MW00453131

Springer Series in
Nuclear
and **Particle Physics**

Springer Series in **Nuclear** and **Particle Physics**

Editors: Mary K. Gaillard · J. Maxwell Irvine · Erich Lohrmann · Vera Lüth
Achim Richter

Hasse, R.W., Myers, W.D.
Geometrical Relationships of Macroscopic Nuclear Physics

Belyaev, V.B.
Lectures on the Theory of Few-Body Systems

Heyde, K.L.G.
The Nuclear Shell Model

Gitman, D.M., Tyutin I.V.
Quantization of Fields with Constraints

Sitenko, A.G.
Scattering Theory

Fradkin, E.S., Gitman, D.M., Shvartsman, S.M.
Quantum Electrodynamics with Unstable Vacuum

Brenner, M., Lönnroth, T., Malik, F.B. (Editors)
Clustering Phenomena in Atoms and Nuclei

M. Brenner   T. Lönnroth   F. B. Malik
Editors

Clustering Phenomena in Atoms and Nuclei

International Conference on Nuclear and Atomic Clusters
1991, European Physical Society Topical Conference,
Åbo Akademi, Turku, Finland, June 3–7, 1991

With 230 Figures

Springer-Verlag

Berlin Heidelberg New York
London Paris Tokyo
Hong Kong Barcelona
Budapest

Professor Dr. Mårten Brenner
Dr. Tom Lönnroth

Department of Physics, Åbo Akademi, Porthansgatan 3,
SF-20500 Turku, Finland

Professor Dr. F. Bary Malik

Department of Physics, Southern Illinois University at Carbondale,
Carbondale, IL 62901-4304, USA

ISBN 3-540-55101-8 Springer-Verlag Berlin Heidelberg New York
ISBN 0-387-55101-8 Springer-Verlag New York Berlin Heidelberg

Library of Congress-Cataloging-in-Publication Data.
Clustering phenomena in atoms and nuclei / [edited by] M. Brenner, T. Lönnroth, F. B. Malik. p. cm. – (Springer series in nuclear and particle physics) Includes bibliographical references and index. ISBN 0-387-55101-8 (U.S.) 1. Cluster theory (Nuclear physics)–Congresses. 2. Nuclear structure–Congresses. 3. Atomic theory–Congresses. 4. Metal crystals–Congresses. I. Brenner, M. (Mårten) II. Lönnroth, T. (Tom) III. Malik, F. B. (Bary) IV. Series. QC793.3.S8C575 1992 539.7–dc20 92-3940

This work is subject to copyright. All rights are reserved, whether the whole or part of the material is concerned, specifically the rights of translation, reprinting, reuse of illustrations, recitation, broadcasting, reproduction on microfilm or in any other way, and storage in data banks. Duplication of this publication or parts thereof is permitted only under the provisions of the German Copyright Law of September 9, 1965, in its current version, and permission for use must always be obtained from Springer-Verlag. Violations are liable for prosecution under the German Copyright Law.

© Springer-Verlag Berlin Heidelberg 1992
Printed in Germany

The use of general descriptive names, registered names, trademarks, etc. in this publication does not imply, even in the absence of a specific statement, that such names are exempt from the relevant protective laws and regulations and therefore free for general use.

Typesetting: Camera-ready by the authors
57/3140-5 4 3 2 1 0 – Printed on acid-free paper

Preface

In these days of specialization it is important to bring together physicists working in diverse areas to exchange and share their ideas and excitement. This leads to cross-fertilization of ideas, and it enriches, as in biological systems, a specialized field with new strength, development and direction derived from another area. Although this might be an uncommon thing, it is an important step in our understanding of the physical world around us, which is, after all, the main purpose of physics.

The seed for this conference was really sowed when one of us (MB) and Mr. Manngård showed some α-scattering data at backward angles to FBM one summer about four years ago. That occasion led to a long research collaboration between the Åbo Akademi physicists and other scientists in several countries. The actual idea to explore the possibility of holding a conference, however, crystallized in the summer of 1989 during a visit of FBM to Åbo Akademi.

The final decision to organize a conference was made after MB visited Professor Ben Mottelson in Copenhagen and Professor Anagnostatos in Athens. At this point it was recognized that there are similarities as well as differences between clustering phenomena in nuclei and systems consisting of atoms. It was therefore conjectured that it could be very stimulating to bring together these groups to exchange their ideas and to learn from each other's fields. A conference along these lines, we hoped, would contribute to an increased mutual understanding.

The idea of some of the participants to hold another conference of a similar nature in two years' time in Greece seems to indicate that we have achieved the goal, at least to some extent. The large number of very personal letters expressing thanks from participants also seems to mean that the conference was successful. It is thus our wish that we continue to search for the unity in the diversity of physics.

The article and reports in these proceedings were presented as talks at the conference in Turku/Åbo, June 3–7, 1991. As many of the topics of the conference are related – if at all – only in an unconventional way, the organization of the material has been difficult. In order to keep the nuclear and atomic aspects close we have by force put the articles under headings that are believed to cover both aspects.

In the first chapter, "Clusters and Nuclei", and in the last chapter, "Post-Conference Thoughts", ideas of magic numbers, periodicity in mass spectra and the electron structure of atomic clusters provide the best evidence for common phenomena in the two subtopics of the conference. The chapter "Theoretical Approaches" is of general interest. In the chapter "Cluster Radioactivity (f-Decay)" the spontaneous decay of nuclei and atomic clusters forms a bridge between the two aspects. Moreover, the cluster decay of nuclei presents an interesting new bridge between the alpha decay and fission of nuclei, which is commented on by B. Mottelson in his concluding remarks. Decay, fragmentation and fission are uniting concepts in this chapter. "Clustering in Light Nuclei" covers a field in which clustering within clusters, i.e. nuclei, displays interesting new evidence for the existence of small and large clusters in nuclei. New aspects related to the dynamical properties of atomic clusters are collected in the chapter "Atomic Clusters and Ions". The production of clusters by sputtering is included as well. The optimism of the editors has brought nuclei and atomic clusters under the same heading "Reactions and Clusters". It is of course probable that experiments on collisions between atomic clusters will be a hot issue in the near future. So far theoretical considerations on this subject can be presented together with results from well-established nuclear reactions.

The ordering of the articles within the chapters was even more difficult than the choice of chapters. Plenary speakers' articles are first. The editors regret the shortcomings of the organization of the material.

Åbo (Turku)   Mårten Brenner
Carbondale   Tom Lönnroth
September 1991   F. Bary Malik

Acknowledgements

We are very grateful for the strong support and encouragement of Professor Mottelson (Copenhagen) and our honorary chairman, Professor Bergström (Stockholm). Professor Mottelson's vision has been a key factor in setting the tone and the format of the conference. Without the active participation of Professors Anagnostatos (Athens), Bargholz (Stockholm), Gridnev (Leningrad), Hefter (Heidelberg), Manninen (Jyväskylä) and Rubchenya (Leningrad) it would have been next to impossible to put together the program with a balanced composition of speakers. We appreciate their help very much.

Of course, the conference would not have taken place without the untiring efforts of the local organizing committee and we are very much indebted to Mrs. K. von Schalien, Mrs. M. Vainio, Mr. K.-M. Källman and Mr. P. Manngård. Furthermore, we are very thankful to Mr. Manngård, Mr. Källman, Mr. J. Martin, Dr. Alam and Dr. Hefter for their assistance in preparing the proper format of some of the manuscripts for this volume.

Last but not least, we thank the members of the International Advisory Committee for their support, which was vital at the formative stage of the conference. Similarly, we are thankful to the European Physical Society for their support and interest. A most important factor for the success of the conference was, no doubt, the financial support, for which we express our sincere thanks. The supporters are: the Academy of Finland, the Ministry of Education, the City of Turku, the Åbo Akademi University, the Foundation of Åbo Akademi Research Institute, and Suomi Mutual Life Assurance Company.

Opening Address

by M. Brenner

Mr Chairman, honorable participants of the conference.

This is the first conference organized to bring together scientists from nuclear and atomic physics to exchange ideas about clustering phenomena. From theoretical standpoints there are already many similarities between nuclear and atomic clusters. The identification of magic numbers of atomic clusters is the most significant example, but there are many others. Hopefully, conferences like this will in the future bring out more interesting new aspects that are common in both areas. The speakers have been asked to make their ideas understandable by every participant, no matter if one is a chemist or physicist. This may of course be difficult in some cases without losing essential aspects, but it is worth trying because new ideas are likely to develop during discussions among scientists with different backgrounds.

The program of these five conference days perhaps looks like an odd mixture or salad of vastly diverse ingredients. In fact, a talk on one topic may be followed by another which seemingly bears no relation to the preceding one. This has been done on purpose in order to make everybody acquainted with different points of view on clustering phenomena in atoms and nuclei, particularly from a theoretical or experimental perspective.

There was a conference on nuclear cluster dynamics in 1988 in Sapporo, Japan, and there will be one on the physics and chemistry of finite systems of clusters in Richmond, Virginia, in October 1991. I hope that this conference will successfully complement these two, and other more specialized conferences.

Let me mention the circumstances which have encouraged the organizers to arrange this conference. Three years ago, Professor Malik suggested that the research at the Åbo Akademi accelerator would profit from a conference on molecular states in nuclei. Åbo Akademi has cooperative research with the Leningrad State University, the V.G. Khlopin Radium Institute in Leningrad and the I.V. Kurchatov Atomic Energy Institute in Moscow. These projects are also related to cluster phenomena in nuclei. Thus, physicists from these laboratories became interested in such a conference.

It is 32 years since our chairman, Prof. Ingmar Bergström, organized in Fiskebäckskil, Sweden, the first Nordic conference on the use of accelerators in nuclear physics. Following the tradition of Fiskebäckskil, the sixth of these conferences was held in Kopervik, Norway, two years ago. At that conference Prof. Ben Mottelson showed in a most fascinating talk the close relation between nuclear and atomic physics in the study of cluster phenomena. The talk inspired us to join the nuclear and atomic aspects in one conference. We have been lucky to have Professor Mottelson's support in organizing it. After all, we would not be here today without Professor Mottelson's enthusiasm and encouragement.

The members of the organizing committee are spread over an area of 23 million square kilometers, and hence no committee meeting has been held. Therefore, please do not blame individual members of the committee for mistakes and bad organization. I wish to welcome you all wholeheartedly.

Contents

Part VII   **Post-Conference Thoughts**

Part I

Clusters and Nuclei

Shell Structure in Nuclei and in Metal Clusters

S. Bjørnholm[1], *J. Borggreen*[1], *K. Hansen*[1], *T.P. Martin*[2],
H.D. Rasmussen[1], *and J. Pedersen*[1]

[1]The Niels Bohr Institute, University of Copenhagen,
  Blegdamsvej 17, DK-2100 Copenhagen Ø, Denmark
[2]Max-Planck-Institut für Festkörperforschung,
  Heisenbergstr. 1, W-7000 Stuttgart 80, Fed. Rep. of Germany

The conduction electrons in ultrasmall metal drops form ordered quantum states in analogy to the nucleons in the nucleus. This becomes visible in experiments with clusters produced by expansion of sodium vapour. The droplets form an entire periodic system of "giant" nuclei that can be understood by applying Bohr's quantization rule to electrons moving classically inside the droplets in triangular and square orbits. Drops with 2 to 3000 atoms are ordered in 22 periods divided between two main groups, or supershells.

Introduction

Free and independent electrons bonded by the spherical surface of a small metallic drop will quantize into a total quantum state that exhibits shell structure[1-5]. There is here an analogy to the atomic electrons surrounding the nucleus and even more, an analogy to the ordering of protons and neutrons in the nucleus itself as described by the nuclear shell model[6]. The most significant difference is the absence of any spin-orbit splitting in the metal clusters. In the study of how metallic clusters grow, one therefore leans heavily on theories and experience from atomic and in particular from nuclear physics [7,8]. In addition, since metal clusters readily exceed nuclei in the number of elementary constituents they contain, the metal clusters provide a basis for significant extensions of the quantum many body concepts developed to describe nuclear shell structure.

We shall here present measurements of the size distribution of sodium metal clusters, with up to 3000 atoms per cluster, obtained by expanding hot sodium vapour through a fine nozzle. Steps in the abundance distributions occur locally at sizes that differ in radius by 0.61 r_{ws}, where r_{ws} is the Wigner-Seitz radius (i.e. radius of sphere representing the volume available to each atom or electron). For a cold electron gas the Pauli exclusion principle imposes a unique relation between r_{ws} and the most energetic (Fermi) electron. The action of a classical electron moving with this energy will increase by one Planck unit h for each increment in radius of the above magnitude, provided it moves inside the cluster surface in something intermediate between a closed triangular and a closed square orbit. Quantization of such classical closed orbits can therefore explain the observed periodicity. The experimental data further show clear evidence of a quantum beat mode, reflecting alternating constructive and destructive interference between quantized triangular and square orbits, respectively. The altogether 22 shell periods observed are divided into two groups separated by a minimum in the amplitude of the shell effect. In addition one sees a shift in phase in going from group one to group two, as expected for a beat mode.

The experiment

Metal droplets are produced by expansion of metal vapour through a fine nozzle, the same way water droplets are formed by letting steam through a valve, Fig. 1.

The sodium vapour is generated in a stainless steel vessel held at 700-800 °C and the expansion is assisted by pressurizing the vessel with a large

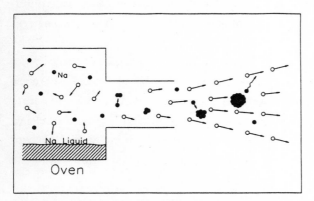

Fig. 1 - **Production of sodium clusters, i.e. giant atoms**. *In the expansion of an inert gas through a 0.1 -0.2 mm diameter nozzle, random thermal motion is converted to uniform translational motion, resulting in strong cooling of the inert gas. Introducing atomic sodium vapor into this medium, the sodium atoms aggregate into clusters with a broad, uniform size distribution.*

surplus of argon or xenon gas. The mixture is expanded into a three meter long, differentially pumped flight tube, resulting in a narrow beam of free-flying metal clusters with velocities close to the initial thermal speed of the argon or xenon atoms. In the expansion and clustering process, lasting about 100 nanoseconds, the noble gas medium cools to a few tens of Kelvin. In the sodium clusters, on the other hand, there will be a competition between heating, due to condensation, and cooling, due to collisions. We believe that the resulting internal temperature in the freshly formed clusters is just some 100-200 °C below the oven temperature. In the ensuing free flight for about one millisecond the droplets will therefore loose sodium atoms by stepwise evaporation, cooling to about 100-200 °C. This evaporation process is sensitive to shell-like variations in the atomic separation (free) energies; and these variations are thought to be responsible for the step-like modifications of the experimentally observed size distributions, while presumably the pre-evaporation size distributions are completely uniform.

One meter downstream the size distribution is sampled by time-of-flight mass spectrometry. Ultraviolet photons with energies just about the ionization threshold and an energy spread of 1 eV (compared to a total Fermi energy in sodium of 3.24 eV), are used to produce a representative sample, in the form of ions, from the otherwise neutral size distribution. Figure 2, upper part, shows an example of an abundance distribution I_N vs. N obtained in this way. The quantity N is the number of atoms or conduction electrons in the cluster.

Magic numbers

One sees clearly in Fig. 2 how the global bell-shaped distribution is scarred with saw-tooth or s-shaped irregularities at certain "magic" sizes.

Since the interest focusses on these, it is convenient to display the experimental result in terms of relative intensity changes, i.e.:

$$\Delta_1 \ln I_N = \ln(I_{N+1}/I_N) \approx 2\frac{(I_{N+1} - I_N)}{(I_{N+1} + I_N)} \ . \tag{1}$$

As one sees from Fig. 2, lower part, this makes the magic numbers stand out very clearly as dips in the curve. At the same time, one sees how the dips become smaller and broader with increasing magic number, N_0. For

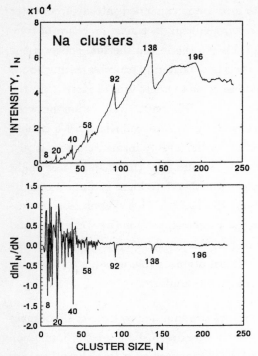

Fig. 2 - Mass spectra. *Top panel: Abundance distribution for sodium clusters produced by adiabatic expansion and measured after a one meter free flight by time-of-flight mass spectrometry. In this measurement the individual mass peaks are fully resolved. Each point represents the integrated mass peak, corrected for background. Bottom panel: Logarithmic derivative $\Delta_1 \ln I_N$ of the results in top panel according to eq. (1).*

higher N_0-values the dips signalling shell closures will drown in noise due to finite counting statistics. In this situation we have found it useful to make a compromise between "band pass" in terms of the size interval sampled and statistical noise, by computing a properly weighted logarithmic derivative for properly spaced mass points N,

$$\langle \Delta_1 \ln I_N \rangle_{K_0} = \frac{\displaystyle\sum_{K=2K_0/3}^{K_0} \frac{(I_{N+1+K} - I_{N-K})(2K+1)}{I_{N+1+K} + I_{N-K}}}{\displaystyle\sum_{K=2K_0/3}^{K_0} (2K+1)^2} , \qquad (2)$$

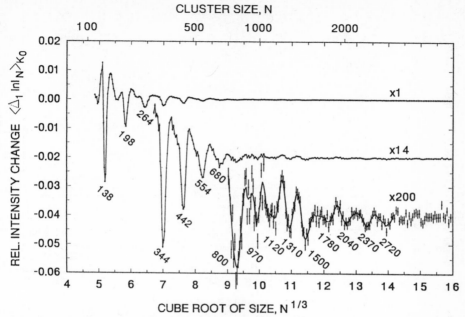

CLUSTER SIZE, N

Fig. 3 - **Magic numbers.** *Relative changes in the cluster abundance distributions,* $\langle \Delta_1 \ln I_N \rangle_{K_o}$, *calculated from the measured mass abundance spectra* I_N *according to eq. (2) and plotted as a function of* $N^{1/3}$. *In these measurements the individual masses are not resolved in the time-of-flight spectra. Time-of-flight is converted to mass by extrapolation from the low-mass region, where peaks are resolved, and the intensity per mass* I_N *calculated without background correction. (Some background originates from doubly ionized clusters. This has been kept at a minimum by adjustment of the ionizing light intensity to a relatively low level).*

choosing values of $K_o = 0.03\,N$. The derivative is thus sampled over intervals of $2K_o + 1$, i.e. $\pm 3\,\%$ of the actual size. In this way it becomes realistic to scale up the measured derivatives in order to display very small irregularities in the intensity pattern. Figure 3 shows the results, plotted as a function of the linear dimensions of the clusters, $\sim N^{1/3}$. At each dip the magic number N_o is indicated. The dips are very nearly equidistantly spaced in the plot (Fig. 3), indicating that each new period require the same increment in drop radius $\Delta R = r_{ws}\,\Delta N_o^{1/3}$ to be completed.

A closer test of the linearity is made in Fig. 4, where $N_o^{1/3}$ is plotted against a running index n, accounting for the period (or shell) number.

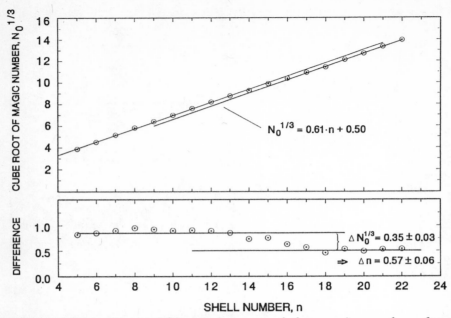

*Fig. 4 - **The phase shift**. Cube root of the magic numbers from Fig. 3, signifying shell closures, plotted against shell number n. (Top panel). For higher shell numbers the points fall on two straight lines, phase-shifted by one half shell number. This is shown more clearly in the panel below where the points result from subtraction of a straight line from the data above.*

From the slope of the straight lines in Fig. 4 one finds that the increment in radius per period is $\Delta R = (0.61 \pm 0.01)\, r_{ws}$. The increment ΔL in circumference $2\pi\Delta R$ is then $(3.83 \pm 0.06)\, r_{ws}$. The de Broglie wavelength of an electron with the Fermi energy[9] is $\lambda_F = (32\pi^2/9)^{1/3}\, r_{ws} = 3.28\, r_{ws}$, which is not so far from the increment in the circumference or, equivalently, in the length of a full circular orbit. However, triangular and square orbits are in much closer agreement with the de Broglie quantization rule $\Delta L = \lambda_F$, having length increments of $3\sqrt{3}\Delta R = (3.17 \pm 0.06)\, r_{ws}$ and $4\sqrt{2}\Delta R = (3.45 \pm 0.06)\, r_{ws}$, respectively. This is the first very significant observation resulting from simply measuring the positions N_0 of the dips in Fig. 3.

The second result is the observation, Fig. 4, that the magic numbers, or rather the $N_0^{1/3}$-values, fall on two straight lines displaced by one-half unit of n, the shell index. The sum of two cosine functions of n with some 7-10 %

difference in wave numbers $k_\triangle$ and $k_\square$, respectively would show such a phase shift:

$$\cos(k_\triangle n) + \cos(k_\square n) = 2\cos\left(\frac{k_\triangle + k_\square}{2}n\right) \ \cos\left(\frac{k_\triangle - k_\square}{2}n\right) . \quad (3)$$

Combined with the first result, this strongly suggests that quantized triangular and square orbits together are responsible for the observed periodicities in the abundance distributions.

Intensities

An interference pattern of the type eq. (3) should show up as a long wavelength modulation of the short wavelength periodicities visible in Figs. 2 and 3. Instead, one sees there a seemingly monotonous decrease in the amplitudes of the shell dips. There are two reasons for this decrease.

Shell structure is due to periodic variations in an otherwise uniform single particle electron level density. In a constant density medium with constant Fermi energy, or, equivalently, constant width of the conduction band, the spacing $\hbar\omega_{shell}$ between consecutive modulations must decrease as $1/n$. This will cause an $N_o^{-1/3}$ decrease in the expected dips. If furthermore the amplitudes of the shell modulations decrease as $N_o^{-1/6}$, one arrives[8] at a decrease varying as $N_o^{-1/2}$ – even at zero absolute temperature.

The other, more important, reason for the strongly decreasing dip amplitudes is temperature[10−12], which tends to wash out the observable shell structure – exponentially[8] in the effective temperature parameter τ,

$$\tau = k_B T \frac{2\pi^2}{\hbar\omega_{shell}} , \quad (4)$$

with

$$\hbar\omega_{shell} \approx 2\varepsilon_F/n \ \propto \ N^{-1/3} . \quad (5)$$

Here k_B is Boltzmann's constant, T is the real temperature at the time of sampling the abundance distributions, estimated[13] to be 400-500 K, while $\varepsilon_F = 3.24$ eV is the Fermi energy.

In order to compensate for these effects we have scaled the experimental logarithmic derivatives, $\langle\Delta_1 \ln I_N\rangle_{K_o}$ cf. Fig. 3, with the factor $N^{1/2}\exp(cN^{1/3})$, setting $c = 0.65$. The result is presented in Fig. 5.

9

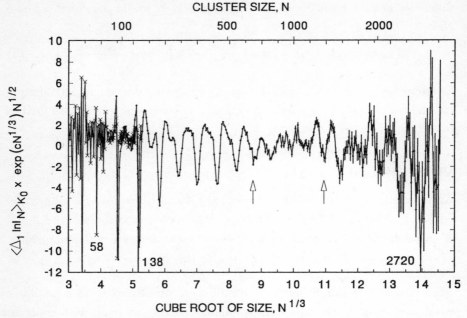

Fig. 5 - **The quantum beat.** *Relative changes* $\langle \Delta_1 \ln I_N \rangle_{K_0}$ *in experimental cluster abundance* I_N, *corrected for the effect of temperature and shell compression, as a function of linear cluster dimensions* $N^{1/3}$. *The model considerations behind these corrections are described in refs.[8,10−12]. The highest shell in the first group (supershell) according to the phase shift plot, Fig. 4, and the first shell in the second group are indicated by arrows. The measurements from Fig. 2, lower part, are also shown (crosses). The logarithmic amplitudes* $\Delta_1 \ln I_N$ *are here summarily scaled by an extra factor of 0.2 to account for the superior mass resolution and the use of eq. (1) instead of eq. (2) for these data.*

Since the scaling function is monotonously increasing, it cannot by itself introduce the large scale modulations seen in the plot. They indeed have the character of a beat mode as expected from eq.(3) and from the results of the previous section.

Correcting the experimental abundance modulations for the influence of finite temperature and for the compression of the shells with increasing size thus corroborates the previous results, obtained solely by examining the sequence of magic numbers.

The quantum beat

From this analysis one obtains an understanding of the periodic abundance variations with cluster size as the result of quantization of triangular and square electron orbits inside the sodium droplets, in the spirit of Bohr[14] and de Broglie[15]. The presence of two orbits, and hence two slightly differing periods, results in a beat pattern superposed on the "fundamental", primary shell periodicity. The beat mode has a minimum around shell no. 16, where there are about 1100 conduction electrons in the cluster. An illustration of how the combination of "preferred" cluster radii for electrons orbiting with equal energy in triangular and square orbits, respectively, can lead to an interference pattern[16] is given in Fig. 6.

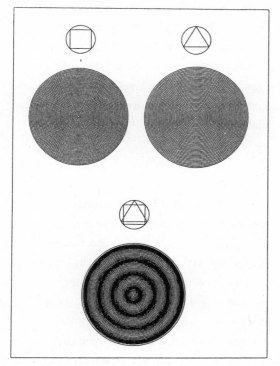

Fig. 6 - **Correspondence principle.** *For a particle with Fermi-energy, triangular orbits will have integer action values for a set of discrete drop radii (top right). Analogously, square orbits require another set of discrete radii in order for the action to be an integer multiple of h (top left). The drop sizes that meet both requirements exhibit a beat pattern (bottom center).*

Modern theories

The attempt to discuss the experiment in the simplest possible terms is deliberate. It does not imply that modern quantum theories have overlooked the effects shown or are inadequate for their description. Indeed, we were inspired by several newer theories of shell structure in nuclear systems, in particular refs.[8,17], in first mounting the experiment. Also, the prior theoretical treatment of size and temperature effects[8], have guided the analysis of the admittedly very weak abundance modulations. Some theoretical developments[10,11,18] have occurred in parallel with the experimental campaign, helping to sharpen and focus the effort.

The modern theories[18] are particularly helpful in answering why just two orbits, triangles and squares, suffice to understand the experiment, and why pentagons, hexagons, etc., etc. can be neglected. In a more complete theory, all closed orbits indeed contribute to the final spectrum of discrete quantal energy eigenstates. But the experiment averages over this forest of discrete states, being only sensitive to the gross features in the level density, and these features are found to be dominated by contributions from the two shortest closed orbits (with non-zero angular momentum).

Particles and waves

Periodicities due to quantum shell structure have been identified in a large number of experiments[1,2,19,20], the most recent one[21] extending to cluster sizes almost as large as the ones described here. Of special interest is ref.[19]. Working with cold sodium clusters at about $100 \, K$, the authors find quantum shell periodicities for clusters with 1400 atoms or less, while periodicities in the size interval from $N = 1400$ all the way up to $N = 20000$ give evidence of geometrical packing of sodium ions into icosahedral quasicrystals. This last aspect has no analogue in nuclear systems and is an example of the coexistence in one experiment of wave-order and particle-order, respectively. Combined with the present, high-temperature experiment it also illustrates how the periodicities that one encounters in connection with the growth of clusters towards macroscopic dimensions can depend rather strongly on temperature, reflecting basically the difference between soft (perhaps liquid) and solid end products.

12

The authors wish to thank O. Echt, W.D. Knight and B.R. Mottelson for most valuable advice and encouragement.

References

1. W.D.Knight, K. Clemenger, W.A. de Heer, W.A. Saunders, M.Y. Chou, and M.L. Cohen, *Phys. Rev. Lett.* **52**, 2141 (1984), and W.A. de Heer, W.D. Knight, M.Y. Chou, and M.L. Cohen, *Solid State Physics* **40**, 93 (1987)

2. I. Katakuse, T. Ichihara, Y. Fujita, T. Matsuo, T. Sakurai, and H. Matsuda, *Int. J. Mass Spectrom. Ion Proc.* **67**, 229 (1985)

3. W. Ekardt, *Phys. Rev.* **B29**, 1558 (1984)

4. O.E. Beck, *Solid State Commun.* **49**, 381 (1984)

5. M.Y. Chou, and M.L. Cohen, *Phys. Lett.* **A133**, 420 (1986)

6. M.G. Mayer, and J.H.D. Jensen, *Elementary Theory of Nuclear Shell Structure*, 1 (Wiley, New York, 1955)

7. A. Bohr, and B.R. Mottelson, *Nuclear Structure*, Vol. **I**, 2 (Benjamin, New York, 1969).

8. A. Bohr, and B.R Mottelson, *Nuclear Structure*, Vol. **II**, 2 (Benjamin, London, 1975)

9. N.W. Ashcroft, and N.D. Mermin, *Solid State Physics*, 1 (Saunders College, Philadelphia, 1976).

10. M. Brack, O. Genzken, and K. Hansen, *Z. Phys.*, to appear (1991)

11. M. Brack, O. Genzken, and K. Hansen, *Z. Phys.*, to appear (1991)

12. S. Bjørnholm, J. Borggreen, O. Echt, K. Hansen, J. Pedersen, and H.D. Rasmussen, *Z. Phys.* to appear (1991)

13. C. Brechignac, Ph. Cahuzac, J. Leygnier, and J. Weiner, *J. Chem. Phys.* **90**, 1492 (1989)

14. N. Bohr, *Phil. Mag.* 26, 1 (1913)

15. L. de Broglie, *C.R. Acad.Sci.Paris*, **179**, 39 and 676 (1924)

16. S. Bjørnholm, in *Clusters of Atoms and Molecules* (ed. Haberland, H.) Ch. 2.4, pp.24 (Springer, Berlin, to appear 1991)

17. R. Balian, and C. Bloch, *Ann. Phys.* **69**, 76 (1972)

18. H. Nishioka, K. Hansen, and B.R. Mottelson, *Phys. Rev.* **B42**, 4377 (1990)

19. T.P. Martin, T. Bergmann, H. Göhlich, and T. Lange, *Chem. Phys. Lett.* **172**, 209 (1990)

20. S. Bjørnholm, J. Borggreen, O. Echt, K. Hansen, J. Pedersen, and H.R. Rasmussen, *Phys. Rev. Lett.* 65, 1627 (1990)

21. T.P. Martin, S. Bjørnholm, J. Borggren, C. Bréchignac, P. Cahuzac, K. Hansen, and J. Pedersen, *Chem.Phys.Lett.* submitted (1991)

Simple Models for Metal Clusters

W.A. de Heer

Institut de Physique Expérimentale, Ecole Polytechnique
Federale de Lausanne, PHB 1015, Lausanne, Switzerland

The electronic properties of simple metal clusters are examined in several simple models. The electronic shell effects, polarizabilities, ionization potentials and plasma resonances are calculated using the Sommerfeld model and the Drude model in an ellipsoidal potential well with a shape derived in the Nilsson-Clemenger model. Despite their simplicity these models describe and predict electronic properties surprisingly well, far better than might be expected considering the simplifications, and in some cases even surpass much more sophisticated treatments.

INTRODUCTION

The electronic properties of simple metal clusters have been investigated intensively for more than a decade and are reasonably well understood. It is now clear that many properties of simple metal clusters may be described using very simple models. [1,2] It is well known that simple models for bulk metals, like the Drude model, already describe the optical response of metals.[3] The Sommerfeld model quantizes the electron gas and gives more reasonable estimates of the electronic heat capacity and the electron mean free paths than the Drude model. In all these models the important simplification is that only the conduction electrons contribute to the electronic properties. Furthermore these electrons are considered to be free, that is to say that the ionic cores hardly perturb their motion.[3]

Clusters have been studied theoretically in various models, ranging from the most sophisticated configuration interaction (CI) quantum chemical calculations,[4] to the self-consistent jellium model where the ionic core structure is ignored,[1,5] and the independent electron approach with an *ad hoc* effective single particle potential.[1] Even though the CI calculations are usually considered to be most exact, this approach tends to be so complex that the final results are difficult to generalize.

Simple models are necessary for several reasons. Firstly, they highlight the basic physics and furthermore, generalization is practically built in since the

Springer Series in Nuclear and Particle Physics **Clustering Phenomena in Atoms and Nuclei**
Editors: M. Brenner · T. Lönnroth · F.B. Malik   © Springer-Verlag Berlin, Heidelberg 1992

cluster size and electronic density are usually the parameters in the model. Here we present several models, each which describe certain aspects simple metal clusters. Of course these models do injustice to the enormous theoretical efforts to understand these clusters from first principles. Nevertheless it is certainly helpful to recognize that ultimately these extremely complex many body systems have very simple and understandable physical properties.

Here the usefulness of simple models is demonstrated. We focus primarily on properties of small sodium clusters and discuss the mass spectra, the polarizabilities, the ionization potentials and the plasma resonances.

ELECTRONIC SHELL STRUCTURE

The first experimental indication for the underlying simplicity was clear in the cluster mass spectra of sodium cluster beams in Berkeley by Clemenger, de Heer, and Saunders under supervision of W. D. Knight.[6] The clusters were produced by condensing the metal vapor in an inert gas in an adiabatic expansion through a small nozzle. In order to detect them they were first photo-ionized and the masses were measured with a quadupole mass analyzer. A typical mass spectrum is shown in Fig.1, and it is clear that the the Na_N clusters with N= 8, 20, 40, 58,

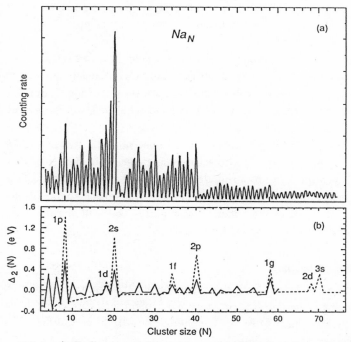

FIG.1. a) Sodium mass spectrum. b) Calculated second differences in the spherical jellium model (see Ref 1) (dashed), and in the Nilsson-Clemenger model (bold).

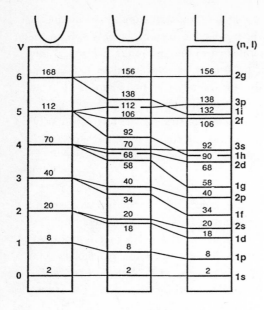

FIG.2. Energy levels in three effective single particle wells. From left to right: harmonic, rounded square well, and square well. The rounded square well potential most closely fits the observed mass spectrum.

92,... are especially abundant. It was immediately recognized that these numbers correspond to the electronic shells closing numbers in a spherical potential well.[6] Further experiments have shown that this progression continues up to the many thousands of atoms per cluster.[7]

From these observations a semi-empirical cluster model was formulated. The valence electrons in the the cluster were considered to be confined in a rounded square well potential. The energy levels in this well are shown in Fig. 2, and the electronic shell closing numbers are seen to correspond with the observed special peaks in the mass spectra.

Subsequently the corresponding numbers in the spectra of Li, K, Cs and Rb clusters have been found. The noble metals also have electronic shell structure with shell closings corresponding to a rounded square well potential.[8] Of course to reproduce the angular momentum shell closings the exact shape of the potential is not as important as the spherical symmetry. The well shape affects only the ordering of the angular momentum shells.

THE NILSSON-CLEMENGER MODEL

Already from the mass spectra it is clear that the spherical approximation is inadequate to account for the rich fine structure. To explain this, corrections must be made and an obvious one is to relax the spherical constraint and allow more general shapes as shown below.

The analogy between the electronic shell model proposed for the alkali clusters and the nuclear shell model stimulated Clemenger to extend the spherical cluster model[9] as had been done for nuclei by Nilsson years ago.[10] He suggested that the cluster shapes could be determined by minimizing the total electronic energy of an idealized spheroidal system. To lowest order the total electronic energy is assumed to be proportional to the sum of the single electron energies within the spheroidal well. Clemenger constrained the cluster shapes to spheroids as Nilsson had done, for the nuclei however it is not much more difficult to extend the model to ellipsoids.

The Nilsson-Clemenger (NC) single particle hamiltonian is similar to the Nilsson hamiltonian except that the spin-orbit term is neglected.

$$H = \frac{p^2}{2m} + \frac{1}{2} mw_0{}^2 r_0{}^2 \left(x_0{}^{-2} + y_0{}^{-2} + z_0{}^{-2}\right) - U \, hw_0 \left(l^2 - \langle l^2 \rangle_v\right)$$

$$\langle l^2 \rangle_v = \frac{1}{2} v \, (v+3)$$

$$x_0 \, y_0 \, z_0 = 1$$

(1)

Here v is the harmonic oscillator quantum number, l the electronic angular momentum and x_0 etc. the semi axes of the ellipsoid normalize to the radius of a sphere r_0, which has the volume of the cluster. The anharmonic parameter U is adjusted so that the energy levels conform with those from self consistent calculations and is about 0.04 for sodium clusters up to N=40.

The energy levels are plotted as a function of the spheroidal distortion in the Nilsson-Clemenger diagram, Fig. 3. The cluster shapes are indicated by a dot at the highest occupied level as shown.

Although the energy scale, hw_0 is not important for the cluster shapes, Clemenger demonstrated, adapting the discussion of deShalit and Feshbach[11] that hw_0 is almost exactly $E_f/N^{1/3}$, where E_f is the Fermi energy.

Since the anharmonic correction term is small for small clusters, it may be neglected to first order so that the effective potential reduces to that of an anisotropic harmonic oscillator. This form is so simple that the energy the problem can be solved trivially and Fig. 4 shows the energy levels of sodium clusters up to N=20. These are compared with those from self-consistent spheroidal jellium calculations.[12] It is clear that this very simple model already reproduces many of the features of much more sophisticated methods.

For comparison with experiment, in Fig.1b the second differences of the total energies in the self consistent spherical jellium model and from the NC model are compared with the features in the sodium cluster abundance spectra. Peaks in the second differences should correspond with clusters with enhanced stabilities, and it is clear that the NC model reproduces the fine structure features in the mass spectra.

17

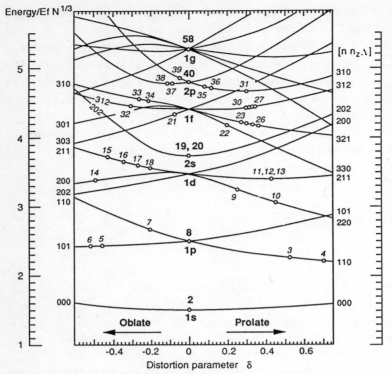

FIG. 3. Nilsson-Clemenger diagram for spheroids, $x_0 = y_0$. The distortion parameter $\delta = 2(z_0 - x_0)/(z_0 + x_0)$.

POLARIZABILITIES

Classically the polarizability of a metal sphere is the cube of the radius so that the polarizability per atom is constant:

$$A/N = R^3/N. \tag{2}$$

The measured polarizabilities are found to be larger than the classical value,[13] and this is because the radius of the electronic cloud is slightly greater than that of the ionic cores. This spill-out effect causes externally applied electric fields to be screened beyond the classical radius.[14] An analogous effect also occurs in the bulk, as has been demonstrated by Lang and Kohn.[15] Hence to approximate the experimental values, the radius used in the classical expression should be R+d, where d represents the spillout. For a sodium surface d is 1.3 a.u.. With this value the experimental curve is reasonably well reproduced (except for electronic shell effects), showing that the semi-classical expression applies even for small clusters, when the electronic spillout effect is included.

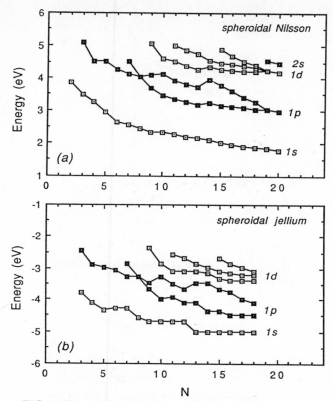

FIG. 4. Comparison between the energy levels from the N-C model with those from self consistent spheroidal jellium calculations (Ref. 5)

IONIZATION POTENTIALS

The ionization potential of a classical metal sphere is given by

$$IP = WF + e^2/2R. \tag{3}$$

where WF is the bulk work function. The second term is the classical charging energy of a sphere of radius R. Fig. 5 shows the ionization potentials of potassium clusters up to N=20.[16,1] It is clear that the measured ionization potentials have much more structure than predicted by the classical expression, reflecting among other things the cluster size dependent variations of the highest occupied levels due to shell effects.

These shell corrections can be calculated in the NC model. Furthermore, as for the polarizabilities, the classical radius should be enhance by a spillout factor. Hence a semi classical expression for the IP may be written as

$$IP = WF + e^2/2(R+d) + SC \tag{4}$$

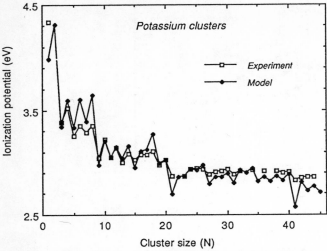

FIG. 5. Measured ionization potentials of potassium clusters and calculated ionization potentials from equation 4.

where SC is the shell correction, obtained from the highest occupied levels in the NC model.

Fig. 5 shows the semi classical ionization potentials and correspond very well with the observed values using the bulk workfuntion and d=1.7 a.u. Furthermore, plotting the IPs of closed shell clusters as a function of $e^2/2(R+d)$, yields nearly a straight line that intercepts the y axis at the bulk work function, as predicted by the model. Hence it follows that the bulk work function is already reflected in the ionization potentials of even the smallest clusters.

PLASMA RESONANCES

The classical optical response of a very small metal sphere is given by the Mie equation (in the Drude model) in the limit that the radius is much smaller than the wave length of the light. A single resonance is predicted with frequency

$$\omega^2 = e^2/m\alpha. \tag{5}$$

where $\alpha = A/N$ is the static polarizability per atom.[2,17]

This is a giant collective resonance (also known as the surface plasma resonance) that exhausts the dipole sum rule. The resonance may be though of a coherent oscillation of the valence electron cloud. Before examining experimental data it is useful to first consider obvious modifications of this expression for small sodium clusters. Very large resonances have been first observed in small sodium clusters by de Heer et al,[2,17] and examples are shown

20

in Fig. 5. The closed shell clusters, for example Na$_8$ and Na$_{20}$, in fact have very strong resonances very near to the values from the classical expression, using measured values for the polarizabilities.

It is also clear that the open shell clusters have more structure. The reason is that these clusters are not spherical but ellipsoidal. Since the polarizability along a long axis in an ellipsoid is greater than along a short axis, it is expected that the plasma resonance is split accordingly. The ellipsoidal NC model is used to calculate the shapes and the corresponding depolarization factors. From these shapes and the experimental polarizabilities, the resonance can be predicted.[17] Correspondence with experiment is remarkable. The splitting of the resonance into 2 or 3 depends on whether the cluster is spheroidal or ellipsoidal in the NC model and agrees with experiment even for very small clusters.[17]

The widths of the plasmons is recently under theoretical investigation and understood to be primarily caused by the Frank-Condon factors involved in the optical transition.[18] It is clear that the plasma frequency is strongly dependent on the cluster shape so that when a photon near the plasma frequency is absorbed the cluster typically will be in a highly excited vibrational state. This effect is demonstrated in Fig. 6 where the total energy curves are calculated from the NC model and the excited state includes a plasmon. Note that the shape dependence of the plasmon causes a distortion of the cluster excited by a plasmon. Assuming that the the cluster was at about 300 K gives approximately a width of 0.3 eV to the plasmon. Widths of this magnitude are experimentally observed.

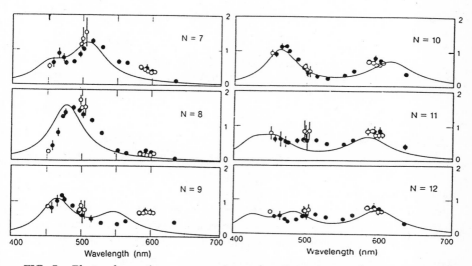

FIG. 5a. Photo-absorption cross sections of sodium clusters from Ref. 2. Solid curves are calculated in the model explained in the text. The area under the curves corresponds with 70% of the total oscillator strength. Absorption is measured in Å^2 per valence electron.

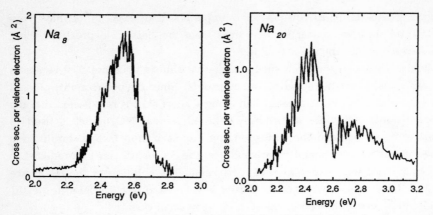

FIG. 5b . The plasmons of Na$_8$ and Na$_{20}$ (after Ref. 20). Note the fragmentation of the Na$_{20}$ plasmon.

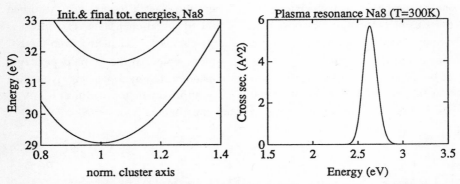

FIG. 6. The shape of the plasmon of Na$_8$ as calculated in the model. Note the shift of the total energy curve in the excited state, causing broadening of the resonance of an initially hot cluster.

It is seen however that the plasmons of closed shell clusters are also split. The effect becomes more pronounced as the cluster size increases and for N=40 there is considerably more structure than predicted by the simple model. This fragmentation of the plasma resonance is understood to be caused by interference effects of the plasmon with single particle excitations, and is not described in the simple model above, but in a model where the single particle excitations are determined from the shell model, and the collective excitations from the model above, it is clear that the fragmentation effect can be described.

Besides the fragmentation of the plasmons the line shapes are also remarkable. For example, examining the line shape of the plasmon of Na$_{20}$ (Fig 5b) it is seen that there seems to be a hole with very low cross sections in the resonance.

Theoretically the fragmentation effect has been described in the random phase approximation.[19] In these calculations the resonance frequencies are calculated together with their oscillator strengths. Using an *ad hoc* smoothing function (usually a lorenzian with a given width) the experimental curves are approximately reproduced. However it is always observed that the experimental shapes which often show that the fragmentation features that look more like "holes" than "bumps". Here we show that these features are in fact easily understood in the simple model.

A FINAL EXAMPLE, THE HOLES IN THE PLASMA RESONANCES

To demonstrate how the interference of a single particle excitation with a plasmon can produce a hole in the plasmon within the simple models presented above, we assume that the coupling between the single particle excitation with the plasmon has a strength α. Hence the hamiltonian which describes the coupled state is

$$H = \begin{bmatrix} E_{pl} & \alpha \\ \alpha & E_{sp} \end{bmatrix}$$

(6)

where E_{pl} and E_{sp} are the unperturbed plasmon and single particle energies. It is further assumed that in the unperturbed system, the plasmon carries all the oscillator strength, so that the single particle excitation gets oscillator strength only by virtue to the mixing, i.e. if the coupled wave functions are

$$\phi_1 = a \ \phi_{pl} + b \ \phi_{sp}$$
$$\phi_2 = c \ \phi_{pl} + d \ \phi_{sp}$$

(7)

then the oscillator strength of ϕ_1 is a^2 and of ϕ_2 it is c^2.

As shown above, both the plasmon and the single particle level energies depend on the shape of the cluster, so that to lowest order

$$E_{pl} (r) = E_{pl}(0) + K_{pl} \, dr$$
$$E_{sp} (r) = E_{sp}(0) + K_{sp} \, dr$$

(8)

where dr is a small change in the shape of the cluster along a normal axis. The constants K_{pl} and K_{sp} can be estimated from the Drude model and the NC model. Inserting these values into the H, and assuming that there is a single particle state near the plasmon the energies of the coupled system as a function of dr are as in Fig 7. Note the avoided crossing.

Furthermore the total energy of the cluster also depends on the shape, and near the equilibrium shape:

$$E_t = E_t (0) + 1/2 \ K_t \, dr^2$$

(9)

23

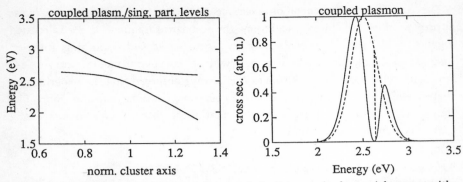

FIG. 7. Left : the energy levels of the coupled plasmon-single particle state, with a=0.1 eV. Right: the resulting plasmon with the hole caused by the avoided crossing (bold). For very small a, the hole reduces to a notch at the unperturbed single particle energy (dashed).

When the cluster is hot it undergoes thermal shape oscillations as shown above. By thermally populating this mode and following the eigen values and the oscillator strengths of resonances gives rise to shapes as in Fig. 7. The effect of the avoided crossing is to produce the hole in the resonance, approximately as observed experimentally. Although (to my knowledge) this feature has not been described in the literature, it is obviously a direct consequence of the coupling between single particle levels and the plasmon.

CONCLUSION

It is not possible to give a detailed exposition of simple models for metal clusters in a short review, however it is clear that these models are indispensable to understand the properties of metal clusters. In fact as shown in the final section, effects which have not yet been described in thorough calculations are nevertheless easily understood in the simple models. Further examples, will be given in a more extensive review in progress.[21]

1. For a review, see W. A. de Heer, W. D. Knight, M. Y. Chou, and M. L. Cohen, in *Solid State Physics*, edited by F. Seitz and D. Turnbull (Academic New York, 1987), Vol. 40.
2. a. W. A. de Heer, K. Selby, V. Kresin, J. Masui, M. Vollmer, A. Châtelain, and W. D. Knight, Phys. Rev. Lett. **59**, 1805 (1987).
3. N. Ashcroft and Mermin, *Solid State Physics*, (Holt, Rinehart, Winston, 1976) ch. 1,2 and 3.
4. see for example, V. Bonacic-Koutecky, M. M. Kappes, P. Fantucci, and J. Koutecky, Chem. Phys. Lett. **170**, 26 (1990)
5. W. Ekardt, Phys. Rev. **B 29**, 1558 (1984).

6. W. D. Knight, K. Clemenger, W. A. de Heer, W. A. Saunders, M. Y. Chou, and M. L. Cohen, Phys. Rev. Lett. **52**, 2141 (1984).

7. a. S. Bjornholm, J. Borggreen, O. Echt, K. Hansen, J. Pederson, and H. D. Rasmussen, Phys. Rev. Lett. **65**, 1627 (1990); b. T. P. Martin, T. Bergman, H. Görlich, and T. Lange, Phys. Rev. Lett. **65**, 748 (1990).

8. I. Katakuse, T. Ichihara, T. Matsuo, T Sakurai, and H. Matsuda, Int. J. Mass Spectrom. Ion Processes **67**, 229 (1985).

9. a. K. Clemenger Phys. Rev. **B 32**, 1359 (1985); b. K. Clemenger, Ph. D Thesis, University of California, Berkeley (1985).

10. a. S. G. Nilsson Mat-Fys. Medd. K. Dan. Vidensk. Slesk. **29**, 16 (1955); b. Å. Bohr and B. R. Mottelson, *Nuclear Structure*, Vol. II (Bejamin , Reading, Mass. , !975).

11. A. de Shalit and H. Feshbach, *Theoretical Nuclear Physics*, Vol 1, (Wiley, New York, 1974) pp194-199

12. W. Ekardt and Z. Penzar, Phys. Rev. **B 38**, 4273 (1988).

13. W. D. Knight, K. Clemenger, W. A. de Heer, and W. A. Saunders, Phys. Rev. **B 31**, 445, (1985).

14. D. R. Snider and R. S. Sorbello, Phys. Rev. **B 28**, 5702 (1983).

15. N. D. Lang and W. Kohn, Phys. Rev. **B 3**, 1215 (1971).

16. W. A. Saunders, K. Clemenger, W. A. de Heer, and W. D. Knight, Phys. Rev. **B 32**, 1366 (1985).

17. a. K. Selby, M. Vollmer, J. Masui, V. Kresin, W. A. de Heer, and W. D. Knight, Phys. Rev. **B 40**, 5417 (1989); b. K. Selby, V. Kresin, J. Masui, M. Vollmer, W. A. de Heer, A. Scheidemann, and W. D. Knight, Phys. Rev. **B 43**, 4565 (1991).

18. G. F. Bertsch and D. Tománek, Phys. Rev. **B 40**, 2749 (1989).

19. J. M. Pacheco, Y. Yannoleas, and R. A. Broglia, Phys. Rev. Lett. **61**, 294 (1988).

20. S. Pollack, C. R. C. Wang, and M. M. Kappes, J. Chem. Phys. **94**, 2496 (1991).

21. W. A. de Heer and M. Brack, Rev. Mod. Phys. (in preparation).

Electronic Shell Structure in Metal Clusters

T.P. Martin, T. Bergmann, H. Göhlich, T. Lange, and U. Näher

Max-Planck-Institut für Festkörperforschung,
Heisenbergstr. 1, W-7000 Stuttgart 80, Fed. Rep. of Germany

Introduction

In 1949 Maria Goeppert-Mayer [1] and Haxel, Jensen and Suess [2] suggested a shell model to explain magic numbers of stability for atomic nuclei. Recently, a similar model has been used to sucessfully describe another fermion system - the electrons in metallic clusters [3-18].

If it can be assumed that the electrons in metal clusters move in a spherically symmetric potential, the problem is greatly simplified. Subshells for large values of angular momentum can contain hundreds of electrons having the same energy. The highest possible degeneracy assuming cubic symmetry is only 6. So under spherical symmetry the multitude of electronic states condenses down into a few degenerate subshells. Each subshell is characterized by a pair of quantum numbers n and ι. Under certain circumstances the subshells themselves condense into a smaller number of highly degenerate shells. The reason for the formation of shells out of subshells requires more explanation.

The concept of shells can be associated with a characteristic length. Every time the radius of a growing cluster increases by one unit of this characteristic length, a new shell is said to be added. The characteristic length for shells of atoms is approximately equal to the interatomic distance. The characteristic length for shells of electrons is related to the wavelength of an electron in the highest occupied energy level (Fermi energy). For the alkali metals these lengths differ by a factor of about 2. This concept is useful only because the characteristic lengths are, to a first approximation, independent of cluster size.

The concept of shells can also be described in a different manner. An expansion of N, the total number of electrons, in terms of the shell index K will always have a leading term proportional to K^3. One power of K arises because we must sum over all shells up to K in order to obtain the total number of

Springer Series in Nuclear and Particle Physics   Clustering Phenomena in Atoms and Nuclei
Editors: M. Brenner · T. Lönnroth · F.B. Malik   © Springer-Verlag Berlin, Heidelberg 1992

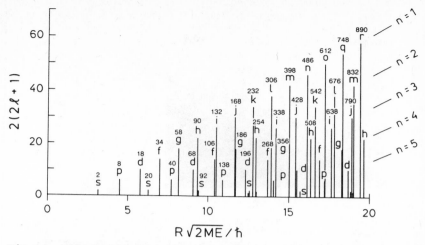

Fig. 1: The degeneracy of states of the infinitely deep spherical well on a momentum scale. The total number of fermions needed to fill all states up to and including a given subshell is indicated above each bar.

particles. One power of K arises because the number of subshells in a shell in-creases approximately linearly with shell index. Finally, the third power of K arises because the number of particles in the largest subshell also increases with shell index. Expressing this slightly more quantitatively, the total number of particles needed to fill all shells, k, up to and including K is

$$N_K = \sum_{k=1}^{K} \sum_{\ell=0}^{L(k)} 2\,(2\,\ell+1) \sim K^3 \tag{1}$$

where L(k) is the highest angular momentum subshell in shell k.

Shell structure is not necessarily an approximate and infrequent bunching of states as in the example of the potential well, Fig. 1. Clearly, almost none of the subshells are exactly degenerate for this potential. However, shell structure can also be the result of exactly overlapping states. Such degeneracies signal the presence of a symmetry higher than spherical symmetry. Subshells of hydrogen for which $n + \ell$ have the same value are degenerate. This additional degeneracy in the states of hydrogen is a result of the form of its potential, $1/r$, which bestows on hydrogen O(4) symmetry. Subshells of the spherical harmonic oscillator for which $2n+\ell$ have the same value are also degenerate, due to the form of the potential, r^2, and the resulting symmetry, SU(3). For this reason it is said that these systems,

27

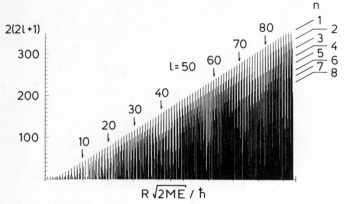

Fig. 2:. The states of the infinitely deep spherical well for very large values of ℓ. Notice the periodic bunching of states into shells. This periodic pattern is referred to as supershell structure.

hydrogen and oscillator, have quantum numbers $n + \ell$ and $2n+\ell$ that determine the energy. We will show that $3n+\ell$ is an approximate energy quantum number for alkali metal clusters [16]. As the cluster increases in size, electron motion quantized in this way would finally be described as a closed triangular trajectory [19].

The grouping of large subshells into shells is illustrated in Fig. 2 for the spherical potential well. Here, it can again be seen that in certain energy or momentum regions the subshells bunch together. However, the states are so densely packed in this figure that the effect is perceived as an alternating light-dark pattern. That is, for the infinite potential well, bunching of states occurs periodically on the momentum scale. The periodic appearance of shell structure is referred to as _supershell_ structure [20,21]. Although supershell structure was predicted by nuclear physicists more than 15 years ago, it has never been observed in nuclei. The reason for this is very simple. The first supershell beat or interference occurs for a system containing 1500 fermions. There exist, of course, no nuclei containing so many protons and neutrons. It is possible, however, to produce metal clusters containing such large numbers of electrons.

Experimental

The technique we have used to study shell structure is photo-ionization time-of-flight (TOF) mass spectrometry. The mass spectrometer has a mass range of 600 000 amu and a mass resolution of up to 20 000. The cluster source is a low pressure, rare gas, condensation cell. Sodium vapor was quenched in cold He gas having a pressure of about 1 mbar. Clusters condensed out of the quenched vapor were transported by the gas stream through a nozzle and through two chambers of intermediate pressure into a high vacuum chamber. The size distribution of the clusters could be controlled by varying the oven-to-nozzle distance, the He gas pressure, and the oven temperature. The clusters were photoionized with a 1 μJ, 2x1 mm, 15 nsec dye laser pulse. The high resolution mass spectra showed no evidence of cluster fragmentation.

Since phase space in the ion optics is anisotropically occupied at the moment of ionization, a quadrupole pair is used to focus the ions onto the detector. All ions in a volume of 1 mm^3 that have less than 500 eV kinetic energy at the moment of ionization are focused onto the detector [22].

Observation of Electronic Shell Structure

Knight, Clemenger, de Heer, Saunders, Chou and Cohen [3] first reported electronic shell structure in sodium clusters in 1984. Electronic shell structure can be demonstrated experimentally in several ways: as an abrupt decrease in the ionization energy with increasing cluster size, as an abrupt increase or an abrupt decrease in the intensity of peaks in mass spectra. The first type of experiment can be easily understood. Electrons in newly opened shells are less tightly bound, i.e. have lower ionization energies. However, considerable experimental effort is required to measure the ionization energy of even a single cluster. A complete photoionization spectrum must be obtained and very often an appropriate source of tunable light is simply not available. It is much easier to observe shell closings in photoionization, TOF mass spectra. However, depending upon the intensity and wavelength of the ionizing laser pulse, the new shell is announced by either an increase or a decrease in mass peak height.

Cluster intensities can sometimes be increased by a factor of ten by using a seed to nucleate the cluster growth. For example, by adding less than 0.02% SO_2 to the He cooling gas, Cs_2SO_2 molecules form which apparently promote further cluster growth. Mass spectra of $Cs_{n+2}(SO_2)$ clusters obtained [15] using four different dye-laser photon energies are shown in Fig. 3. Although it is not possible to distinguish the individual mass peaks in this con-

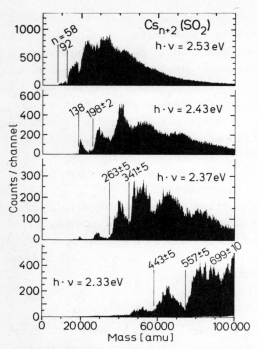

Fig. 3: Mass spectra of $Cs_{n+2}(SO_2)$ clusters with decreasing photon energy of the ionizing laser from 2.53 eV (top) to 2.33 eV (bottom). The values of n at the steps in the mass spectra have been indicated.

densed plot, it is evident that the spectra are characterized by steps. For example, a sharp increase in the mass-peak intensity occurs between n = 92 and 93. This can be more clearly seen if the mass scale is expanded by a factor of 50 (Fig. 4). Notice also that the step occurs at the same value of n for clusters containing both one and two SO_2 molecules. In addition to the steps for n = 58 and 92 in Fig. 3, there are broad minima in the 2.53 eV spectrum at about 140 and 200 Cs masses. These broad features become sharp steps if the ionizing photon energy is decreased to 2.43 eV. By successively decreasing the photon energy, steps can be observed for the magic numbers n = 58, 92, 138, 198±2, 263±5, 341±5, 443±5, and 557±5 [15,16]. However, the steps become less well defined with increasing mass. We have studied the mass spectra of not only $Cs_{n+2}(SO_2)$ but also $Cs_{n+4}(SO_2)_2$, $Cs_{n+2}O$, and $Cs_{n+4}O_2$. They all show step-like features for the same values of n.

30

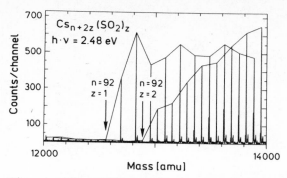

Fig. 4:. Expanded mass spectra of $Cs_{n+2z}(SO_2)_z$ clusters for an ionizing photon energy of 2.48 eV. The lines connect mass peaks of clusters containing the same number z of SO_2 molecules. Notice that the steps for clusters containing (SO_2), and $(SO_2)_2$ are shifted by two Cs atoms.

First, we would like to offer a qualitative explanation for these results and then support this explanation with detailed calculation. Each cesium atom contributes one delocalized electron which can move freely within the cluster. Each oxygen atom, and each SO_2 molecule, bonds with two of these electrons. Therefore, a cluster with composition $Cs_{n+2}(SO_2)$, for example, can be said to have n delocalized electrons. The potential in which the electrons move is nearly spherically symmetric, so that the states are characterized by a well-defined angular momentum. Therefore, the delocalized electrons occupy subshells of constant angular momen-tum which in turn condense into shells. When one of these shells is fully populated with electrons, the ionization energy is high and the clusters will not appear in mass spectra obtained using sufficiently low ionizing photon energy.

In another experiment [9] the closing of small subshells of angular momentum was shown to be accompanied by a sharp step in the ionization energy for Cs-O clusters having certain sizes, namely, for $Cs_{n+2z}O_z$ with n = 8, 18, 20, 34, 58, and 92. The closing at n = 40 seen in all other alkali-metal clusters could not be observed, neither in the experiments nor in the calculations. The steps were observed for clusters containing from one to seven oxygen atoms.

Density Functional Calculations

Self-consistent calculations have been carried out applying the density functional approach to the spherical jellium model

Main quantum number

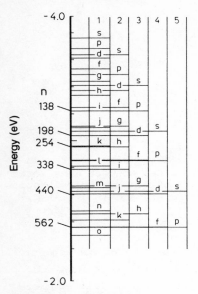

Fig. 5: The self-consistent, one-electron states of a 600 electron cesium cluster calculated using a modified spherical jellium background.

[10,11]. We used an exchange correlation term of the Gunnarsson-Lundqvist form and a jellium density $r_S = 5.75$ corresponding to the bulk value of cesium. This model implies two improvements over the hard sphere model. Firstly, electron-electron interaction is included. Secondly, the jellium is regarded to be a more realistic simplification of the positive ion background than the hard sphere. The O^{2-} ion is taken into account only by omitting the cesium electrons presumably bound to oxygen. The calculations were performed on Cs_{600} clusters.

We found, that if a homogeneous jellium was used, the grouping of subshells was rather similar to the results of the infinite sperical potential well. However, a nonuniform jellium yielded a shell structure in better accordance to experimental results. We found that the subshells group fairly well into the observed shells only if the background charge distribution is slightly concentrated in the central region. This was achieved, for example, by adding a weak Gaussian (0.5% total charge density, half-width of 6 a.u.) charge distribution to the uniform distribution (width 48 a.u.). Figure 5 shows the ordering of subshells obtained from this potential. This leads to the rather surprising

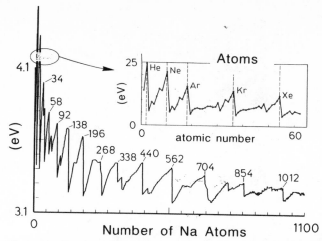

Fig. 6: Ionization potentials calculated as a function of n for Na_n clusters. A positive background charge distribution slightly concentrated in the central region has been used. Notice the similar behaviour of the ionization energies of the chemical elements (inset).

result that the Cs^+ cores seem to have higher density in the neighbourhood of the center, perhaps due to the existence of the O^{2-} ion. All attempts to lower the positive charge density in the central region led to an incorrect ordering of states.

This first calculation addressed the problem of the grouping of low lying energy levels in one large Cs_{600} cluster. However, in the experiment the magic numbers were found by a rough examination of ionization potentials of the whole distribution of cluster sizes. A more direct way to explain magic numbers is to look for steps in the ionization potential curve of Cs-O clusters. Therefore, we calculated the ionization potentials of $Cs_{n+2}O$ for $n \leq 600$ and of Na_n for $n \leq 1100$ using the same local-density scheme described above, Fig. 6. Starting from a known closed shell configuration for $n = 18$, electrons were succesively added. Three test configurations were calculated for each cluster size testing the opening of new subshells. The configuration with minimum total energy was choosen for the calculation of the ionization potential.

Transition from Shells of Electrons to Shells of Atoms

Recently, two types of shell structure have been observed in the same mass spectrum of large sodium clusters, Fig. 7. For

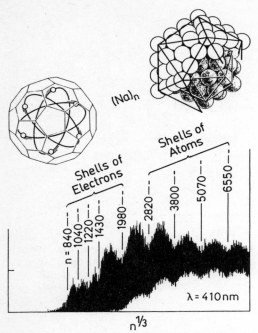

Fig.7: Mass spectrum of $(Na)_n$ clusters photoionized with 3.02 eV photons. Two sequences of structures are observed at equally spaced intervals on the $n^{1/3}$ scale - an electronic shell sequence and a structural shell sequence.

small clusters ($n \leq 1500$) the pattern appears to be due to the filling of electronic shells. For large clusters the shells seem to be composed of atoms.

Why might one expect a transition from electronic shell structure to shells of atoms? For very small clusters the atoms are highly mobile. There is no difficulty for the atoms to arrange themselves into a sphere-like conformation if this is demanded by the closing of an electronic shell. At a size corresponding to about 1500 atoms under our experimental conditions, the clusters become rigid. There after, each newly added atom condenses onto the surface and remains there. Further growth takes place by the accumulation of shells of atoms.

References

1. M.G. Mayer; Phys. Rev. **75**:1969L (1949).
2. O. Haxel, J.H.D. Jensen, and H.E. Suess, Phys. Rev.**75**:1766L (1949).

3. W.D. Knight, K. Clemenger,W.A. de Heer, W.A. Saunders, M.Y. Chou, and M.L. Cohen, Phys. Rev. Lett., **52**:2141 (1984).

4. M.M. Kappes, R.W. Kunz, and E. Schumacher, Chem. Phys. Lett. **91**:413 (1982).

5. I. Katakuse, I. Ichihara, Y. Fujita, T. Matsuo, T. Sakurai, and T. Matsuda, Int. J. Mass Spectrom. Ion Processes **67**:229 (1985).

6. C. Brechignac, Ph. Cahuzac, J.-Ph. Roux, Chem. Phys. Lett. **127**:445 (1986).

7. W. Begemann,.S. Dreihofer, K.H. Meiwes-Broer, and H.O. Lutz, Z. Phys.D **3**:183 (1986)

8. W.A. Saunders, K. Clemenger, W.A. de Heer, W.D. Knight, Phys. Rev.B **32**:1366 (1986).

9. T. Bergmann, H. Limberger, T.P. Martin, Phys. Rev. Lett. **60**:1767 (1988).

10. J.L. Martins, R. Car, J. Buttet, Surf. Sci. **106**:265 (1981).

11. W. Ekardt, Ber. Bunsenges. Phys. Chem **88**:289 (1984).

12. K. Clemenger, Phys. Rev. B **32**:1359 (1985) .

13. Y. Ishii, S. Ohnishi, and S. Sugano, Phys. Rev B **33**:5271 (1986).

14. T..Bergmann, T.P. Martin, J. Chem. Phys **90**:2848 (1989)

15. H. Göhlich, T. Lange, T. Bergmann, and T.P. Martin, Phys. Rev. Lett. **65**:748 (1990); to be published in Z. Phys. D.

16. T.P. Martin, T. Bergmann, H.Göhlich, and T. Lange, Chem. Phys. Lett..**172**:209 (1990);.to be published in Z. Phys. D.

17. S. Bjørnholm, J.Borggreen, O. Echt, K. Hansen, J. Pederson, and H.D.Rasmussen, Phys. Rev. Lett. **65**:1627 (1990).

18. J.L. Persson, R.L. Whetten,.Hai-Ping Cheng, and R.S. Berry, to be published; E.C. Honea, M.L. Horner, J.L. Persson and R.L. Whetten, Chem. Phys. Lett. **171**:147 (1990).

19. R. Balian, and C. Bloch, Ann. Phys, **69**:76 (1971).

20. A. Bohr, B.R. Mottelson, Nuclear Structure, Benjamin, London (1975).

21. H. Nishioka, K. Hansen, B.R. Mottelson, Phys. Rev. B **42**:9377 (1990).

22. T. Bergmann, H. Göhlich, T.P. Martin, H. Schaber, and G. Malegiannakis, Rev. Sci.Instrum. **61**:2585 (1990).

23. B.A. Mamyrin, V.I. Katataer, D.V. Shmikk, and V.A. Zagulin, Sov. Phys.JETP **37**:45 (1973).

24. T. Bergmann, T.P. Martin, H. Schaber,Rev. Sci Instrum. **61**:2592 (1990).

25. T. Lange, H. Göhlich, T. Bergmann, and T.P. Martin, to be published in Z. Phys. D.

Quantum Chemical Interpretation of Absorption Spectra of Small Alkali Metal Clusters

V. Bonačić-Koutecký[1], *P. Fantucci*[2], *and J. Koutecký*[1]

[1]Institut für Physikalische und Theoretische Chemie,
 Freie Universität Berlin, W-1000 Berlin 33, Fed. Rep. of Germany
[2]Dipartimento di Chimica Inorganica e Metallorganica,
 Università di Milano, Centro CNR, I-20133 Milano, Italy

Comparison of the depletion spectra of small alkali metal clusters with the results obtained from the quantum chemical investigation of excited states determined for the equilibrium geometries makes possible the structural assignments. The knowledge gained about specific geometrical and electronic properties responsible for excitations in small clusters is used for explanation of the predicted and measured spectroscopic patterns. Similarities and differences between the optical response properties of Na_n and Li_n clusters have been found and discussed.

Introduction

Theoretical investigation of excited states of simple s^1 metal clusters offers an excellent opportunity to study a development of specific electronic and structural properties as a function of the cluster size[1]. Moreover, a comparison of quantum chemical predictions of absorption spectra[1-11] with the recorded ones by depletion spectroscopy[12-27,10,11] represent a new tool for structural assignments. The interplay between electronic and geometrical structure gives rise to very characteristic spectroscopic patterns with small number of very intense transitions and a large number of weak ones.

It will be clearly shown that small alkali metal clusters exhibit several of their own specific features, although formal similarities, due to common physical laws, might be present for systems which possess a number of substantially different physical properties. For example, some features of absorption spectra resemble a giant resonances found for nuclei. Similarly, individual cluster properties might resemble metallic characteristics but others can be far from those typical for the metallic state. Important is to find the reasons for such similarities in order to distinguish the physical nature of these parallels.

The ab initio quantum chemical methods which take into account electronic correlation (e.g. configuration interaction — CI) are well suited for this purpose, since they avoid a priori assumptions which can be well established and justified in other fields. Consequently, a part from their predictive power, they can serve to

Springer Series in Nuclear and Particle Physics   **Clustering Phenomena in Atoms and Nuclei**
Editors: M. Brenner · T. Lönnroth · F.B. Malik   © Springer-Verlag Berlin, Heidelberg 1992

extract the essentials which are necessary to include in simplified models used to determine ground and excited states properties of elemental clusters.

Therefore, the quantum molecular approach will be introduced first by presenting the atomization energies for the most stable ground state geometries of the neutral and charged Li_n and Na_4 clusters[28-33]. The measurements of molecular beam abundances of alkali metal clusters[34-39] yielded information about special stabilities of clusters which survived fragmentation[40-45]. Except of very recent development of cold cluster sources[46], the most experiments used hot cluster sources and therefore it has been argued that the positions of nuclei are of no importance and can be ignored as assumed in the framework of the jellium model[34]. In contrast to abundance experiments, the depletion spectroscopy offers more structural information although the exact temperature is not known. The first photodissociation experiments on clusters larger than tetramers have been carried out for Na_n (n=8,9,10,12,16 and 20)[12-14], but with low resolution and only for selected visible wavelengths, so that some spectral regions have not been covered. These measurements have been interpreted in terms of classical collective electronic oscillations in spherical, spheroidal (or elipsoidal) metal droplets, since they exhibited from one to three broad maxima. In the case of cationic clusters K_n^+ (n=9,21)[24,25] a pronounced single intense transition was observed, resembling a "giant resonance" found for the interaction of atomic nuclei with radiation. Similarly, the photoabsorption spectra of two closed–shell clusters Cs_8 and $Cs_{10}O$[26] have been interpreted as due to plasmon or plasmon–enhanced excitations. The high–resolution spectra for Na_n (n=3–21,8,20)[17-20,27], Li_n (n=2–4,6–8)[22,23,10,11] and Li_yNa_x[21] also became available, covering a good part of the spectral region. These experimental findings called for theoretical interpretations using more sophisticated quantum molecular models, since the presence of quantum effects became evident. It will be shown that the recently calculated transition energies and oscillator strengths using ab initio quantum chemical methods[1-11] for the lowest energy structures determined in the earlier work on the ground state properties[29,31-33] are in excellent agreement with the recorded absorption spectra[17-23,25,10,11,27]. This illustrates unambiguously that the position of nuclei is important in spite of relatively large mobility of alkali metal clusters due to their delocalized many center bonds[47,28]. Although our calculations are directly relevant for T=0 their reliability can be extended for low temperatures as well. This is supported by the recent work on Na_n[48] using Car–Parrinello method also for T>0 in which relative rigidity of Na_n cluster geometries up to T $\approx$ 300 K has been reported. Moreover, the successful structural assignment of topologies determined at T=0 to the recorded depletion spectra suggests that i) certain rigidity of geometries must be present since the experiments are not carried out at T=0 and that ii) the experimental temperature cannot be substantially higher than T=300 K.

Finally, a comparison of the CI predictions and different simplified approaches such as RPA based on the Hartree—Fock approximation[1,49] or on jellium potential[50] applied for determination of absorption spectra will be made. The results of the ab initio CI calculations will be analyzed also in a qualitative manner and used for the physical interpretation of the specific features characteristic for the excitation in small alkali metal clusters.

II. Methods Employed for Interpretation of Absorption Spectra

There are five approaches presently available which have been used for description of absorption spectra of alkali metal clusters:

1) Classical Mie—Drude theory has been frequently employed to estimate the surface plasmon frequency[12-14]. Of course, the concept of the volume plasmon which is a well defined collective mode in quantum mechanics of very large systems, yields also, in the classical limit a good estimate of the plasmon frequency, which depends only on the electron density. The surface plasmon involves additional assumptions upon the nature of the restoring force and can yield only a rough estimate of the interval in which the frequencies of the absorbed light can be expected. Using measured static polarizabilities and assumption upon spherical,spheroidal or elipsoidal shape of clusters, one, two or three frequencies, respectively, can be calculated. The predictions about the positions of intense bands using this simple model are in some cases in agreement (e.g. Na_8) and in some not (e.g. Na_{20} and Na_{40}) with the experimental findings[13,14]. It is also remarkable that the disagreement is more pronounced for larger clusters. In summary, it is not surprising that additional "quantum effects" are present in clusters of finite small dimensions and that the appearance of one, two or three broad intense bands in absorption spectra of small clusters is not a proof for the existence of the surface plasmon in spherical metal droplets due to a collective motion.

2) The time dependent local density approximation (TDLDA) based on jellium model[51-54] has been applied to the calculations of photoabsorption cross sections. In this type of method the quantum mechanical treatment of electronic correlation is approximately taken into account, since the calculations of the imaginary part of the dynamic polarizability involve interactions among single particle—hole transitions. Kohn—Sham orbitals have been used and the calculated maxima have been attributed to "surface plasmons" or to volume plasmons at higher energies. Although the quantum effects are included many approximations are involved which makes the analysis of the results difficult, since the effects produced by partial

treatment of correlation effects cannot be separated from those due to the jellium potential.

3) In the framework of the random phase approximation (RPA) based on jellium model[50], the electron correlation has been taken into account using the variational density formalism and approximate exchange correlation potential due to Gunnarson and Lundquist[55]. In the RPA procedure the interaction among single particle—hole transitions is explicitly taken into account and in addition the matrix elements between forwards and backwards transitions have a similar form as those used for the interaction between the ground state and two particle—two holes transitions in the CI treatments (cf. Ref. 1).

4) The RPA calculations based on the ab initio Hartree—Fock ground state orbitals[49,1,8] account for the correlation effects in the same manner as the RPA with jellium background[50]. Therefore a comparison of results obtained from both RPA approaches makes possible to analyze the deficiencies due to the jellium approach (e.g. approximate potential and neglect of position of nuclei).

5) The ab initio multireference CI approach used in our work[1—11] accounts for correlation effects at higher level of approximation. The interaction among two—particles two—holes transitions (double excitations) are treated at the same footing as interactions among single excitations and higher order excitations are taken into account at least partly in the truncated CI. Therefore a comparison between results obtained from ab initio RPA and CI results makes possible to find out which kind of correlation treatment is necessary for successful interpretation of absorption spectra. Notice, that the accuracy of ab initio calculations in addition to the truncation of the CI expansions is determined by the choice of the atomic basis sets (AO). The quality of AO basis sets used in our work have been carefully checked comparing a number of calculated spectroscopical properties with the available experimental values (for details compare Ref. 1 and references therein).

III. Ground State Structures of Li_n and Na_n Clusters

The binding energy per atom $E_b/n = (nE_1 - E_n)/n$ where E_n is the CI energy of the cluster with n atoms calculated at the SCF optimized geometry as a function of the cluster size exhibits increasing tendency for neutral and charged Li_n and Na_n clusters as shown in Figure 1. Pronounced maxima can be depicted for systems with 2 and 8 valence electrons. The topologies of the optimized structures obtained in the framework of the analytical gradient method drawn in Figure 1 illustrate that the small clusters assume planar geometries, due to the Jahn—Teller deformation, which can be viewed as deformed sections of the fcc lattice. Since a number of valence electrons is not sufficient to fill up the degenerate one—electron levels of highly

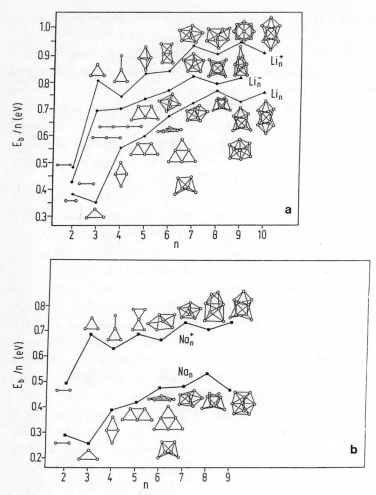

Fig. 1: The CI binding energy per atom E_b/n as a function of the nuclearity n for a) neutral, anionic and cationic Li_n clusters and for b) neutral and cationic Na_n clusters. The topologies of the H–F optimized geometries are shown. Small, but appropriate AO basis sets are used. For details cf. Refs. 29,31,32.

symmetrical structures the lowering of the symmetry takes place. Therefore, the optimal structure of the neutral Li_4 or Na_4 tetramer is a rhombus and not a tetrahedron. In contrast, Li_8 and Na_8 assume compact symmetrical T_d structures since eight valence electrons are sufficient to occupy also three degenerate MO's or in the superatom picture the ground state configuration crresponds to $1s^2 1p^6$ (cf. Ref. 1).

40

Hexamers represent an interesting case for which the transition between the two–dimensionality (2D) to the three–dimensionality (3D) might occur. The energy differences between the planar D_{3h} structure and the flat pentagonal pyramid (C_{5v}) or the C_{2v} geometry built from the tetrahedral subunits are very small. However, the C_{2v} structure has the lowest energy in the case of Li_6 but for Na_6 both the planar and almost planar C_{5v} structure have lower energies than the C_{2v} structure. Also only the planar Na_6 structure is a local minimum on the SCF energy surface while all three Li_6 structures represent local minima. Notice, however, that determination of local minima and saddle points on such flat energies surfaces is dependent on the details of calculations.

A possible structural difference between Li_6 and Na_6 will be apparent when comparing the depletion spectra with the quantum chemical predictions. Topologies of the energetically low lying neutral structures for Na_{2-8} and Li_{2-8} are identical. It is worth of mentioning that the geometries of small cations and anions can substantially differ from the neutral geometries. Notice also that E_b/n curves calculated for clusters representing sections of the fcc lattice obtained without geometry optimization exhibit different behaviour indicating that certain classes of geometries have definitly lower energies than others. Also E_b/n for cationic clusters obtained from the SCF calculations do not increase with the cluster size. The correlation effects contribute more than 50% to the E_b/n (for details cf. Ref. 1 and 27).

IV. Quantum Molecular Assignment of the Absorption Spectra

The first absorption spectra have been recorded several years ago for Na_3 and Li_3 using two–photon ionization technique and they have been interpreted in terms of dynamical Jahn–Teller effect[56,15,16,35]. Additional electronic bands with considerable intensities have been later recorded via depletion spectroscopy for both Na_3 [18] and Li_3 [57]. The ab initio CI calculation for the optically allowed excited states of Na_3 in the energy interval up to 3.1 eV at the optimal ground state geometry (obtuse isosceles triangle) and corresponding oscillator strengths f_e are compared with the depletion spectra of Na_3 [18] in Figure 2. There is a large number of optically allowed states in the energy interval between 2–3 eV and the calculations of f_e helps to clarify the assignment. The seven vertical transitions with

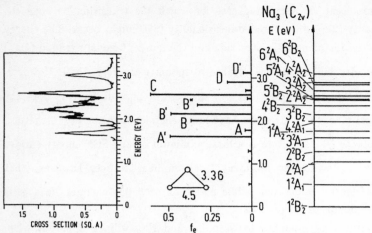

Fig. 2: Comparison of the photodepletion spectrum[18] and the CI optically allowed transitions (eV) and oscillator strengths f_e for Na_3 (AE–MRDCI treatment)[4].

respect to the ground state 1^2B_2 of the isosceles triangle together with their oscillator strengths have been assigned to the recorded bands exhibiting a complete agreement with dominant as well as fine features. The spectroscopy of trimers represents a special case since the resolved vibrational fine structure (i.e. progression in the excited states and some hot bands) has been measured and some indirect evidence about ground state geometry has been obtained.

Recent progress in the experimental field has opened the new possibility to investigate the size dependent features of absorption pattern especially the high resolution spectra of Na_n (n=3–8,20)[17–20,27], Li_n (n=3,4,6–8)[22,23,10,11] and Li_yNa_x[21] covering the spectral region up to 3.2 eV. In these cases the structural assignment is possible only from a comparison of theoretical predictions and experimental spectra. Therefore the optically allowed states and the corresponding oscillator strengths have been calculated not only for the most stable structures, but also for energetically close lying ones in order to investigate how dependent is the spectroscopic pattern on topology, symmetry, dimensionality and the number of valence electrons.

For tetramers Na_4 and Li_4 the planar rhombic structures are the most stable geometries and the deformed tetrahedron forms ($C_{2v} \approx D_{2d}$) are the best geometries for the triplet ground states with energies which are substantially higher with respect to the rhombic singlet ground states (> 0.25 eV). A comparison of predicted and recorded spectra for both singlet (2D) and triplet (3D) structures of Na_4 is

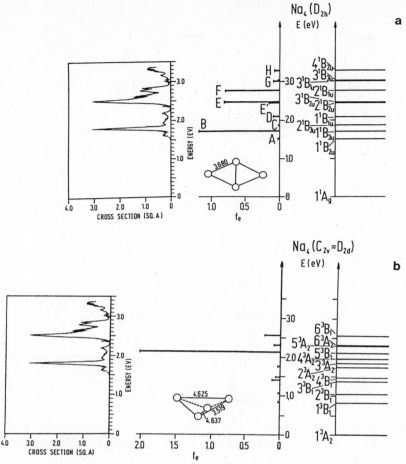

Fig. 3: Comparison of the depletion spectrum of Na_4[17,18] and the CI energies of optically allowed states (eV) and the oscillator strengths f_e for a) the best neutral rhombus (singlet state) and b) the deformed tetrahedron (best triplet state) structures[2,4].

shown in Figure 3. This illustrates clearly that the all eight measured bands can be entirely assigned to the rhombic structure. The transition energies and oscillator strengths are in excellent agreement with the rich spectrum including three intense and five weaker transitions. In contrast 3D–triplet structure gives rise to the spectrum dominated by one intense transition which location and intensity do not agree with the experiment. This finding illustrates clearly that only rhombic structure is responsible for the recorded spectrum and that the 3D structure of such small cluster as Na_4 with 4 valence electrons gives rise to only one dominant

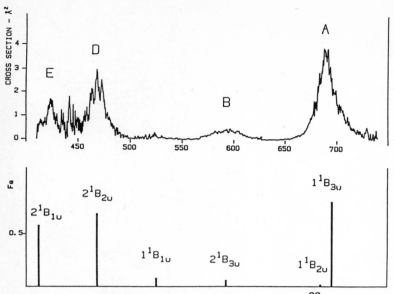

Fig. 4: Comparison of the photodepletion spectrum[22] and the CI optically allowed transitions (nm) and oscillator strengths f_e for Li_4[5,10].

transition. The assignment of the rhombic structure to the measured Li_4 spectrum as shown in Figure 4 is straight forward. The similarity with predicted as well as recorded spectra of Li_4 and Na_4 is striking. In the latter case just several additional weak transitions have been found.

The analysis of the CI wavefunctions for each state of Na_4 or Li_4 in terms of leading configurations shows that linear combinations of the same type of excitations can lead to very intense or very weak transitions illustrating interference phenomena known in molecular spectroscopy. This analysis is also useful for a qualitative comparison of the CI results with simplified methods. Notice, however, that the contributions of a large amount of configurations which are included in the CI in addition to the leading ones is essential for a reliable prediction of locations of transitions and their oscialltor strengths.

It is interesting to compare the CI and the RPA predictions for optically allowed states of Li_4 in order to find out to which extend the simplified electron correlation treatment can be reliable. Fig. 5 illustrates that there is discrepancy between the CI and the RPA results for transition energies and oscillator strengths except for the $^1B_{3u}$ states. In a fact the single excitations play the leading role in the CI wavefunctions of the $^1B_{3u}$ states and therefore they are reliably described in the

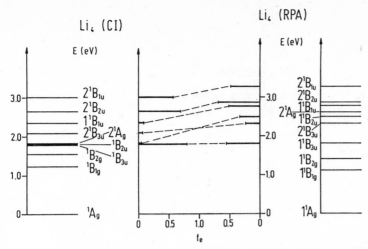

Fig. 5: Comparison of ab initio (AE)–CI and RPA transition energies (eV) and oscillator strengths f_e for optically allowed states of Li_4 (D_{2h}) rhombic structure.

framework of the RPA as well. However, the inclusion of double and single excitations at the same footing in the correlation treatment is a necessary condition for acceptable description of optically allowed states in the energy interval up to 3.0 eV for both Li_4 and Na_4 tetramers (cf. Ref. 1) which are needed for the structural assignment.

The optically allowed transitions have been calculated for two Na_5 structures: the planar (C_{2v}) and the deformed trigonal bipyramid (C_{2v}). The CI energy difference of their ground states is 0.25 eV in favour of the planar structure. A comparison of predicted CI absorption spectra for both structures is given in Figure 6. The calculated vertical spectrum for the planar Na_5 exhibits large oscillator strengths for transitions located between 2.0 and 2.5 eV. In the energy interval 2.5–3.0 eV no intense transitions have been calculated. In the case of bipyramidal structural there are more transitions with considerable oscillator strengths than for the planar structure and they are distributed over the whole energy interval 2.0–2.8 eV. Although absorption spectra recorded at high resolutions are not yet completed, it is likely that the planar structure will be responsible for the spectrum.

Hexamers represent a challenging case for theoretical interpretation. The planar structure and the flat pentagonal pyramid differ topologically only by the position of one atom and the latter one is almost two–dimensional (2D). In a fact the calculated CI spectroscopic patterns for both structures are similar and both of

45

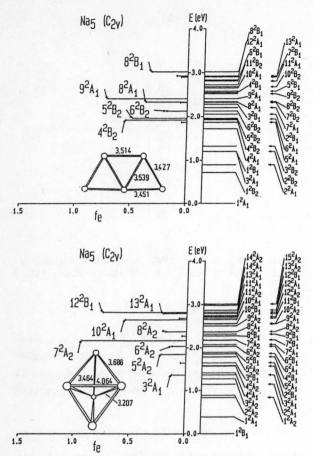

Fig. 6: The CI optically allowed transitions (eV) and oscillator strengths f_e for planar and bipyramidal structure of Na_5 [8].

them substantially differ from the pattern predicted for the three–dimensional C_{2v} structure as shown in Figure 7. For both the 2D planar and the C_{5v} structure the CI oscillator strengths have large values for transitions located at $\sim$ 2.1 eV. The transitions with considerably lower intensity have been predicted at $\sim$ 3.0 eV. In addition two weak transitions red shifted with respect to the dominant feature have been obtained for the planar structure. Dominant as well as all fine features of the absorption spectra obtained from the CI procedure for the planar Na_6 structure are in a good agreement with the recorded spectrum (Figure 7). The C_{5v} Na_6 structure is not a local minimum on the SCF energy surface, but as mentioned earlier the

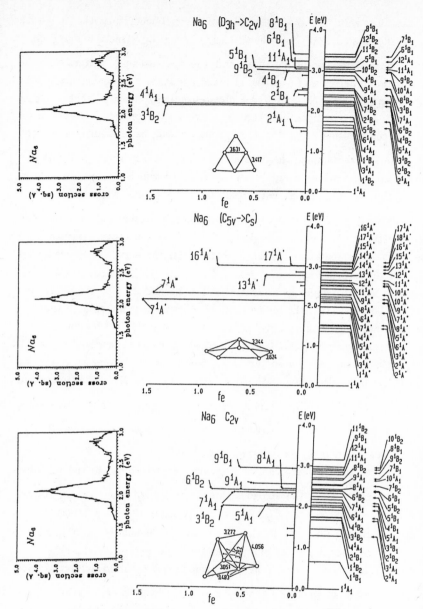

Fig. 7: Comparison of the photodepletion spectrum[27] and the CI optically allowed transitions (eV) and oscillator strengths f_e for three Na_6 structures[8,1]: $D_{3h} \rightarrow C_{2v}$, $C_{5v} \rightarrow C_s$ and C_{2v}. Notice, that the labels of the states correspond to the C_{2v} point group for both planar and pentagonal pyramid since the geometries deviate slightly from the D_{3h} and C_{5v} symmetry, respectively.

47

energy surface are very flat and the results of the frequency analysis should be considered with some caution. Since the CI energy of the C_{5v} structure is negligibly higher with respect to the planar form and the spectroscopic pattern for the pentagonal pyramid agrees well with the recorded dominant and weaker features in the energy interval $\sim$ 2.0 eV and $\sim$ 3.0 eV, respectively, it is difficult to rule completely out a contribution from this structure to the recorded spectrum. In contrast, the contribution from the C_{2v} structure can be excluded, since the predicted pattern with dominant features at $\sim$ 2.5 eV is not in agreement with the experimental finding and the energy of this structure which is also not a local minimum is considerably higher with respect to the planar Na_6. This finding demonstrates clearly the importance of the positions of nuclei and therefore the structural aspects should not be neglected due to the unknown experimental conditions such as temperature. If the average of all three structures would be responsible for the recorded spectrum, provided they are given comparable weights, three dominant features would be present at $\sim$ 2.2, 2.5 and 2.8 eV. Therefore, the conclusion can be drawn that the temperature of experiment is not so high that the interconversion between structures occurs via vibrations. This is also supported by molecular quantum dynamic (QMD) calculations using density functional theory[48]. In this work a competition between the flat pyramid and the planar structure for Na_6 at T=0 has been also found. Moreover, at $\sim$ 200K the 2D related structure remains rigid, at $\sim$ 350K the C_{5v} structure oscillates between two mirror structures and only at 600 K the D_{3h} and C_{5v} structures do start to interconvert. The 3D C_{2v} structure has been found as higher lying transition state.

The finding that the C_{2v} structure of Na_6 does not contribute to the recorded spectrum is interesting since the anlogous three–dimensional form has the lowest energy for Li_6[33]. Moreover, the C_{2v} Li_6 structure gives rise to a spectroscopic pattern with dominant features located at $\sim$ 2.5 eV which is in a complete agreement with the recorded spectrum[10,11] as shown in Figure 8. The other two considered structures of Li_6: planar and flat pentagonal pyramid exhibit similar spectroscopic patterns as analogous topologies of Na_6. From a comparison of the experimental data and the CI results for all three Li_6 structures the contributions from the planar geometry and the flat pyramid can be excluded. The conclusion can be drawn that there is a structural difference between Li_6 and Na_6. The three–dimensionality of Li_6 might be connected with the stronger tendency of the Li atom (in comparison with Na) to assume the s–p type of bonding which leads to the compact structure of Li_6.

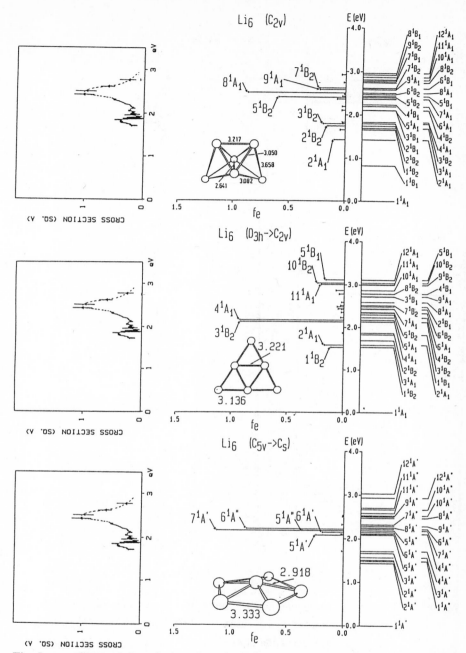

Fig. 8: Comparison of the photodepletion spectrum and the CI optically allowed transitions (eV) and oscillator strengths f_e for three Li_6 structures: $D_{3h} \to C_{2v}$, $C_{5v} \to C_s$ and C_{2v}[10,11]. For explanation of labels compare Figure 7.

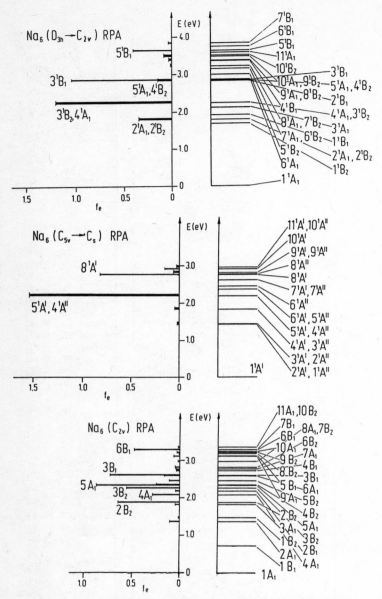

Fig. 9: The RPA (based on the SCF procedure) transition energies (eV) and oscillator strengths f_e for three Na_6 structures[8,1]: $D_{3h} \rightarrow C_{2v}$, $C_{5v} \rightarrow C_s$ and C_{2v}. For explanation of labels compare Figure 7.

A comparison of the results obtained from the CI and the RPA treatments for the spectrum of the three structures of Na_6 (Figures 8 and 9) offers a good opportunity to point out that one has to be aware of approximations involved in different treatments in order to judge the reliability of their predictions. The dominant pair of transitions located at 2.1–2.2 eV present for both, the planar and the flat pentagonal pyramid occur to states which contain as leading configurations two single excitations arising among pairs of almost degenerate one–electron levels. Although the CI wavefunctions contain few additional leading configurations, two single excitations represent ~ 70% of the wavefunctions. Consequently, the transition energies obtained by the RPA technique are only negligibly higher with respect to the CI ones. However, there is a substantial difference between the CI and the RPA results for the higher lying excited states concerning their location and values of oscillator strengths. For example, the RPA yields a considerably larger f_e value for a transition to the $3\,^1B_1$ state of the planar Na_6 structure and its location is ~ 0.3 eV higher with respect to the corresponding CI state (cf. Figures 8 and 9). This is due to the leading role of single and double excitations (two–particle, two–hole transitions) in the CI wavefunctions. Since the RPA does not take into account the interaction among double excitations it does not describe these type of states adequately. However, the location of the $3\,^1B_1$ state obtained by the RPA method coincides fortuitously with the region of the measured weak broad band, although the calculated oscillator strength is too large with respect to the most intense transitions. The discrepancy between the RPA and the CI results are even larger for the C_{2v} structure of Na_6.

From a comparison between the CI and the RPA results a rough estimate can be made that a dominant intense transition can arise due to an interaction of very few (two or three) particle–hole transitions among degenerate or almost degenerate pairs of one–electron levels. In this case the RPA will reproduce this part of spectrum relatively well. However, the approximate location of the intense transition depends roughly upon the energy difference between the one–electron levels among which excitations occur. This one–electron scheme is determined by the symmetry, geometry and the potential. Therefore, differences between described RPA results and those based on jellium model[50] are to be expected[1]. Of course, if the symmetry of the systems plays important role a discrepancy among the results of the simplified treatments will be less severe.

For heptamers and octamers the structural assignment of Na_n and Li_n is again very similar. The pentagonal bipyramids are the most stable structues of Na_7 and Li_7 and they give rise to a dominant feature located at ~ 2.5 eV[8,10] in a good agreement with the recorded spectra[27,10].

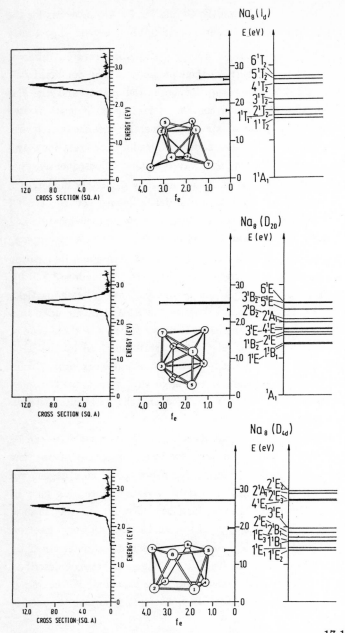

Fig. 10: Comparison of the photodepletion spectrum[17,18] and the CI optically allowed transitions (eV) and oscillator strengths f_e for three Na_8 structures[3,5]: T_d, D_{2d} and D_{4d}.

The spectra of Na_8 [17,18] and Li_8 [10] are characterized by a dominant intense band located at ~2.5 eV. The dominant features of Li_7 and Li_8 almost coincide in the location, and in the case of Na_8 the intense band is only slightly blue shifted with respect to the one measured for Na_7 [27].

A comparison of the CI results for three structures of Na_8 with the experimental spectrum is given in Figure 10. As already pointed out the most stable structures for Li_8 and Na_8 are highly symmetrical compact T_d structures (cf. Section I). The transition energies up to 3.0 eV and oscillator strengths have been calculated for two-additional structures of Na_8 with D_{2d} and D_{4d} symmetry. The D_{2d} geometry is closely related to the T_d structure and represents a deformed section of the fcc lattice. The D_{2d} and D_{4d} strcutures have been considered since they have been found as local minima in the LSD[30] and LSD–QMD work[48] and they belong to a class of compact and symmetrical geometries.

For the T_d structure, the transition to the 4^1T_2 state located at 2.49 eV, has dominant oscillator strength and coincides with the maximum of the observed intense band (cf. Fig. 10). There are only three leading configurations in the wavefunction of the 4^1T_2 state arising from single excitations from triply degenerate $1t_2$ to $3t_2$ and $1e$ MO's. There are also two close lying transitions with considerable intensities slightly blue shifted with respect to the dominant transitions, but also lying in the energy interval of the observed intense band. The leading features of their wavefunctions are determined by the same type of single excitations as in the case of 4^1T_2 illustrating again the interference phenomenon which gives rise to a larger or smaller transition dipole results. Notice, that in the case of Na_8 the RPA and the CI results are in relatively good agreement for these intense transitions[49,1] since they are dominated by the single excitations. The fine structure which is red shifted with respect to dominant features calculated for the T_d Na_8 structure is also in a good agreement with the experimental finding (cf. Figure 10). Notice, that two other structures D_{2d} and D_{4d} of Na_8 give rise to similar spectroscopic patterns as the T_d structure does which is the only equilibrium geometry in our treatment. It is important to realize that the consideration of dynamical averaging of geometries to yield "spherical" Na_8 in which the position of the nuclei can be ignored is an inadequate assumption[50] for this level of response measurements. The vibrations might be rather responsible for the broading of the intense absorption band. However, the QMD work[18] suggests again relative rigidity of the local minima found for Na_8 at T=0 (D_{2d}, T_d and D_{4d}) at not very high temperature (200 K).

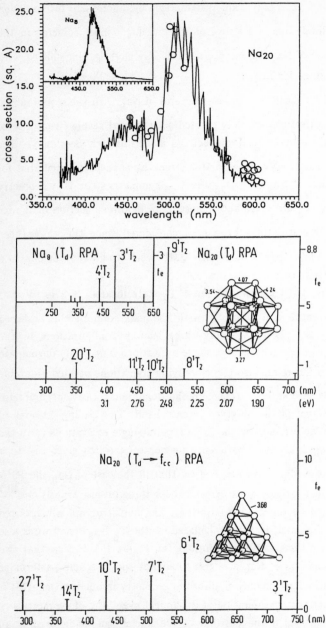

Fig. 11: Comparison of the photodepletion spectrum of Na_{20}[19] and Na_8[18] and the RPA optically allowed transitions (nm) and oscillator strengths f_e a) for the highly symmetrical T_d structures of Na_{20} and Na_8 and b) for the T_d structure of Na_{20} which is a section of fcc lattice[1,58].

A comparison of ab initio–CI results carried out for the optically allowed states for the T_d equilibrium structure of Li_8 with the experimental data makes possible the assignment of this very compact and symmetrical structure to the redcorded spectrum[10].

It is instructive to mention that the description of the excited states of clusters in terms of interaction of individual excitations is just a way of improving the wrong starting point which is the one electron picture. However, this discussion is useful since we can distinguish between a) individual particle–hole excitations and b) many–electron effects which can be qualitatively determined by i) interaction among very few leading excitations as is the case in many molecular excited states and those ii) for which a very large number of excitations with comparable weights is involved so that the notion of collective excitations can be introduced.

In this context, the question can be raised at which cluster size such a molecular picture as presented here will disappear. Therefore, the interpretation of the absorption spectrum of Na_{20} is of particular interest. Although the structural assignment of Na_{20} absorption spectrum[19] is not yet completed, a comparison of the RPA predictions based on the HF SCF calculations for the two T_d structures with the experimental data shown in Figure 11 illustrates two important points. First, the "spherical" structure yields one dominant transition which location is in a good agreement with the maximum of the recorded intense band. There are several other weaker transitions blue shifted coinciding with the weaker band. The second T_d structure which is the section of the fcc lattice gives rise to entirely different spectroscopic pattern which is not in agreement with the experimental finding. Second, the analysis of the RPA results for the intense transition obtained for the "spherical" T_d structure contains contributions with comparable weights from the five single particle–hole transitions illustrating that there might be a long way to go to reach the collective type of excitations.

Summary

The predictive power of quantum chemical methods, although still limited to small clusters with a small number of valence electrons, has been clearly demonstrated by successful structural assignments and interpretation of the optical response properties. A comparison between accurate theory and experiment makes it possible to gain the information about the geometries and specific electronic properties such as nature of excitation responsible for the spectroscopic pattern. As illustrated on a number of examples, in spite of a relatively large mobility of alkali–metal atoms in small clusters the positions of their nuclei cannot be ignored.

We have shown that a number of valuable general rules which govern the electronic and geometrical structure of small alkali metal clusters have resulted from the quantum chemical investigations.

Acknowledgement

This work has been supported by the Deutsche Forschungsgemeinschaft (Sfb 337, Energy transfer in molecular aggregates) and the Consiglio Nazionale delle Ricerche (CNR). We thank our co—workers I. Boustani, J. Pittner, C. Scheuch and colleagues C. Gatti, M. F. Guest and S. Polezzo who have substantially contributed to the work included in this contribution. We extend our thanks to M.M. Kappes, M. Broyer, J. P. Wolf and L. Wöste for providing us with their data prior to publication.

References

1   V. Bonačić–Koutecky, P. Fantucci, and J. Koutecky, Chem. Reviews, **91**, 1035 (1991).
2   V. Bonačić–Koutecky, P. Fantucci, and J. Koutecky, Chem. Phys. Lett., **166**, 32 (1990).
3   V. Bonačić–Koutecky, M.M. Kappes, P. Fantucci, and J. Koutecký, Chem. Phys. Lett., **170**, 26 (1990).
4   V. Bonačić–Koutecky, P. Fantucci, and J. Koutecky, J. Chem. Phys., **93**, 3902 (1990).
5   V. Bonačić–Koutecky, P. Fantucci, and J. Koutecky, Chem. Phys. Lett., **146**, 518 (1988).
6   V. Bonačić–Koutecky, P. Fantucci, J. Gaus, and J. Koutecky, Z. Phys. D — Atoms, Molecules and Clusters, **19**, 37 (1991).
7   V. Bonačić–Koutecky, J. Gaus, M.F. Guest, and J. Koutecky, J. Chem. Phys., in press.
8   V. Bonačić–Koutecky, J. Pittner, C, Scheuch, M.F. Guest, and J. Koutecky, to be published.
9   V. Bonačić–Koutecky, P. Fantucci, and J. Koutecky, Z. f. Phys. Chemie, Neue Folge **169**, 35 (1990).
10 J. Blanc, V. Bonačić–Koutecky, M. Broyer, J. Chevaleyre, P. Dugourd, J. Koutecky, C. Scheuch, J.–P. Wolf, and L. Wöste, J. Chem. Phys., in press.
11 P. Dugourd, J. Blanc, V. Bonačić–Koutecky, M. Broyer, J. Chevaleyre, J. Koutecky, J. Pittner, J.–P. Wolf, and L. Wöste, Phys. Chem. Lett., in press.
12 a) W.A. de Heer, K. Selby, V. Kresin, J. Masui, M. Vollmer, A. Chatelain, and W. D. Knight, Phys. Rev. Lett., **59**, 1805 (1987).
    b)K. Selby, M. Vollmer, J. Masui, V. Kresin, W.A. de Heer, and W. D. Knight, Phys. Rev. **B40**, 5477 (1989).
13 K. Selby, V. Kresin, J. Masui, M. Vollmer, A. Scheidemann, and W. D. Knight, Z. Phys. D — Atoms, Molecules and Clusters **19**, 43 (1991).
14 K. Selby, V. Kresin, M. Vollmer, W. A. de Heer, A. Scheidemann, and W. D. Knight, Phys. Rev. **B43**, 4565 (1991).
15 M. Broyer, G. Delacretaz, P. Labastie, J.–P. Wolf, and L. Wöste, Phys. Rev. Lett., **57**, 1851 (1986).

16 M. Broyer, G. Delacretaz, N. Guoquon, J.–P. Wolf, and L. Wöste, Chem. Phys. Lett., **145**, 232 (1988).

17 C. Wang, S. Pollack, and M.M. Kappes, Chem. Phys. Lett., **166**, 26 (1990).

18 C. Wang, S. Pollack, D. Cameron, and M.M. Kappes, J. Chem. Phys., **93**, 3787 (1990).

19 S. Pollack, C. Wang, and M.M. Kappes, J. Chem. Phys., **94**, 2496 (1991).

20 C. Wang, S. Pollack, J. Hunter, G. Alameddin, T. Hoover, D. Cameron, S. Liu, and M.M. Kappes, Z. Phys. D – Atoms, Molecules and Clusters, **19**, 13 (1991).

21 S. Pollack, C.R.Ch. Wang, T. Dahlseid, and M.M. Kappes, J. Chem. Phys., submitted.

22 M. Broyer, J. Chevaleyre, P. Dugourd, J.–P. Wolf, and L. Wöste, Phys. Rev. A**42**, 6954 (1990).

23 J. Blanc, M. Broyer, J. Chevaleyre, P. Dugourd, H. Kühling, P. Labastie, M. Ulbricht, J.–P. Wolf, and L. Wöste, Z. Phys. D – Atoms, Molecules and Clusters, **19**, 7 (1991)

24 C. Brechignac, P. Cahuzac, F. Carlier, and J. Leygnier, Chem. Phys. Lett., **164**, 433 (1989).

25 C. Brechignac, P. Cahuzac, F. Carlier, M. de Frutos, and J. Leygnier, Z. Phys. D – Atoms, Molecules and Clusters, **19**, 1 (1991).

26 a) H. Fallgren and T. P. Martin, Chem. Phys. Lett., **168**, 233 (1990).
b) H. Fallgren, K. M. Bowen, and T. P. Martin, Z. Phys. D – Atoms, Molecules and Clusters, **19**, 81 (1991).

27 C. R. Ch. Wang, S. Pollack, T. Dahlseid, and M.M. Kappes, to be published.

28 J. Koutecky and P. Fantucci, Chem. Rev. **86**, 539 (1986).

29 a) I. Boustani, W. Pewestorf, P. Fantucci, V. Bonačic–Koutecky, and J. Koutecky, Phys. Rev. B**35**, 9437 (1987);
b) I. Boustani and J. Koutecky, J. Chem. Phys. **88**, 5657 (1988).

30 J. L. Martins, J. Buttet, and J. Car; Phys. Rev. B**31**, 1804 (1985).

31 V. Bonačic–Koutecky, P. Fantucci, and J. Koutecky; Phys. Rev. B**37**, 4369 (1988).

32 V. Bonačic–Koutecky, I. Boustani, M. F. Guest, and J. Koutecky; J. Chem. Phys., **89**, 4861 (1988).

33 J. Koutecky, I. Boustani, and V. Bonačic–Koutecky; Int. J. Quant. Chem.; **38**, 149 (1990).

34 W. A. de Heer, W. D. Knight, M. Y. Chou, and M. L. Cohen, Solid State Phys., **40**, 93 (1987).

35 E. Schumacher, Chimia, **42**, 357 (1988).

36 W. D. Knight, E. Clemenger, W. A. de Heer, W. A. Saunders, and M. Y. Chou; Phys. Rev. Lett., **52**, 2141 (1984).

37 M. L. Cohen, M. Y. Chou, W. D. Knight, and W. A. de Heer, J. Chem. Phys., **91**, 3141 (1987).

38 M. M. Kappes, R. Kunz, and E. Schumacher, Chem. Phys. Lett.; **91**, 413 (1982).

39 M. M. Kappes, P. Radi, M. Schär, C. Yeretzian, and E. Schumacher, Z. Phys. D – Atoms, Molecules and Clusters, **3**, 115 (1986).

40 C. Brechignac, Ph. Cahuzac, and J.–Ph. Roux, Chem. Phys. Lett., **127**, 445 (1986).

41 C. Brechignac, Ph. Cahuzac, and J.–Ph. Roux, J. Chem. Phys., **87**, 229 (1987).

42 C. Brechignac, Ph. Cahuzac, R. Pflaum, and J.–Ph. Roux, J. Chem. Phys., **88**, 3732 (1988).

43 C. Brechignac, Ph. Cahuzac, and J.–Ph. Roux, J. Chem. Phys., **88**, 3022 (1988).

44 C. Brechignac, Ph. Cahuzac, J. Leygnier, and J. Weiner, J. Chem. Phys., **90**, 1492 (1989).

45   C. Brechignac, Ph. Cahuzac, F. Carlier, and M. de Frutos, Phys. Rev. Lett., **64**, 2893 (1990).

46   K. E. Schriver, J. L. Persson, E. C. Honea, and R.L. Whetten, Phys. Rev. Lett. **64**, 2539 (1990).

47   H.–O. Beckmann, J. Koutecky, and V. Bonacic–Koutecky, J. Chem. Phys., **73**, 5182 (1980).

48   U. Röthlisberger and W. Andreoni, J. Chem. Phys., **94**, 8129 (1991)..

49   C. Gatti, S. Polezzo, and P. Fantucci, Chem. Phys. Letters, **175**, 645 (1990)

50   C. Yannouleas, R. A. Broglia, M. Brack, and P. F. Bortignon, Phys. Rev. Letters, **63**, 255 (1989); C. Yannouleas, J. M. Pacheco, and R.A. Broglia; Phys. Rev. **B41**, 6088 (1990); P.–G. Reinhardt, M. Brack, and O. Genzken, Phys. Rev. **A41**, 5568 (1990); C. Yannouleas and R.A. Broglia; to be published.

51   W. Eckardt, Phys. Rev. Letters, **52**, 1925 (1984).

52   M. J. Pruska, R. M. Nieminen and M. Manninen; Phys. Rev. **B31**, 3486 (1985).

53   W. Eckardt, Phys. Rev. **B31**, 6360 (1985).

54   W. Eckardt, Z. Penzar, and M. Sunjic; Phys. Rev. **B33**, 3702 (1986).

55   O. Gunnarson and B.I. Lundquist; Phys. Rev. **B13**, 4274 (1976).

56   A. Herrmann, M. Hofmann, S. Leutwyler, E. Schumacher, and L. Wöste, Chem. Phys. Letters, **62**, 216 (1979).

57   J.–P. Wolf, G. Delacretaz ,and L. Wöste; Phys. Rev. Lett., **63**, 1946 (1989).

58   C. Gatti, P. Fantucci, S. Polezzo, C. Scheuch, V. Bonačić–Koutecky, and J. Koutecky, to be published.

Effects of Crystal Field Splitting and Surface Faceting on the Electronic Shell Structure

M. Manninen, J. Mansikka-aho, and E. Hammarén

Department of Physics, University of Jyväskylä,
P.O. Box 35, SF-40351 Jyväskylä, Finland

1. Introduction

The shell structure of the valence electrons is clearly observed in all alkali and noble metal clusters containing up to hundreds of atoms[1 − 4]. It is seen in the abundances of the clusters, in the ionization potential and in the polarizability. The shell structure of the valence electrons is closely related to the shell model of nuclei, but is simpler owing to the negligibly small spin-orbit interaction. The ability to produce all sizes of metal clusters has made the metal clusters a test ground for the super-shell structure[5].

In atomic clusters the electrons move in an effective potential provided by the ion pseudopotentials and electron-electron interactions. *Ab initio* electron structure computations for the smallest clusters[6, 7] and for the bulk metals[8] have shown that the mean field theory, i.e. the local density approximation of the density functional theory[9] is accurate enough to make quantitative predictions. In alkali metals the ion pseudopotential is weak and in the bulk metal the electron system is nearly a homogeneous electron gas which moves in a compensating homogeneous positive background. The obvious model for a spherical cluster is then a potential well. This simple model leads to the observed shell structure. Refinements can be made by introducing a rounded edge for the potential well, e.g. Wood-Saxon potential[5], or using the self-consistent jellium model[10, 11]. Moreover, the departure of the geometry from the spherical well can be taken into account by using a spheroidal potential[12] or a spheroidal jellium model[13].

In the case of large clusters two questions arise. First, the weak pseudo-potentials of the ions reduce the symmetry of the potential. This will split the highly degenerate large angular momentum states. If this *crystal field splitting* becomes larger than the energy difference between neighbouring energy shells, the shell structure effects will diminish. Secondly, in large clusters the ion arrangement will be the same as in the infinite solid. The ground state cluster geometry will then be determined by minimizing the surface energy and the resulting geometry will be the so-called Wulff polyhedron. The question arising now is what is the effect of the polyhedral shape to the shell structure.

We have studied the effects of the crystal field splitting and the polyhedron formation on the electronic shell structure using several simple models. In the case of alkali metals the crystal field splitting was observed to be very small[14] and to be able to destroy the shell structure only when the cluster had about a million atoms. On the other hand the polyhedron formation has a more drastic effect on the shell structure. Cubo-octahedral clusters are expected to loose their shell structure already when the cluster has of the order of 100 atoms, whereas in the icosahedral structure the shell model is visible in much larger clusters.

2. Crystal field splitting in simple metal clusters

The spherical symmetric potential of the jellium model leads to a large degeneracy of the electron states corresponding to high angular momentum eigenvalues (in atomic clusters the effect of the spin-orbit interaction to the energy levels can be ignored). In real clusters the effective potential is not strictly spherical while it is affected by the ionic structure. The symmetry is reduced and consequently the electronic levels will split. This is called crystal field splitting. Group theory dictates the maximum degeneracy a state can have in a given symmetry. For the cubic lattices the maximum degeneracy is three and for the icosahedral symmetry it is five[15]. If the splitting of the individual energy levels is larger than the gap between different energy shells, the shell structure effects will disappear. Most of the observed shell structure effects measure the shell structure at the Fermi energy of the cluster (magic numbers, reactivity, ionization potential).

In a bulk metal the crystal field splitting can be seen as the deviation of the energy bands from the free electron energy levels. In metals with nearly free electrons this happens only close to the Brillouin zone boundaries and in monovalent metals the Fermi surface is nearly spherical. However, there is always a small crystal field splitting at the Fermi energy and this gives the limits to the size of the cluster that can exhibit the shell structure.

We have estimated the effect of the crystal field splitting in simple metals using the perturbation theory[14]. The energy change due to the ion pseudopotentials can be written as

$$\epsilon_{\mathbf{k}} - \epsilon_{\mathbf{k}}^0 = \sum_{\mathbf{G}} \frac{|U_{\mathbf{G}}|^2}{\epsilon_{\mathbf{k}}^0 - \epsilon_{\mathbf{k-G}}^0}, \tag{1}$$

where $U_{\mathbf{G}}$ is the Fourier component of the screened pseudopotential and $\epsilon_{\mathbf{k}}^0 = \hbar^2 k^2/2m$. The crystal field splitting can be understood as the variation of $\epsilon_{\mathbf{k}}$ across the Fermi sphere. For spherical clusters we can estimate that the shell structure effects will disappear when the crystal field splitting becomes comparable to the energy difference between different energy shells in a spherical potential box. In a large spherical potential box the energy difference between the shells can be approximated to be[14]

$$\Delta E \approx \frac{\hbar^2 \ell}{mR^2}, \tag{2}$$

where ℓ is the angular momentum eigenvalue and R the cluster radius. By equalizing ΔE to the crystal field splitting we can estimate largest cluster that can exhibit shell structure effects. For alkali metals this limit is of the order of a million atoms[14], corresponding to a cluster diameter of hundreds of Ångströms.

3. Hückel model - nonspherical potential boxes

Apart from the cube the energy levels in nonspherical potential boxes in three dimensions cannot be easily solved analytically. A numerical solution for the lowest energy levels is straightforward by discretizing the space in a proper mesh. We have shown[16] that the discretation of the wave function in an fcc, bcc or simple cubic mesh leads to the Hückel model for the same structure. Using the fcc mesh we can then easily compute the electronic structure of the free electrons in a octahedral, cubooctahedral or any truncated octahedral potential box. The solution of the free electron model is then equivalent of solving the Hückel model with

$$H_{ii} = \frac{\hbar^2 c}{2ma^2}$$
$$H_{ij} = -\frac{\hbar^2}{2ma^2} \quad , \tag{3}$$

where c is the coordination number, a the lattice constant and the nondiagonal elements are nonzero only for the neareast neighbour atoms. Figure 1. shows the density of states for the spherical potential box and for the cubooctahedral potential box calculated numerically using the Hückel model. The density of the discrete states are smoothened with a Lorenzian. All the energy levels obtained with the about 1000 lattice points are included in the figure. Only the lowest energy states give an accurate estimate for the free electrons. At higher energies an increase of the density of states typical to the density of states for the Hückel model in bulk fcc is seen. In the case of the spherical potential box the discretation with about 1000 points gives the shell structure correctly up to the shell closing 338 as seen in Fig 1. A more detailed comparison is made in Ref.[17]. In the case of the cubooctahedral geometry the shell structure is from the very beginning different from that of the spherical potential box. Only the first two shell closings, namely 2 and 8, are similar as in the case of the spherical potential box. No electronic shell structure effects are then expected for cubooctahedral clusters, if described with an infinite potential box.

We have also studied cubic, octahedral and truncated octahedral (Wulff polyhedra) structures[17], but in all these the spherical shell structure has completely vanished when the cluster has more than about 50 electrons. Icosahedral symmetry can not be constructed from any cubic lattice. Nevertheless, we have studied also the icosahedral clusters using the Hückel model. The results seem to indicate that there the spherical shell structure persists to much larger clusters than in any other faceted geometry.

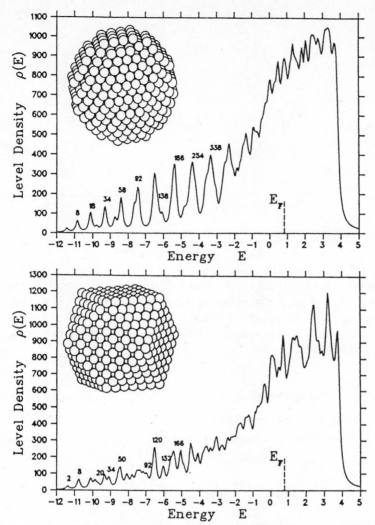

Figure 1. Density of states calculated using the Hückel model for a spherical potential box and for a cubooctahedron. The insets show the atomic arrangement in the fcc Hückel clusters. The numbers above the peaks indicate the total number of atoms at various shell closings.

4. Icosahedral and cubooctahedral potential wells

The use of the Hückel model is well suited for calculating the electronic shell structure of clusters where the boundary conditions are that of an infinite potential wall and the cluster can be obtained by cutting from a cubic lattice. The icosahedral alkali metal clusters do not fulfill any of these

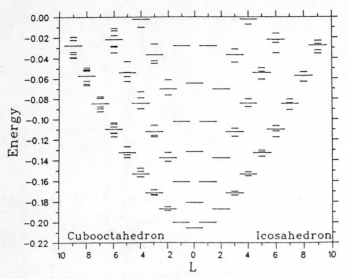

Figure 2. The level structure of a 147 atom sodium cluster approximated with a potential well. L is the angular momentum eigenvalue. The long lines show the results for the spherical well, the short lines on the left side show the results for the cubooctahedral well and the short lines on the rigth the results for the icosahedral well.

requirements. Since the icosahedral cluster shape is fairly spherical it is possible to use the perturbation theory to study how the electron energy levels change when the spherical potential well is replaced with an icosahedral well. We start from a potential well

$$V(r) = -V_0 \theta(r_0 - r), \tag{4}$$

where θ is a step function. The depth of the potential V_0 can be estimated as the sum of the Fermi energy and the ionization potential of the bulk metal. The potential radius $r_0 = r_s N^{1/3}$, r_s being the Wigner-Seitz radius and N the number of atoms in the cluster. For the icosahedral potential well we assume the same depth and the same volume as for the corresponding sphere. The difference of the icosahedral potential well and the spherical potential well leads to a the perturbation potential ΔV which is nonzero only close to the surface of the sphere (radius r_0). The first order perturbation theory now gives the energy change of the energy level $E_{\ell,m}$

$$\Delta E_{\ell,m} \approx |R_\ell(r_0)|^2 V_0 r_0^3 B_{\ell m}, \qquad (5)$$

where R_ℓ is the radial wave function and $B_{\ell m}$ a constant which depends only on the symmetry of the potential well. The approximation (5) is based on the fact that the radial wave function R_ℓ changes only slightly over the distance where the perturbation ΔV is nonzero. The constant $B_{\ell m}$ in Eq. (5) has to be computed for each ℓ and m only once.

Figure 2 shows the electronic shell structures for a spherical, icosahedral and cubooctahedral potential wells. The parameters of the well correspond to a sodium cluster with 147 atoms. In the icosahedral cluster all levels with ℓ larger than 2 will split but the splitting is still small compared to the average level separation in the spherical potential well. In the case of the cubooctahedral potential well the level splitting is about 50 % stronger. Figure 3 shows the density of states for the 147 atom sodium cluster of spherical, icosahedral and cubooctahedral geometry. The discrete levels are smoothened with a Lorenzian. Compared to the the spherical case the peaks become wider in the cases of icosahedral and cubooctahedral symmetries, but in the icosahedral case the the shell structure is still clearly visible. The results show that when the cluster grows the shell structure effects will disappear faster if the cluster is cubooctahedral than if it is icosahedral. A further study will be published elsewhere[18].

The comparison of Fig. 1 and 3 reveals that the surface faceting disturbes the shell structure less in the case of the finite potential well than in the case of the infinite potential box. The reason for this is that in the finite well the wave functions can penetrate outside the well where the potential again has a spherical symmetry, whereas in the infinite well the wave functions are forced to be zero at the box boundary.

5. Conclusions

In alkali metal clusters the crystal field splitting is so small that the spherical cluster can have of the order of million atoms before the crystal field splitting will disturb the electronic shell structure of the valence electrons. The faceting of the crystal surface will have a much stronger effect. It seems

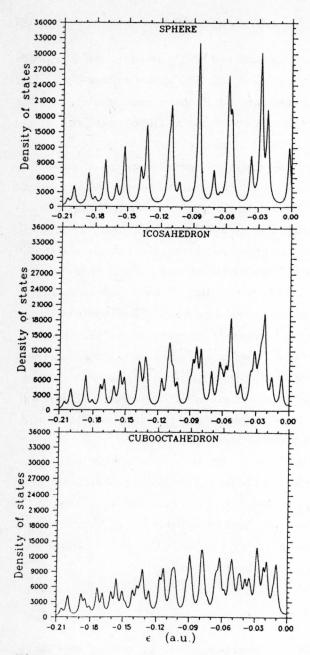

Figure 3. Density of states for a 147 atom sodium cluster calculated by approximating the potential by spherical, icosahedral and cubooctahedral potential wells.

that the icosahedral cluster can preserve the shell structure better than any simple polyhedron cutted from an fcc lattice. The icosahedral cluster can contain at least hundreds of atoms and still show the shell structure of a sphere.

Acknowledgments

We would like to thank Prof. P. Jena and Dr. J. Suhonen for useful discussions. This work vas supported by the Academy of Finland.

References

1. W. D. Knight, K. Clemenger, W. A. de Heer, W. A. Saunders, M. Y. Chou, and M. Cohen, Phys. Rev. Lett. **52**, 2141 (1984).

2. M. M. Kappes, M. Schär, P. Radi, and E. Schumacher, J. Chem. Phys. **84**, 1863 (1986).

3. W. A. de Heer, W. D. Knight, M. Y. Chou, and M. L. Cohen, Solid State Physics **40**, 93 (1987).

4. H. Göhlich, T. Lange, T. Bergmann, and T. P. Martin, Phys. Rev. Lett. **65**, 748 (1990).

5. H. Nishioka, K. Hansen, and B. R. Mottelson, Phys. Rev. B **42**, 9377 (1990).

6. J. Martins, J. Buttet, and R. Car, Phys. Rev. B **31**, 1804 (1985).

7. P. Ballone, W. Andreoni, R. Car, and M. Parrinello, Europhys. Lett. **8**, 73 (1989).

8. V. L. Moruzzi, J. F. Janak, A. R. Williams, *Calculated Electronic Properties of Metals* (Pergamon, New York 1978).

9. *Theory of Inhomogeneous Electron Gas*, ed. by S. Lundqvist and N. H. March (Plenum, New York 1983).

10. W. Ekard, Phys. Rev. B **29**, 1558 (1984).

11. M. J. Puska, R. M. Nieminen, and M. Manninen, Phys. Rev. B **31**, 3486 (1983).

12. K. Clemenger, Phys. Rev. B **32**, 1359 (1985).

13. Z. Penzar and W. Ekardt, Z. Phys. D **17**, 69 (1990).

14. M. Manninen and P. Jena, Europhys. Lett. **14**, 643 (1991).

15. *Symmetries in Physics*, W. Ludvig and C. Falter (Springer, Berlin 1988).

16. M. Manninen, J. Mansikka-aho, and E. Hammaren, Europhys. Lett. (in press).

17. J. Mansikka-aho, M. Manninen, and E. Hammarén, Z. Phys D (submitted).

18. J. Mansikka-aho, M. Manninen, and E. Hammarén, to be published.

Fermion-Boson Classification in Microclusters

G.S. Anagnostatos

Institute of Nuclear Physics, National Center for Scientific
Research "Demokritos", GR-153 10 Aghia Paraskevi Attiki, Greece

Microclusters composed of atoms with *non* delocalized odd number of
valence electrons possess the usual magic numbers for fermions in a
central potential and those with an even number of valence electrons
possess the magic numbers for bosons coming from the packing of
atoms in nested icosahedral or octahedral or tetrahedral shells. On the
other hand, microclusters composed of atoms *with* delocalized valence
electrons, either with an odd or with an even number of electrons,
exhibit electronic magic numbers (according to the jellium model) but
also magic numbers coming from the (same, as above) packings of
their bosonic ion cores. Finally, through the present work, an
alternative approach to study atomic nuclei as quantum clusters
appears possible and promising.

1. Introduction

Magic numbers (intensity anomalies in the mass spectra) in microclusters mainly have
been interpreted as coming from two distinct and basically different origins which refer to
separate categories of elements. That is, magic numbers are considered either as a result of
electron structure (as in jellium model), e.g. in alkali clusters [1], or as a result of close
packing of atoms, e.g. in rare gas clusters [2-3]. The magic numbers for these two categories
are 2,8,20,40,58,.... and 1,13,55,147,309,...., respectively. Recently however, it has been found
that the magic numbers in certain mass spectra include numbers from both of the above sets
[4-6]. Further complexity in understanding magic numbers comes from the fact that besides
those for alkali and rare gas clusters, different categories of magic numbers have been
established for different elements or combinations of them, e.g., magic numbers for
semiconductor [7] or alkali-halide clusters [8], or for clusters made of mixed rare gases [9] or
mixed alkalis [10,11], etc.

The question raised by the present work is whether all these different sequences of
magic numbers are indeed independent of each other or whether there is something
fundamental, out of which one may derive them. In this paper, we propose a new
concept that the magic numbers are mainly determined by the nature of particles
involved in forming clusters. Depending on their odd or even number of electrons, the
constituent atoms or ion cores behave as heavy fermions or heavy bosons. It is this
property of constituent particles, together with the delocalized electrons, whenever they
exist, that determine the structure of a microcluster [6].

Springer Series in Nuclear and Particle Physics   **Clustering Phenomena in Atoms and Nuclei**
Editors: M. Brenner · T. Lönnroth · F.B. Malik   © Springer-Verlag Berlin, Heidelberg 1992

2. The Model

2.1 Conceptualization of the model

The starting point of the present model is the comparative study of small-size-clusters of neutral and ionized alkali atoms shown in figures 1a, 1b, and 1c, which show, respectively, the celebrated experimental mass spectrum of sodium clusters [1], the prediction of the same in the jellium model, and the observed mass spectrum of sodium cluster cations [5].

The magic-number sequence in Figure 1(a) is 2,8,20,40,58,92,...., while the predicted magic numbers according to the jellium model (Figure 1(b)) are 2, 8, 18, 20, 34, 40, 58, 68, 70, 92,..... Thus, between the experimental data of Figure 1(a) and the jellium-model predictions there are differences.The model predicts additional peaks at N= 18,34,68,70,..... However, these missing numbers from Figure 1(a) are present in Figure 1(c), e.g., the peak at N=19 sodium cations which corresponds to 18 delocalized electrons ($N_e=N-1$) which is predicted by the jellium model. In addition, the spectrum in Figure 1(c) exhibits peaks at N=13, (19), 25 atoms, which are very well known as magic numbers of rare gas clusters [2-3]. Similar comments can be made when comparing other born neutral and born ionized small size alkali clusters.

One may therefore, infer that either a sodium mass spectrum does not include all predictions of jellium model (specifically the numbers 18,34,68,70,...) or it includes them but additional magic numbers, familiar from the close-packing of spheres structure (e.g., from the rare gas clusters), also exist. This is a general conclusion of all similar examples on alkali or alkali-like clusters.

The application of the jellium model implies that the valence electron from each alkali atom is delocalized and that all such electrons in the cluster move in a common central potential somehow created by the ion cores, a fact which leads to the electron magic numbers [1]. Indeed, there are experimental conditions which can secure the delocalization of the valence electrons, e.g., those valid for the experiment [5] of Figure 1(c). However, this is not necessarily the case in all experiments. Thus, if we do not have delocalization of valence electrons, one does not fullfill the assumption underlying the jellium model and electron magic numbers. Hence, Figure 1(a) can be seen as an example of localized electrons and thus the atoms themselves are the constituents of the cluster . In that case the jellium model is inapplicable. Also, the extra numbers (i.e., N=13,19,25) appearing in Figure 1(c) can be seen as magic numbers of the ion cores which are formed after the delocalization of valence

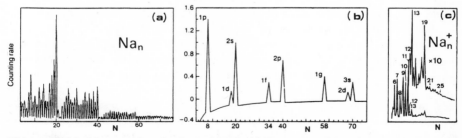

Figure 1(a)-(c).. Mass spectra of sodium clusters. (a) Experimental data for neutral atoms, (b) jellium model predictions, and (c) experimental data for cations

electrons. Indeed, such ion cores do not exist when the constituents of the cluster are the neutral atoms themselves, a fact which is consistent with the absence of additional peaks in Figure 1(a).

2.2 Development of the Model

We present here a model where the nature of an atom or its ion core, taken as composite particles, is incorporated. That is, a neutral alkali atom (or a neutral atom of another element possessing an odd number of electrons) is considered as a heavy fermion due to its half integer total spin, while a neutral atom or an ion core possessing an even number of electrons is considered as a heavy boson due to its integer total spin [6].

Thus, in the model we are concerned not only with delocalized electrons and left-over ion cores, but also with atoms as particles. It is the Fermion or Boson nature of these particles rather than the forces among them that are emphasized in the model. The physical properties of Fermi- and Boson-like particles are different and they follow different statistics. Bosons usually occupy (if possible) the lowest available energy level, while Fermions occupy different energy levels, according to the Pauli principle.

In the model, each atom (or ion core) feels an average central potential created by all atoms (or ion cores) of a shell in the microcluster [6-12]. Inside this potential an atom (or ion core) moves independently from the motions of the other atoms (or ion cores). Of course, this model is analogous to that in nuclear physics. The difference is that the quantum constituent there is nucleons , whereas here it is the atoms (or ion cores) themselves which are either Fermions or Bosons, depending on their spin, determined by the number of electrons attached to them. This model should be distinguished from the jellium model [1], where the quantum constituent is the delocalized electrons and the central potential is somehow created by the ion cores. In this sense the jellium model is similar to the atomic shell model.

The model proposed herein is applicable, by itself, to a microcluster of neutral atoms. For the case of a microcluster composed of ionized atoms, the relevant model is a combination of the present model and the jellium model because the atomic-ion-cores are discribed by this model, whereas the electron motion is given by the jellium model. In this case the potential of the jellium model in general does not have, as usually assumed, an infinite spherical symmetry, but a reduced symmetry, determined by the structure of the ion cores. When the constituent atoms (either fermionic or bosonic) are neutral the magic numbers come from the structure of these atoms alone, and when the constituent atoms possess delocalized valence electrons, these numbers come from both the structure of the (always bosonic) ion cores *and* from the structure of the (fermionic) electrons [6].

In the model, the atoms are in continuous motion determined by their wave functions. However, the viscosity of the fluid formed by the atoms in the microclusters is much larger than that of the nucleons in the atomic nucleus, due to the much larger mass of the atoms compared to that of the nucleons. This large viscosity, of course, implies relatively slow motion of the atoms in the clusters. However, a geometrical structure of the cluster always results, if for each atom in the cluster one considers its average position. Such positions are discussed in [7-11,3,13-15] and are employed, e.g., in [6,12,15] in order to evaluate parameters of the relevant central potential needed for additional quantitative predictions by the present model. In refs [3,10,13] , dealing

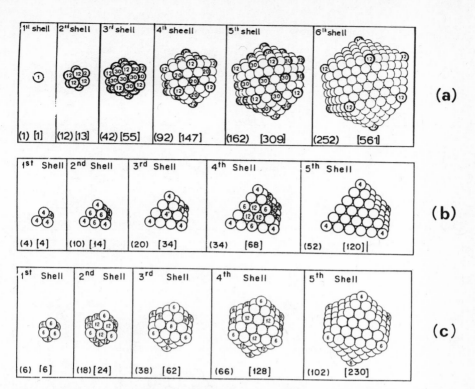

Figure 2(a)-(c). Close packing of *soft spheres* standing for atomic bosons (either as neutral atoms or as ion cores) in nested polyhedral shells. (a) icosahedra, (b) tetrahedra, and (c) octahedra.

with fermionic clusters, the aforementioned average positions are simply useful *representations*, but in [3,7-9,14] dealing with bosonic clusters, these average positions form a structure which closely approximate the real structure of the relevant microcluster.

In Figure 2 the close packing of soft spheres standing for either neutral atoms or ion cores with an even nember of electrons (bosons) is presented. In Figure 2(a) the first five successive shells of rare gas clusters are shown as nested icosahedral shells [3,9], while in Figure 2(b) and (c) those of semiconductor [7] and alkali-halide [8] clusters, as nested tetrahedral and nested octahedral shells, respectively, are presented. The relevant magic numbers are as follows: Figure 2(a) : N=1, 13, 55, 147, 309,...; Figure 2(b) : N=4, 6, 10, 14, 18, 22,..., also 5, 7, 11, 15, 19, 23,.... ; Figure 2(c) : N=6, 14, 18, 20, 24,...., also 7, 10, 13, 17, 19, 25,...., [3,7-8]

The initial choice for each specific cluster to assume one of the above three packing structures depends on the softness of spheres presenting the relevant atoms at each case. The softness of a sphere presenting an atom is a measure of the degree of completion of the outermost electronic shell of this atom and takes its minimum value (~10%) when the outermost shells are complete, as in the case of rare gases [3], and larger values otherwise, e.g. , for semiconductors (~40%) and alkali-halides (~30%) .

The final choice among the three possibilities considered in Figure 2, dealing with bosonic constituent atoms (or ion cores), does not depend only on the softness of the

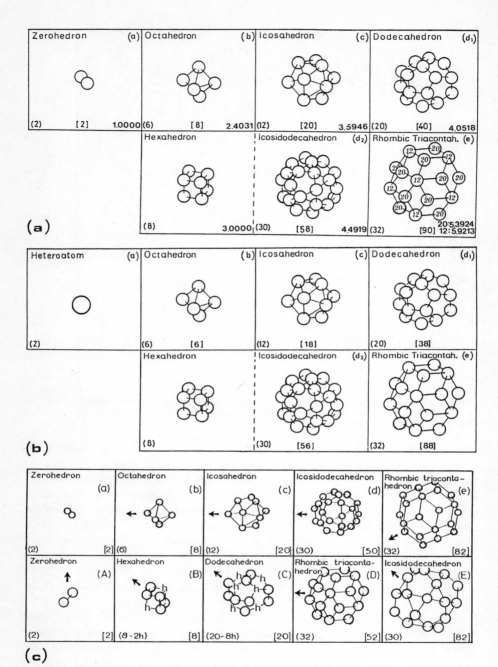

Figure 3(a)-(c). Close packing of *shells* composed of hardspheres standing for atomic fermions in nested equilibrium polyhedra. (a) Alkali homoclusters, (b) clusters of alkali heteroatom, and (c) two alkali clusters.

relevant spheres mentioned above, but also on the temperature (or size) of the cluster. Indeed, a higher temperature or (very closely related) a larger size of a cluster corresponds to an excitation of the cluster which can alter the initial choice of structure and thus leads to a metastable structure. Thus, depending on the temperature and size of a cluster, a mixture of all magic numbers corresponding to all three parts of Figure 2 can be obtained in one and the same mass spectrum [6].

It is satisfying that shell structure based on nested tetrahedral, octahedral, and icosahedral packing simultaneously possess stable equilibrium (which is necessary for the stability of a cluster) and minimization of electrostatic energy among atoms [16] of bosonic type occupying the lowest energy levels.

In Figure 3 the close packing of *shells* composed of hard spheres standing for neutral atoms with an odd number of electrons (fermions) is presented. In Figure 3(a) the first successive shells of alkali homoclusters are shown [3], while in Figure 3(b) and (c) those of alkali-heteroatoms [13] and of two kinds of alkali clusters [10] are shown. All three parts of Figure 3 are made up from nested equilibrium polyhedra as shown. It is satisfying that all such polyhedra possess an equilibrium of the *average* positions of particles assumed on their vertices (middles of their edges or centers of faces) whatever the exact form of the force among fermion particles may be [16].

2.3 Simple Quantitative Treatment

The quantum mechanical treatment of rare gas microclusters (which here are representative clusters of bosonic atoms) has been presented in [17], based on Path - Integral Monte-Carlo algorithm (instead of the wave function formalism), whereas treatment for alkali microclusters (which here are representative clusters of fermionic atoms) based on one body central forces has been studied in [12]. Here only some elements of a simple quantum mechanical treatment (taken from [12]), valid both for bosonic and fermionic atoms , are presented (and compared) taking advantage of the fact that both kinds of clusters form shells (called high fluximal shells) possessing a geometrical representation [3].

Here, the potential previously mentioned is better defined; it is specifically assumed that all atoms in a shell of the cluster taken together create an average central potential, assumed to be harmonic, common for all atoms in this shell and that in this potential each atom performs an independent particle motion obeying the Schrödinger's equation. In other words, we consider a multi-harmonic potential description of the cluster, as follows

$$H\psi = E\psi, \quad H = T + V \tag{2.3.1}$$

$$H = H_{1s} + H_{1p} + H_{1d2s} + \ldots \tag{2.3.2}$$

where

$$H_i = V_i + T_i = -\bar{V} + 1/2m(\omega_i)^2 r_i^2 + T_i \tag{2.3.3}$$

That is, we consider a state-dependent Hamiltonian, where each partial harmonic oscillator potential has its own state-dependent frequency ω_i. All these ω_i's are determined from the harmonic oscillator relation (2.3.4)

$$\hbar\omega_i = (\hbar^2/m\langle r_i^2\rangle) \ (n+3/2), \tag{2.3.4}$$

where n is the harmonic oscillator quantum number and $\langle r_i^2\rangle^{1/2}$ is the average radius of the relevant maximal probability of occupation (hence forth called high fluximal) of a shell made of either bosonic [3,7-9,14] or fermionic [3,10,13] atoms. Before applying (2.3.4), to each of the shells a value (0,1,2,3,.....) of the harmonic quantum number n is assigned and a value of $<r_i^2>^{1/2}$ is derived from the geometry of the shell taking the finite size of the atomic sphere into account. Thus, $\hbar\omega_i$ changes value each time either n or $<r_i^2>^{1/2}$ (or both) changes its value.

In the case of bosonic atoms there is no restriction for the number of atoms in a shell, since any number of such atoms is accepted for the same quantum state (symmetric total wave function). In the case of fermionic atoms, however, the atoms on each shell are restricted by the Pauli principle (antisymmetric total wave function). It is satisfying that all relevant shells for fermionic atoms [3,10,13] fulfil this fundamental requirement, as explained in detail in [12].

According to the Hamiltonian of (2.3.2), the binding energy , BE, of a cluster of N atoms is given by (2.3.5).

$$BE = 1/2 \ (\overline{V}N) - 3/4[\sum_{i=1}^{N} \ \hbar\omega_i \ (n+3/2)], \tag{2.3.5}$$

where $\overline{V}$ is the average potential depth given [12] by (2.3.6)

$$\overline{V} = -\alpha N + b + c/N, \tag{2.3.6}$$

The coefficient c in (2.3.6) expresses the sphericity of the cluster and has the same numerical value everywhere the outermost shell of the structure is completed and otherwise c has a zero value. Of course, one expects that different kinds of atoms will assume different values of parameters a, b, and c in (2.3.6).

The relative binding energy gap for a cluster with N atoms compared to clusters with N+1 and N-1 atoms is given by (2.3.7).

$$\delta(N) = 2E_B(N) - [E_B(N-1) + E_B(N+1)]. \tag{2.3.7}$$

As is apparent throughout the present work and the cited references, the average positions of the atoms (or their ion cores) in the clusters have a shell structure either for fermionic or for bosonic atoms. In this respect the structure of the clusters, to some extent, resembles nuclear structure.

Thus, several well-documented nuclear phenomena, e.g. collective effects, are reasonably expected for the clusters as well. Hence, small clusters of size N far from magic numbers are expected to be deformed. Furthermore, deformed (prolate or oblate) clusters are expected to rotate [18] and spherical (close to magic numbers) clusters are expected to vibrate. Besides these collective excitations, clusters can show single particle excitation either due to their atomic or electronic constituent (partial levels of ionization). All these interesting phenomena are out of the scope of the present work which mainly intends to obtain a classification of microclusters according to the statistics of the constituent atoms (i.e., Fermi or Boson statistics) depending on whether this constituent has half integer or integer spin.

3. Application of the Model

3.1 Alkali atoms without delocalized electrons

This category of clusters has been examined earlier [1] (See Sect. 2.1)

3.2 Alkali atoms with delocalized electrons

For example, in [4-5] dealing with Li_n^+, Na_n^+, and Rb_n^+ clusters, the appeared magic numbers are N=3,5,7, 9,11,13, (15),19, 21, 23,(25),35,41,..... Out of these numbers, due to electron structure alone, magic numbers are predicted to be at N=3,9,19,21,35,41 (since for cations N_e=N-1). The remaining magic numbers, according to the present model, should come from the structure of the bosonic ion cores. Indeed, the numbers 13,19,(25) come from nested icosahedral packing [3] and the numbers 5,7,11,(15),19,23 from nested tetrahedral packing [7] with a central atom. In addition, it is satisfying that for negatively charged alkali clusters the magic numbers due to the ion cores remain the same [19].

3.3 Alkali-like atoms (Cu, Ag, Au)

These clusters are almost identical to alkali clusters. Because of the proximity of magic number 55 (due to ion cores) and to magic number 58 (due to electron structure), dramatic behavior is observed between 55 and 58 in all mass spectra of such atoms [19].

3.4 Even-valence atoms without delocalized electrons

In [20] for Pb_n the magic numbers are 10,13,15,17,19,23,25,..... It is satisfying that these numbers are almost identical to those discussed and explained previously for the ion cores of alkali clusters. Additional examples are in [21] and [22] for Co, Ni, and Ba.

3.5 Even-valence atoms with delocalized electrons

For born-ionized Zn_n^+ and Cd_n^+ [23] magic numbers appear at N=10, 18, 20,28,30, 32,35,41,46,54,57,60,69,..... Out of them the numbers N=10, 18, 20, 30,35,46,57,69,... (with number of electrons 19,35,39,59,69,91,113,137,....) may be explained exactly or closed to the numbers predicted by the jellium model. With the exception of 41, all remaining magic numbers are interpreted by the nested octahedral [8] packing (e.g., the numbers 28,32, also 18,30) and by the nested tetrahedral [7] packing (e.g., the number 60).

3.6 Odd-valence atoms without delocalized electrons

In [24] for Nb clusters, the expected magic numbers for fermions 2 and 8 show up for light clusters. while for heavier clusters delocalization of electrons occurs (due to the higher temperature of the cluster) and the magic numbers 10,13,16,25,.... are exhibited [25]. With the exception of 16 all other numbers can be explained as close packing of ion cores [8].

3.7 Odd-valence atoms with delocalized electrons

Here, Al_n^+ is taken as an example [26]. Enhancements in mass spectra which appear at N=3,7,14,20,23,.... (with number of electrons 8,20,41,59,68,...) are explained by the jellium model, while enhancements at N=5,10,15,18,... by the close packing of ion cores. Specifically, the first three are interpreted as nested tetrahedral shells [7], while the last one as nested octahedral shells [8].

3.8 Rare gas atoms

Since no valence electron exists here, all magic numbers come from the well known nested icosahedral shells [2,3].

3.9 Large size alkali clusters

In [27] large alkali clusters have been reported up to N≃22000 atoms. According to this reference, the magic numbers for alkali clusters come, up to the size N≃1500, from the electronic structure alone, while beyond this number, from shells of atoms alone. However, a closer examination of the experimental data of this reference, within the context of the present model, leads to different conclusions. Specifically, as shown in Figure 4, shells of atoms exist even below N≃1500 and electronic shells exist after this number as well [28].

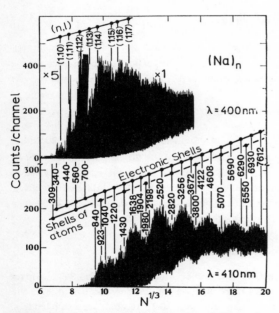

Figure 4. Coexistence of electronic shells and shells of atoms for sodium clusters. Positions marked by (n,l) values demonstrate the contribution of electronic subshells on the fine structure of the mass spectrum

The shells of atoms are estimated by using (3.9.1)

$$N_{cluster} = 1/3\ (10K^3 - 15K^2 + 11K - 3) \tag{3.9.1}$$

(where K is a shell index) and the electronic shells (as groups of subshells having the same energy) by using the 3n+l approximate energy quantum number for alkali [27]. All these are supported by substantial minima in Figure 4. Even secondary minima all over the spectrum of Figure 4 (fine structure of the spectrum) are attributed to the electronic subshells. e.g., $(n,l)=(1,10), (1,11), (1,12), (1,13), (1,14), (1,15), (1,16), (1,17),...$

Thus, change of phase from the electronic shells to shells of atoms proposed in [27] is not suported by the present model. Such change, however, is supported here [12,29] in going from stochastic shells of atoms alone ($N \lesssim 70$) for born-neutral alkali clusters to the coexistence of electronic shells and shells of atoms ($N \gtrsim 70$).

4. Extension of the Model to Nuclear Physics

The identity of light magic numbers in two independent branches of physics, alkali clusters and nuclear physics, obeying two basically different types of forces, electromagnetic and strong force, respectively, does not seem to be coincidental. This remark is in agreement with the fundamental premise of the present model, wich emphasizes the statistical properties of the constituents (i.e., fermionic, bosonic nature) rather than the forces among them. The model clearly demonstrates that many properties can be understood directly from general consideration of the statistical properties rather than the strength of the particular force [10].

Many concepts and methods of treatment in cluster physics come from nuclear physics. However, the above remarks may be seen as a hint to reverse the flow of knowledge, now, from cluster physics to nuclear physics. The consideration of the size of nucleons via the sizes of their bags is essential, since we cannot speak about point nucleons in a structure resembling that of small clusters. We now apply specifically the model to nuclear structure employing 0.974 fm for the neutron bag and 0.860 fm for that of a proton [30] . These values are consistent with our knowledge from particle physics [31] that supports their relative size as well [32]. These different sizes of bags imply a weak isospin symmetry, or in other words they imply that a nucleus consists of two almost different (distinct) kinds of fermions. Thus, the nucleus resembles those of clusters which are made up of two kinds of alkalis [10]., i.e., those presented by Figure 3(c).

The close packing of average sizes of shells assumed by this figure permits the determination of the average radial sizes of all nuclear shells with respect to the sizes of the nucleon bags alone. The necessary formula is [30]

$$R_x = <r^2>^{1/2}_{shell}\ =\ R\cos\ \alpha + (\ d^2 - R^2\sin^2\alpha)^{1/2}, \tag{4.1}$$

where R_x is the average radius of the shell to be determined, R the average radius of the previous shell in contact, d the distance of the centers of two nucleon bags in contact, and α an angle defined by the symmetry and relative orientation of both shells involved each time in the calculation according to [33].

Table 1. Charge root mean square radii in units Fermi

NUCL.	MOD.	EXP.	NUCL.	MOD.	EXP.
H		0.8	^{98}Mo	4.40	4.391 (26)
^{4}He	1.71	1.71(4)	^{98}Tc	4.43	
^{7}Li	2.06	2.39(3)	^{102}Ru	4.46	4.480(22)^d
^{9}Be	2.22	2.50(9)	^{103}Rh	4.49	4.510(44)
^{11}B	2.31	2.37	^{106}Pd	4.52	4.541(33)
^{12}C	2.37	2.40(56)^b	^{107}Ag	4.55	4.542(10)^d
^{14}N	2.54	2.540(20)	^{114}Cd	4.57	4.624(8)
^{16}O	2.70	2.710(15)^c	^{115}In	4.60	4.611(10)^d
^{19}F	2.84	2.85(9)^b	^{120}Sn	4.63	4.630(7)
^{20}Ne	2.98	3.00(3)	^{121}Sb	4.65	4.63(9)
^{23}Na	2.95	2.94(4)^b	^{130}Te	4.67	4.721(6)
^{24}Mg	3.06	3.08(5)	^{127}I	4.72	4.737(7)
^{27}Al	3.14	3.06(9)	^{132}Xe	4.77	4.790(22)^d
^{28}Si	3.21	3.15(5)	^{135}Cs	4.82	4.801(11)^d
^{31}P	3.27	3.24	^{138}Ba	4.85	4.839(8)^d
^{32}S	3.33	3.263(20)	^{139}La	4.91	4.861(8)
^{35}Cl	3.37	3.335(18)	^{140}Ce	4.95	4.883(9)
^{40}Ar	3.40	3.42(4)	^{141}Pr	4.99	4.881(9)
^{39}K	3.44	3.436(3)^c	^{142}Nd	5.03	4.993(35)
^{40}Ca	3.47	3.482(25)	^{146}Pm	5.06	
^{45}Sc	3.51	3.550(5)^c	^{152}Sm	5.10	5.095(30)^d
^{48}Ti	3.55	3.59(4)	^{153}Eu	5.13	5.150(22)^d
^{51}V	3.59	3.58(4)	^{158}Gd	5.16	5.194(22)^d
^{52}Cr	3.62	3.645(5)^c	^{159}Tb	5.19	
^{55}Mn	3.65	3.68(11)	^{164}Dy	5.22	5.222(30)^d
^{56}Fe	3.68	3.737(10)	^{165}Ho	5.25	5.210(70)^d
^{59}Co	3.71	3.77(7)	^{166}Er	5.28	5.243(30)^d
^{58}Ni	3.73	3.760(10)	^{169}Tm	5.30	
^{63}Cu	3.81	3.888(5)^c	^{174}Yb	5.32	5.312(60)^d
^{64}Zn	3.87	3.918(11)	^{175}Lu	5.35	
^{69}Ga	3.93		^{180}Hf	5.37	5.339(22)^d
^{72}Ge	3.99	4.050(32)^d	^{181}Ta	5.40	5.500(200)^d
^{75}As	4.04	4.102(9)^d	^{184}W	5.42	5.42(7)
^{80}Se	4.08		^{187}Re	5.44	
^{79}Br	4.13		^{192}Os	5.46	5.412(22)^d
^{86}Kr	4.17	4.160^c	^{193}Ir	5.48	
^{87}Rb	4.21	4.180^c	^{195}Pt	5.50	5.366(22)^d
^{88}Sr	4.25	4.26(1)	^{197}Au	5.52	5.434(2)
^{89}Y	5.29	4.27(2)	^{202}Hg	5.54	5.499(17)^d
^{90}Zr	4.32	4.28(2)	^{205}Tl	5.56	5.484(6)
^{93}Nb	4.36	4.317(8)^d	^{208}Pb	5.58	5.521(29)

aThe experimental radii come from [34] except as noted below in b-d.
b See [39] ; c See [40] ; d See [41].

Now, the knowledge of the average radial size of all shells permits the determination of the average values of all nuclear radii (e.g., charge radii) by using (4.2), noted below, and assuming the filling of subshells according to the simple shell model [30].

$$<r^2>^{1/2}_{nucleus}= [\sum_1^Z <r^2_i>/Z + (0.8)^2 - (0.116)N/Z]^{1/2},$$ (4.2)

where the $<r^2_i>^{1/2}$ values are given by (4.1) and the constants $(0.8)^2$ and (-0.116) are the ms charge radii accounting for the proton and the neutron finite sizes, respectively [34]. One can consult Table 1 for predictions of the model for all nuclei from H to Pb, where the only two parameters involved are the sizes of the neutron bag and the proton bag (specified above).

In Hamiltonian (2.3.3), besides the nuclear dimentions, we are concerned with the potential whose depth is taken from (4.3) and (4.4) noted below for neutron and protons, repsectively [35].

$$- N\bar{V} = - N\bar{V}_0 + (27.2) (N-Z)/A$$ (4.3)

and

$$- Z\bar{V} = -Z\bar{V}_0 - (27.2)(N-Z)/A + 2E_C/Z.$$ (4.4)

where the second term in each equation stands for the simplest possible isotope effect [36], N,Z and A have their usual meaning, and E_C stands for the Coulomb energy [37]. according to (4.5) below for $R=1.25 A^{1/3}$,

$$E_C= e^2/R [0.6Z(Z-1) - 0.46 Z^{4/3}],$$ (4.5)

and

$$N\bar{V}_0 = Z\bar{V}_0 = 79.26 - 0.0879 |A-74|, \quad for A=16-74$$ (4.6)

or

$$N\bar{V}_0 = Z\bar{V}_0= 79.26 - 0.0313 |A-74|, \quad for A=74-208$$ (4.7)

The seven closed-shell nuclei in Table 2 are used for the determination of the three constants (parameters) in (4.6) and (4.7), while the nine open-shell nuclei of Table 3 constitute a sample of nuclei spread all over the table of isotopes for which the model makes real predictions. Specifically, nuclear charge radii come from (4.2) by using

Table 2: Binding energies and rms charge radii of closed-shell nuclei.

	^{16}O	^{40}Ca	^{58}Ni	^{90}Zr	^{120}Sn	^{142}Nd	^{208}Pb
E_C mod	12	65	123	223	331	447	757
EC emp	12	68	123	224	324	445	744
BE mod	125	350	495	782	1031	1185	1626
BE exp[a]	128	342	506	784	1021	1185	1637
$<r^2>^{1/2}$ mod	2.70	3.47	3.73	4.32	4.63	5.03	5.58
$<r^2>^{1/2}$ exp	2.710	3.482	3.760	4.28	4.630	4.993	5.521
	(15)	(25)	(10)	(2)	(7)	(35)	(29)

a See [42] ; b See [34], [40], and [41]

Table 3 : Predicted binding energies in MeV and rms charge radii in fm of a sample of ten open-shell nuclei close and far from magic numbers

	^{28}Si	^{36}Ar	^{40}Ar	^{56}Fe	^{104}Pd	^{110}Pd	^{126}Te	^{136}Ba	^{138}Ba	^{202}Hg
E_C emp	30	50	55	107	280	280	345	392	390	713
BE mod	234	310	354	494	863	953	1067	1143	1157	1621
BE exp[a]	237	307	344	492	893	940	1066	1143	1159	1595
$\langle r^2 \rangle^{1/2}$ mod	3.21	3.41	3.40	3.68	4.52	4.51	4.67	4.86	4.85	5.54
$\langle r^2 \rangle^{1/2}$ exp	3.15[b]	3.396[c]	3.42[b]	3.737[b]	4.581[d]	4.595[c]	4.721[b]	4.833[b]	4.836[b]	5.499[d]
	(5)	(7)	(4)	(10)	(22)	(3)		(10)		(17)

[a] See [42] ; [b] See [34] ; [c] See [40] ; [d] See [41]

$\langle r_i^2 \rangle^{1/2}$ values from (4.1), while nuclear binding energies are calculated (2.3.5) by using $\hbar\omega_i$ values from (2.3.4) with the help of (4.1) and $\bar{V}$ values from (4.6 - 4.7).

All predictions of the model on radii and binding energies (see Tables 2 and 3) are satisfactory. This implies that an alternative method of studying atomic nuclei via quantum small-cluster concepts is possible and highly promising. Of course, a lot of work is necessary for the refinement of the method and its application to all spectrum of nuclear properties.

5. Concluding Remarks

The model introduced by the present paper, which is based on the nature of the constituent atoms or their ion cores, seems to be justified by all experimental data known to us. Thus, the concept of fermionic and bosonic nature for atoms (or ion cores) with an odd and an even number of electrons, respectively, appears to combine the two views of electronic structure and atom-packing origin of magic numbers and at the same time to unify the comprehension of magic numbers in many kinds of clusters.

Specifically, clusters composed of atoms with non delocalized valence electrons and with an odd number of electrons have stochastic atom magic numbers alone at N=2,8,20,40,.... and those with an even number of electrons possess magic numbers coming from the packing of atoms alone in icosahedral or octahedral or tetrahedral form or mixed. On the other hand, clusters composed of atoms with delocalized valence electrons either with an odd or with an even number of valence electrons exhibit magic numbers due to the structure of their delocalized valence electrons but also magic numbers due to the packing of their (always bosonic) ion cores in forms similar to those discussed above.

Depending on the temperature or/and the size of the clusters, the forms (and thus the relevant magic numbers) of clusters assumed by bosonic atoms or bosonic ion cores (i.e., nested tetrahedra, or octahedra, or icosahedra) may change from the one (ground state) into the other form (excited or metastable structure). In a mixture of cluster sizes, i.e., in clusters with different temperatures, one may expect a coexistence of different forms and related magic numbers.

The state of matter of microclusters is apparently related to the present explanation. Specifically, bosonic clusters with no delocalized valence electrons are expected to closely resemble solid state of matter (as it is known, e.g. for rare gas clusters), while fermionic clusters are expected to closely resemble gas phase of matter(as it is believed, e.g. for alkali clusters) [38]. On the other hand, clusters with delocalized valence electrons (e.g. clusters born ionized) either bosonic or fermionic are initially expected to have structure close to the solid state phase. However, due to the appearence of the ion cores in the cluster, a greater mobility of the constituent atoms exists, a fact which could shift the phase towards the structured liquids.

The equilibrium geometry of the average alkali shells in Figure 3 is not a fixed geometry like the one we are familiar with in solid state physics, but it simply is a geometrical representation of high fluximal shells like those we are familiar with from molecular orbitals.

Besides the novel quantum mechanical explanation of magic numbers, the present paper underlines the idea that *new* , as yet *unobserved* properties of microclusters should be investigated. Perhaps, the most important of them is the orbiting properties of atoms implying a series of properties due to orbital angular momentum, i.e., definite spin properties, independent particle and collective modes of excitation of individual species etc. For an experimental verification of such properties nuclear methods should be employed.

Finally, an alternative method of studying atomic nuclei via concepts of quantum clusters seems possible and promising. It seems that the clusters made up of two kinds of alkali atoms (two kinds of fermions) assume a structure close to nuclear (neutron and proton) structure. However, a lot of work towards this direction is still needed.

Acknowledgements

I express my sincere appreciation to Professors P. Jena, B.K. Rao and S.N. Khanna of the Virginia Commonwealth University, Department of Physics, for their support to the initial ideas of this work, and to Professors J.W. Negele and J. Goldstone of the M.I.T. Department of Physics for their invitations to join the highly stimulating scientific environment at their Center for Theoretical Physics in 1988-89 and 1990.

References

1. Ekardt, W: Phys. Rev. **B 29**, 1558 (1984); Knight, W.D., Clemenger, K., de Heer, W.A., Saunders, W.A., Chow, M.Y., Cohen, M.L.: Phys. Rev. Lett. **52**, 2141 (1984).
2. Echt, O., Sattler, K., Recknagel, E.: Phys. Rev. Lett. **47**, 1121(1981).
3. Anagnsotatos, G.S.: Phys. Lett. A **124**, 85(1987)
4. Saito, Y., Watanabe, M., Hagiwara, T., Nishigati, S., Noda, T.: Jpn J. Appl. Phys. **27**, 424(1988); Saito, Y., Minami, K., Ishida, T., Noda, T.:Z.Phys. D-Atoms, Molecules and Clusters **11**, 87(1989).
5. Bhaskar, N.D., Frueholz, R.P., Klimcak, C.M., Cook, R.A.: Phys. Rev. B **36**, 4418(1987).
6. Anagnostatos, G.S.: Phys. Lett. **A 157**, 65 (1991).
7. Anagnostatos, G.S.: Phys. Lett. A **143**, 332(1990
8. Anagnostatos, G.S.: Phys. Lett. A **150**, 303(1990)
9. Anagnostatos, G.S.: Phys. Lett. A **133**, 419(1988)
10. Anagnostatos, G.S.: Phys. Lett. A **128**, 266(1988)
11. Anagnostatos,G.S.: Z. Phys D - Atoms. Molecules and Clusters **19**, 121 (1991).
12. Anagnostatos, G.S.: Phys. Lett. A **154**, 169 (1991).
13. Anagnostatos, G.S.: Phys. Lett. A **142**, 146(1989).
14. Anagnostatos, G.S.: Phys. Lett. A **148**, 291(1990).
15. Anagnostatos, G.S.: Z. Phys. D-Atoms, Molecules and Clusters **19**, 125 (1991).

16. Leech, J.: Math. Gazette **41**, 81(1957).
17. Franke, G., Hilf, E., Palley, L.: Z. Phys. D.-Atoms, Molecules and Clusters **9**, 343(1988).
18. Lipparini, E., Stringari, S.: Phys. Rev. Lett. **63**, 570(1989).
19. Katakuse, I., Ichihara, T., Fujita, Y., Matsuo, T., Sakurai, T., Matsuda, H.: Int. J. Mass Spectr. Ion Proc. **74**, 33(1986).
20. Mühlbach, J., Pfau, P., Sattler, K., Recknagel, E.: Z. Phys. B **47**, 233(1982).
21. Klots, T.D., Winter, B.J., Parks, E.K., Riley, S.J.: J.Chem. Phys. **92**, 2110(1990).
22. Rayane, D., Melinon, P., Cabaud, B., Hoareau, A., Tribollet, B., Broyer, M.: Phys. Rev. D-Atoms, Molecules and Clusters **39**, 6056(1989).
23. Katakuse, I., Ichihara, T., Fujita, Y., Matsuo, T., Sakurai, T., Matsuda, H.: Int. J. Mass Spectr. Ion Proc. **69**, 109(1986).
24. Geusic, M.E., Morse, M.D., Smalley, R.E.: J. Chem. Phys. **82**, 590(1985).
25. Whelten, R.L., Zakin, M.R., Cox, D.M., Trevor, D.J., Kaldor, A.: J. Chem. Phys. **85**, No.3(1986).
26. Begemann, W., Driehöfer, S., Meiwes-Broer, K.H., Lutz, H.O.: Z. Phys. D-Atoms, Molecules and Clusters **3**, 183(1986).
27. Martin, T.P., Bergmann, T., Göhlich, H., Large, T.: Chem. Phys. Lett. **172**, 209(1990).
28. Anagnostatos, G.S.: To appear.
29. Bjornholm, S., Borggreen, J., Echt, O., Hansen, K., Pedersen, J., Rasmussen, H.D.: Phys. Rev. Lett. **65**, 1627(1990).
30. Anagnostatos, G.S.: Int. J. Theor. Phys. **24**, 579(1985).
31. Thomas, A.W.: Adv. Nucl. Phys. **13**, 1(1984).
32. Celenza, L.S., Shakin, C.M.: Phys. Rev. C **27**, 1561(1983).
33. Coxeter, H.S.M.: Regular polytopes, 3rd Ed. (Macmillan, New York, 1973).
34. de Jager, C.W., de Vries, H., de Vries, C.: At. Data Nucl. Data Tables **14**, 479(1974).
35. Anagnostatos, G.S.: To appear.
36. Hornyak, W.F.: *Nuclear Structure*, Academic, New York, 1975.
37. Hill, D.L.: In *Encyclopedia of Physics* , Flugge, S. ed., Springer-Verlag, Berlin, vol. XXXIX, p. 211.
38. Gspann, G.: Z. Phys. D-Atoms, Molecules and Clusters **3**, 143(1986).
39. Engfer, R., Schneuwly, H., Vuilleumier, J.L., Valter, H.K., Zehnder, A.: Atomic Data and Nuclear Data Tables **14**, 509 (1974).
40. Brown, B.A., Bronk, C.R., Hodgson, P.E.: J. Phys. G **10**, 1683 (1984).
41. Wesolowski, E.: J. Phys. G **10**, 321 (1984).
42. Wapstra, A.H., Gove, N.B.: Nuclear Data Tables **9**, 267 (1971).

Multipole Response of ^{3}He Clusters

J. Navarro[1], *Ll. Serra*[2], *M. Barranco*[3], *and Nguyen Van Giai*[4]

[1]Departament de Física Atòmica, Molecular i Nuclear,
  Universitat de Valencia, E-46100 Burjassot, Spain
[2]Departament de Física, Universitat de les Illes Balears,
  E-07071 Palma de Mallorca, Spain
[3]Departament d'Estructura i Constituents de la Matèria,
  Universitat de Barcelona, E-08028 Barcelona, Spain
[4]Division de Physique Théorique, Institut de Physique Nucléaire,
  F-91406 Orsay, France

Ground state properties of normal ^{3}He drops have been studied using either a correlated wave function in conjunction with a realistic potential of Aziz type [1] or a mean-field description based on an effective potential [2,3]. In general, an overall good agreement between both methods has been found. The second one has the advantage of being rather easily applicable to both static and dynamic calculations, although being less fundamental than the first one. In this work we are concerned with the description of the collective modes of normal ^{3}He drops within the self-consistent Random-Phase Approximation (RPA), in which the same effective interaction is used to generate both the mean-field and the residual interaction. We restrict ourselves to surface excitations induced by external multipole fields of $r^l Y_{l0}$ type and to the monopole volume mode generated by the field r^2.

Our starting point is a zero-range, density and velocity dependent, effective ^{3}He-^{3}He interaction [2] whose parameters have been fixed so as to reproduce some selected bulk properties (saturation density, binding energy, incompressibility coefficient, specific heat and liquid surface tension). With this phenomenological interaction the Hartree-Fock (HF) problem for the cluster ground state can be solved and then the self-consistent mean-field and single-particle orbitals ϕ_i are obtained. It is found that the magic numbers are those of the harmonic oscillator and that the threshold in the number of atoms to produce bound clusters is near 30 atoms per cluster [1-3].

Such an effective interaction allows to write the total energy of the cluster as a local density functional. The residual particle-hole interaction is obtained as the second functional derivative of the cluster energy with respect to the particle density [4]. To calculate the excited states of ^{3}He clusters, we have constructed a discrete particle-hole basis and diagonalized the Hamiltonian $H = H_0 + V$, sum of the HF Hamiltonian H_0 and the residual particle-hole interaction V. The RPA equations are written in terms of angular-momentum coupled matrix elements of the residual interaction. This calculation closely follows those carried out to describe giant resonances in nuclei[5] or plasma resonances in metal clusters[6].

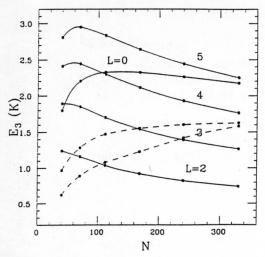

Figure 1. ^{3}He $L=2$-5 surface mode and $L=0$ volume mode energies as a function of the number N of atoms in the cluster.

The strength function $S(\omega)$ corresponding to an operator Q is defined as $S(\omega) = \sum_n |\langle n|Q|0\rangle|^2 \delta(\omega - E_n)$, where $|0\rangle$ and $|n\rangle$ are respectively the ground and excited state wave functions and E_n are the excitation energies. Before solving the RPA equations one can estimate the energies of multipole modes by evaluating the m_1 and m_3 RPA sum rules, defined as $m_k = \int_0^\infty dE E^k S(E)$. These RPA sum rules can be evaluated with the only knowledge of the HF ground state wave function [7-9]. From m_1 and m_3 we define the average energy $E_3 = (m_3/m_1)^{1/2}$, which gives a measure of the excitation energy provided the strength be concentrated in a narrow energy region (resonance state). In fact, E_3 provides an upper bound to the centroid of the strength, whereas the energy $E_1 = (m_1/m_{-1})^{1/2}$ provides a lower bound.

Figure 1 shows the E_3 energies corresponding to the $L = 2-5$ surface modes and to the $L = 0$ volume mode. The calculations have been done for several magic clusters. The lines are drawn only to guide the eye. One can see that these energies show a smooth, well defined N- and L-dependence. For $N \geq 150$ they decrease with N, qualitatively following a $N^{-1/3}$ law, which can be analitically deduced considering a big cluster of constant density [7]. The behavior changes for small clusters, especially in the monopole case, for which the average excitation energy increases when going from $N = 40$ to 112. Dashed lines represent the absolute value of the single particle energies corresponding to the last occupied and the first unoccupied states. One can see that, in contradistinction with the nuclear case, for small L-values and not too small sizes, the collective energies are in the discrete part of the spectrum.

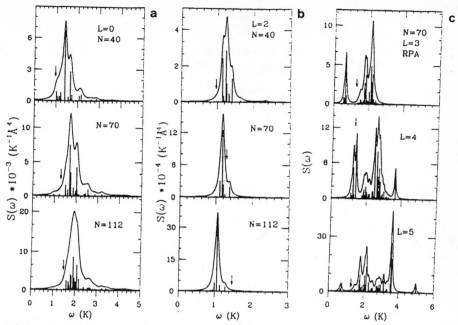

Figure 2. (a) Monopole $S(\omega)$ (in $10^{-3}K^{-1}\mathring{A}^4$) for $N=40$, 70 and 112. (b) Quadrupole $S(\omega)$ (in $10^{-4}K^{-1}\mathring{A}^4$) for $N=40$, 70 and 112. (c) $N=70$ $S(\omega)$ for multipoles $L=3$ (in $10^{-6}K^{-1}\mathring{A}^4$), $L=4$ (in $10^{-8}K^{-1}\mathring{A}^4$) and $L=5$ (in $10^{-10}K^{-1}\mathring{A}^4$).

Let us now comment on the detailed results obtained [10] from the solution of the RPA equations for the magic clusters $N=40$, 70 and 112. Figure 2 displays the strength distributions $S(\omega)$. Vertical lines represent the discrete states whose height gives the corresponding transition probability. For the sake of presentation we also display the curves obtained by folding the discrete lines with Lorentzian functions having an artificial width (solid lines).

One can notice that the RPA strength of the monopole (Fig. 2a) is rather fragmented (more fragmented, in fact that the HF strength). This is due to the strong repulsion effect of the residual interaction which shifts the strength above the particle emission threshold and produces a resonance in the continuum, similarly to the giant resonances in atomic nuclei. The distribution of collective states is broad (around $0.6K$). This width is larger than the parameter Γ and therefore it is a genuine width, reflecting the combined effects of particle escape and Landau damping.

Quadrupole results are displayed in Figure 2b. The RPA distributions are narrower than the monopole ones. This is especially marked for the $N=70$ and 112 clusters, for which a single line exhausts 60% and 72% of the quadrupole strength respectively. We have found that the quadrupole mode for $N \geq 70$ lies in the discrete part of the spectrum. This explains why this mode is less Landau damped than the monopole mode. According to the sum rule estimation, one finds that the evolution with N of the collective mode is opposite to the monopole case, now slowly decreasing in energy with increasing size.

Finally, in Fig. 2c the results for the cluster $N=70$ and the multipoles $L=3$, 4 and 5 are displayed. One can see that for higher multipolarities the response function is strongly fragmented over a wide energy range, thus indicating that the collective character of the excitation is lost for multipoles $L \geq 3$.

References

1. V.R. Pandharipande, S.C. Pieper and R.B. Wiringa, Phys. Rev. **B34** (1986) 4571.
2. S. Stringari, Phys. Lett. **A107** (1985) 36.
3. S. Stringari and J. Treiner, J. Chem. Phys. **87** (1987) 5201.
4. G.F. Bertsch and S.F. Tsiai, Phys. Reports **18** (1975) 125.
5. J.P. Blaizot and D. Gogny, Nuc. Phys. **A284** (1977) 429.
6. C. Yannouleas, R.A. Broglia, M. Brack and P.F. Bortignon, Phys. Rev. Lett. **63** (1989) 255.
7. Ll. Serra, F. Garcias, M. Barranco, J. Navarro and Nguyen Van Giai, Z. Phys. **D** (1991) in press.
8. O. Bohigas, A.M. Lane and J. Martorell, Phys. Reports **51** (1979) 267.
9. E. Lipparini and S. Stringari, Phys. Reports **175** (1989) 103.
10. Ll. Serra, J. Navarro, M. Barranco y Nguyen Van Giai, UIB preprint (1991).

Size-Dependent Plasmons in Small Metal Clusters

W. Ekardt and J.M. Pacheco

Fritz-Haber-Institut der Max-Planck-Gesellschaft,
Faradayweg 4–6, W-1000 Berlin 33, Fed. Rep. of Germany

The collective motion of the system of loosely bound valence electrons is studied within the framework of the TDLDA (time dependent local density approximation) applied to spherical jellium clusters for the magic clusters, or to spheroidal jellium clusters for the non-magic ones. For Na_n Clusters the main results are the following:

— 1. A well-defined surface plasmon does exist in the size-range $100 < N < 200$. Here the surface plasmon line exhausts the Thomas-Reiche-Kuhn sum rule of the oscillator strength to nearly 100%.

— 2. In the size range $N < 40$ the surface plasmon is ruined by *Landau fragmentation*, i.e. the line splits in many lines which have their origin in individual particle-hole-pair excitations.

— 3. *Landau damping* does not exist, simply because the surface plasmon frequency is always lower than the first ionization potential of the cluster.

— 4. In the optical spectra the intensity of the volume plasmon excitation is smaller by 3 orders of magnitude. In order to excite volume plasmons one should performe inelastic electron scattering.

— 5. In order to take into account the effect of the Coulomb tail we have modified the TDLDA to the SICTDLDA, which amounts to the inclusion of corrections of the self-interaction effects within the framework of PERD-DEW and ZUNGER.

This enables us to study the charge dependence of the fragmentation, and the results are as follows. The neutral Na_{20} has a split plasmon consisting of two components. The 20-electron system Na_{21}^+ has no fragmentation and the corresponding anion Na_{19}^- is heavily fragmented. The predictions for the neutral Na_{20} and for the cation are experimentally confirmed. Unfortunately, we are not aware of any absorption spectra of the anions.

Springer Series in Nuclear and Particle Physics   **Clustering Phenomena in Atoms and Nuclei**
Editors: M. Brenner · T. Lönnroth · F.B. Malik   © Springer-Verlag Berlin, Heidelberg 1992

Nucleonic Orbitals in the Valence Particle Plus Core Model of Nuclear Reactions

W. von Oertzen[1] and B. Imanishi[2]

[1]Hahn-Meitner-Institut and Freie Universität Berlin,
 W-1000 Berlin 39, Fed. Rep. of Germany
[2]Institute of Nuclear Study, University of Tokyo, Tokyo 188, Japan

Reactions between nuclei characterised by a valence nucleon and two inert cores are discussed. Several experimental cases will be given where molecular orbital aspects, hybridisation and Landau-Zener transitions are observed. The two-center shell model and the rotating molecular orbital approach are illustrated and the relation to the structure of nuclei is demonstrated.

1. Introduction. Reactions between nuclei close to the Coulomb barrier exhibit in many cases a very distinct dependence on the structure of the nuclei; the most pronounced aspects are connected to the single particle states if nuclei with one or a few valence particles outside closed shells are involved [1,2]. In such cases the cores are considered as clusters, whose intrinsic structure can be neglected. For reactions, where the two cores (clusters) stay at larger distances only their collective inelastic excitations must be considered, and in most cases of weak inelastic transitions the core-core interaction potential can be obtained in the framework of the optical model. The motion of the valence particles can then be considered separately in detail. Such systems can be very effectively described by a two-center basis of the valence particle [3,4,5]. In the following section a general description of the properties of valence orbitals in a two-center system is given. In section III a few specific examples are given for which a comparison of experiments with theoretical calculations has been done. They mainly involve ^{12}C and ^{16}O as cores: $^{12}C + ^{13}C$ [6,7,8], $^{16}O + ^{17}O$ [9,10,11], $^{16}O + ^{13}C$ [12,13], $^{13}C + ^{14}C$ [14]. Recently also proton exchange in a heavier system $^{36}S + ^{37}Cl$, where ^{36}S acts as a core cluster has been studied [15]. The dynamics of the reaction in which nucleon transfer and single particle excitations occur, is described in a coupled reaction channel (CRC) approach which is well known also in atomic and molecular collisions [16].

II. Nucleon orbits in a Two-Center System. We consider the interaction of two nuclei defined by M_1 and M_2 (core 1 and core 2) and a valence nucleon m (see fig. 1). The eigenstates of the asymptotically defined nuclei consisting of M_1 and M_2 plus a valence nucleon are given by the shell model and are defined by wave functions $\Psi_{i,k}$ (R_i) i = 1,2; k = (nlj). From these basis states the eigenstates ϕ_α of the instrinsic two-center Hamiltonian H_α are constructed by a linear combination of nucleon orbitals [4]. For each total angular momentum and parity n the total wave function of the system, $\Phi^{JM\Pi}$ is expanded,

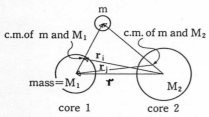

Fig. 1. Definition of coordinates for the core plus valence particle model.

Springer Series in Nuclear and Particle Physics **Clustering Phenomena in Atoms and Nuclei**
Editors: M. Brenner · T. Lönnroth · F.B. Malik   © Springer-Verlag Berlin, Heidelberg 1992

$$\Phi^{JM_{II}} = \sum_{\alpha} \Phi_{\alpha}^{JM_{II}} u_{\alpha}^{J_{II}}(r) / r$$

With the total Hamiltonian we obtain the CRC equation for the radial relative motion, which in matrix form can be written as

$$\{E = (T + R + \epsilon + U^D + K^t)\} u = 0$$

The operator K^t contains all transfer processes and is nonlocal but is approximated by a local form with a first order inclusion of recoil effects; U^D causes the direct excitations. The operator R describes the angular part of relative motion; the kinetic energy, T, is completely separated from the other terms unlike in the conventional treatment of CRC; we can thus discuss the local effective interaction $U^{tot} = R + \epsilon + U^D + K^t$.

With this interaction U^{tot}, which can be diagonalized by using molecular orbital states, we obtain the <u>adiabatic potential</u> $V_p(r)$ as function of the core-core distance

$$V_p(r) = \left(A^{-1}(r) \; U^{tot} \; A(r) \right)_{pp} ,$$

and for the <u>rotating molecular orbitals,</u> Φ_p, RMO (because we included the rotational terms) we have

$$\Phi_p = \sum_{\alpha} \Phi_{\alpha} A_{\alpha p}(r).$$

The transformation matrix (amplitudes) is identical to the unitary operator I at large distances. With these molecular orbital states Φ_p a new set of coupled equations is obtained with the adiabatic potentials $V_p(r)$. Transitions are now induced by radial couplings only (described by interactions $\Delta V_{pq}(r)$) and are particularly strong for cases of avoided crossings - i.e. situations where the energies of two states Φ_p as a function of the two-center distance would cross.

To illustrate the properties of nucleonic orbitals of the nuclear two-center problem fig. 2 and fig.3 show examples of correlation diagrams.

Fig.2 shows the case of $^{17}O + ^{12}C$ from ref.[3] where the energies of all orbitals (not only of the valence orbits but also those contained in the cores) are shown as obtained in a two-center shell model approach. These energies are obtained without diagonalizing the rotational and radial interaction terms, therefore transitions in this correlation diagram, which corresponds to those used in atomic collisions, will be induced by <u>rotational</u> and <u>radial</u> couplings.

In fig. 3 we show the adiabatic potential $V_p(r)$ as defined above for the valence neutron orbits in the $^{12}C + ^{13}C$ system. In this RMO approach for each total spin and parity (here $J^{\pi} = 7/2^-$) the energy of the molecular orbits are obtained; asymptotically these orbits merge into the states of ^{13}C (gs, $\frac{1}{2}^-$; $s\frac{1}{2}$ at 3.08 MeV and $d_{5/2}$ state at 3.85 MeV). An avoided crossing is observed at a radial distance of 6 fm between the $p=1$ and $p=2$ states (asymptotically $p\frac{1}{2}$ and $s\frac{1}{2}$) which gives rise to a sharp peak in the radial coupling shown in the middle part of the figure. Transitions induced by coupling at an avoided crossing are named Landau-Zener transitions [4,5]. The observation of features connected to Landau-Zener transitions in nuclear physics is a subject which has been discussed extensively in the past 5 years [4,5]. In the next section we will show some examples, where evidence for nuclear Landau-Zener transitions has been obtained. Generally, the quantum mechanical properties of the nuclear systems and the strong rotational coupling makes it very improbable that resonances in integrated cross sections can be observed due to Landau-Zener transitions. Experimental data relevant to this claim [17,18], where shown, by the same authors, to be due to target

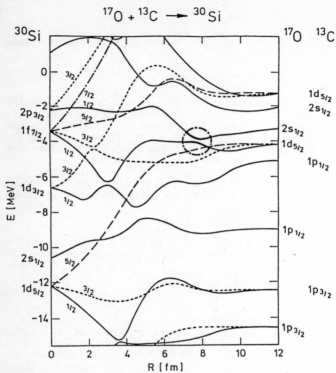

Fig. 2. Correlation diagram for all valence particles in the $^{17}O + ^{12}C$ system (ref. 3).

contaminations [19]; similarly schematical calculations, which showed such resonant behaviour, were shown to be incorrect in later work [20].

The lower part of fig.3 shows the mixing coefficients for different states labelled by the index p, which are classified by the projection K of the nucleon spin along the molecular axis. We note the strong mixing between the $p\frac{1}{2}$, $s\frac{1}{2}$ and $d_{5/2}$ components which sets in at a distance of r=6.5 fm. The effect of the dynamical mixing of configurations is a very specific feature of the molecular orbital approach. For the case of mixing of different parities a dramatic distortion of the wave function and of the densities is observed, which is known as hybridization in atomic physics [21].

Fig. 4 shows two-center density distributions of the valence orbital for four different distances together with the adiabatic potential for the p=1 state. The lowering of the core-core potential due to the molecular orbital formation is a typical effect of the present approach which is connected to the experimentally observed enhancement of the fusion reaction cross section below the barrier [13, 22].

III.   Experimental Examples In this section examples are shown which have been analysed in the CRC-approach and in the RMO-approach. The experiments involve measurements of angular distribution at incident energies close to the Coulomb barrier, where the valence particle plus core (cluster) model is particularly adequate. The decisive information is obtained from the energy variation of the angular distributions of simple reactions - one nucleon transfer

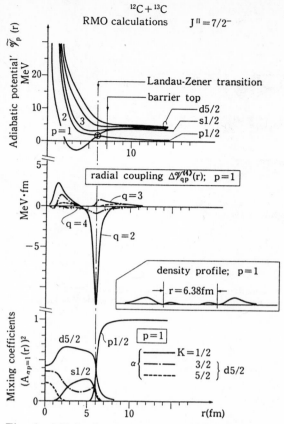

Fig. 3. Adiabatic potentials, coupling elements and mixing coefficients for the system $^{12}C + n + ^{12}C$ for $J^n = 7/2-$.

and single particle excitation. The shapes of the angular distributions and the sometimes dramatic changes with incident energy generally can not be reproduced in a first-order pertubation calculation, - however, often they cannot be reproduced by the CRC calculation if the radial coupling is smeared out due to strong absorption, or because the basis was still to small.

a) $^{12}C + ^{13}C$ The $^{12}C + ^{13}C$ system is characterised by two ^{12}C cores which have a closed p3/2 shell and a binding energy for neutron of 18.7 MeV; the neutron in ^{13}C is bound only by 4.95 MeV. Other favourable cases for the valence particle plus core model are summarised in fig. 5 (binding energies are: $(n + ^{15}O) \rightarrow ^{16}O$, $E_B = 15.6$ MeV; $(^{35}P + p) = > ^{36}S$, $E_B = 13.0$ MeV.

Strong hybridisation effects occur in $^{12}C + ^{13}C$ because of the close vicinity of orbitals for different major shells. Thus it occurs at the same element as in chemistry because of the similarity of the shell structure of atoms and nuclei. An extensive study [8] of angular distributions as function of energy has been done from the vicinity of the Coulomb barrier E_B up to energies two times E_B. They showed a fast change in the shapes for the s$\frac{1}{2}$ state, which could not be described by second order calculations at lower energies (Fig. 6). At higher energies the data are reasonable well reproduced. The fast change in the barrier region is explained

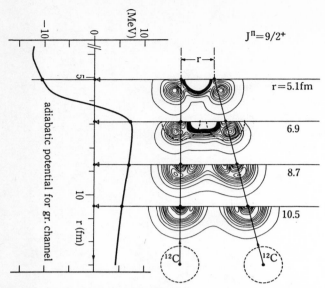

Fig. 4. Density distributions of the valence neutron in the lowest molecular orbital in $^{12}C + ^{13}C$ for $J^{\Pi} = 9/2 +$ (grazing angular momentum for 8.8 MeV $= E_{CM}$)

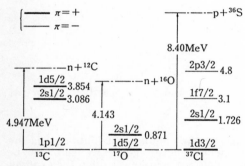

Fig. 5. Levels of valence particles with strongly bound cores, $^{12}C, ^{16}O, ^{36}S$.

by the hybridization effect and the Landau-Zener transition between the ground state and the first excited state ($s\frac{1}{2}$), which is shown in fig. 3 to occur at a radial distance just behind the Coulomb barrier. Therefore it becomes active as soon the $\frac{1}{2}^+$ state moves above the barrier in the energy region between $E_{CM} = 7.80$ to 10.7 MeV. For other states, the $d_{5/2}$ state, the transition is generated by smooth radial coupling at distances larger than the grazing distance. Thus the transition shows a smooth energy dependence.

b) $^{14}C + ^{13}C$ For the $^{14}C + ^{13}C$ system we have two non-identical cores ^{12}C and ^{14}C. The ^{14}C nucleus is very similar to ^{16}O, the closed neutron shell is responsible for the spherical shape and the absence of excited states below 6 MeV. Avoided crossings are observed for the transition $^{14}C + ^{13}C \rightarrow ^{12}C + ^{15}C$ populating the $s_{1/2}$ state in ^{15}C.

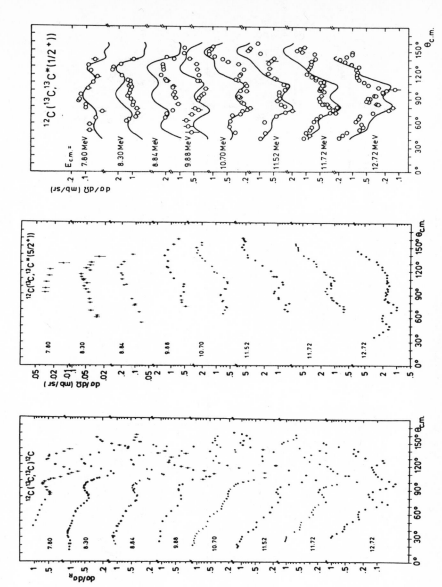

Fig. 6. Angular distributions of elastic and inelastic scattering of $^{12}C + ^{13}C$ The full curves for the $\frac{1}{2}^+$ state show CRC calculations with a basis: $p\frac{1}{2}$, $s\frac{1}{2}$, $d5/2$.

93

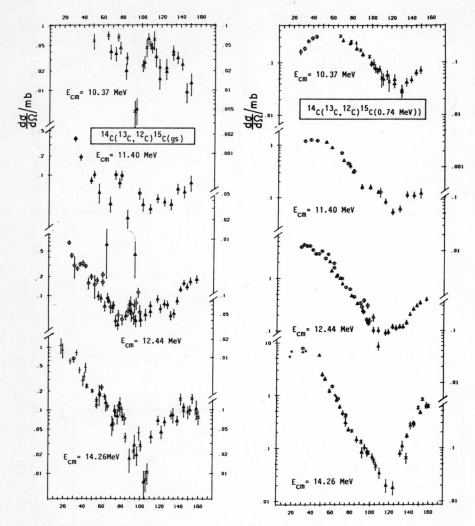

Fig. 7. Angular distributions of the reaction, $^{14}C(^{13}C,^{12}C),^{15}C$ showing fast changes of angular distributions for the $s\frac{1}{2}$ state (gs state of ^{15}C)

For this system again angular distributions for several energies between $E_{CM} = 10.37$ to 14.5 MeV have been measured for inelastic transitions and for transfer channels. In the lower energy steps large changes in the shapes of the angular distributions are observed - and only for those ($s_{1/2}$ states in ^{15}C and ^{13}C) which participate in the strong coupling induced by the avoided crossings. Fig. 7 shows some examples of the energy dependence of angular distributions. We find that the behaviour of some states is actually <u>chaotic</u> as a function of incident energy. From these two examples, and other, we may conclude that fast changes in the shapes of angular distributions are the most significant signature of Landau-Zener transitions in the valence-particle plus core systems. The strong coupling at avoided crossings actually may often create situations, which exhibit chaotic behaviour.

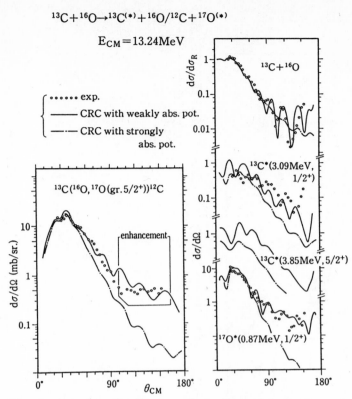

$$^{13}C + ^{16}O \rightarrow ^{13}C^{(*)} + ^{16}O / ^{12}C + ^{17}O^{(*)}$$

$E_{CM} = 13.24 MeV$

•••••• exp.

——— CRC with weakly abs. pot.

—·—· CRC with strongly
        abs. pot.

$^{13}C(^{16}O, ^{17}O(gr. 5/2^+))^{12}C$

enhancement

$^{13}C + ^{16}O$

$^{13}C^*(3.09MeV, 1/2^+)$

$^{13}C^*(3.85MeV, 5/2^+)$

$^{17}O^*(0.87MeV, 1/2^+)$

Fig. 8. Angular distribution of elastic and inelastic scattering and transfer channels in the reaction of $^{16}O + ^{13}C$.

c) $^{16}O + ^{13}C$ A typical expample for the valence particle plus cores model is the system $^{16}O + ^{13}C$, for which a variety of data exists, including fusion cross sections as function of energy around the Coulomb barrier. In this case the neutron transfer $^{16}O + ^{13}C \rightarrow ^{12}C + ^{17}O$ shows an avoided crossing at rather small distances between the ^{13}C (gs, $p_{1/2}$) the $s_{\frac{1}{2}}$ configuration in ^{17}O. The $^{16}O + ^{12}C$ scattering itself is well known as a weakly absorbing system. CRC-calculations for the $^{13}C + ^{16}O \rightarrow ^{12}C + ^{17}O$ system show a pronounced effect due to the Landau-Zener transitions at small distances. The result of the CRC calculations is shown in fig. 8.

Here calculations with strong absorption and weak absorption are compared. In the case of strong absorption the structure at large angles in the elastic scattering is not reproduced and the cross sections for the neutron transfer channels are too small at large angles. The cross sections at large angles are strongly enhanced by the contributions from small distances in the case of the weakly absorbing potentials. Another important effect, the enhancement of the fusion cross sections at the threshold (below the Coulomb barrier), is also reproduced in these calculations [13].

d) $^{36}S + ^{37}Cl$ In this system the ^{36}S nucleus is a reasonably well defind cluster core due to the closed neutron shell (N = 20). The proton orbits in ^{37}Cl span configurations at the end of the sd-shell and the beginning of the fp-shell. This fact gives rise to strong configuration mixing of the proton orbits due to the presence of

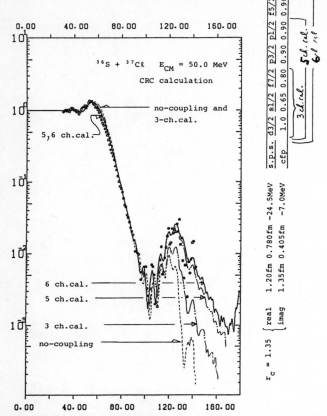

Fig. 9. Elastic ground state transfer in the system $^{37}Cl + {}^{36}S$. The backwardrise shows strong enhancement if other single particle levels in ^{37}Cl are included.

the field of the second nucleus, which is enhanced by the presence of the Coulomb interaction. The experimental data show strong excitations of $s\frac{1}{2}$ state (1.73 MeV), the $f7/2$-state (3.10 MeV) and the $p3/2$ state (4.27 MeV). In the analysis the full onfiguration space of the valence proton spanning the sd shell and the fp shell (6 configurations, see fig. 9) has been included. For the case that only the $d_{3/2}$ ground state is included the elastic transfer channel is underpredicted by a factor of 2. With the inclusion of the higher orbits of the valence proton the ground state transfer cross section rises, the data are well reproduced (fig. 9).

In order to understand the effect of the increasing cross section the density distribution of the valence particle in the intrinsic frame (RMO) can be calculated. For a distance of 8.64 fm of the two cores fig. 10 shows the density distribution for a given total spin and parity. For the state with positive parity the mixing of configurations with different parity induces a strong concentration of the valence particle between the two centers (hybridisation effect). This effect explains the increase in the cross section of the ground state transition in the molecular orbital approach. A mixing with configurations of the same parity would have produced only a rather small effect.

1-channel (d3/2) calculation   3-channel (d3/2, s1/2, f7/2) calculation

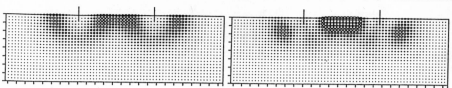

6-channel (d3/2, s1/2, f7/2, p3/2, p1/2, f5/2) calculation

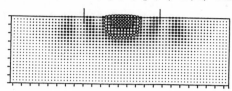

Fig. 10. Proton density distribution in the two center system $^{36}S + ^{37}Cl$ for grazing angular mometum ($J^{\pi} = 63/2+$) and a core-core distance of $\Upsilon = 8.64$ fm.

IV.   Conclusions The fast evolution of the last ten years, in both experimental possibilities as well as in the theoretical analysis (with the use of larger computers), has made it possible that the physics of molecular valence orbitals outlined more than 10 years ago, has reached now a quantitative level, which explains a rich variety of phaenomena as they are known partially from atomic physics. The valence particle plus cluster-core approach gives for adequately chosen situations (strong binding of the nucleons in the cores, smaller binding energy of the valence particle) an excellent description of various cross sections in nucleus-nucleus collisions. The physical phenomena (hybridisation, Landau-Zener transitions etc.) are in these cases analoguous to the cases of ion-atom collisions [16], where the motion of the valence particles can be described in a molecular orbital approach. The cluster properties of nulcei are particularly strongly developed for ligther masses and closed shells. For heavier nuclei it becomes increasingly difficult to single out valence particles (even at closed shells) because of the much weaker overall binding energy of nucleons. However, the future evolution of the experimental and computational possibilities may render the study of collisions between heavy nuclei possible in similar details as described here for light systems.

References

1.   G. Breit, M.E. Ebel, Phys. Rev **103** (1956) 679
2.   L.J.B. Goldfarb, W. von Oertzen, in "Heavy Ion Collisions" Vol. 1. ed. R. Bok, North Holland, 1979
3.   Y. Park, W. Scheid, W. Greiner, Phys. Rev. **C20** (1979) 188; **C12** (1980) 1958
4.   B. Imanishi, W. von Oertzen, Phys. Reports **155** (1987) 29
5.   A. Thiel, J.Phys. **G16** (1990) 867
6.   H. Fröhlich et al. Nucl. Phys. **A420** (1984) 124
7.   B. Imanishi, W. von Oertzen, H. Voit, Phys. Rev. **C35** (1987) 359
8.   H. Voit et al. Nucl. Phys. **A476** (1988) 491
9.   D. Kalinski et al. Nucl. Phys. **A250** (1975) 364
10.   G. Bauer, H.H. Wolter, Phys. Letters **B51** (1974) 205

11. B. Imanishi, S. Misono, W. von Oertzen, Phys. Letters B210 (1988) 35
12. W. Bohne et al. Nucl. Phys. A332 (1979) 501
13. B. Imanishi, S. Misono, W. von Oertzen, Phys. Letters B241 (1990) 13
14. N. Bischoff et al. Nucl. Phys. A490 (1988) 485
15. R. Bilwes, B. Bilwes et al. to be published
16. R.E. Johnson, Introduction to Atomic and Molecular Coillsions, Plenum Press (NY), 1983
17. R.M. Freeman et al., Phys. Rev. C28 (1983) 437
18. Y. Abe and J.Y. Park, Phys. Rev. C28 (1983) 2316
19. R.M. Freeman et al. Phys. Rev. C38 (1988) 1081
20. T. Tazawa and Y Abe, Phys. Rev. C41 (1990) R17
21. L. Pauling, The Nature of Chemical Bound (Cornell Uni. Press, 1969)
22. M. Beckermann, Rep. Prog. Phys. 51 (1988) 1047

Magic Numbers in Atomic Clusters: Energetic, Electronic and Structural Effects

A.W. Castleman, Jr.

Pennsylvania State University, Department of Chemistry,
University Park, PA 16802, USA

Introduction: The subject of magic numbers in clusters has been one of long-standing interest since early observation of pronounced intensity anomalies seen in the mass spectral distributions of molecular and atomic systems (1-10). Early observations in the case of water (2,3)revealed a particularly stable entity involving 21 molecules bound to a proton, which was first seen in the electron impact ionization of neutral water clusters, and later in proton expansion experiments and upon the bombardment of ice surfaces with fast atomic ions (11). The underlying commonality strongly pointed to the origin of magic numbers involving a particularly stable ion, a fact which has recently been proven (12-15). During the last decade there has been extensive experimental and theoretical interest in atomic systems (6,7,15-18), particularly for clusters of rare gas atoms and those involving metallic-type constituents. In view of the fact that the magic number patterns observed in the case of rare gas clusters coincided almost exactly with compact structures calculated for the neutrals, they were first interpreted in terms of especially stable neutral aggregates, despite the fact that fragmentation could obviously play a role following the ionization and concomitant fragmentation often involved.

The literature has been replete with suggestions concerning the origin of these pronounced anomalies seen in the mass spectral distributions of clusters, with explanations ranging from transient intermediates and frozen-in fragments formed in a molecular beam, to especially stable neutral or ionic species. In this paper we draw on several examples from work conducted in our own laboratory which point to the origin of magic numbers in several important atomic systems. Those involving rare gas systems are considered first, where new studies have definitively established the interrelationship of thermochemical stability and magic numbers in the xenon cation system. Secondly, we turn to some simple metallic systems followed by a consideration of metallic/nonmetallic zintl-like species. Finally, we end up with a consideration of atomic clusters comprised of a mixture of anions and cations, namely MgO clusters.

Springer Series in Nuclear and Particle Physics **Clustering Phenomena in Atoms and Nuclei**
Editors: M. Brenner · T. Lönnroth · F.B. Malik © Springer-Verlag Berlin, Heidelberg 1992

Xenon Cluster Ions: The Interrelationship of Bonding Energies and Magic Numbers:

Magic numbers have been observed in all of the rare gas cluster systems, and here we consider new data (15) for the magic numbers in xenon. Magic numbers have been observed for a vast number of cluster sizes, the smaller ones of which coincide with known icosahedra. In view of the dielectric relaxation which occurs following ionization, and the fact that the overlap between the initial neutral and final ion states do not allow good Frank-Condon overlap, adiabatic ionization may not obtain for moderate size atomic rare gas clusters. Hence, following ionization, evaporative dissociation commences leading to what is termed "metastable decay" (19,20). As pointed out by Klots (21), such systems are not microconical ensembles and hence the evaporation leads to a change in decay fraction which varies with the time of observation and shows an increase with cluster size.

We have recently demonstrated (19,20) through a detailed investigation of ammonia clusters whose bonding is well known from other measurements, that a time-of-flight mass spectrometer equipped with a laser-based ionization system and reflectron enable quantification of the metastable decay fractions, the average kinetic energy released upon dissociation, and a determination of the thermochemical bond energies for the cluster ions as a function of size. Various statistical theories can be used to treat the data including those developed by Engelking (see Ref. 19) and others formulated by Klots termed the evaporative canonical ensemble model.

Through use of the reflectron technique, we have recently measured decay fractions defined as $D=I_d/(I_d+I_p)$ where I_d and I_p are the daughter and parent ions, respectively. Through a high precision measurement enabling an integration of the parent and daughter ion peaks, we have determined (15) the decay fractions of the xenon cluster ions over a range of size. Based on the evaporative canonical ensemble approach, the daughter ion population can be calculated as follows:

$$Xe_n^+ \longrightarrow Xe_{n-1}^+ + Xe, \quad n=5-40 \tag{1}$$

$$D = 1-(\alpha W_n)^{-1}\ln[1+(\exp(\alpha W_n)-1)t_0/t] \tag{2}$$

where

$$\alpha W_n = \gamma^2(W_n/\Delta E_n)/[C_n(1-\gamma/2C_n+(\gamma/C_n)^2/12...)^2 \tag{3}$$

$$W_n = \Delta E_n\{1+[(dE/d\Delta E_n)_k-1](\Delta E_n-\Delta E_{n-1})/\Delta E_n\} \tag{4}$$

$$(dE/d\Delta E_n)_k = (C_n/\gamma)(1 + \gamma/2C_n + (\gamma/C_n)^2/12...) \tag{5}$$

This method enables a determination of the ratio $\Delta E_n/\Delta E_{n+1}$. While the resulting values are somewhat sensitive to the choice of parameters in the equations for the case of very small ions, for those in excess of n=10 the deduced bond energy ratios are insensitive to the choices of heat capacity and the Gspann parameter γ (which has been determined to be about 25 for a wide range of systems). The heat capacity C_n is estimated from the bulk heat capacity and the required time ratio, t_0/t, is deduced experimentally. Employing

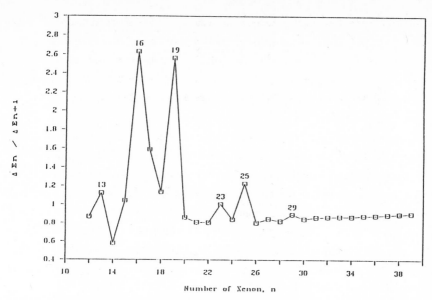

Figure 1. Relative binding energies of $\Delta E_n/\Delta E_{n+1}$ deduced by fitting the calculated values to the measured ones are plotted as a function of cluster sizes, n, assuming $\gamma=25$ and $C_n=(2.77n-6)k$. [Taken from Ref. 15]

information on the molar bulk heat capacity for xenon, we deduce an expression for C_n equal to $(2.77(n-6)k$.

Figure 1 shows a plot of the relative bond energies versus cluster size. Importantly, the especially stable peaks marked at n=13, 16, 19, 23, 25, and 29 coincide exactly with the magic numbers observed for this cluster system (15,18). To the best of our knowledge, this is the first time that data on bond energies and magic numbers have been available for the same system, and hence these experiments provide the first definitive evidence for an exact one-to-one correspondence between cluster ion stability and magic numbers.

The Role of Tunneling in Effecting the Metastable Decay of Rare Gas Cluster Ions:

Metastable decay arising through the evaporative dissociation of "hot" clusters shows an interesting trend of an increasing "rate" with cluster size, obtained when the dissociation fraction is put in terms of rate terminology by dividing its value by the observational time of dissociation. More appropriate treatment follows from a consideration of the decay fraction, which also displays a similar trend. Although molecular systems can show some degree of metastability even for very small clusters, this is in general not possible for atomic clusters. The reason is that for small clusters there are few vibrational modes in which energy can be stored, and any excess energy above the dissociation limit leads to rapid dissociation and rather small fractions of metastability in any observational time window. Nevertheless, in careful investigations of rare gas systems we have observed metastability on a very long time scale (22) extending to as much as 40 microseconds in length for the case of Ar_3^+ dissociating to $Ar_2^+ + Ar$. These observations are incompatible with the usual statistical theory and point to some other factor enabling energy above the dissociation limit to be stored.

We considered this problem following experiments which revealed not only long term metastability, but an increase in metastable fraction arising from collision induced dissociation with variations in experimental parameters which were believed to substantially increase the rotational excitation of these systems. The observations strongly pointed to the influence of high angular momentum states, and to the importance of trapping behind a rotational barrier. Although this problem is very difficult to treat theoretically for a multi-dimensional system, we investigated it (22,23) using the WKB approximation and a consideration of one dimensional tunneling probabilities which depend on the reduced mass of the system μ and the effective potential V_{eff},

$$P \propto \exp \frac{-2}{\hbar} \int_{r_2}^{r_1} [2\mu(V_{eff} - E)]^{1/2} \, dr \,, \tag{6}$$

where r is the internuclear distance and E is the total internal energy of the ion. the effective potential energy is given by

$$V_{eff} = V(r) + \frac{\hbar^2 L (L + 1)}{2\mu r^2} \,, \tag{7}$$

where the second term on the right-hand side is the familiar rotational barrier, $V(r)$ is the rotationless potential function, and L is the total angular momentum quantum number. At large r and L the second term becomes relatively more important and when it is dominant the reduced mass in the denominator tends to offset the reduced mass in Eq. (6).

Further quantification follows from a consideration of an appropriate potential for the system which we have parameterized in terms of the standard Morse potential function for

$$V(r) = D_o \{1 - \exp[-\alpha(r - r_o)]\}^2 \,. \tag{8}$$

The parameters employed in the calculations are deduced from theoretical and experimental considerations (22,23). Although the rotational barrier is lowered due to the deeper attractive well, this is somewhat offset by a large range of rotational states that are bound; see Figure 2. Computations have been done where bolts on distribution of rotational levels was assumed. More recently we have carried out computations for somewhat larger systems (23) using the same approach as for the trimer system. Figure 3 reveals that for realistic rotational temperatures, a fairly broad distribution of lifetimes can be obtained in agreement with the experimental observations. Explicit knowledge of the metastable

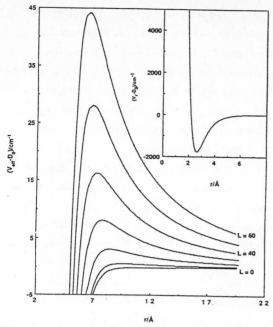

Figure 2. Rotational barrier used in the calculation of the WKB tunneling lifetimes. The insert shows the pure Morse potential employed. [Taken from Ref. 22]

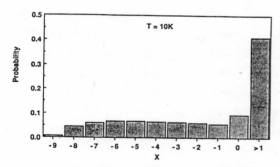

Figure 3. Distribution of tunneling lifetimes of Ar_3^{+*} decaying into Ar_2^+ calculated within the WKB approximation. Each bar in a histogram contains the integrated probability of having a lifetime between 10^x and $10^{(x+1)}$ s. The last bar on the right contains the total probability of observing lifetimes longer than 1 s. [Taken from Ref. 22]

lifetimes for a particular experimental condition requires knowledge of the energy levels, cluster structure and temperature distribution. Despite this shortcoming, our findings do strongly point to the role of rotational tunneling as one source of long term metastability, particularly in small atomic cluster systems.

Metastability of Metals and Metal Alloys: The work of Knight and co-workers (24) provided strong evidence for the inter-relationship of magic numbers and the electronic states of metal clusters for the simplest alkaline metal systems. Among the possible free-electron systems, aluminum is a very good sample. Studies of ionization potentials and photoelectron detachment spectroscopy have suggested that there are small variations in the neighborhood of predicted shell closings based on the jellium model, though the error bars for such experiments are comparable in magnitude to the small variations in the energy levels at the sizes corresponding to the expected shell closings.

An alternative approach elucidating shell closings is to investigate variations in chemical reactivity with the numbers of metal atoms. The reason that this approach is so successful hinges on the fact that chemical reactions are dramatically affected by shell closings, and can reveal extremely small differences in energy levels due to electronic effects on reactions. In this context, we have investigated in detail anions of aluminum (25,26), including alloys of aluminum (27) where one atom has been substituted with either niobium or vanadium. These experiments have been conducted using a fast flow reactor and a laser vaporization source to produce the anion clusters. Our experiments reveal that aluminum anions readily react with O_2, through an etching mechanism which proceeds to selectively remove one aluminum atom at a time from the cluster; the anion cluster retains the charge and is merely diminished in size. The expected odd/even alternation in reactivity is observed in accordance with expectations based on pairing and unpairing of electrons for even and odd electron containing systems. Particularly notable were observations of the unreactivity for aluminum 13⁻,23⁻ and 37⁻, all of which correspond to the jellium shell closings in view of the electronic character of aluminum being a $(3s^23p^1)$ atom. Most interestingly, not only are the clusters mentioned comparatively unreactive, they are actually formed due to the successive etching reactions producing the aluminum 13, 23, and 37 from the etching of higher order clusters; see Figure 4(a). More recent experiments with niobium and vanadium, each having five valence electrons $(4d^45s^1$, and $3d^34s^2$, respectively) reveal similarly interesting reaction patterns; see Figure 4(b). The results forniobium are also in accord with the jellium model showing an unusually unreactive species, $NbAl_4^-$, which is evidently an 18 electron system. Nevertheless, another species $NbAl_6^-$ and (only) a VAl_6^- species also are observed to be formed in abundance. If every valence electron is counted as in the previous example, these latter species would be 24 electron entities, in discrepancy with the jellium model. Failure is not unexpected in view of the fact that niobium and vanadium are not likely to be free electron systems. It is possible that hybridization of the electrons occurs or that one of the four d electrons of niobium is promoted and that both systems may be 20 electron species. Further work is needed to elucidate the exact electronic nature of these systems.

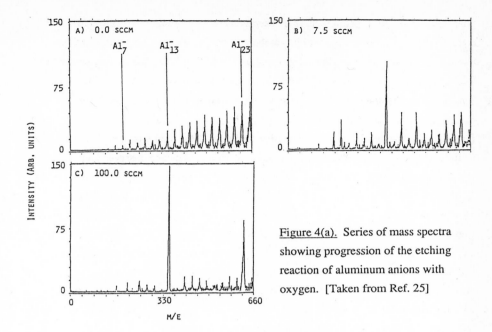

Figure 4(a). Series of mass spectra showing progression of the etching reaction of aluminum anions with oxygen. [Taken from Ref. 25]

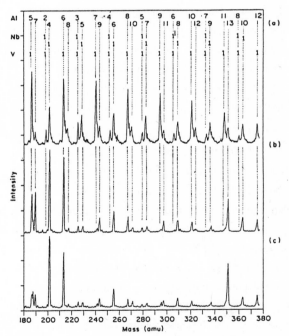

Figure 4(b). Reaction of metal clusters with O_2: (a) 0.0 sccm of O_2, (b) 30.6 sccm of O_2, and (c) 60.0 sccm of O_2. Intensity scales are arbitrary. [Taken from Ref. 27]

Gas-Phase Zintl Ions: Zintl ions, first observed nearly one hundred years ago in condensed phases, provide further hints at the influence of electron structure on magic numbers in cluster systems. Several groups have investigated these, including Duncan and Recknagle and Martin. Other than work involving cesium with lead and cesium with tin, most prior studies other than those from our own group have involved systems where neither constituent is very likely to display any free electron behavior. In recent investigations, we have conducted extensive studies of the sodium/bismuth (31) and sodium/antimony (32) systems. In the case of the former, we have observed a number of prominent magic numbers in the cluster spectrum for the cations $Bi_xNa^+_{y+1}$ correspondence to $Bi_x{}^{y-}$ for every zintl ion of bisumth known except $Bi_2{}^-$. Instead for this system, we observe $Bi_2Na_3{}^+$ which reveals the anion character of $Bi_2{}^{2-}$ and suggests the electronic octet rule determines the stability of this particular cluster. Clusters of sodium have revealed $Bi_4{}^{4-}$, $Bi_6{}^{4-}$, $Bi_7{}^{3-}$, $Bi_8{}^{4-}$, $Bi_9{}^{5-}$, and $Bi_{14}{}^{6-}$. The observations of these and the antimony sodium system are in general accord with Wades rules for "hypo" compounds. Nevertheless, we do observe some particularly stable species for which current theories (31,32) do not account, pointing to the interesting character of these species and the fact that more can be learned from a study of metal alloy cluster systems.

Co-Clusters of Atomic Anions and Cations: The MgO System:

In general, one may think of MgO as a molecule in view of its known existence as an isolatable entity in the gas phase. Nevertheless, in the condensed phase, it displays very ionic character and the bulk solid of MgO is made up of individual doubly charged anions and cations of O and Mg, respectively. The intriguing aspect of these atomic systems is that their interaction can be well described by a pair-wise Born-Mayer potential of the form

$$V_{ij} = Z_iZ_je^2/4\pi\epsilon_oR_{ij} + \lambda exp(R_{ij}/\rho) \qquad (9)$$

where Z_i and Z_j are the charges (in units of e) on the two interacting ions, e is the elementary charge, ϵ_o is the permitivity of vacuum, R_{ij} is the distance between the ions, and ρ and λ are two parameters that can be obtained from properties of the monomer and/or bulk. The cluster binding energy relative to the ions at infinite separation is found by summing V_{ij} over all ions ($i \neq j$).

Calculations (33) were made with charges of ± 1 or $+2$ on the constituent ions and either of two sets of parameters: ρ=0.301Å, λ=391 eV. In the first set ρ was taken from the Gilbert-type exponential repulsive term in a more complex pair-potential used to study a number of properties of solid MgO. This value has also been recommended for MgO lattice energy calculations with a Born-Mayer potential. The corresponding value of λ was calculated by using ρ, ionic charges of ± 2, the crystalline MgO bond length of 2.10Å, and equations relating these quantities. The second parameter set was obtained similarly, except that ρ was first calculated by using a bulk compressibility of 6.09×10^{-7} atm^{-1}. Calculations were performed on $(MgO)_n$ clusters using single and double charges and both parameter sets, whereas only single charges and the first parameter set were used in the $(MgO)_n Mg^+$ calculations.

Importantly, calculations assuming doubly charged anions and cations revealed that rings and stacked rings of clusters would be the most stable structures for small sizes. But cubes and cubes containing ledges are found to be the most stable structures if the assumption is made that the small clusters contain charges of plus and minus 1 rather than plus and minus 2. In every case, the calculated cubic structures (including those with steps and ledges) fit exactly the magic number patterns observed (33) in our experiments.

The experiments were made using a gas aggregation source and studied via a laser ionization time-of-flight technique similar to that referred to earlier. The data, in conjunction with the calculations reveal the interesting fact that the small clusters readily adopt bulk structures from the smallest observational sizes, but evidently only evolve to the electronic character of the bulk at much larger degrees of aggregation. Such findings provide further impetus for cluster research on these and related systems.

We also have investigated (34) doubly charged clusters of the MgO system, where we have observed compositions corresponding to $(MgO)_n Mg^{++}$ (n=4-172) and $(MgO)_n Mg_2^{++}$ (n=12-24). The findings for the doubly charged single excess Mg atom containing systems show that the clusters also have cubic structures resembling pieces of the bulk cubic lattice of MgO, with the most stable structures being cuboids and cuboids with an atom vacancy or an attached terrace. In the case of the $(MgO)_n Mg_2^{++}$ clusters, these are also explainable in terms of cubic structures, but for the situation where the O atom vacancies may be occupied by one or two excess electrons analogous to a solid state F-center. The large clusters are observed to congruently vaporize MgO, while for small sizes the doubly charged aggregates appear to undergo a Coulomb explosion leading to $(MgO)_n^+$ and $(MgO)_n Mg^+$.

Conclusions: The subject of magic numbers of small atomic systems is seen to be a rich one for investigation. Findings interrelate to a number of fields including an investigation of transitions from the gas to the condensed state, the onset of metallic-like character which can be attributed to changes in electronic states of small systems, the nature of the electronic states and evolution with composition and size for alloys, and insight into the onset of bulk geometric structure and electronic states for ionic systems. We can expect new and exciting findings for a wide variety of systems in the future, but most important will be the interplay between experiment and theory, where progress has been impeded by insufficient attention to the latter.

Acknowledgments: Financial support by the Department of Energy, Grant Numbers DE-FGO2-88ER60658 and DE-FGO2-88ER60668, the National Science Foundation, Grant No. ATM-90-15855, and the Wentworth Institute of Technology are gratefully acknowledged. Thanks to Ziyun Chen, Andreas Hartmann, Amy Harms, Robert Leuchtner, Kent Shi, Shiqing Wei, Yasuhiro Yamada, and Paul Ziemann.

Literature References

1. S. S. Lin, Rev. Sci. Instrum. 44, 516 (1973).
2. J. Q. Searcy and J. B. Fenn, J. Chem. Phys. 61, 5282 (1974).
3. P. Holland and A. W. Castleman, Jr., J. Chem. Phys. 72, 5984 (1980); V. Hermann, B. D. Kay, and A. W. Castleman, Jr., Chem. Phys. 72, 2031 (1982).
4. D. Dreyfuss and H. Y. Wachman, J. Chem. Phys. 76, 2031 (1982).
5. A. J. Stace and C. Moore, Chem. Phys. Lett. 96, 80 (1983).
6. O. Echt, K. Sattler and E. Recknagel, Phys. Rev. Lett. 47, 1121 (1981).
7. I. A. Harris, R. S. Kidwell, and J. A. Northby, Phys. Rev. Lett. 53, 2390 (1984).
8. A. W. Castleman, Jr. and R. G. Keesee, Chem. Rev. 86, 589 (1986).
9. A. W. Castleman, Jr. and R. G. Keesee, Ann. Rev. Phys. Chem. 37, 525 (1986).
10. A. W. Castleman, Jr. and R. G. Keesee, Science 241, 36 (1988).
11. G. M. Lancaster, F. Honda, Y. Fukuda and J. W. Rabalais, J. Am. Chem. Soc. 101, 1951 (1979).
12. S. Wei, Z. Shi, and A. W. Castleman, Jr., J. Chem. Phys. 94, 3268 (1991).
13. X. Yang and A. W. Castleman, Jr., J. Am. Chem. Soc. 111, 6845 (1989).
14. X. Yang and A. W. Castleman, Jr., J. Phys. Chem. 94, 8500 (1990); Errata, J. Phys. Chem. 94, 8974 (1991).
15. S. Wei, Z. Shi, and A. W. Castleman, Jr., "Elucidating the Origin of Magic Numbers: Trends in the Relative Binding Energies of Xenon Cluster Ions and Their Implications," J. Chem. Phys., accepted for publication.
16. D. Kreisle, O. Echt, M. Knapp, and E. Recknagel, Phys. Rev. A. 33, 768 (1986).
17. T. D. Märk and P. Scheier, Chem. Phys. Lett. 137, 245 (1987); and J. Chem. Phys. 87, 1456 (1987).
18. O. Echt, M. C. Cook and A. W. Castleman, Jr., Chem. Phys. Lett. 135, 229 (1987).
19. S. Wei, W. B. Tzeng and A. W. Castleman, Jr., J. Chem. Phys. 92, 332 (1990).
20. S. Wei, K. Kilgore, W. B. Tzeng, and A. W. Castleman, Jr., "Evaporative Dissociation of Ammonia Cluster Ions: Quantification of Decay Fractions and Isotope Effects," J. Phys. Chem., accepted for publication.
21. (a) C. E. Klots, J. Chem. Phys. 83, 5854 (1985); (b) C. E. Klots, Nature 327, 222 (1987); (c) C. E. Klots, Acc. Chem. Res. 21, 16 (1988); (d) C. E. Klots, J. Chem. Phys. 58, 5364 (1973) and 90, 4470 (1989).
22. E. E. Ferguson, C. R. Albertoni, R. Kuhn, Z. Y. Chen, R. G. Keesee, and A. W. Castleman, Jr., J. Chem. Phys. 88, 6335 (1988).
23. C. R. Albertoni, A. W. Castleman, Jr., and E. E. Ferguson, Chem. Phys. Lett. 157 159 (1989).
24. W. A. de Heer and W. D. Knight, in *Elemental and Molecular Clusters* (G. Benedek, T. P. Martin, and G. Pacchioni, Eds.) Springer-Verlag, Berlin Heidelberg, pp 45-63 (1988).
25. R. E. Leuchtner, A. C. Harms and A. W. Castleman, Jr., J. Chem. Phys. 91, 2753 (1989).
26. R. E. Leuchtner, A. C. Harms, and A. W. Castleman, Jr., J. Chem. Phys. 94, 1093 (1991).
27. A. C. Harms, R. E. Leuchtner, S. W. Sigsworth, and A. W. Castleman, Jr., J. Am. Chem. Soc. 112, 5672 (1990).

28. R. G. Wheeler, H. K. LaiHing, W. L. Wilson, and M. A. Duncan, J. Chem. Phys. 88, 2831 (1988).
29. D. Schild, R. Pflaum, K. Sattler, and E. Recknagel, J. Phys. Chem. 91, 2649 (1987).
30. T. P. Martin, J. Chem. Phys. 83, 78 (1985).
31. R. W. Farley and A. W. Castleman, Jr., J. Am. Chem. Soc. 111, 2734 (1989).
32. A. Hartmann and A. W. Castleman, Jr. (to be published).
33. P. J. Ziemann and A. W. Castleman, Jr., J. Chem. Phys. 94, 718 (1991).
34. P. J. Ziemann and A. W. Castleman, Jr., "Mass Spectrometric Study of the Formation, Evaporation, and Structural Properties of doubly Charged MgO Clusters," Phys. Rev. B., in press.

Part II

Theoretical Approaches

Medium Dependence in Heavy Ion Collisions

A. Faessler

Institut für Theoretische Physik, Universität Tübingen,
W-7400 Tübingen, Fed. Rep. of Germany

Abstract: One of the proclaimed goals of heavy ion collisions with 100 MeV
up to 2000 MeV per nucleon is the determination of the equation of state of
nuclear matter, which one needs for example for neutron stars, supernova
explosions and the early universe. But the situation in heavy ion collisions is
quite different from thermal equilibrium with a spherical momentum distribution and a fixed temperature. The effective nucleon-nucleon interaction
as determined for example by the solution of the Bethe-Goldstone equation depends through the Pauli operator and through the single particle
energies on the surrounding nuclear matter. This dependence is especially
pronounced since the nucleon-nucleon interacting is highly momentum dependent: It is attractive at small relative momenta and repulsive at higher
values. Thus the effective nucleon-nucleon interaction in heavy ion collisions
depends on the distribution of the surrounding nuclear matter in orbital and
momentum space. Here results are presented using for the description of
heavy ion reactions at intermediate energies Quantum Molecular Dynamics
(QMD) in a non-relativistic and in a completely covariant (RQMD) form.
We show that the production of gamma-rays, pions and the inclusive spectra of nucleons and light nuclei are not sensitive to the equation of states.
The most sensitive observable is the perpendicular momentum distribution
in heavy ion collisions.

1. Introduction

Properties of nuclei and their extrapolation to infinite nuclear matter have
been studied in the past at zero temperature and for the saturation density.
Heavy ion collisions at intermediate energies from about 100 to 2000 MeV
per nucleon allow to get information on nuclei and nuclear matter at finite

Springer Series in Nuclear and Particle Physics   Clustering Phenomena in Atoms and Nuclei
Editors: M. Brenner · T. Lönnroth · F.B. Malik   © Springer-Verlag Berlin, Heidelberg 1992

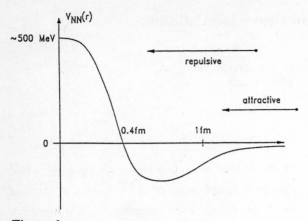

Figure 1
Qualitative sketch of the radial dependence of the nucleon-nucleon interaction in a S wave. The interaction is highly momentum dependent. At high relative momentum it is strongly repulsive and at low momenta it is attractive.

temperature and also at densities other than the saturation density of about 0.17 nucleons per fm^{-3}. Thus it has often been claimed that one of the main aims of heavy ion collisions at intermediate energies is the determination of the equation of state (EOS) of nuclear matter. The equation of state is not only an interesting relation for nuclear physics, but is also needed in astrophysics to describe supernova explosions, neutron stars and phases in the early universe. But at least in the earlier phase of heavy ion collisions one is far away from the thermal equilibrium, which is assumed to be present in nuclear matter for the EOS. The momentum distribution is unisotropical and this is influencing the effective nucleon-nucleon interaction during the collision. The bare nucleon-nucleon interaction as determined by the scattering between two nucleons in vacuum is highly momentum dependent (see qualitative sketch of this dependence in fig. 1).

The strong momentum dependence of the nucleon-nucleon interaction does not allow to use perturbation theory to calculate the nucleon-nucleon matrix elements in nuclear matter, but the distortion of the nucleon-nucleon wave function due to the interaction has to be taken into account. This is done with the help of the Brueckner-Theory in solving the Bethe-Goldstone

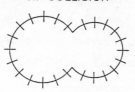

Figure 2

Local distributions in momentum space for the equation of state (EOS) and for the initial states of a heavy ion collision (right-hand side). The slashes perpendicular to the Fermi surface indicate that the surface is smeared out due to finite temperatures. The volume inside the Fermi sphere is proportional to the density of nuclear matter at the spatial point considered. The momentum distribution in a heavy ion collision (right-hand side) is not spherical.

equation. In a heavy ion collision the Pauli operator of the Bethe-Goldstone equation is not spherical in momentum space, but reflects the bombarding energy of the incoming heavy ion.

$$G(\rho_t, \rho_p, E/A, \epsilon_1 + \epsilon_2) = V + V \frac{Q}{\epsilon_1 + \epsilon_2 - H_o + i\eta} G \qquad (1)$$

The Brueckner-reaction matrix depends as the free nucleon-nucleon interaction on the momenta of the two incoming and the two outgoing nucleons. But in addition it depends also on the distribution of the surrounding nucleons in orbital and momentum space for the early stage of the nucleus-nucleus collision that means on the densities of the target ρ_T and the projectile ρ_P at the interaction point, on the bombarding energy E/A and on the starting energy $W = \epsilon_1 + \epsilon_2$. The dependence on the densities and the bombarding energy enters the Bethe-Goldstone equation through the Pauli operator Q. A further dependence on the medium comes through the single particle energies ϵ_1 and ϵ_2, which are modified from the free kinetic energies due to the interacting with the neighbouring nucleons [4,5,6].

Opposite to the situation in heavy ion collisions one needs for the equation of state of nuclear matter a spherical Fermi distribution in momentum space, where only the surface can be smeared out due to a finite temperature. Figure 2 shows the situation for the EOS and a heavy ion collision in momentum space.

To extract from a heavy ion collision (right-hand side of figure 2) information on the EOS (left-hand side of figure 2) one needs a detailed knowledge of the effective medium dependent interaction in heavy ion collisions. In the present contribution we will discuss three points:

(i) By solving the Bethe-Goldstone equation for the momentum distribution shown on the right-hand side of figure 2 one can calculate the local energy density, which is complex. By subtracting the energy of two heavy ions at large distance one can obtain the optical potential between the two nuclei. This optical potential allows to describe in a quantitative way elastic and inelastic scattering of two heavy ions.

(ii) At intermediate energies we describe heavy ion collisions with the help of Quantum Molecular Dynamics (QMD), which is a special version of the Vlasov-Uehling-Uhlenbeck approach. The collision term is derived from the Brueckner reaction matrix G obtained from solving equation (1).

$$d\sigma_{NN}/d\Omega \; \propto \; |G|^2$$
$$U(i) = \sum_j < i,j|G|i,j > \rho_j \qquad (2)$$

The selfconsistent potential U(i) for particle i is also determined by the Brueckner reaction matrix solved for the density and momentum distribution derived from the Brueckner reaction matrix G determined including the special Pauli operator and the selfconsistent single particle energies in the surrounding nuclear medium.

(iii) Finally we extend QMD to a fully covariant version into Relativistic Quantum Molecular Dynamics (RQMD).

2. Elastic and inelastic heavy ion collisions

In this chapter we calculate the optical potential for the collision of two nuclei using a local density approximation. We solve the Bethe-Goldstone equation (1) for each spatial point in a heavy ion collision, where the momentum distribution is characterized as indicated on the right-hand side of figure 2. The radii of the two spheres are given by the densities of the target

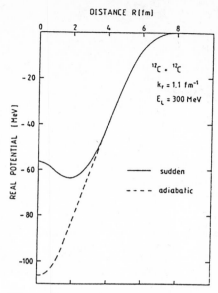

Figure 3

Microscopically calculated real part of the optical potential for ^{12}C at a bombarding energy of 1016 MeV. The solid line is the microscopically calculated potential using the Reid-Soft core interaction. The dashed line is a fit to the data by Bunere et al [7].

and the projectile at the spatial point considered. The distance of the two centres of the two momentum space spheres are given by the bombarding energy and indicate the average relative momenta of the nucleons in the targets relative to the nucleons in the projectile. The Brueckner reaction matrix obtained in this way by solving equation (1) is complex, because the two interacting nucleons can be scattered into states on shell allowed by the Pauli operator Q. A complex energy density is then calculated at each spatial point using the Hartree-Fock approach. By integrating over the two heavy ions at a given distance and by subtracting the energy of the two heavy ions at large distance one obtains a good approximation for the real and the imaginary part of the optical potential. The optical potential and the scattering cross section obtained in this way for ^{12}C on ^{12}C at 1016 MeV bombarding energy in the lab system is shown in figures 3 and 4.

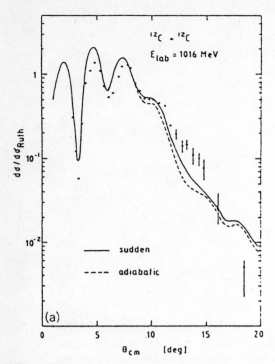

Figure 4

Elastic $^{12}C-^{12}C$ scattering cross section in units of the Rutherford cross section as a function of the centre of mass scattering angle. The data are from reference 7, the two theoretical curves assume for the density distribution of the two ^{12}C nuclei a sudden and an adiabatic approximation. The total reaction cross section is experimentally 996 (+50 -250) mb and theoretically one obtains 993 mb.

3. Quantum molecular dynamics with medium dependence

In quantum molecular dynamics like also in the Vlasov-Uehling-Uhlenbeck approach one is not solving the Vlasov equation with the collision term directly

$$\delta_t f_i(\mathbf{p}, \mathbf{r}, t) - \nabla_{\mathbf{r}_i} U_i(\mathbf{p}, \mathbf{r}, t) \cdot \nabla_{\mathbf{p}_i} f_i(\mathbf{p}, \mathbf{r}, t) + \mathbf{r}_i \nabla_{\mathbf{r}_i} f_i(\mathbf{p}, \mathbf{r}, t) =$$
$$\sum_2 \sum_3 \sum_4 d2 \cdot d3 \cdot d4 \ \sigma_{i2;34} \ |\mathbf{v}_{rel}|[f_3 f_4(1 - f_i)(1 - f_2) - f_i f_2(1 - f_3)(1 - f_4)] \tag{3}$$

but one follows the time evolution of each nucleon with the help of Hamilton equations [2,3].

$$\dot{\mathbf{r}}_i(t) = \nabla_{p_i} H(1,...A,t)$$
$$\dot{\mathbf{p}}_i(t) = -\nabla_{r_i} H(1,...A,t) \tag{4}$$

$f_i(\mathbf{p}, \mathbf{r}, t)$ is the Wigner transform giving the momentum and spatial distribution of nucleon i. The collision term on the right-hand side of equation (3) contains a gain and a loss term, describing scattering into the state i and out of the state i. The factors in the square brackets take into account the Pauli principle. For the Hamiltonian in equation (4) we use the Hartree-Fock approximation.

$$H = \sum_{n=1}^{A} [\frac{P_n^2}{2M_N} + U_n(t)] + V_{SU} + V_{SY} + V_{coul}$$

$$U_n(t) = \sum_2 \int d2\, G(n,2) f_2(\mathbf{r}_2, \mathbf{p}_2, t)$$

$$U_n(t) = \alpha\rho(n,t) + \beta\rho^\gamma(n,t) + [U_p(\mathbf{p}_n)] \tag{5}$$

$$V_{SU} = \frac{\eta}{8M_N}(\nabla\rho)^2$$

$$V_{SY} = C(\rho_n - \rho_p)^2/\rho^{1/3}$$

$$V_{coul} = Coulomb$$

$$\sigma_{i2;34} \propto |<3,4|G|i,2>|^2$$

For the selfconsistent potential $U_n(t)$ for the nucleon n we use a microscopic expression calculated as the Hartree-Fock potential from the Brueckner reaction matrix

$G(n,2) = < n,2\,|G|n,2 >$. Sometimes we use also a phenomenological expression from a Skyrme force with the parameters α and β adjusted to give the right binding energy per nucleon and the right saturation density in nuclear matter. In this expression we sometimes also take into account a momentum dependence $U_p(\mathbf{p}_n)$. The potential contains also corrections for the surface (V_{SU} = Weizsäcker surface correction term) and for the symmetry energy depending on the neutron ρ_n and the proton ρ_p density.

Figure 5 shows the photo production cross section of ^{12}C on ^{12}C with 84 MeV per nucleon. The left-hand side shows the calculation using the Skyrme potential and the nucleon-nucleon cross section of Cugnon [8]. On the right-

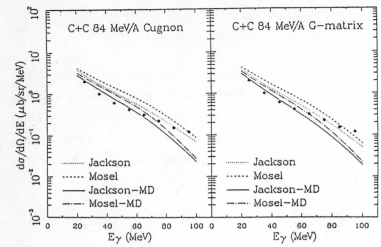

Figure 5

Photon production cross section calculated with the Soft EOS (incompressibility $K = 200$ MeV) at the emission angle of 90° with respect to the beam axes in $^{12}C + ^{12}C$ collisions at $E_{lab} = 84$ MeV/A. Results are obtained with the Cugnon [8] (left-hand side) and the G-matrix (right-hand side) in-medium NN cross section. Dotted (solid) and dashed (dashed-dotted) curves are obtained using the momentum independent (dependent) interaction with the Jackson formula and the microscopic expression of reference 10 for the pnγ cross section, respectively. The experimental data are taken from reference 11. The two lowest curves contain the momentum dependence in the selfconsistent field (5). At higher energies the production cross section is smaller including the momentum dependence due to the fact that the momentum dependence is repulsive and reduces the probability for nucleon-nucleon collisions and by that also the production cross section for gamma rays.

hand side we show the gamma ray production cross section as a function of the energy of the produced gammas using the microscopic collision term from the Bethe-Goldstone equation (1) and (2). The selfconsistent potential is here again calculated with a Skyrme force (5). The four different curves have been calculated with and without momentum dependence in the self-consistent potential $U_n(t)$ and with a different photo production cross sections.

120

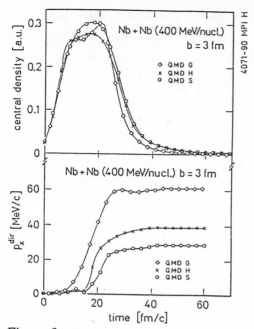

Figure 6

Central density and directed transverse momentum as a function of the reaction time for the reaction Nb on Nb at 400 MeV/A with an impact parameter b=3 fm. We compare the results obtained using the Skyrme interaction with the hard equation of state (QMD H) with those obtained using the soft equation of state (QMD S) and the Brueckner reaction matrix (QMD G) for the collision term and the selfconsistent potential.

Figure 6 shows the perpendicular momentum flow as a function of the time for Nb on Nb at 400 MeV/A for an impact parameter of b=3 fm. The three curves shown are calculated with the soft (circles) equation of state (incompressibility K=200 MeV), with the hard equation of state (crosses; K=380 MeV) and with the collision term and the selfconsistent potential calculated from the Brueckner reaction matrix (Reid-soft core potential; squares). The directed transverse momentum is defined by

$$< p_x^{dir} >= \frac{1}{N} \sum_{i=1}^{N} sign \ (y_i) p_x(i) \tag{6}$$

as a function of time. The signum of the rapidity y_i of the different emitted

121

particles in a heavy ion collision guarantees that the directed transverse momentum is different from zero. The transverse momenta of the target particles scattered to one side and of the projectile particles scattered to the other side are counted by expression (6) all with the same sign.

The results shown here in figure 5 for the gamma ray production cross section and corresponding results for the pion production cross section indicate that these production cross sections are not sensitive to the equation of state. The most sensitive quantity is the transverse momentum as defined in (6) and shown in figure 6.

4. Relativistic quantum molecular dynamics

A classical covariant treatment of the many body problem was first attempted by Dirac [12]. But his approach cannot include interactions between the nucleons. An extension which can take interactions fully into account was given by Samuel [13] and first used by Sorge, Stöcker and Greiner [14] to describe nuclear collisions. We follow here the approach of Samuel [13]. The relativistic phase space has 8 dimension for each nucleon. We reduce this to a six dimensional phase for each nucleon by constraining the momenta and energies of each nucleon on shell and by fixing the different times of each nucleon in a covariant fashion[15]. The time evolution parameter is defined as the average time of all the nucleons in the total centre of mass system.

$$\tau = Q \cdot P/|P| \Rightarrow \frac{1}{A} \sum_{i=1}^{A} t_i;$$

$$Q = \frac{1}{A} \sum_{i=1}^{A} (t_i, \mathbf{r}_i) \tag{7}$$

$$P = \sum_{i=1}^{A} (E_i, \mathbf{p}_i)$$

The interaction between the nucleons is assumed to be a Lorentz scalar and is so defined that their non-relativistic reduction is the Skyrme force.

$$V_{NN}(1,2) = t_1 \delta(\mathbf{r}_{12}) + t_3 \rho^{\gamma-1} \delta(r_{12})$$

$$U(i) = \alpha \rho(i, \tau) + \beta \rho^{\gamma}(i, \tau) \tag{8}$$

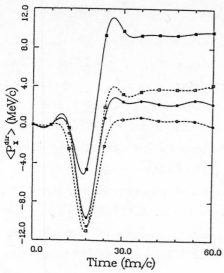

Figure 7
Directed transverse momentum as defined in equation (6) for the collision ^{40}Ca on ^{40}Ca with E/A=1050 MeV and an impact parameter b=2.6 fm as a function of time. The non-relativistic results (QMD) are indicated by open symbols, while the relativistic results are indicated by filled symbols. The soft equation of state (K=200 MeV) is characterized by circles and the hard equation of state (K=380 MeV) by squares.

Figure 7 shows the collision of ^{40}Ca on ^{40}Ca for E/A=1050 MeV for an impact parameter p=2.6 fm. The directed transverse momentum defined in equation (6) is given as a function of time for the hard and the soft relativistic (RQMD) and non-relativistic (QMD) treatment. One sees an appreciable difference between the relativistic and the non-relativistic treatment. RQMD gives a smaller transverse momentum than QMD. As expected the transverse momentum is also larger for the hard equation of state (K=380 MeV) than for the soft one (K=200 MeV).

We have also calculated relativistically the production of gamma rays and pions and found no dependence on the equation of state and also practically no dependence on a relativistic and a non-relativistic treatment. Thus the strongest differences depending on the EOS and on QMD and RQMD can be seen in the directed transverse momentum (6).

5. Summary

The free nucleon-nucleon interaction in vacuum is strongly momentum dependent. This strong momentum dependence of the free nucleon-nucleon interaction translates in a strong medium dependence if the nucleon-nucleon collision is considered within a nuclear medium as in a collision between two heavy ions. If one wants to extract the equation of state (EOS) from data of heavy ion collisions one has to take into account this medium dependence. In the present work we have solved the Bethe-Goldstone equation for the local density and momentum distribution and obtained in this way a medium dependent nucleon-nucleon interaction the G-matrix. With this Brueckner reaction matrix we calculate the collision term and the self-consistent potential for a quantum molecular dynamics (QMD) calculation to describe heavy ion collisions.

Due to the strong repulsion at high relative momenta one needs in this more realistic and microscopic approach no high densities as for Skyrme forces to get a repulsive potential between two nuclei. It is enough to have high relative momenta.

Presently we calculated the collision term using the G-matrix calculated by solving the Bethe-Goldstone equation approximating the local momentum and density distribution by two spheres in momentum space for each spatial collision point. But we have not yet taken into account the smearing of the Fermi surface due to temperature effects. The self-consistent potential has been calculated in the Hartree-Fock approach using Brueckner reaction matrix elements calculated again for temperature zero. Since in the potential one averages over the second nucleon we simplified here further by taking for the Pauli operator only a spherical momentum distribution. In addition we compared with results calculated with Skyrme forces.

We also extended QMD to a fully covariant treatment (Relativistic Quantum Molecular Dynamics =RQMD).

The production cross section for gamma rays and pions turned out to be not sensitive to the EOS or to relativistic and non-relativistic treatments. The most sensitive quantity turned out to be the directed transverse momentum defined in equation (6).

I would like to thank Drs. Ohtsuka, Linden, Khoa, Ismail, Maruyama, Li and the Ph. D. students Huang and Lotfy and the colleagues from Heidelberg Drs. Aichelin, Jaenicke and Bohnet, with whom this work has been performed.

References

1. G. F. Bertsch, H. Kruse, S. D. Gupta, Phys. Rev. **29** (1984) 673
2. H. Stöcker, W. Greiner, Phys. Rep. **137** (1986) 277
3. J. Aichelin, H. Stöcker, Phys. Lett. **163B** (1985) 59
4. M. Trefz, A. Faessler, W. H. Dickhoff, Nucl. Phys. **A428** (1985) 499 and S. Krewald, A. Faessler, Nucl. Phys. **A341** (1980) 319
5. N. Ohtsuka, R. Linden, A. Faessler, F. B. Malik, Nucl. Phys. **A465** (1987) 550
6. A. Bohnet, N. Ohtsuka, J. Aichelin, R. Linden, A. Faessler, Nucl. Phys. **A494** (1989) 349
7. M. Buenerd et al, Phys. Rev. **C26** (1982) 1299
8. J. Cugnon, T. Mitzutani, J. Vandermeulen, Nucl. Phys. **A352** (1981) 505 and Phys. Rev. **C22** (1980) 1885
9. D. T. Khoa, N. Ohtsuka, S. W. Huang, M. Ismail, A. Faessler, M. El Shabshiry, J. Aichelin, to be published in Nucl. Phys. (1991)
10. M. Schäfer, T. S. Biro, W. Cassing, U. Mosel, UGI- preprint 89-1
11. E. Grosse, P. Grimm, H. Heckwolf, W. F. J. Müller, H. Noll, A. Oskarsson, H. Stelzer, Europhys. Lett **2** (1986) 555
12. P. A. M. Dirac, Proc. Roy. Soc. **A246** (1958) 326
13. J. Samuel, Phys. Rev. **D26** (1982) 3475 and 3482
14. H. Sorge, H. Stöcker, W. Greiner, Ann. Phys. **5** (1989) 266
15. T. Maruyama, S. W. Huang, N. Ohtsuka, A. Faessler, J. Aichelin, submitted to Nucl. Phys. A

Energy-Density Functional Formalism in Nuclear Physics

F.B. Malik[1] *and I. Reichstein*[2]

[1]Physics Department, Southern Illinois University,
 Carbondale, IL 62901, USA
[2]School of Computer Science, Carleton University,
 Ottawa, Ontario, K1S 5B6, Canada

The nature of the energy-density functional (EDF) relevant to the calculation of various macroscopic properties of nuclei is discussed. Nuclear masses calculated by the EDF method using density distribution functions deduced from electron scattering and μ-mesic data are in very good agreement with the observed masses. The potentials between two nuclei calculated from the EDF method are in agreement with those needed to account for angular distributions in eleastic scattering. A simple prescription to scale the potential using the EDF approach as a guide is given and found successful in many cases. The EDF calculations indicate the existance of an external barrier between the saddle and scission points. Incorporation of such a barrier in the fission process can account for the observed half-lives, mass, charge, and total kinetic energy distributions and provides an understanding of cold fission.

1. Introduction

In recent years, the use of energy-density functional (EDF) formalism to calculate various observed macroscopic properties pertinent to nuclear physics has become popular. In particular, the method has been applied to calculate heavy-ion heavy-ion interaction in a number of ways. In this talk, I shall review the application of the EDF calculated from realistic two-nucleon potential to account for various observed properties such as (a) total binding energy (b) potential between two elastically scattered nuclei and, (c) the external barrier in radioactive decay which are relevant to nuclear physics. Furthermore, understanding of the approach allows one to scale the parameters of the potential between two nuclei as a function of mass number A, and thereby, one can, in many instances, predict the general trend and magnitude of the elastic scattering and fission cross sections as well as mass distributions in radioactive decay.

Springer Series in Nuclear and Particle Physics   **Clustering Phenomena in Atoms and Nuclei**
Editors: M. Brenner · T. Lönnroth · F.B. Malik   © Springer-Verlag Berlin, Heidelberg 1992

2. The Formalism

The energy density formalism in nuclear physics has been initially developed to calculate total binding energy of nuclei and has its roots in Thomas-Fermi Statistical Model [1-4]. However, the formalism could also be viewed to have its origin in many body theory. The corner stone of the latter approach is embedded in the Hohenberg-Kohn's [5] observation that the total energy of the ground state of a system of Fermion can always be expressed as a functional of its density. The proof is based simply on the fact that the total ground state energy, E(exact), is the expectation value of the Hamiltonian. The Raleigh-Ritz variational principle then allows one to show that any energy E, calculated from a density functional is an upper bound of the exact ground state energy;

$$E(exact) \leq E(\rho) \tag{1}$$

where

$$E(\rho) = \int dr...dr_N \psi^*[\sum_i T(i) + \frac{1}{2}\sum_{i \neq j} v(ij)]\psi \tag{2}$$
$$= \int dr\ \epsilon[\rho(r)]$$

In (2), T(i) is a one-body operator which may contain a one-body potential u(i), in addition to the kinetic energy operator $(-\hbar^2/2m)\Delta(i)$. The crux of the energy-density formalism is to find an appropriate $\epsilon[\rho(r)]$. One can get some idea about the procedure to find such $\epsilon[\rho(r)]$ from the following considerations.

The exact wavefunction $\psi(1,2.....N)$ can be expanded in a complete set of determinants of one body orthonormal wavefunctions $\phi(i)$, det $[\phi(i) .... \phi(N)]$:

$$\psi(1,2....N) = \Sigma_{mn} A_{mn} [\det[\varphi(i)...\varphi(N)]]_{mn} \tag{3}$$

Because of the orthonormality condition

$$(\varphi_m \varphi_n) = \delta_{mn} \tag{4}$$

(2) leads to

$$E(\rho) = \sum_{pm} A_{pp}^* A_{mm} I_{mp} + (1/2) \sum_{mnpq} A_{pq}^* A_{mn} [J_{mnpq} - K_{mnpq}] \tag{5}$$

where

$$I_{mp} = \int dr_i\ \varphi_p^*(r_i)\ T(i)\ \varphi_m(r_i) \tag{6a}$$

$$J_{mnpq} = \int dr_i\ dr_j\ \varphi_p^*(r_i)\ \varphi_q^*(r_j)\ v(ij)\ \varphi_m(r_i)\ \varphi_n(r_j) \tag{6b}$$

127

and

$$K_{mnpq} \equiv \int dr_i \, dr_j \, \varphi_p^*(r_i) \, \varphi_q^*(r_j) \, v(ij) \, \varphi_m(r_j) \, \varphi_n(r_i) \tag{6c}$$

In case T(i) and v(ij) contain only central potential, (5) reduces to

$$E(\rho) = \sum_{mm} [|A_{mm}|^2 I_{mm}] + (1/2)\sum_{mn} |A_{mn}|^2 [J_{mnmn} - K_{mnnm}] \tag{7}$$

For a single determinantal expansion of ψ, (7) reduces to Slater's expression for atoms. It is clear that for central potentials, (6a), (6b) and (6c), involve only the density of the system, since

$$\rho = \sum_{mm} |A_{mm}|^2 \int dr_i \, \varphi_m^*(r_i) \, \varphi_m(r_i) \tag{8}$$

In particular, one may choose the basis set to be plane waves normalized in a box in the expansion (3). In that case the Pauli principle dictates that one can put only two particles of a particular momentum in a volume of h^3. In principle, the expression (5) or (7) remains still valid but in atomic as well as in nuclear physics the observed density distributions can only be generated by a large superposition of plane waves.

At this point, one has to acknowledge that the many body problem cannot be solved exactly but one is to resort to various approximations, the nature of which depends on the physicsal nature of the problem in consideration. In the Hartree-Fock theory used in atomic and nuclear physics, one considers only one determinant in the expansion (3) and gets a set of coupled integro-differential equations for single particle orbitals φ_m by applying variational method. This leads to a self-consistant way to determine the ground state energy and its density.

In nuclear physics the realistic nature of two-body interaction v(ij) is too complicated to allow for a proper solution of the coupled set of equation. On the other hand, the short range nature of this two-body interaction may allow us to represent the wave function as modulated plane waves. However, the estimation of Gomes, Walecka and Weisskopf [6] indicates a healing distance of a few Fermi and as such one may, in the first approximation , omits the modulation in order to calculate the binding energy. On the other hand, this cannot reproduce the density distribution function. One may, thus, rewrite $E[\rho(r)]$ as follows:

$$E[\rho(r)] = \int dr \, \rho(r) \, [T(\rho) + V(\rho)] \tag{9}$$

where $T(\rho)$ and $V(\rho)$ are contributions to the total energy, respectively, from the one-and two-body operators T(i) and v(ij). One may, then, calculate $T(\rho)$ and $V(\rho)$ using plane waves and use observed density distribution or determine ρ from a variational principle either solving the equation of ρ obtained by varying E or making an ansatz for ρ and determining its parameters by variation. In effect, $[T(\rho) + V(\rho)]$, are effective operators that may concoct the actual situation. The dependence of kinetic energy $T(\rho)$ on density has been calculated by Brueckner, Coon and Dabrowski (BCD) [7] for systems having unequal numbers of neutrons and protons and is given by

$$T(\rho) = C_1(\alpha)\rho^{2/3} \tag{10}$$

with

$$C_1(\alpha) = (3/5)\,(\hbar^2/2M)\,(3\pi^2/2)^{2/3}(1/2)[(1-\alpha)^{5/3} + (1+\alpha)^{5/3}] \tag{11}$$

For $V(\rho)$, BCD calculated the non-Coulomb part of the interaction using a K-matrix approach. Thus, the non-Coulomb part of the neutron and proton potential $V_n(k)$ and $V_p(k)$ are given by

$$V_n(k) = \sum_{q<k_{nf}} [(kq|K_{nn}|kq) - Exchange] + \sum_{q<k_{pf}} (kq|K_{np}|kq) \tag{12a}$$

$$V_p(k) = \sum_{q<k_{pf}} [(kq|K_{pp}|kq) - Exchange] + \sum_{q<k_{nf}} (kq|K_{np}|kq) \tag{12b}$$

where k_{nf} and k_{np} are, respectively, neutron and proton Fermi momenta related to nuclear matter Fermi momentum k_f by

$$K_{nf} = (1+\alpha)^{1/3}k_f \quad ; \quad K_{pf} = (1-\alpha)^{1/3}k_f \tag{13}$$

$\alpha = (N - Z)/A$, the neutron excess. The dependence of this part of potential energy V (nucl) on α and ρ are shown in Fig. 1. and it can be parametrized by

$$V(nucl) = b(1 + a_1\alpha^2)\rho + b_2(1 + a_2\alpha^2)\rho^{4/3} + b_3(1 + a_3\,\alpha^2)\rho^{5/3} \tag{14}$$

The calculated V(nucl)shown in Fig. 1 corresponds to the realistic two-nucleon potential of Brueckner and Gammel[8].

To correct for the non-compatibility of the plane wave function with the observed density distribution, BCD propose to add a term of the form $\gamma(\hbar^2/8M)\,(\nabla\rho)^2/\rho$ with γ, a free parameter (M: nucleon mass) to be adjusted to reproduce binding energy per nucleon. This term is somewhat different from the inhomogeneity term of Weissäcker which is given by $(\hbar^2/8M)\,(\nabla\rho)^2$. In addition, $V(\rho)$ includes the Coulomb interaction among protons, $(1/2)\,\Phi_c$ with the Pauli correction to it which is taken to be $(-0.738e^2)\,(\rho_p^{4/3})/\rho$ from the work of Peaslee [9]. Here

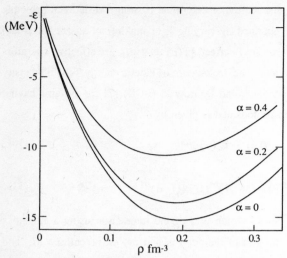

Fig. 1. Energy per nucleon in MeV versus density in fm^{-3} taken from [7] for nuclear matter ($\alpha = 0$), and neutron excesses $\alpha = 0.2$ and 0.4.

$$\Phi_c = e \int dr' \rho_p(r')/|r-r'|$$

Thus,

$$\rho V(\rho) = \rho V(nucl) + (1/2)\Phi_c\rho - (3/4)(3/\pi)^{1/3}e^2\rho_p^{4/3} + (\hbar/8M)\gamma(\nabla\rho)^2 \quad (15)$$

We may note the following:

(a) In principle, one may, instead of plane wave, choose any other form of one-body wave function e.g., the Hartree-Fock or elementary shell model wave function as done in the work of Danilov, Gridnev, Subotin and Malik [10]. In that case, the density-dependency and the nature of the effective operators $T(\rho) + V(\rho)$ would be different.

(b) Certainly, one can impose a variational condition and determine the density function from the derived equations. Such a density function would only be close to the actual density of a system, be it atomic, solid-state or nuclear, only if the operators $T(\rho) + V(\rho)$ are close to the actual one. In atomic physics this leads to the Thomas-Fermi equation and such attempts have also been done in nuclear physics [11,12]. (c) For a single determinant in the expansion (3) and $\gamma = 0$, one can derive from the functional, the Hartree-Fock set of equations. Similarly, an expansion of ψ in an antisymmetrized multi-pair or multi-cluster or Jastrow functions would lead to the appropriate equations relevant to that particular expansion [14], and (d) Whereas, Weisäcker attributes the gradient term to the corrections to the kinetic energy, Brueckner and his collaborators [11, 12]

130

claim that the first order correlation energy also leads to such a term. The nature of this term is still uncertain and one may look upon it as an empirical term.

Despite some uncertainties, the energy density functional approach has been successful in describing macroscopic properties such as the total binding energies of the nuclei. Refining the nature of the functional may lead to more reliable calculations of other properties.

3. Total Binding Energies of Nuclei

Contrary to popular belief, nuclei have quite an extensive surface and the number of nucleons residing at the surface is at least one half of those located in the constant density zone, a point made by Reichstein and me [13] about a decade and a half ago. The point is best illustrated by approximating the observed nuclear density, ρ, by a trapezoidal function

$$\rho = \begin{cases} \rho_o & \text{for } r < c \\ \rho_o \dfrac{d - r}{d - c} & \text{for } c \leq r < d \\ 0 & \text{for } d \leq r \end{cases} \qquad (16)$$

ρ_o is either the number of nucleon or amount of energy per unit volume. The total number of nucleons is then simply given by

$$(\pi \rho_o/3) \ (c+d) \ (c^2 + d^3) \qquad (17)$$

and the total number of nucleons in the constant density zone and at the surface are, respectively

$$(4\pi \rho_o/3)c^3 \qquad (18a)$$

$$(\pi \rho_o/3) \ [(c + d) \ (c^2 + d^2) - 4c^3] \qquad (18b)$$

In table 1, we have indicated the total numbers of nucleons $N(A)$, the number of nucleons in the constant density zone, $N(o)$, and the numbers of nucleons at the surface, $N(s)$, for $c = 1.2 \ A^{1/3}$ fm. and d-c = 2.5 and 3.0 fm. in the unit of $(\pi \rho_o/3)$. Both the number of nucleons in the surface region and the contribution to the total energy by them are significant. (18b) indicates that the surface energy may not be proportional to $A^{2/3}$ and as such, the coefficient of A in the liquid-drop mass formula may not reflect, the energy per nucleon in nuclear matter [14].

Since a large fraction of nucleons in a nucleus resides on the surface, it is necessary to derive total nuclear masses using proper density distribution functions. The

131

Table 1: Fractions of nucleons in the surface N(s), and interior N(o) regions for mass number A = 125 and 238. N(A) is the total number of nucleons in arbitary unit. d-c is defined in (10) and (15).

A=Mass No.	d - c = 2.5 fm.			d - c = 3.0 fm.		
	N(A)	N(o)	N(s)	N(A)	N(o)	N(s)
125	1016	500	516	1157	500	657
238	2676	1647	1032	2938	1647	1291

EDF approach provides an appropriate method for that. In case the choce of the effective operator $[T(\rho) + V(\rho)]$ is a good one, one should be able to reproduce the nuclear masses by choosing appropriate density distribution functions. Using density distribution functions obtained from Hartree-Fock type of calculations Lombard [15] has been able to reproduce the observed masses for $\gamma = 15.2$. Ngô and Ngô [16] and Reichstein and Malik [13, 17] have also reproduced the masses reasonably well using, respectively, Fermi and trapezoidal distribution functions.

In this paper, we present in table 2 calculated nuclear masses using $\gamma = 8$ and 9 and the density distibution functions derived either from the electron scattering or μ-mesic data and compare them with the observed values as well as those obtained from Myers and Swiaticki's liquid drop mass formula [18]. Calculated masses can be well reproduced with proper density distributions i.e., also with proper root mean squared radii.

The expressions (10) and (15) indicate that the compressibility has a complicated dependence on local density. At the saturation density, defined to be the minimum in the E/A versus ρ plot, the compressibility is about 180 in the usual unit.

In future, it would be interesting to evaluate γ from a more fundamental theory and obtain the density distribution either from an equation obtained from $\delta E = 0$ or making an ansatz for ρ and determining its parameter from the minimum condition on the total energy of a nucleus. Ohtsuka et al. [19] have made an initial attempt in the latter direction using a slightly different functional.

Table 2: Calculated total binding enrgies (col. 4 and 5) using EDF for $\gamma = 8$ and 9. Experimental values given in col. 3 are from [37]. 2pf and 3 pf in col. 2 are two and three point Fermi distribution functions from [38]. The last col. are results of liquid drop [18].

ELEMENT	DENSITY FUNC	B. E(MeV) EXPT	B. E(MeV) $\gamma = 8$	B. E(MeV) $\gamma = 9$	B. E(MeV) M + S
^{12}C	2pf	92.2	92.5	88.0	
^{14}N	3pf	104.7	111.0	106.9	
^{16}O	3pf	127.6	125.2	121.3	123.0
^{24}Mg	2pf	198.3	194.1	189.3	
	3pf		194.5	189.9	
^{28}Si	2pf	236.5	234.3	228.5	
	3pf		239.1	233.3	
^{40}Ca	3pf	342.1	340.6	333.8	340.0
^{51}V	2pf	445.8	461.5	451.5	
^{58}Ni	3pf	506.5	516.5	506.7	
^{70}Ge	2pf	610.5	609.1	599.1	
^{88}Sr	2pf	768.4	793.5	778.7	
^{114}Cd	2pf	972.6	984.1	969.3	
^{139}La	2pf	1164.8	1184.1	1166.1	
^{148}Sm	2pf	1225.4	1229.6	1212.6	
^{165}Ho	2pf	1344.2	1339.4	1321.9	
^{197}Au	2pf	1559.4	1592.8	1568.9	
^{206}Pb	2pf	1622.3	1630.1	1607.7	
^{208}Pb	2pf	1636.4	1667.8	1642.3	1627.0
^{238}U	2pf	1801.7	1808.6	1785.1	1805.0

4. Nuclear-Nuclear Interaction

On the basis of a simple analysis of the elastic scattering data of ^{16}O by ^{16}O Block and Malik in 1967 [20] proposed that the nuclear-nuclear potential is non-local and can be concocted by a non-monotonic local potential having a soft short range repulsion. This idea was soon corroborated by Scheid, Ligensa and Greiner [21]. The energy density formalism is reasonably suitable to calculate the macroscopic structure of the nuclear-nuclear interaction from a realistic two nucleon potential and the initial calculation by Brueckner, Buchler and Kelly [22], indeed, supports the non-monotonic nature of the interaction. However, the detailed nature of this interaction depends on the competition between the collision time of two nuclei and the relaxation time of the nucleon-nucleon collision in a medium [23].

In the London-Heitler approximation, the potential $V(R)$ between two nuclei separated by a distance R is given by

$$V(R) = E[\rho(R)] - E[\rho_1(R=\infty)] - E[\rho_2(R=\infty)] \tag{19}$$

where $E[\rho(R)]$, $E[,\rho_1(R=\infty)]$, and $E[\rho_2(R=\infty)]$ are, respectively, the total energies of the colliding system, and of particles (1) and (2) when they are far apart. One can either obtain $E[\rho_1(R=\infty)]$ and $E[\rho_2(R=\infty)]$ from the observed masses or calculate them with reasonable certainty following the method of the previous section. The calculation of $E[\rho(R)]$ needs, however, a knowledge of the density distribution function ρ of the composite system which is not easy to obtain. One could, however, obtain this in two extreme cases noted below:

(A) ρ can be calculated by adding the densities of two colliding nuclei i.e. $\rho(R) = \rho_1 + \rho_2$. This is called the sudden approximation and invariably leads to a non-monotonic potential with a reasonable repulsion at a short distance.

(B) ρ can be calculated by minimizing the energy with respect to characteristic parameters defining ρ at every separation distance with the auxiliary condition that at no point the density exceeds the saturation density of the nuclear matter [13]. This approach which is very important for the external barrier in fission, is called the adiabatic approximation. This is a very time consuming process and a large number of calculations indicates that it is quite often sufficient to use the following ansatz for $C_i(i=1,2)$, the half-density radius and $t_i(i=1,2)$ the surface thickness parameter [13,17,23]

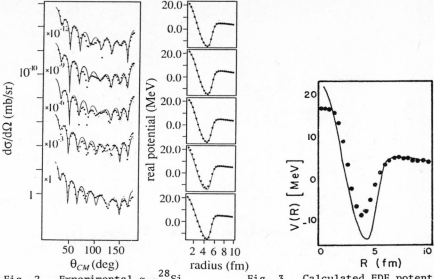

Fig. 2. Experimental α -^{28}Si angular distributions [39] (solid dots) and the fit to them by two molecular potentials. Solid lines and dots on the left correspond to solid line (called standard potential) and dotted potential to the right [24].

Fig. 3. Calculated EDF potential for the α -^{28}Si system in sudden approximation (dots) compared to the standard potential of Fig. 2.

$$
\begin{aligned}
C_i(R) &= C_c \exp[\ln(C_1/C_2) \cdot (R/R_{cut})^n] & \text{for} R \le R_{cut} \\
&= C_i & \text{for} R \le R_{cut}
\end{aligned}
\qquad (20)
$$

(and the same for t_i).

Here C_c is the half-density radius of the composite system and R_{cut} is usually taken to be (3/4) of the half-density radii of two nuclei. The ansatz (20) is referred to as the special adiabatic approximation (SAA) and n is usually 2. The parameter ρ_o, the central density of the composite system is determined from the condition of mass conservation

$$
A = A_1 + A_2 = \int \rho(r)dr \qquad (21)
$$

In Fig.2 we show the fit to the elastic scattering data of alpha particle by ^{28}Si at 25.0, 26.0, 26.5, 27.0 and 28.0 MeV (lab) and the real part of the potential which is basically found to be energy independent [24]. In Fig. 3 the empirically determined real part of the potential is compared with the one calculated from the energy density formalism using the sudden approximation. The agreement is quite good. In general,

135

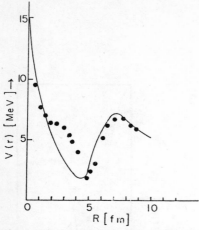

Fig. 4. Calculated EDF potential for the ^{12}C - ^{12}C system in the SAA (solid dots) is compared to the Alam potential derived from the inverse scattering theory [25] which fits the data.

it seems that the sudden approximation is a reasonable one for the alpha scattering at energies twice the Coulomb barrier height and for targets heavier than oxygen.

One of the most definitive potentials derived to date is that between two ^{12}C nuclei, and presented by Dr. Alam in this conference [25]. ^{12}C has a large surface and may have a reasonable deformation which smears the surface further. In addition, the analysis of Alam is at energies which are less than twice the Coulomb barrier height. Thus, we expect here an adiabatic situation. In Fig. 4 the calculated potential in the SAA with n = 1.6 and R_{cut} = 6 fm. are noted by dots and reproduces the general trend of the Alam potential shown by a solid line. The potential calculated in the sudden approximation has a core of 20 to 50 MeV [19].

Before concluding the section, we may note that the imaginary part of the potential can also be determined using the energy density functional by calculating the contribution from off the energy shell T-matrices [19] or calculating the virtual excitation probabilities [27].

5. Predicting Potential by Scaling

From the general structure of the formalism, it is possible to scale the potential from one system to another and thereby predict the general trend of the scattering cross sections provided the energy range in both cases are close to one another. The scaling is best done in the energy range where the sudden approximation holds approximately. To understand the scaling we adopt a trapezoidal density distribution.

In order to scale the real part of the potential we note that it can be decomposed in three essential segments: (a) the Coulomb interaction, (b) the attractive part of the interaction originating from the unsaturated nuclear matter at the surface and (c) the repulsive core originating from the region where the density becomes larger than its saturated value.

The Coulomb part, V_{Coul} is simply given by

$$
\begin{aligned}
V_{Coul} &= (Z_1 Z_2 e^2/2R_c)\ (3 - r^2/R_c^2) \quad \textit{for } r \leq R_c \\
&= Z_1 Z_2 e^2/r \qquad\qquad\qquad \textit{for } r > R_c
\end{aligned}
\tag{22}
$$

This part is governed by R_c which is that distance where two charged nuclei just touch each other. Thus, $R_c = R_{c1} + R_{c2}$ where R_{c1} and R_{c2} are the maximum extend of charge densities of two colliding nuclei. Except for very light nuclei such as α-particles, R_{c1} and R_{c2} can simply be scaled as $r_o A_1^{1/3}$ and $r_o A_2^{1/3}$, respectively ($R_{c\alpha}$ is the radius of alpha-particle)

$$
R_c = R_{c\alpha} + r_{oc} A_2^{1/3} \quad \textit{(for } \alpha\textit{-nucleus)}
\tag{23a}
$$

$$
= r_{oc}(A_1^{1/3} + A_2^{1/3}) \quad \textit{(for two nuclei)}
\tag{23b}
$$

One could probably refine this further but at this stage, it is unwarranted.

The attractive part of the interaction, V(attr) may be written as

$$
V(attr.) = -V_o[1 + \exp\{(r - R_o)/a\}]^{-1}
\tag{24}
$$

The parameter, 'a' is not expected to depend strongly on mass number. R_o is that distance between two nuclei when about one quarter of the central density of each nuclei overlaps. Hence,

$$
\begin{aligned}
R_o &= R_{o1} + R_{o2} \\
&= R_{\alpha1} + r_{oo} A_2^{1/3} \qquad \textit{(for } \alpha\textit{-nucleus)} \\
&= r_{oo}(A_1^{1/3} + A_2^{1/3}) \qquad \textit{(for nucleus-nucleus)}
\end{aligned}
\tag{25a}
$$

The repulsive part of interaction $V_R(R)$ may be parametrized as

$$
V_R(R) = V_R \exp[-(r/R_R)]
\tag{26}
$$

In this case R_R determines the range where the potential drops to e^{-1}. This is expected to occur when about $(3/4)$ of the central density of each colliding nuclei overlap. For a trapezoidal density distribution, this point also scales as $A^{1/3}$. Thus,

$$
\begin{aligned}
R_R &= R_{R1} + R_{R2} \\
&= R_{R\alpha} + r_{0R} \, A_2^{1/3} \qquad for \, (\alpha-nucleus \, case) \\
&= r_{0R}(A_1^{1/3} + A_2^{1/3}) \qquad for \, (nucleus-nucleus \, case)
\end{aligned}
\tag{27a}
$$

The strength of the repulsive part V_R is simply governed by the sum of the central density of two nuclei which does not depend on the mass numbers significantly and can be kept unchanged in the first approximation.

The strength of the attractive part V_0 in (24) can also be scaled by noting that V_0 is the difference between the surface energy of the composite system and that of the individual colliding nucleus. The surface energy, E_s, for a trapezoidal density distribution (16) is given by

$$
\begin{aligned}
E_s &= \frac{\pi \rho_o}{3}[(c + d) \, (c^2 + d^2) - 4c^2] \\
&= \frac{\pi \rho_o}{3} c^2 x \, [6 + (4x/c) + (x/c)^2]
\end{aligned}
$$

where $x = d-c$ which is 2 to 3 fm. Hence, one may neglect the last two terms in the parenthesis and obtain $E_s \sim \pi \, \rho_o \, 6xc^2/3 = $ constant $c^2 = $ constant $A^{2/3}$. Therefore,

$$
V_c = b_s[A_1^{2/3} + A_2^{2/3} - (A_1 + A_2)^{2/3}]
\tag{28}
$$

The same expression has been derived by Bass [26] for the interaction between two nuclei via a proximity force.

In Fig. 5 we present the potential of the α-^{32}S and α-^{34}S systems that are obtained by scaling the parameters of the α - ^{28}Si system. We note that they are indeed very close to the ones calculated from the EDF formalism in the sudden approximation [28].

The scaling of the imaginary part W(R), of the potential is more uncertain. However, the model of Alexander and Malik [27] indicates that

$$
W(R) = f(R) \, x \, (a \, function \, depending \, on \, energy/nucleon)
$$

In case the energy per nucleon in two systems is approximately equal, one may use the same energy dependence and scale R as $R_{\alpha i} + r_{oi} A_1^{1/3}$ (for the alpha-nucleus system) and $R = r_{oi} (A_1^{1/3} + A_2^{1/3})$ for two colliding nuclei.

In Fig. 6, we have shown by solid curve and dots the calculated and observed angular distributions, respectively, of alpha particle scattered by ^{32,34}S [28]. The calculations done using scaled potentials, has been performed before the data were taken and the agreement is good.

138

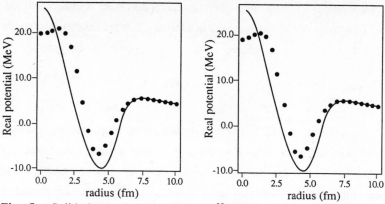

Fig. 5. Solid dots and lines are α = ³²S (left) and α = ³⁴S potentials obtained, respectively, from the EDF and by scaling.

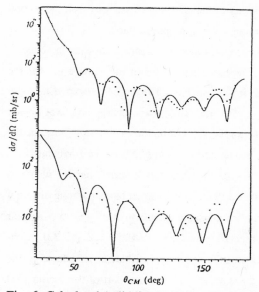

Fig. 6. Calculated (solid lines) angular distributions using the scaled potentials for the α = ³²S (upper) for 18.1 MeV and α = ³²S (lower) for 18.6 MeV systems are compared to the data [40, 41].

Haider and Malik [29] could obtain a good fit to the 90° excitation function and some angular distributions for the ¹²C - ¹²C system at energies greater than about 37 MeV (cm) by scaling the potential used for the ¹⁶O + ¹⁶O system.

Recently it has also been possible to account for the angular distribution of elastically scattered ³²S by ⁶²Ni by scaling the potential from the ¹²C-¹²C case [30].

Thus, one is now in a position to predict the angular distributions of two colliding nuclei from the knowledge of the potential used in analyzing data for another (preferably nearby) system. This has not been achieved in the standard optical model approach.

6. Radioactive Decays

In 1971 it was recognized [31] that the molecular nature of the potential between two nuclei in the exit channel of the final state may have very serious consequences for our understanding of both spontaneous and induced fission processes. Extensive works have been done in this direction with various collaborators and these are summarized in a paper presented at the 50 years of nuclear fission [32].

The radioactive decay of clusters may be conceived as a two step process [13,31-36]. First, there is the formation probability of the decaying cluster and then the second step is the tunneling of that cluster through the appropriate barrier. (This model is similar to the one reported by Dr. Rubchenya in this conference). In case the clusters have similar masses, the pre-formation probabilities are about the same e.g., for the fission if ^{238}U to its asymmetric modes, it is about 1.6×10^{-5} (to be compared to about 10^{-2} to 10^{-3} for the alpha-particle decay). For half-lives, mass, charge and total kinetic energy distributions the determining factor is, therefore, the penetration through the external barrier that exists between the saddle and scission points in the final state.

The EDF provides a suitable way to calculate this external barrier which is different for different daughter pairs. This difference in external barrier among different pairs of the daughter products along with the available excitation energy for a particular decay mode is responsible for the observed mass, charge and TKE [13,33]. For radioactive decay one has an adiabatic situation and a typical barrier calculated in the SAA is shown in Fig. 7 [13]. In Fig. 8 we present the external part of the barrier [33] pertinent for the decay of ^{240}Pu to two different decay modes, namely ^{142}Ba + ^{98}Sr and ^{122}Cd + ^{118}Pd. The calculated barriers in two cases differ which is an important factor in determining the differences in mass and charge yields in two cases.

Actual calculation of this barrier is a very time consuming job. Fortunately, this external barrier can again be scaled easily. Hooshyar, Compani and Malik [34-36] have been able to account for the charge and mass distributions in spontaneous and particle induced fission in a number of cases using barrier obtained by scaling. The decay of ^{258}Fm to predominantly symmetric mode has also been predicted properly [35].

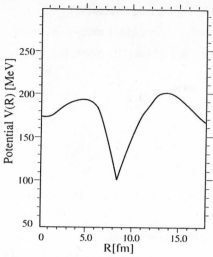

Fig. 7. A typical fission barrier in MeV for the decay of [234]Pu to
[142]Xe and [92]Sr calculated by the EDF in adiabatic approximation as a
function of separation distance R in fm. [13] .

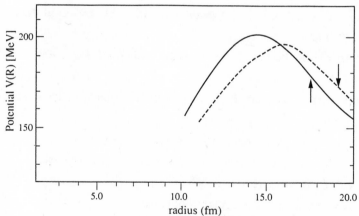

Fig. 8. The potential in the exterior region for the decay of [240]Pu
to [142]Ba + [98]Sr (solid line) and to [122]Cd + [118]Pd (dashed line)
using the EDF in adiabatic approximation by superposing two spheroidal density distribution.

This external barrier is important in explaining the cold fission [32].

It is my pleasure to acknowledge with many thanks various fruitful discussions with Professor M. Brenner, Dr. M. M. Alam, and Messr. P. Manngard and Z. Shehadeh.

References

1.   P. Gombas, Ann. Physik 10, 253 (1952)

2.   T. H. R. Skyrme, Phil. Mag. 1, 1093 (1956)

3.   L. Wilets, Phys. Rev. 101, 1805 (1956); Rev. Mod. Phys. 30, 542 (1958)

4.   H. A. Bethe, Phys. Rev. 167, 879 (1968)

5.   P. Hohenberg and W. Kohn, Phys. Rev. 136, B864 (1964)

6.   L. C. Gomes, J. D. Walecka and V. F. Weisskopf, Ann. Phys. (N.Y.) 3, 241 (1958)

7.   K. A. Brueckner, S. A. Coon and J. Dabrowski, Phys. Rev. 168, 1184 (1968)

8.   K. A. Brueckner and J. L. Gammel, Phys. Rev. 109, 1023 (1958)

9.   D. C. Peaslee, Phys. Rev. 95, 717 (1959)

10. P. B. Danilov, K. A. Gridnev, V. B. Subotin and F. B. Malik, Izv. Akad. Nauk, (Fiz) 53, 2220 (1989)

11. K. A. Brueckner, J. R. Buchlar, S. Jorna and R. J. Lombard, Phys. Rev. 171, 1188 (1968)

12. K. A. Brueckner, J. R. Buchlar, R. C. Clark and R. J. Lombard, Phys. Rev. 181, 1543 (1969)

13. I. Reichstein and F. B. Malik, Ann. Phys. (N.Y.) 98, 322 (1976)

14. F. B. Malik and J. Y. Shapiro, Condensed Matter Theories, 3, 11 (1988)

15. R. J. Lombard, Ann. Phys. (N.Y.) 77, 380 (1973)

16. H. Ngô and Ch. Ngô, Nucl. Phys. A348, 140 (1980)

17. I. Reichstein and F. B. Malik, Condensed Matter Theories 1, 291 (1985)

18. W. D. Myers and W. J. Swiatecki, Nucl. Phys. 81, 1 (1966)

29. N. Ohtsuka, R. Linden, A. Faessler and F. B. Malik, Nucl. Phys. A465, 550 (1987)

20. B. Block and F. B. Malik, Phys. Rev. Lett. 19, 239 (1967)

21. W. Scheid, R. Ligensa and W. Greiner, Phys. Rev. Lett. 21, 1479 (1968)

22. K. A. Brueckner, J. R. Buchler and M. M. Kelly, Phys. Rev. 173, 944 (1968)

23. I. Reichstein and F. B. Malik, Phys. Lett. B 37, 344 (1971)

24. P. Manngard M. Brenner, M. M. Alam, I. Reichstein and F. B. Malik, Nucl. Phys. A504, 130 (1989)

25. M. M. Alam and F. B. Malik, This proceeding

26. R. Bass, Phys. Lett. 47B, 139 (1973)

27. D. R. Alexander and F. B. Malik, Phys. Lett. 42B, 412 (1972)

28. P. Manngard, M. Brenner, I. Reichstein and F. B. Malik, The Åbo Akademi Accl. Lab. Triennial Report (1987-1989)

29. Q. Haider and F. B. Malik, J. Phys. G 7, 1661 (1981)

30. Z. Shehadeh and F. B. Malik, Nucl. Phys. (To be submitted)

31. B. Block, J. W. Clark, M. D., High, R. Malmin and F. B. Malik, Ann. Phys. (N. Y.) 62, 464 (1971)

32. B Campani-Tabrizi, M. A. Hooshyar and F. B. Malik, 50 Years With Neclear Fission, eds. J. W. Behrens and A. D. Carlson (American Nuclear Society 1989) P. 643.

33. I. Reichstein and F. B. Malik, Superheavy Element ed. M. A. K. Lodhi (Gordon and Breach 1976)

24. M. A. Hooshyar and F. B. Malik, Phys. Lett. 38B, 495 (1972); Helv.. Phys. Acta 46, 720 (1973)

35. M. A. Hooshyar and F. B. Malik, Helv. Phys. Acta 46, 724 (1973)

36. M. A. Hooshyar, B. Compani-Tabrizi and F. B. Malik, Proc. Int'l Conf. on Interactions of Neutrons with Nuclei ed. E. Sheldon (ERDA: Conf. - 7607115-Pl, 2976) and Proc. V. Int'l. Conf. on Nucl. Reaction Mech. ed. E. Gadioli (Univ. of Milano Press 1988) P. 385

37. A. H. Wapstra and G. Audi, Nucl. Phys. A432, 1 (1985)

38. H. DeVries, C. W. De Jaeger and C. DeVries, At. Data Nucl. Data Tables 36, 495 (1987)

29. L. Jarczyk, B. Macius, M. Simenko and W. Zipper, Acta Phys. Pol. B7, 53 (1976)

40. J. C. Corelli, E. Bleuler and J. Tendam, Phys. Rev. 116, 1184 (1959)

41. Å. Bredbacka and Z. Maté, (Private Communication)

An Inverse Scattering Theory for the Identical Particles and the ^{12}C–^{12}C Potential

M.M. Alam and F.B. Malik

Physics Department, Southern Illinois University,
Carbondale, IL 62901, USA

We have applied the recently developed inverse scattering method of identical particles to determine the nature of the ^{12}C $-^{12}$C potential between 14 and 33 MeV. The real part of the potential is found to be of non-monotonic molecular type and nearly energy independent. The imaginary part of the potential depends smoothly on the energy.

1. INTRODUCTION

Almost a quarter century ago, Block and Malik[1] proposed that the nature the nucleus-nucleus interaction is of molecular type and the local version of that has a short range repulsion. Such a potential has been derived from realistic two nucleon interaction using an energy-density formalism[12] and by Scheid et al.[2] using a kind of proximity model.

On the other hand, attempts have been made to analyze the heavy-ion heavy-ion elastic scattering data using a very deep strongly energy dependent real potential that is expected from a folding model[3,4] which neglects the Pauli principles among nucleons.

Our ideal is to determine the general nature of this interaction using an inverse scattering method which uses phase shifts as inputs. Almost all inverse scattering methods at a fixed energy including that of Newton[5] and Hooshyar and Razavy[6,7] cannot be applied to the scattering of two identical particles, since they require a knowledge of phase shifts at successive partial waves. On the other hand, the scattering of two identical particles provides information only on even partial waves.

We, therefore, first present here an outline of a recently developed[9] new inverse scattering method that requires only the information on even partial waves in order to construct the potential. The method is then, applied to determine the potential between two ^{12}C nuclei using Ledoux et al.'s[8] phase shifts analysis.

Springer Series in Nuclear and Particle Physics   Clustering Phenomena in Atoms and Nuclei
Editors: M. Brenner · T. Lönnroth · F.B. Malik   © Springer-Verlag Berlin, Heidelberg 1992

2. THE METHOD

Let us consider the scattering of two identical particles with reduced mass μ. The radial part of the Schrödinger's equation for the lth partial wave for a spherically symmetric potential can be written as

$$\frac{d^2\psi_l}{dr^2} + \left[k^2 - U(r) - \frac{l(l+1)}{r^2}\right]\psi_l = 0 \tag{1}$$

where

$$k^2 = \frac{2\mu E}{\hbar^2}, \qquad U(r) = \frac{2\mu V(r)}{\hbar^2}$$

In the above relation, E is the energy in the center of mass system and $V(r)$ is the sum of Coulomb and non-Coulomb potentials. In general $V(r)$ is complex and after a finite range R, known as the range of nuclear interaction, it is given by

$$V(r) = \frac{(Ze)^2}{r} \qquad \text{for} \quad r \geq R \tag{2}$$

here Z is the atomic number of the scattering particles. The solution $\psi_l(r)$ of (1) satisfies the boundary condition

$$\lim_{r \to 0} \psi_l(r) \sim (kr)^{l+1} \tag{3}$$

To solve the inverse problem, we define the function $\phi_l(r)$ as follows:

$$\phi_l(r) = (kr)^{-(l/2+1)}\psi_l(r) \tag{4}$$

Substitution of (4) in (1) leads to

$$\left(\frac{d^2}{dr^2} + \frac{l+2}{r}\frac{d}{dr} + k^2 - U(r) - \frac{l(3l+2)}{4r^2}\right)\phi_l = 0 \tag{5}$$

For the inverse scattering of non-identical particles Hooshyar and Razavy[6] have used a different substitution and the counterpart of the relation (5) for non-identical cases are different from that of the identical cases. It is obvious from the relation (3) and (4) that the solution $\phi_l(r)$ of (5) must satisfy the following boundary condition

$$\lim_{r \to 0} \phi_l(r) \sim (kr)^{l/2} \tag{6}$$

which means $\phi_l(r)$ is not a singular function at the origin.

145

The inverse scattering problem for the identical particles can be stated as follow: at a particular energy k^2, one has the information on each *even* partial wave contributing to the scattering amplitude and from this information one is to construct the relevant potential. To solve this problem, we convert the differential equation (1) to a difference equation. For that purpose first we divide R in N equal parts of length Δ, use $n\Delta$ for variable r and central differences for the derivatives. We get the following difference equation

$$\phi_{n+1} = A_n(l)B_n(l)\phi_n + C_n(l)\phi_{n-1} \tag{7}$$

$$n = 1, 2, \ldots, N$$

where

$$A_n(l) = 2 - \Delta^2 k^2 + \Delta^2 U_n + \frac{l(3l+2)}{4n^2} \tag{8}$$

$$B_n(l) = \frac{2n}{l+2+2n} \tag{9}$$

and

$$C_n(l) = \frac{l+2-2n}{l+2+2n} \tag{10}$$

The structure of equation (7) is the same as that of the scattering of two non-identical particles determined by Hooshyar and Razavy but the expression for the coefficients $A_n(l)$, $B_n(l)$ and $C_n(l)$ are different in the two cases. Now, we, may define the function $Z_N(l)$ and the matrix $\Re(l)$ as follows:

$$Z_N(l) = \frac{N}{2}\left(\frac{\phi_{N+1} - \phi_{N-1}}{\phi_N}\right) \tag{11}$$

$$\Re(l) = \frac{\psi_l}{\psi_l'}\bigg|_{r=R} = R[l/2 + 1 + R(\phi_l'/\phi_l)]^{-1} \tag{12}$$

Using central difference for the derivative, the $\Re(l)$ matrix and function $Z_N(l)$ can be related as follows:

$$Z_N(l) \simeq \frac{R}{\Re(l)} - \frac{l}{2} - 1 \tag{13}$$

The relation between function $Z_N(l)$ and matrix $\Re(l)$ are different for identical and non-identical cases. Beyond the range of nuclear interaction, ψ_l can be expressed as

$$\psi_l = F_l(\gamma, kr) + T_l[G_l(\gamma, kr) + iF_l(\gamma, kr)] \qquad \text{for } r \geq R \qquad (14)$$

Here F_l and G_l are the regular and the irregular Coulomb wave function respectively. The matrix T_l is related to the phase shift δ_l and the attenuation coefficient η_l or to the S-Matrix S_l as follows:

$$T_l = \frac{\eta_l \exp(2i\delta_l - 1)}{2i} = \frac{S_l - 1}{2i} \qquad (15)$$

γ is the Sommerfeld parameter and is given by

$$\gamma = \frac{\mu(Ze)^2}{\hbar^2 k} \qquad (16)$$

In the following we proceed to show that one can construct the potential from the knowledge of the even partial wave phase shifts only. Using (12) – (14) we can calculate $Z_N(l)$ for $l = 0, 2, 4, \ldots, L$, where $N \equiv L/2 + 1$. The potential is known at $r = R$ by using (2) in (8), $A_N(l)$ can be calculated for all even l. Using (7) for $n = N$ and (11) we get

$$\frac{\phi_N}{\phi_{N-1}} = [C_N(l) - 1] / \left[\frac{2}{N} Z_N(l) - A_N(l) B_N(l) \right] \qquad (17)$$

Now, we define

$$l_n = 2(N - n - 1), \qquad n = 1, 2, \ldots, N - 1 \qquad (18)$$

Science both N and n are integer l_n is an even number and according to (10), $C_n(l_{N-n}) = 0$. Hence, from (7), we get

$$A_{N-1}(l_1) = \frac{\phi_N(l_1)}{\phi_{N-1}(l_1)} \frac{1}{B_{N-1}(l_1)} \qquad (19)$$

We can continue this process and find A_{N-j} for $j = 2, 3, \ldots, N - 1$ as a continued fraction

$$A_{N-j}(l_j) = \frac{1}{B_{N-j}(l_j)} \left[\frac{C_{N+1-j}(l_j)}{-A_{N+1-j}(l_j) B_{N+1-j}(l_j) +} \cdots \right.$$
$$\left. \frac{C_{N-2}(l_j)}{-A_{N-2}(l_j) B_{N-2}(l_j) +} \frac{C_{N-1}(l_j)}{-A_{N-1}(l_j) B_{N-1}(l_j) + \phi_N(l_j)/\phi_{N-1}(l_j)} \right]$$
$$j = 2, 3, \ldots, N - 1 \qquad (20)$$

After calculating $A_{N-j}(l_j)$, we can determine U_{N-j} and hence, V_{N-j} using

$$U_{N-j} = \frac{1}{\Delta^2}\left[A_{N-j}(l_j) - 2 + \Delta^2 k^2 - \frac{l_j(3l_j + 2)}{4(N-j)\Delta}\right] \tag{21}$$

and

$$V_{N-j} = \frac{\hbar^2}{2\mu}U_{N-j} \tag{22}$$

The analytic form $V(r)$, for the potential can be found using a suitable fitting method for the set of N determined potential points $\{V_n\}$. We note that the method is valid for energy dependent potentials.

3. THE NATURE OF ^{12}C $-^{12}$C POTENTIAL

The method has been tested for a number of complex potentials pertinent to heavy-ion heavy-ion interaction and found to be effective and satisfactory.[9]. As it is well known, sometime one may obtain a set of potentials rather than a unique one from the phase shifts at a single energy. However, the analytic behavior of the set of phase shifts corresponding to different potentials as a function of energy is not the same. Therefore, the analysis of data in few neighboring energies reduces this ambiguity and in most cases, selects an unique potential.

The phase shift analysis of the observed angular distributions of the elastic scattering of ^{12}C by ^{12}C have been done by Ledoux et al.[8] at several energies. In this paper we have attempt to determine the nature of the potential using the phase shifts at 20.9 and 24.5 MeV(cm). The method determines only $(l_{max}/2 + 1)$ points of the potential at a given energy, where l_{max} is the largest partial wave. The real and imaginary parts of the determined potential points for both energies are shown in Fig. 1. One may conclude that the potential is definitely not deep but non-monotonic with a small short range repulsion and its real part is approximately energy independent. As a further check we take the following analytic potential function and determine the parameters to yield approximately the determined points of the potential and investigate whether such a potential could reasonably account for the observed angular distributions. In Fig. 2 we have ploted the calculated angular distribution using this analytic potential at 20.9 and 24.5 MeV together with the experimental values. The analytic form of the real part of the potential at both the energies is the same and given by

$$V(r) = \frac{-10}{1 + \exp\left(\frac{r-5.48}{.53}\right)} + 13\exp(-r/2.36) + V_c(r) \tag{23}$$

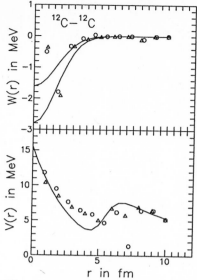

Fig. 1 Derived complex potential $V(r) + iW(r)$ from phase shifts. Triangles and circles are derivation at 20.9 and 24.5 MeV center of mass energy. Solids line are the analytic potentials used to calculate the angular distribution.

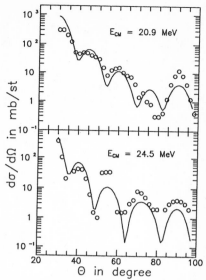

Fig. 2 Calculated angular distribution of ^{12}C $-^{12}$C scattering at 20.9 and 24.5 MeV are shown by solid lines. The experimental points (circles) are taken from ref. 5. (Both cross section and θ are in center of mass system.)

where $V_c(r)$ is the Coulomb potential due a uniformly charged sphere. The analytic form of the imaginary part of the potential is given by

$$W(r) = W_I(E) \exp[-(r/2.36)^2] \qquad (24)$$

where W_I is the energy dependent depth of the imaginary part of the potential, it is -1.64 Mev and -2.8 MeV respectively for 20.9 MeV and 24.5 MeV of energy. At this stage we may inquire the predictability of cross section using this potential. In the first approximation, we expect the real part to be nearly energy independent and adjust $W_I(E)$ to reproduce the magnitude. $W_I(E)$ at 32.3 and 14.6 are taken, respectively to be -4.8 and $-.04$ MeV. The calculated angular distributions shown as solid lines are compared to the data at 31.3 and 14.6 MeV in Fig. 3. Calculations do account for the data at 31.3 MeV and the general trend of the data at 14.6 MeV reasonably. The agreement at 14.6 MeV improves if 5.48 and 13 in the

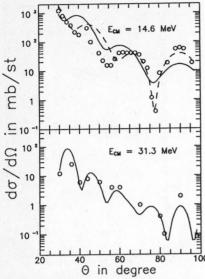

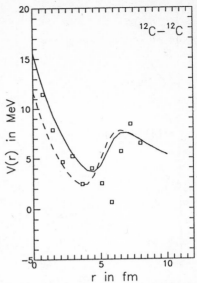

Fig. 3 Solids lines are the calculated angular distribution using (23) as the real part of the potential and the dashed line is the distribution, calculated with the parameter changed potential. The experimental points (circles) are taken from ref. 4. θ and cross section both are in center of mass system.

Fig. 4 The solid line shows the potential relation (23). The dashed line is the parameter changed potential used to calculate the cross section for the angular distribution at 14.6 MeV. The potential points calculated from the inverse scattering method at 14.6 MeV are shown as squares.

first and second expression of (23) are changed to 5.00 fm and 12 MeV, respectively. The calculated cross sections using this latter set of parameters are shown by dashed line in Fig. 3. In Fig. 4, we have plotted the potential points determined from the phase shifts at 14.6 at 14.6 MeV using our inverse scattering method along with the two potentials corresponding to the calculations represented by the solid and dashed lines in Fig. 3. Both potentials are consistent with the points determined by the inverse scattering method.

The depth of the imaginary part of the potential $W_I(E)$ has the following functional relationship with the center of mass energy, E which has been determined by a polynomial fitting.

$$W_I(E) = -3.22(-1 + .06E + .007E^2) \qquad (25)$$

It basically increases linearly with E and plotted in Fig. 5. The domain of the validity of (25) is detected by the fact that $W_I(E)$ must be negative.

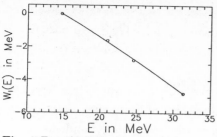

Fig. 5 Depth of the imaginary part of the potential is shown as a function of center of mass energy. The solid line is the polynomial fitting and is given by the relation (25).

4. COMMENTS AND CONCLUSION

Our analysis indicates that the $^{12}C-^{12}C$ interaction in the energy range considered here is non-monotonic i.e., of the molecular types first suggested by Block and Malik[1] for the $^{16}O-^{16}O$ case. The real part of the potential is weekly energy dependent.

Gobi et al.[10] attempt to fit the low energy data with a shallow potential without much success. Weiland et al.'s[3] attempt to fit the angular distribution between 35 and 65 MeV using deep real part of the potential generated from folding model has had only limited success. Stockstad et al.[4] latter has to normalize the depth of the folding model potential at every energy to improve the fit. Our determined potential is at variance with these deep and strongly energy dependent potentials.

Haider and Malik[11] on the other hand could reproduce the general trend of the 90° excitation function between 13 and 62 MeV center of mass energy and angular distribution at six energies between 70.7 and 112.0 MeV using complex molecular potential having nearly an energy dependent real part.

Our potential is similar to the Haider-Malik potential but is shallower and the height of the repulsive core somewhat smaller. Our analyses of angular distribution is, however, done at lower energy where one might have the onset of adiabaticity discussed by Reichstein and Malik[12].

A microscopic calculation done within the frame work of an energy density functional using a realistic two nucleon interaction by Ohtsuka et al.[13] indicates that the real part of the potential at the energies considered here is of molecular type with significant absorption. Their real part of the potential at zero incident energy which is pertinent for our case is closer to the potential deduced here.

The inverse scattering method which we have presented here is an effective method to determine the general pattern of the nuclear interaction. This method is a simple one and almost free from any complicated numerical procedures. Though this method may generates some errors, with careful step by step approach these errors can be reduced and it is possible to determine the overall functional form of the potential reasonably well. Our inverse scattering method should not be confused to be a procedure to determine the nuclear interaction with high accuracy; rather should be considered as a mean to get a preliminary idea about the macroscopic nature of the interaction.

REFERENCES

1. B. Block and F.B. Malik, Phys. Rev. Lett. 19, 239 (1967).
2. W. Scheid, R. Ligensa and W. Greiner Phys. Rev. Lett. 21, 1479 (1968).
3. R.M. Wieland, R.G. Stokstand. G.R. Satchler and L.D. Rickertsen, Phys. Rev. Lett. 37, 1458 (1976).
4. R.G. Stokstand, R.M. Wieland, G.R. Satchler, C.B. Fulmer, D.C. Raman, L.D. Rickertsen, A.H. Snell and P.H. Stelson, Phys. Rev. C 20, 655 (1979).
5. R.G. Newton, J. Math. Phys. 3, 75 (1962).
6. M.A. Hooshyar and M. Razavy, Can. J. Phys. 59, 1627 (1981).
7. M.M. Alam and F.B. Malik, Phys. Lett. B 237, 14 (1990).
8. R.J. Ledoux, M.J. Bechara, C.E. Ordonez, H.A. Al-Juwair and E.R. Cosman, Phys. Rev. C 27, 1103 (1983); also private communication.
9. M.M. Alam and F.B. Malik, Nucl. Phys. A 524, 88 (1991).
10. A. Gobi, R.M. Wieland, L. Chua, D. Shapira and D.A. Bromley, Phys. Rev. C 7, 30 (1973).
11. Q. Haider and F.B. Malik, J. Phys. G 7, 1661 (1981).
12. I. Reichstein and F.B. Malik, Phys. Lett. B 37, 334 1971).
13. N. Ohtsuka, R. Linden, A. Faessler and F.B. Malik, Nucl. Phys. A 465, 550 (1987).

The Role of a Repulsive Core in the Investigation of Clusters in Light Nuclei

K.A. Gridnev

Leningrad State University, St. Petersburg, Russia

What are clusters? Clusters are spatially correlated nucleons. What kind of forces act between clusters, and between clusters and nuclei? Japanese theoreticians showed [1], that there are four typical "di-molecules" in nature (molecules consisting of atoms, nuclei, nuclei plus hyperon, nucleons) with the same kind of potential. We consider the typical potential of the interaction between ^{4}He atoms (Fig.1).

One can approximately consider ^{4}He atoms as hard spheres with the radius 2.7 Å. The minimum implies that atoms have a tendency to form crystals. But the energy of heat does not permit it. In a different energy scale the situation is very similar. In the case of a nuclear potential [2] the origin of the attractive part of the potential is due to the exchange of nucleons.

Let us now consider the repulsive core. The best understanding can be obtained in the micro- and macro- approaches. The existence of a repulsive core in the potential interaction of two complex nuclei follows from the Pauli principle which forbids their mutual interpenetration. As an example we consider the interaction of two alpha particles. In this interaction the repulsive core is manifested from the formation of quasimolecular states of the nucleus. Figure 2 gives the $\alpha + \alpha$ phase shifts obtained from different experiments. The phase shift δ_2 increases rapidly from an energy about 1 MeV and δ_4 from an energy about 6 MeV. This indicates that the core has an extension with a quite sharp boundary (in the interior region) with R $\leq$ 4 fm. Note, that the impact parameter $b = \lambda L$ is approximately 4 fm at E_{cm}=1 MeV and 3 fm at E_{cm}=6 MeV. This character of the scattering can be well described by a phenomenological potential with a repulsive core (Fig. 1).

With the above potential quasibound 0^+, 2^+ and 4^+ levels arise and form the well known rotational band of the ^{8}Be nucleus with states at 0, 2.9, and 11.4 MeV. The levels of this band have alpha-particle widths approximately equal to the Wigner limit. The distance between the levels corresponds to a moment of inertia given by $m_\alpha R_\alpha^2$ with R=4.5 fm. It was shown in [3] that a phenomenological alpha + alpha potential with the repulsive core can be obtained microscopically as a result of the Pauli principle. In this approach the connection to the ordinary shell model is not lost. The repulsive core at a radius of about 2 fm arises as a node of a

Springer Series in Nuclear and Particle Physics **Clustering Phenomena in Atoms and Nuclei**
Editors: M. Brenner · T. Lönnroth · F.B. Malik   © Springer-Verlag Berlin, Heidelberg 1992

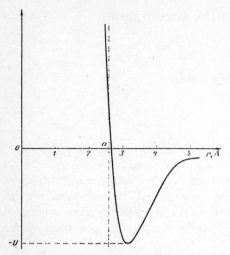

Fig. 1. The potential of the interaction between ^{4}He atoms

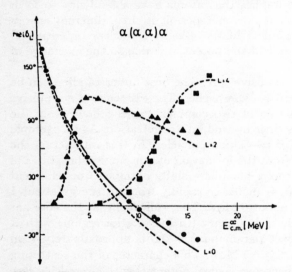

Fig. 2. The phases δ_0, δ_2 and δ_4 for $\alpha - \alpha$ scattering plotted as function of the energy of the particles in the c.m.-system. The broken lines were obtained using a repulsive ℓ—dependent potential in the form of a hard wall plus a long-range attractive potential. The full lines correspond to a soft core.

wave function which oscillates in the internal region and rapidly increases in amplitude in the external region, i.e. for R $\geq$ 2 fm.

The development of heavy ion beams has made it possible to pursue systematic and intensive investigations of light nuclei by transfer reactions. The studies have shown that there are many pure alpha-cluster states in light nuclei [4] in the region of high excitation energies (10-30 MeV). The

154

reduced alpha-widths for the decay to the ground state are of the order of the Wigner limit. A very good explanation of this phenomenon can be obtained with the help of an Effective Surface Potential (ESP). The ESP interaction comprises of an attractive part, which is usually of the Woods-Saxon type, and an ℓ—dependent repulsive hard or soft core in the region of the nuclear surface. Such a potential reproduces extremely well not only the positions and the widths of the alpha-cluster levels but also the angular distributions of alpha-particle scattering by light nuclei [5].

What is the nature of this hard core in terms of the Pauli principle? There are two points which are very important [6]:

1. The so-called Firsov splitting, which amounts to using different potentials for different parities: $V_0^{even} = V_0 + J_0$ and $V_0^{odd} = V_0 - J_0$, where J_0 is the exchange interaction. In short

$$V = V_{00} + (-1)^L \cdot V.$$

2. The splitting of the potential due to symmetry. It appears in light nuclei when there is a Wigner SU(4) supersymmetry as:

$$V = V_0 + (-1)^L \cdot V(S+1),$$

where $S = 0$ for singlet and $S = 1$ for triplet in the case of the deuteron. Different symmetry applies to even and odd states [7].

Now we know that a significant part of the energy dependence of the real part of the optical potential for the system alpha + nucleus is due to internucleus antisymmetrization [8]. In spite of the good results obtained by the aid of the ESP we would like to compare it to the microscopic Orthogonality Condition Model (OCM), which is a simplified version of the Resonating Group Method (RGM). This will hopefully give further support to our model. The nucleus ^{20}Ne will be given as an example. We have solved the OCM equation

$$\lambda_\ell P_\ell \lambda_\ell U_\ell = 0,$$

where the projection operator λ_ℓ eliminates the redundant solutions ($U_{N\ell}(r)$; $M = 2N + \ell \leq 8$), which are prohibited by the Pauli principle, and it is given by the expression

$$\lambda_\ell(r, r') = \delta_\ell(r, r') - \sum_{N=0}^{M} U_{N\ell}(r) U_{N\ell}(r').$$

The symbol P_ℓ stands for

$$P_\ell = -\frac{\hbar^2}{2m} \frac{d^2}{dr^2} + \frac{\hbar^2}{2mr^2} \ell(\ell+1) + V_D^n(r) + V_D^c - E.$$

Here $V_D^n(r)$ and $V_D^c(r)$ denote the direct nuclear and the Coulomb poten-

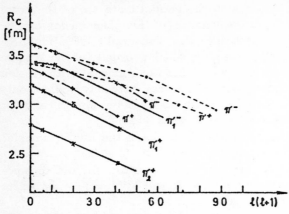

Fig. 3. The radius R_c plotted versus $\ell(\ell+1)$ for the nucleus ^{20}Ne obtained by the ESP (dashed lines), by Laguerre polynomials (dashed-dotted lines) and the OCM (full lines).

tials, and $U_{N\ell}(r)$ is the harmonic oscillator wave function. The elimination of the forbidden states gives rise to energy independent oscillations of the wave function of the relative motion in the nuclear interior. The resulting behaviour of the OCM wave functions suggest that standing waves are formed in the region of the nuclear surface.

In ^{20}Ne the even rotational band I_1^+ ($0^+, 2^+, 4^+, 6^+, 8^+$) consists of $2s-1d$ nucleons, and consequently the number of oscillator quanta is 8. According to Talmi's rule $2(N-1)+\ell=8$ the first allowed states for $I^\pi = 0^+, 2^+, 4^+, 8^+$ are characterized by the quantum numbers $N=5,4,3,2,1$. On the contrary, the states with $N=1,2,3,4$ for $I^\pi = 0^+$, with $N=1,2,3$ for $I^\pi = 2^+$, with $N=1,2$ for $I^\pi = 4^+$, and with $N=1$ for $I^\pi = 6^+$ are forbidden by the Pauli principle.

We can see that the harmonic-oscillator wave functions are proportional to the Laguerre polynomials

$$L_{N-1}^{\ell+\frac{1}{2}}(a^2 r^2) = L_{N-1}^{\ell+\frac{1}{2}}(\xi^2),$$

with $a^2 = \mu\nu$. The frequency ν is determined by the relation $\nu^{-1} = 1.01 A^{1/3} fm^2$, and N is the principal quantum number of the shell model. In order to obtain the radius of the repulsive core, R_c, one has to determine this quantum number.

For the band I_1^+ the zeros of the Laguerre polynomial with $(N,\ell) = (5,0),(4,2),(3,4),(2,6)$ are calculated to study the dependence of the radius of the core as a function of $\ell(\ell+1)$. Similarly the zeros for the I_1^- band were obtained. The dashed-dotted lines in Figure 3 demonstrate that the bands are split according to their parity. For a comparison earlier results of the ESP are included in the figure as the dashed lines. Finally, the OCM was used to compute the wave function of the relative motion, and its zeros are represented by the upper two full curves in Figure 3.

All curves clearly illustrate the splitting of even and odd rotational bands and show that the radius of the repulsive core decreases with increasing $\ell(\ell + 1)$. Since we are only interested in a qualitative understanding of these features we have not attempted to modify any of these parameters to reproduce identical results for all methods used. The lowest full curve in Figure 3 corresponds to the zeros of the 6_2^+ band being formed by nucleons of the $(sd)^2(fp)^2$ shells. It is strongly forbidden in comparison to the even band 6_1^+ with $N = 10$.

The main discrepancy between the shell model and the cluster models appears in the deduction of the widths of the cluster states. The shell model does not work in this respect. It has been shown [9] that there is a good description of the widths of ^{20}Ne in the framework of the ESP and the OCM.

Next let us consider how to simulate the Pauli principle in the description of light-ion scattering and bound states of light nuclei. The parent nucleus is assumed to consist of a daughter nucleus core with mass A_1 and a preformed cluster with mass A_2 in relative motion around their mutual center of mass. The resulting potential may be accurately parametrized in the form [10]:

$$V_{nucl} = \frac{-V_0[1 + cosh(R/a)]}{[cos(r/a) + cosh(R/a)]}.$$

As parameters for all cluster-core combinations we use $a = 0.75$ fm and $R = 1.04(A_1^{1/3} + A_2^{1/3})$ fm. The main requirements of the Pauli principle are taken into account by choosing the quantum numbers of relative motion n (the numbers of internal nodes in the radial wave function) and ℓ (the orbital angular momentum) to satisfy the Talmi condition $2n + \ell \leq N$. The integer N is chosen large enough to correspond to the microscopic situation in which the cluster nucleons all occupy orbitals above those already filled by the core nucleons. In Figure 4 we show the agreement between the OCM wave function and the wave function generated by the above potential.

Another way to simulate the Pauli principle is to exploit the Levinson theorem [6] or to use the Darboux transformation [11]. We will consider the latter approach because the former approach is well known. If the solution of the Schrödinger equation

$$\Psi_k'' + [\epsilon_k - \phi(k)]\Psi_k = 0 \qquad (k = 0...n)$$

is known for the set of the $(n + 1)$ lowest eigenvalues $\epsilon_0, \epsilon_1, ...\epsilon_n$, then the set of lowest n eigenvalues of the differential equation

$$U_{k-1}'' - [\epsilon_k - \epsilon_0 - \Psi\{\frac{1}{\Psi_0}\}'']U_{k-1} = 0 \qquad (k = 1, ...n)$$

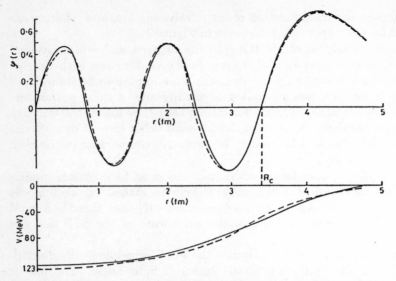

Fig. 4. Wave function and potential, solid line according to OCM, and dashed line after Gyarmati et al. [10]. R_c is the position of the repulsive core.

are given by

$$U_{k-1} = \phi_0(\Psi_k/\Psi_0)',$$

where ϕ_0 is the ground state wave function. Thus for a new potential

$$V^{(1)} = \Psi_0\{\frac{1}{\Psi_0}\}''$$

$\Psi_0(x)$ has no nodes.

This is a method to take into account the forbidden states. It is moreover an interesting way simulate the repulsive and the attractive forces in the corresponding collisions, for instance $\alpha + \alpha$. Using collisions via solitons [12] in the above way the analytical energy dependence of the phase shifts in the $\alpha + \alpha$ collisison has been obtained.

Taking into account the repulsive Springe-type force leads us to the Nonlinear Schrödinger Equation (NOSE) [13]:

$$\frac{\hbar^2}{2\mu}\nabla^2\Psi + V\Psi + V_{NL}\Psi = E\Psi$$

with $V_{NL} = C(\rho - \rho_0)/(2\rho_0)$. The V_{NL} arises due to the short-ranged interaction between the touching n liquid drop nuclei. In terms of the NOSE we can estimate the value at a hard core. For, this purpose it is necessary to rewrite the NOSE as

$$\frac{\hbar^2}{2\mu}\frac{d^2U}{dx^2} + (E - V)U + \beta U^3 = 0.$$

158

After factorization of the wave function U as $U = \Psi_0 f$, with $f = 0$ at $x = 0$, and $f = 1$ for $x \longrightarrow \infty$, we obtain

$$\frac{\hbar^2}{2\mu} \frac{d^2\Psi}{dx^2} + \alpha\Psi + \beta\Psi^3 = 0.$$

Then we have for $\alpha \longrightarrow \infty$, $U = \Psi_0$ and $\Psi_0^2 = -\alpha/\beta$, the equation becomes

$$\frac{\hbar^2}{2\mu\alpha} \frac{d^2\Psi}{dx^2} + (\beta/\alpha)\Psi^3 = 0.$$

In the compact form this equation can be transformed in the following way:

$$-\xi^2 \frac{d^2 f}{dx^2} - f + f^3 = 0.$$

Multiplying this equation by the derivative df/dx we can obtain a solution

$$f = th\left[\frac{x}{\sqrt{2}\xi}\right].$$

This solution is very similar to the correlation function for a Bose gas of hard spheres:

$$f = 0 \qquad \text{for} \qquad r \le \delta,$$
$$f = th[r/\delta - 1] \quad \text{for} \quad r \ge \delta,$$

where δ is the radius of the hard spheres. In terms of this approach the hard core radius is given by the expression

$$\delta = \sqrt{2}\xi = \sqrt{2}\frac{\hbar^2}{\sqrt{2m(E-V)}} \frac{\hbar^2}{\sqrt{m(E-V)}}.$$

Shapes of the potentials given by different relations between the repulsive and attractive parts in relation to the hard core are given in the work of ref. [14]. Microscopic estimates for the value of the hard core were made in the work of V.N. Bragin and D. Donangelo [15] and in the work of A. Tohsaki-Suzuki and K. Naito [16]. The latter authors used the RGM approach. In that case it is very evident to expect the appearance of the hard core. In the first work the authors used the Brueckner theory according to which the total energy of a system of interacting fermions can be written as a functional of the local single-particle energy density $\epsilon(\rho)$. They have computed the heavy-ion potential in the sudden approximation as

$$V(r) = \int \left[\epsilon(\rho_1 + \rho_2) - \epsilon(\rho_1) - \epsilon(\rho_2)\right] d\tau.$$

The potential obtained in this approach is highly repulsive in the nuclear interior, quite in contradiction to the behaviour of the folding model calculations.

Summing up, it is necessary to say that the momentum and energy dependent hard or soft core appears in the macroscopic calculations of two interacting nuclei as due to the Pauli principle. On this base the core can be taken into account also in the macroscopic approaches.

References

1. K. Ikeda: in Proc. Fifth Int. Conf. on Clustering Aspects in Nuclear and Subnuclear Systems, Kyoto, 1988, J. Phys. Soc. Jap. **58** 277 (1989)
2. K. Langanke, I. Leutenantsmeyer, M. Stingl, A. Weiguny: Z. Physik **A291** 267 (1979)
3. K. Ikeda, T. Matumori, R. Tamagaki, H. Tanaka: Prog. Theor. Phys. Suppl. **52** (1972) Chapt. 1
4. K.P. Artemov, V.Z. Goldberg, I.P. Petrov, V.P. Rudakov, I.N. Serikov, V.A. timofeev, P.R. Christensen: Nucl. Phys. **A320** 479 (1979)
5. N.Z. Darwish, K.A. Gridnev, E.F. Hefter, V.M. Semjonov: Nuovo Cimento **42A** 303 (1977)
6. V.I. Kukulin, V.G. Neudachin, Yu.F. Smirnov: Nucl. Phys. **A245** 429 (1975)
7. J. Kondo, S. Nagata, S. Ohkubo, O. Tamimura: Prog. Theor. Phys. **53** 1006 (1975)
8. T. Wada, H. Horiuchi: Phys. Rev. Lett. **58** 2190 (1987)
9. K.A. Gridnev et al.: "Microscopic Optical Potentials" in Lecture Notes in Physics **89**, ed. by H.V. von Geramb (Springer Verlag, Berlin–Heidelberg–New York), 1979, p. 89
10. B. Gyarmati, K.F. Pál, T. Vertse: Phys. Lett. **104B** 177 (1981); A.C. Merchant, B. Buck: Europhysics Lett. **8(5)** 409 (1989)
11. D. Baye: Phys. Rev. Lett. **58** 2783 (1987)
12. E.F. Hefter, K.A. Gridnev: Prog. Theor. Phys. **72** 549 (1984)
13. K. Mikulaš, K.A. Gridnev, E.F. Hefter, V.M. Semjonov, V.B. Subbotin: Nuovo Cimento **93A** 135 (1986)
14. P. Manngåard, M. Brenner, M.M. Alam, I. Reichstein, F.B. Malik: Nucl. Phys. **A504** 130 (1989)
15. V.N. Bragin, R. Donatello: Phys. Rev. C **32** 2176 (1985)
16. A. Tohsaki-Suzuki, K. Naito: Prog. Theor. Phys. **58** 721 (1977)

Microscopic Versus Macroscopic Treatment of Clusters in Nuclei

R.G. Lovas[1], K. Varga[1], and A.T. Kruppa[1],*

[1]Institute of Nuclear Research of the Hungarian Academy of Sciences,
P.O. Box 51, H-4001 Debrecen, Hungary
*Present address: Daresbury Laboratory, Warrington, WA4 4AD, UK

Abstract: We test the limitations of the macroscopic and microscopic versions of the nuclear cluster model by studying their applicability to the description of the fragmentation of the nucleus ^{6}Li into clusters. The models considered are the microscopic α+d breathing cluster model and the macroscopic α+p+n three-particle model. Both models prove reasonably successful, but the minor discrepancies found reveal some important defects of the presently used versions. For the microscopic description to be improved, the intercluster motion should be described in as sophisticated a manner as in the three-particle model. For the macroscopic description to be improved, the spectroscopic amplitude should be re-defined in accord with the microscopic approach. This modified definition contains information on the cluster internal structures through norm operators.

1 Introduction

The existence of cluster structure in nuclei is a highly non-trivial matter. The nucleons' propensity to gather into shell-structured spherical lumps and the rigour of the Pauli principle to deprive them of their individuality seem to act against clustering. Nevertheless, there exist states in which nuclei can be viewed as compositions of interacting clusters. Notable examples among nuclear ground states (g.s.'s) are ^{6}Li as a system of α+p+n clusters [1], ^{7}Li=α+t [2], ^{8}Be=α+α [3], ^{20}Ne=^{16}O+α [4] and ^{44}Ti=^{40}Ca+α [5]. Especially prominent clustering is likely to appear in states of small cluster separation energy [6], and for g.s's this is so for nuclei composed of a closed-shell "core" and another tightly bound cluster. Since, however, shell-model wave functions may overlap significantly with multicluster wave functions, the cluster representation may also work well for states known primarily as shell-model states. Such overlaps can be large because the antisymmetrization tends to wash out differences in configurations, so the Pauli principle does in fact disguise rather than hamper clustering.

The significant antisymmetrization effects seem to require an explicit treatment of the nucleonic degrees of freedom, i.e. a *microscopic* approach. It is, however, well-known that local intercluster potentials not only reproduce scattering data but also nuclear structure properties [4]. The structure models based on such potentials are *macroscopic* in that they only treat the intercluster degrees of freedom. In Sect. 2 we shall briefly review the well-known relationship between the macroscopic and microscopic approaches. From a restricted theoretical point of view they are in fact *equivalent*. A comparison between practically used microscopic and macroscopic models can thus reveal to what extent they embody these abstract ideas and how to extend the equivalence. Our aim is practical. Macroscopic models are obviously less complicated and more plausible, so they can be implemented in a more perfect form. E.g. an exact N-body treatment of the relative motion is less formidable in a macroscopic than in a microscopic framework.

We use the nucleus ^{6}Li as a testing ground. The microscopic α+d breathing model [7,8] and the α+p+n three-particle model should more or less correspond with each other. We shall compare these models in terms of their predictions for cluster fragmentation properties (Sect. 3). In Sect. 4 we shall show that the agreement between the predictions of the two models can be substantially improved by taking the theoretical relationship between the two approaches seriously. In Sect. 5 we shall summarize our findings.

2 Microscopic Foundation of the Macroscopic Approach

A pure-configuration N-cluster model is valid when, behind an intercluster antisymmetrizer $\mathcal{A}$, the nuclear wave function Ψ, written in the intrinsic frame, factors into a product of cluster internal wave functions $\Phi(\xi) = \Phi_1(\xi_1)...\Phi_N(\xi_N)$ and a function φ, which depends solely on relative intercluster variables $\{\mathbf{r}_1, ..., \mathbf{r}_{N-1}\}$, which can always be chosen to be Jacobian coordinates. Such a wave function can be written as

$$\Psi(\xi, \mathbf{r}_1, ..., \mathbf{r}_{N-1}) = \mathcal{A}\{\Phi(\xi)\varphi(\mathbf{r}_1, ..., \mathbf{r}_{N-1})\} \tag{1a}$$

$$= \int d\mathbf{R}_1 ... \int d\mathbf{R}_{N-1} \; \varphi(\mathbf{R}_1, ..., \mathbf{R}_{N-1}) \Psi_{\mathbf{R}_1...\mathbf{R}_{N-1}}, \tag{1b}$$

where, in the second line,

$$\Psi_{\mathbf{R}_1...\mathbf{R}_{N-1}} = \mathcal{A}\{\Phi(\xi)\delta(\mathbf{R}_1 - \mathbf{r}_1)...\delta(\mathbf{R}_{N-1} - \mathbf{r}_{N-1})\} \equiv \Psi_R \tag{2}$$

was introduced. In this way Ψ takes the form of an expansion in terms of a basis $\{\Psi_R\}$ spanned by the continuous indices $R \equiv \{\mathbf{R}_1, ..., \mathbf{R}_{N-1}\}$. With $\Phi(\xi)$ fixed, the functions $\varphi(\mathbf{R}_1, ..., \mathbf{R}_{N-1})$ can indeed be regarded as expansion coefficients. Because of the antisymmetrization involved, the basis $\{\Psi_R\}$ is non-orthogonal: $\langle\Psi_R|\Psi_{R'}\rangle \neq \delta(R - R')$. Nevertheless, the Schrödinger equation of the total system,

$$H\Psi = E\Psi, \tag{3}$$

can be reduced to an equation for the intercluster variables by projecting it onto the subspace $\{\Psi_R\}$:

$$\mathcal{H}\varphi(R) = E\mathcal{N}\varphi(R), \tag{4}$$

where $\mathcal{H}$ and $\mathcal{N}$ are multiple-integral operators whose kernels are

$$H(R; R') = \langle\Psi_R|H|\Psi_{R'}\rangle, \tag{5a}$$

$$N(R; R') = \langle\Psi_R|\Psi_{R'}\rangle. \tag{5b}$$

[E.g., $\mathcal{N}\varphi(\mathbf{R}_1, \mathbf{R}_2) \equiv \int d\mathbf{R}_1' \int d\mathbf{R}_2' \; N(\mathbf{R}_1, \mathbf{R}_2; \mathbf{R}_1', \mathbf{R}_2')\varphi(\mathbf{R}_1', \mathbf{R}_2').$] In Eq. (4) we had to have recourse to operators acting on the parameter coordinates R. In this way we managed to introduce relative variables without violating the Pauli principle. The effect of the Pauli principle is incorporated in the definitions of the kernels. The parameter coordinates are the variables to be identified with the macroscopic coordinates.

Because of the appearance of $\mathcal{N}$ on the right-hand side, Eq. (4) cannot readily be considered an intercluster Schrödinger equation. There are, however, many ways of casting it into the required form. To single out the physically interesting form, one should keep in mind that the Hamiltonian of a macroscopic cluster model is invariably energy-independent and hermitean. One can get such an equation by multiplication of (4) from the left by $\mathcal{N}^{-1/2}$ and re-defining the wave function and the Hamiltonian as

$$\chi(R) = \mathcal{N}^{1/2}\varphi(R), \tag{6a}$$
$$h = \mathcal{N}^{-1/2}\mathcal{H}\mathcal{N}^{-1/2}, \tag{6b}$$

with the result

$$h\chi(R) = E\chi(R). \tag{7}$$

For the inversion operation to be sensible, one should exclude the subspace $\{\varphi_i^{(0)}\}$ of the eigenfunctions of $\mathcal{N}$ that belong to eigenvalue zero. Since $\mathcal{N}\varphi_i^{(0)} = 0$ implies $\sum_i c_i \mathcal{A}\{\Phi\varphi_i^{(0)}(\mathbf{r}_1, ..., \mathbf{r}_{N-1})\} = 0$, this restriction does not go beyond the effect of the Pauli principle.

It is this transformation that implies, for the intercluster wave function, the same normalization as is valid for Ψ:

$$\langle\Psi|\Psi\rangle = (\varphi|\mathcal{N}|\varphi) = (\chi|\chi), \tag{8}$$

where the round brackets denote matrix elements involving integrations over the parameter coordinates. Thus it is this function that is qualified for a probability interpretation by its formal properties [9]. This transformation [10] is the only one that produces an energy-independent hermitean effective Hamiltonian, apart from unitary transformations U applied to χ and h. The function χ is distinguished from the other possible functions $\bar{\chi} = U\chi$ by its representation of the microscopic wave function in matrix elements being more faithful [11]. Unlike the other Hamiltonians derivable from H, the h of Eq. (6b) can be well approximated by a local-potential Hamiltonian. In particular, for systems of more than two clusters, it is only in h of (6b) that the multibody terms caused by the antisymmetrization are negligible [12,13], a property invariably assumed in macroscopic multicluster models.

Thus, whenever the nuclear wave function is fully contained in a cluster-model subspace $\{\Psi_R\}$, corresponding to the microscopic Schrödinger equation (3), there holds a Schrödinger equation (7) for a macroscopically meaningful intercluster relative-motion wave function χ. This equation is valid for a Pauli-restricted space. It is in this sense that the microscopic and macroscopic models are equivalent. Whether the macroscopic wave function represents the microscopic one properly in the probability distribution of a particular dynamical variable depends both on the dynamical variable and on the state itself [14,15], and the correspondence is only exact in very special, though not necessarily unrealistic, cases [9].

The correspondence between the models can be tested more directly by comparing their predictions for cluster fragmentation as the fragmentation (or spectroscopic) amplitudes do not involve dynamical variables. They are just overlaps between the wave functions of the full system and of its fragments. The microscopic amplitude $g(\mathbf{R}) \equiv g_{12}(\mathbf{R})$ of fragmentation into $1 + 2$ and its conventional macroscopic counterpart $G(\mathbf{R})$ are

163

$$g(\mathbf{R}) = \langle \mathcal{A}\{\Psi_1\Psi_2\delta(\mathbf{R} - \mathbf{r}_{12})\}|\Psi\rangle, \tag{9a}$$

$$G(\mathbf{R}) = \langle \chi_1\chi_2\delta(\mathbf{R} - \mathbf{R}_{12})|\chi\rangle, \tag{9b}$$

where $\mathbf{r}_{12}$ is the displacement of the nuclear centre of masses, and $\mathbf{R}_{12}$ is the corresponding macroscopic variable. Quantities more directly comparable with experiment are the spectroscopic factors,

$$s = \int d\mathbf{R}\,|g(\mathbf{R})|^2, \qquad S = \int d\mathbf{R}\,|G(\mathbf{R})|^2, \tag{10}$$

and the so-called "momentum distributions",

$$f(\mathbf{q}) = \left|\frac{1}{(2\pi)^{3/2}}\int d\mathbf{R}\,e^{i\mathbf{q}\mathbf{R}}g(\mathbf{R})\right|^2, \qquad F(\mathbf{q}) = \left|\frac{1}{(2\pi)^{3/2}}\int d\mathbf{R}\,e^{i\mathbf{q}\mathbf{R}}G(\mathbf{R})\right|^2. \tag{11}$$

All analyses we have performed go back essentially to these quantities.

3 Illustrative Examples

We report on our study of the fragmentation of ^{6}Li into α+d and ^{5}He+p fragments. The model we have used for the microscopic description of ^{6}Li and of its fragments is a breathing cluster model. We contrast the predictions of this with the Faddeev-type α+p+n model of Lehman *et al.* [16] and the variational α+p+n model of Kukulin *et al.* [17]. These are macroscopic in that they treat the α particle as structureless.

3.1 The models

Our microscopic models are dynamical. They are based on variational solutions of multinucleon Schrödinger equations with a *central* effective nucleon–nucleon interaction [9]. The clusters involved are the d, t and α clusters. Whether considered individually or as ingredients of larger nuclei, these systems are described as superpositions of 0s harmonic-oscillator (h.o.) Slater determinants of different size parameters, made translation-invariant by omission of the centre-of-mass factors. The sets of size parameters have been chosen by energy minimization. For the individual clusters this model produces the g.s.'s and some breathing modes built upon the g.s.'s. For the deuteron g.s. the wave function provided may be considered numerically exact.

The nucleus ^{6}Li is assumed to consist of a superposition of α+d-type two-cluster states, which contain all α and deuteron states generated by the model for the individual clusters. By including excited cluster states, this model allows for a major part of the distortion that the clusters suffer in each other's field. Being almost exact in the p+n relative motion, the breathing model for ^{6}Li may be considered a sort of microscopic α+p+n model. However, short of non-central interaction terms, an α+d model results in pure p–n and α–(pn) relative orbital momentum and summed nucleon spin values of $L_{pn} = L_{\alpha d} = 0$ and $S = 1$. The ^{5}He system is described as a superposition of α+n and t+d *continua*. In addition to cluster distortions, this model allows for the coupling of these two overlapping clusterization channels. In the interaction region for each pair of cluster states the intercluster motion is expanded in terms of angular-momentum projected shifted Gaussians. For scattering solutions this is matched with relative asymptotic waves [18].

The force parameters were adjusted so that all cluster and two-cluster binding energies as well as cluster sizes be exact [8], and all resonance energies and (s- and p-wave) phase shifts available be in agreement with the data [18]. With no spin-orbit force involved, the latter requirement could only be satisfied by adjusting the force for the partial waves $p_{1/2}$ and $p_{3/2}$ independently.

Although none of the three-particle models has been derived from the nucleonic level, both are claimed to take into account the Pauli principle (by an effective repulsion or projection). The potentials used by Kukulin *et al.* are local, which is in accord with the microscopically constructed cluster–cluster interactions [15]. Lehman's use of non-local separable interactions, for the sake of the Faddeev formalism, is nevertheless not likely to make an essential difference. None of the three-particle models takes into account the distortability of the α particle, but that is not a serious shortcoming either. Both reproduce the subsystem properties and the α+p+n binding energy satisfactorily, apart from the neglect of the Coulomb potential enforced by the Faddeev approach on Lehman's calculations. Unlike in the microscopic model, in these models there is practically no angular-momentum truncation. As to the asymptotic region, the Faddeev approach is exact, while the variational wave function of Kukulin *et al.*, being of the form of α+(pn), behaves like the microscopic wave function.

Since in these models the components with $L_{pn} \neq 0$, $L_{\alpha d} \neq 0$ and $S \neq 1$, which are in excess of those included in the microscopic model, add up to at most a few per cent, one may duly expect very small deviations from the microscopic model in their predictions as well. The agreement in their reproduction of the electromagnetic properties is indeed remarkable [19].

3.2 Results

The quantities that we calculate are those extractable from quasifree knockout data. By detecting the reaction products e' and B_1 of an ${}^6\text{Li}(e, e'B_1)B_2$ reaction, the missing energy E_m and missing momentum $\mathbf{p}_m$ can be determined. The missing energy is the difference between the final and initial intrinsic energies, $E_m \equiv E_{B_1} + E_{B_2} - E_{{}^6\text{Li}}$, whereas the missing momentum is the relative B_1–B_2 momentum in the initial state. The missing-energy and missing-momentum dependence of the cross section is given essentially by the spectral function $S(E_m, \mathbf{p}_m)$, which carries all information on the B_1–B_2 relative motion in ${}^6\text{Li}$.

An energy integral of $S(E_m, \mathbf{p}_m)$ gives a momentum distribution, in the final state(s) covered by the integration, which differs from (11) only in wave-distortion effects ["distorted momentum distribution", $\rho(\mathbf{p}_m)$]. The momentum integral of $S(E_m, \mathbf{p}_m)$ gives a distribution of two-cluster states of energy $E_{B_1} + E_{B_2} = E_m + E_{{}^6\text{Li}}$ present in the g.s. of ${}^6\text{Li}$ ["energy distribution", $n(E_m)$]. In the distorted-wave impulse approximation (DWIA) these can be calculated from the spectroscopic amplitudes (9), and from the ratio of the experimental and DWIA "distorted momentum distributions" one can extract experimental spectroscopic factors.

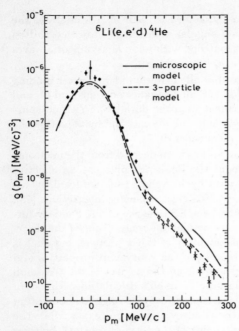

Fig. 1. Distorted α+d "momentum distribution" in the g.s. of ^{6}Li calculated [20] from the microscopic model and from the three-particle model of Lehman *et al.* together with the result of the ^{6}Li(e, e'd)α(g.s.) experiment [21]

Table 1. Cluster-model and experimental ^{5}He+p spectroscopic factors

E_m Region (MeV)	Peak 3.7–5.7	Off-peak 5.7–19.7	Peak 21–22	Off-peak 22–27
Microscopic	0.588	0.648	0.357	0.480
Macroscopic		0.87[a]		
Experiment	0.44±0.05	0.32±0.04	0.30±0.04	0.53±0.07

[a] E_m region: 3.7–19.7 MeV.

The α+d "distorted momentum distributions" calculated from our microscopic model and from the three-particle model of Lehman *et al.* are compared with that deduced from the ^{6}Li(e, e'd)α(g.s.) experiment [21] in Fig. 1. Both models appear to be satisfactory, though we note that in the small-momentum region, where the theoretically uncertain wave-distortion effects are small, the microscopic model is superior. A substantial discrepancy becomes only apparent in the spectroscopic factors: the microscopic model gives 0.93, while the macroscopic model gives 0.632.

The reaction ^{6}Li(e, e'p)^{5}He [22,23] leads to continuum final states punctuated by the two prominent resonances of summed nucleon spins, total orbital momenta and total angular momenta $(S_{^5\text{He}}, L_{^5\text{He}}, J_{^5\text{He}}) = (\frac{1}{2}, 1, \frac{3}{2})$ and $(\frac{3}{2}, 0, \frac{3}{2})$. The "energy

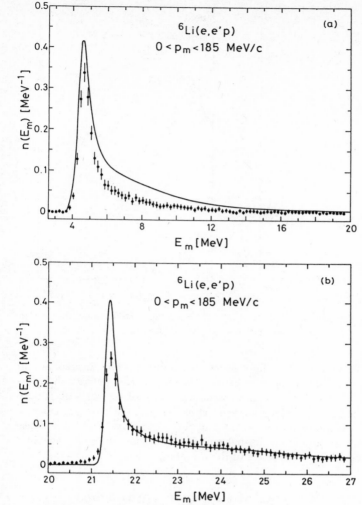

Fig. 2. Experimental [22,23] and microscopic [18,23] ^{5}He–p "energy distributions" below (a) and above (b) the t+d threshold in the g.s. of ^{6}Li

distribution" calculated[1] in the microscopic model [18] is shown together with the experimental data in Fig. 2. While the overall agreement is remarkable, there is a conspicuous discrepancy beyond the first resonance. Detailed analysis [18] shows that the contribution of the partial wave $(\frac{1}{2},1,\frac{1}{2})$ is overwhelming in that region, thus pointing to the probable location of the error [18].

The corresponding "distorted momentum distributions" and spectroscopic factors shown in Fig. 3 and Table 1, respectively, reproduce the experiment with an

---

[1] The DWIA calculations for the reaction ^{6}Li(e, e'p)^{5}He were performed by J. Lanen [18,22,23].

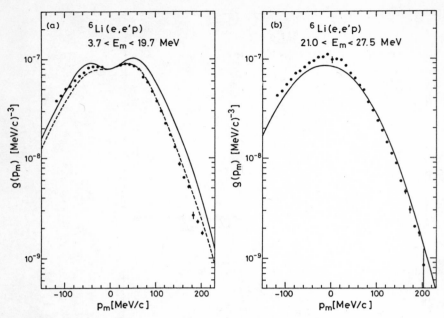

Fig. 3. Experimental [22,23] and theoretical [18,23] ^{5}He–p "distorted momentum distributions" below (a) and above (b) the t+d threshold in the g.s. of ^{6}Li

accuracy camparable with that of the "energy distribution". The discrepancy in Fig. 3a for large momenta corresponds to the one seen in the "energy distribution". In Fig. 3a the prediction of the macroscopic model is also displayed. It does agree with experiment in this region as well. Note that the macroscopic calculation is not extended to the region above the t+d threshold (Figs. 2b, 3b) because in that region the structure of the five-nucleon system is dominated by the t+d configuration, which is out of the scope of the α+p+n model.

4 More about the Relation of the Two Approaches

For being as similar as they are in essentials, the macroscopic and microscopic models seem to differ in their predictions too much. Since the discrepancy appears to be restricted to the fragmentation properties, which hinge on the microscopic and macroscopic amplitudes (9a) and (9b), one feels prompted to re-examine the relation of these two functions.

We shall now cast the α+d and ^{5}He+p microscopic amplitudes $g_{\alpha d}(\mathbf{R})$ and $g_{^5\mathrm{Hep}}(\mathbf{R})$ into a form in which this relation is manifest. Adapting Eq. (1b) to $\Psi = \Psi_{^6\mathrm{Li}}$ and $\Psi = \Psi_{^5\mathrm{He}}$, we have

$$\Psi_{^6\mathrm{Li}} = \int d\mathbf{r} \int d\mathbf{R}\, \varphi_{\alpha\mathrm{pn}}(\mathbf{r},\mathbf{R})\Psi_{\mathbf{rR}} = \int d\mathbf{r} \int d\mathbf{R}\, \varphi'_{\alpha\mathrm{pn}}(\mathbf{r},\mathbf{R})\Psi'_{\mathbf{rR}}, \quad (12\mathrm{a})$$

$$\Psi_{^5\mathrm{He}} = \int d\mathbf{R}\, \varphi_{\alpha\mathrm{n}}(\mathbf{R})\Psi_{\mathbf{R}}, \quad (12\mathrm{b})$$

with

$$\Psi_{\mathbf{rR}} = \mathcal{A}_{\alpha pn}\{\Phi_\alpha \delta(\mathbf{r}-\mathbf{r}_{pn})\delta(\mathbf{R}-\mathbf{r}_{\alpha d})\},$$

$$\Psi'_{\mathbf{rR}} = \mathcal{A}_{\alpha pn}\{\Phi_\alpha \delta(\mathbf{r}-\mathbf{r}_{\alpha n})\delta(\mathbf{R}-\mathbf{r}_{^5Hep})\}, \tag{13a}$$

$$\Psi_{\mathbf{R}} = \mathcal{A}_{\alpha n}\{\Phi_\alpha \delta(\mathbf{R}-\mathbf{r}_{\alpha n})\}. \tag{13b}$$

Following Eq. (5b), with $\Psi_{\mathbf{rR}}$, $\Psi'_{\mathbf{rR}}$ and $\Psi_{\mathbf{R}}$ we can define the norm kernels

$$N_{\alpha pn}(\mathbf{r},\mathbf{R};\mathbf{r}',\mathbf{R}') = \langle\Psi_{\mathbf{rR}}|\Psi_{\mathbf{r'R'}}\rangle, \qquad N'_{\alpha pn}(\mathbf{r},\mathbf{R};\mathbf{r}',\mathbf{R}') = \langle\Psi'_{\mathbf{rR}}|\Psi'_{\mathbf{r'R'}}\rangle, \tag{14a}$$

$$N_{\alpha n}(\mathbf{R};\mathbf{R}') = \langle\Psi_{\mathbf{R}}|\Psi_{\mathbf{R'}}\rangle, \tag{14b}$$

and then the corresponding norm operators $\mathcal{N}_{\alpha pn}$, $\mathcal{N}'_{\alpha pn}$ and $\mathcal{N}_{\alpha n}$. As is seen from (13a), the parameter coordinates in $N_{\alpha pn}$ and $N'_{\alpha pn}$ are related to each other as the physical coordinates $\{\mathbf{r}_{pn},\mathbf{r}_{\alpha d}\}$ are related to $\{\mathbf{r}_{\alpha n},\mathbf{r}_{^5Hep}\}$. The norm operators can be substituted in (9a) adapted to $1+2=\alpha+d$:

$$g_{\alpha d}(\mathbf{R}) = \int d\mathbf{r}\int d\mathbf{r}'\int d\mathbf{R}'\ \Phi_d^*(\mathbf{r})\langle\mathcal{A}_{\alpha d}\mathcal{A}_{pn}\{\Phi_\alpha\delta(\mathbf{r}-\mathbf{r}_{pn})\delta(\mathbf{R}-\mathbf{r}_{\alpha d})\}$$

$$\times|\mathcal{A}_{\alpha pn}\{\Phi_\alpha\delta(\mathbf{r}'-\mathbf{r}_{pn})\delta(\mathbf{R}'-\mathbf{r}_{\alpha d})\}\rangle\varphi_{\alpha pn}(\mathbf{r}',\mathbf{R}') \tag{15a}$$

$$= \int d\mathbf{r}\int d\mathbf{r}'\int d\mathbf{R}'\ \Phi_d^*(\mathbf{r})N_{\alpha pn}(\mathbf{r},\mathbf{R};\mathbf{r}',\mathbf{R}')\varphi_{\alpha pn}(\mathbf{r}',\mathbf{R}') \tag{15b}$$

$$= \int d\mathbf{r}\ \Phi_d^*(\mathbf{r})\mathcal{N}_{\alpha pn}\varphi_{\alpha pn}(\mathbf{r},\mathbf{R}) \tag{15c}$$

$$= \int d\mathbf{r}\ \chi_{pn}^*(\mathbf{r})\mathcal{N}_{\alpha pn}^{1/2}\chi_{\alpha pn}(\mathbf{r},\mathbf{R}). \tag{15d}$$

In the second step we used $\mathcal{A}_{\alpha d}\mathcal{A}_{pn} = \mathcal{A}_{\alpha pn}$, and in the forth step we used $\Phi_d = \chi_{pn}$ and the inverted form of (6a): $\varphi_{\alpha pn} = \mathcal{N}_{\alpha pn}^{-1/2}\chi_{\alpha pn}$.

Similarly, for $1+2=^5$He$+$p, Eq. (9a) can be reformulated as

$$g_{^5Hep}(\mathbf{R}) = \int d\mathbf{r}\int d\mathbf{r}'\int d\mathbf{R}'\ \varphi_{\alpha n}^*(\mathbf{r})N'_{\alpha pn}(\mathbf{r},\mathbf{R};\mathbf{r}',\mathbf{R}')\varphi'_{\alpha pn}(\mathbf{r}',\mathbf{R}') \tag{16a}$$

$$= \int d\mathbf{r}\ \varphi_{\alpha n}^*(\mathbf{r})\mathcal{N}'_{\alpha pn}\varphi'_{\alpha pn}(\mathbf{r},\mathbf{R}) \tag{16b}$$

$$= \int d\mathbf{r}\ \left[\mathcal{N}_{\alpha n}^{-1/2}\chi_{\alpha n}(\mathbf{r})\right]^*\left[\mathcal{N}'^{1/2}_{\alpha pn}\chi'_{\alpha pn}(\mathbf{r},\mathbf{R})\right], \tag{16c}$$

where $\chi_{\alpha n}\equiv\mathcal{N}_{\alpha n}^{1/2}\varphi_{\alpha n}$ and $\chi'_{\alpha pn}\equiv\mathcal{N}'^{1/2}_{\alpha pn}\varphi'_{\alpha pn}$.

In Eqs. (15d) and (16c) the spectroscopic amplitudes are expressed in terms of functions χ identifiable with the relative wave functions of macroscopic models. We see that these formulae differ from Eq. (9b) in $\mathcal{N}_{\alpha pn}^{1/2}$ and in $\mathcal{N}_{\alpha n}^{-1/2}\mathcal{N}'^{1/2}_{\alpha pn}$, respectively. [Note that in (15d) no operator corresponding to $\mathcal{N}_{\alpha n}^{-1/2}$ appears because p and n are treated as elementary particles; at this level $\mathcal{N}_{pn}=1$.] These operators carry information on the cluster internal structure, and they arise because of the Pauli principle.

We are thus forced to assert that *the spectroscopic amplitudes require explicit knowledge of the cluster internal wave functions, and thus the conventional formula, (9b), is, in principle, incorrect.* The quality of (9b) as an approximation depends on the behaviour of the norm operators. The effect of the norm operators

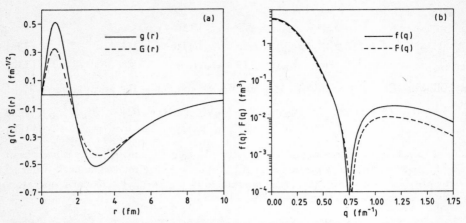

Fig. 4. (a) Radial $\alpha+$d spectroscopic amplitudes $g(r)$ (solid line) and $G(r)$ (dashed line) with $L = 0$ and (b) the corresponding radial "momentum distributions" $f(q)$ (solid line) and $F(q)$ (dashed line) with the $\alpha+$p$+$n and deuteron wave functions of Lehman *et al.* [16]

can be quantified through their eigenvalues. We have found [24] that these are not sensitive to the details of the cluster internal wave functions, and a pure h.o. model with equal cluster sizes gives a good enough estimate. In this limit the eigenfunctions of $\mathcal{N}$'s are h.o. functions [25]. The eigenfunctions belonging to eigenvalues zero span the subspace excluded by the Pauli principle. For the macroscopic model to be sensible, the wave functions χ must be out of this subspace. The additional effect of a norm operator is thus best represented by the the departure of the first non-zero eigenvalue from unity. For $\mathcal{N}_{\alpha pn}$ (and for $\mathcal{N}'_{\alpha pn}$, whose spectrum is the same) this is $\frac{13}{8}$, while for $\mathcal{N}_{\alpha n}$ this is $\frac{5}{4}$ [13]. An enhancement of $\frac{13}{8}$ and $\frac{4}{5}\frac{13}{8} = \frac{13}{10}$ may thus be expected for $\alpha+$d and ^{5}He$+$p, respectively, with respect to the macroscopic spectroscopic factors S.

In Fig. 4a we show the $L = 0$ $\alpha+$d radial spectroscopic amplitude $G(r)$ of Lehman *et al.* and the one, $g(r)$, calculated from the same wave functions via (15d). In Fig. 4b the corresponding (undistorted) radial "momentum distributions" defined in (11) are seen. In Table 2 the spectroscopic factors are collected, together with those calculated with the wave function of Kukulin *et al.* [17].

We see that, as regards the $\alpha+$d fragmentation, the two models are fully reconciled. Not only the $\alpha+$d spectroscopic factors come close to the microscopic estimate (0.93), but also the corresponding amplitudes. The remaining differences could be presumably accounted for by differences in details. The agreement with experiment seems also improved.

For the ^{5}He$+$p fragmentation another estimate of the effect of $\mathcal{N}_{\alpha n}^{-1/2}\mathcal{N}_{\alpha pn}'^{1/2}$ can be obtained from the fact that the microscopic sum rule limit of the removal of a proton from outside the α core of ^{6}Li is 6/5 [22], while the conventional macroscopic picture gives unity [16]. The departure from unity in the microscopic picture is due just to the norm operators in (16). Thus $\mathcal{N}_{\alpha n}^{-1/2}\mathcal{N}_{\alpha pn}'^{1/2}$ causes a 20% enhancement, on an avarage, in the proton spectroscopic factor. Such an increase would also help to bring the predictions of the macroscopic model closer to those of the microscopic

Table 2. Alpha + deuteron spectroscopic factors

Source of $\Psi_{6_{Li}}$	$S_{\alpha d}$	$s_{\alpha d}$
Lehman *et al.* [16]	0.632	0.847
Kukulin *et al.* [17]	0.738	0.970

one. In the region of the peak belonging to the g.s. of ^{5}He this correction could bring the two models in full harmony, with the experiment slightly overshot. But in the smooth continuum beyond this peak a large part of the difference will survive. Nevertheless, $\mathcal{N}_{\alpha n}^{-1/2}\mathcal{N}_{\alpha pn}'^{1/2}$ may still improve the performance of the macroscopic model here, while slightly reducing its disagreement with the microscopic model.

The inferiority of the microscopic to the macroscopic model in reproducing the experimental data here is probably caused by the missing of the $L_{pn} \neq 0$, $L_{\alpha d} \neq 0$ and $S \neq 1$) components. Since the ^{5}He+p fragmentation does contain such components, it is not surprising that their lack in the ^{6}Li wave function causes such errors. The inclusion of such components could reduce the transition by destructive interference with the $L_{pn} = 0$, $L_{\alpha d} = 0$ and $S = 1$ term [18].

5 Conclusion

The comparison of the microscopic and macroscopic models for the fragmentation of ^{6}Li has shed light on hidden aspects of both models.

We have seen that the small $L_{pn} \neq 0$, $L_{\alpha d} \neq 0$ and $S \neq 1$ components of the α+p+n relative motion neglected in the microscopic description are needed to reproduce the proton removal experiment quantitatively. These components are usually ignored because the treatment of the nucleon–nucleon tensor and spin-orbit forces in a microscopic framework for multicluster dynamics is fairly complicated. With their role recognized, these improvements are now under way.

We have come across two breakdowns of the macroscopic model as well. The first one is trivial: a macroscopic α+p+n model cannot describe a (td)+p fragmentation. The second beakdown, however, is rather surprising: the cluster substructure has been found to show up in nuclear processes that involve the splitting of the nucleus into clusters. The α+d spectroscopic factor of ^{6}Li has been shown to be enhanced by more than 30% owing to the non-elementary nature of the α particle. This enhancement improves the agreement of the predictions of the α+p+n models with experiment and restores the accord with the results of the microscopic models. The effect is caused by the Pauli principle. This particular Pauli effect is bound to appear as an extra deviation from the macroscopic models however perfectly the Pauli principle is simulated in the formulation of the macroscopic dynamics.

We have demonstrated that this second breakdown can be cured by the use of the norm operators. The method is, however, more general. The correspondence of the macroscopic wave function $\chi_{\alpha pn}$ to the relative-motion function $\varphi_{\alpha pn}$ appearing in the microscopic model *behind the intercluster antisymmetrizer* is given by $\chi_{\alpha pn} = \mathcal{N}_{\alpha pn}^{1/2}\varphi_{\alpha pn}$. By inverting this relationship, we can always obtain, from the macroscopic wave function χ, a function φ, to be used in formulae of the microscopic approach. E.g. substituting $\varphi_{\alpha\text{pn}} = \mathcal{N}_{\alpha\text{pn}}^{-1/2}\chi_{\alpha\text{pn}}$ in Eq. (12a), the α+p+n three-particle model could be used to describe the (td)+p fragmentation just as well as the microscopic ^{6}Li wave function, whereby the first breakdown is also cured.

From the example of the α+p+n system one can conjecture that deviations should be observable from macroscopic descriptions of cluster removal from (or addition to) other multicluster systems as well. For that matter, such effects should also appear one level lower as a signature of nucleon structure. For instance, the spectroscopic factor of proton removal from triton should also differ from unity, the value obtained with the nucleons treated as structureless.

References

1. P.H. Wackman, N. Austern: Nucl. Phys. **30** 529 (1962)
2. B. Buck (1985): In *Proc. 4th Int. Conf. on Clustering Aspects of Nuclear Structure and Nuclear Reactions*, ed. by J.S. Lilley, M.A. Nagarajan (Reidel, Dordrecht) p. 71;
   H. Walliser, T. Fliessbach: Phys. Rev. C **31** 2242 (1985)
3. A. Săndulescu, E. W. Schmid, G. Spitz: Few-Body Syst. **5** 107 (1988)
4. B. Buck, C.B. Dover, J.P. Vary: Phys. Rev. C **11** 1803 (1975)
5. K.F. Pál, R.G. Lovas: Phys. Lett. **96B** 19 (1980);
   F. Michel, G. Reidemeister, S. Ohkubo: Phys. Rev. Lett. **57** 1215 (1986);
   T. Yamaya, S. Oh-ami, M. Fujiwara, T. Itahashi, K. Katori, M. Tosaki, S. Kato, S. Hatori, S. Ohkubo: Phys. Rev. C **42** 1935 (1990)
6. K. Ikeda, N. Takigawa, H. Horiuchi: Suppl. Prog. Theor. Phys. **Extra Number** 464 (1968)
7. R. Beck, F. Dickmann, A.T. Kruppa: Phys. Rev. C **30** 1044 (1984)
8. R.G. Lovas, A.T. Kruppa, R. Beck, F. Dickmann: Nucl. Phys. **A474** 451 (1987)
9. R. Beck, F. Dickmann, R.G. Lovas: Ann. Phys. (N.Y.) **173** 1 (1987)
10. B. Buck, H. Friedrich, C. Wheatley: Nucl. Phys. **A275** 246 (1977)
11. E. W. Schmid, A. Faessler, H. Ito, G. Spitz: Few-Body Syst. **5** 45 (1988)
12. E. W. Schmid: Z. Phys. A **302** 311 (1981);
    S. Nakaichi-Maeda, E. W. Schmid: Z. Phys. A **318** 171 (1984)
13. K. Varga, R.G. Lovas: Phys. Rev. C **43** 1201 (1991)
14. J. Lomnitz-Adler (1978): University of Illinois (Urbana-Champaign) Report ILL–(TH)–78–58
15. K. Langanke, H. Friedrich (1986): In *Advances in Nuclear Physics, Vol. 17*, ed. by J.W. Negele, E. Vogt (Plenum, New York) p. 223
16. D.R. Lehman, Mamta Rai, A. Ghovanlou: Phys. Rev. C **17** 744 (1978);
    D.R. Lehman, Mamta Rajan: Phys. Rev. C **25** 2743 (1982);
    C.T. Christou, D.R. Lehman, W.C. Parke: Phys. Rev. C **37** 445 (1988)
17. V.I. Kukulin, V.M. Krasnopol'sky, V.T. Voronchev, P.B. Sazonov: Nucl. Phys. **A417** 128 (1984);
    V.I. Kukulin, V.M. Krasnopol'sky, V.T. Voronchev, P.B. Sazonov: Nucl. Phys. **A453** 365 (1986)
18. R.G. Lovas, A.T. Kruppa, J.B.J.M. Lanen: Nucl. Phys. **A516** 325 (1990)
19. A.T. Kruppa, R. Beck, F. Dickmann: Phys. Rev. **C36** 327 (1987);
    A. Eskandarian, D.R. Lehman, W.C. Parke: Phys. Rev. C **38** 2341 (1988);
    V.I. Kukulin, V.T. Voronchev, T.D. Kaipov, R.A. Eramzhyan: Nucl. Phys. **A517** 221 (1990)

20. R. Ent, L. Lapikás (1988): private communication
21. R. Ent, H.P. Blok, J.F.A. van Hienen, G. van der Steenhoven, J.F.J. van den Brand, J.W.A. den Herder, E. Jans, P.H.M. Keizer, L. Lapikás, E.N.M. Quint, P.K.A. de Witt Huberts, B.L. Berman, W.J. Briscoe, C.T. Christou, D.R. Lehman, B.E. Norum, A. Saha: Phys. Rev. Lett. **57** 2367 (1986)
22. J.B.J.M. Lanen, A.M. van den Berg, H.P. Blok, J.F.J. van den Brand, C.T. Christou, R. Ent, A.G.M. van Hees, E. Jans, G.J. Kramer, L. Lapikás, D.R. Lehman, W.C. Parke, E.N.M. Quint, G. van der Steenhoven, P.K.A. de Witt Huberts: Phys. Rev. Lett. **62** 2925 (1989)
23. J.B.J.M. Lanen, R.G. Lovas, A.T. Kruppa, H.P. Blok, J.F.J. van den Brand, R. Ent, E. Jans, G.J. Kramer, L. Lapikás, E.N.M. Quint, G. van der Steenhoven, P.C. Tiemeijer, P.K.A. de Witt Huberts: Phys. Rev. Lett. **63** 2793 (1989)
24. K. Varga, R.G. Lovas: Phys. Rev. **C 37** 2906 (1988)
25. H. Horiuchi: Prog. Theor. Phys. Suppl. **62** 90 (1977)

An Operator Interpolation Method for the Description of Intercluster Dynamics

V.B. Soubbotin

Leningrad State University, St. Petersburg, Russia

A method to describe intercluster dynamics in a "background" of the potential scattering is proposed. For the two interacting clusters, the main idea and results are as follows: The self-adjoint operator H which acts in a Hilbert space $\mathcal{H}$ can be presented as

$$H = H^0 + W \tag{1}$$

where H^0 describes the "background structureless particles" interaction, while the perturbation W arises from the composite nature of the clusters. Let us consider the many particles Hamiltonian H_M and diagonalize it on a finite dimentional subspace of a many body states $\mathcal{J}_M$. It is a typical nuclear structure problem. The operator $\Gamma_M H_M \Gamma_M$ (Γ_M is a projector onto $\mathcal{J}_M$) appears in an "external" channel of relative motion like a $\Gamma H \Gamma$, where Γ -projector, projects the states from $\mathcal{H}$ to subspace $\mathcal{J}$ corresponding $\mathcal{J}_M$ in the "external" channel. Here we refer to the many particles states, which are not orthogonal to the elastic scattering channel (all other channels are closed) and do not take into account the states belonging to the closed channels. The extention of the method to include finite dimensional subspace belonging to the closed channel is straighforward. Because, only the finite number of states $\Psi^i_M \in \mathcal{J}_M$ are considered (and really give some contribution) the contribution of W to the scattering amplitude at high enough energy will be negligible. It is shown that the operator H which satisfies these requirements has the form

$$H = H^0 + \sum_i H^0 \mid u_i > \beta_i < u_i \mid H^0 \tag{2}$$

where u_i is a projection of Ψ^i_M to "external" channel of relative motion and $\{\beta\}=\beta_1,\beta_2,...$ are determined by the solution of a system of q coupled algebraic equations

$$\det\{H^0_{kl} + \sum_m H^0_{km}\beta_m H^0_{ml} - \delta_{kl}E_i\} = 0 \tag{3}$$

here $H^0_{kl} \equiv < u_i \mid H^0 \mid u_l >$, q is the number of different eigenvalues E_i of operator $\Gamma H \Gamma$,which are used as an input data. The contributions from finite dimensional subspace of the closed channels will lead to the similar results, but, in this case, will

Springer Series in Nuclear and Particle Physics **Clustering Phenomena in Atoms and Nuclei**
Editors: M. Brenner · T. Lönnroth · F.B. Malik   © Springer-Verlag Berlin, Heidelberg 1992

be energy dependent. Nevertheless our method gurantees that if $\{\beta\}$ is real, the dynamics in a wider space $\mathcal{H}' = \mathcal{H} \oplus \mathcal{J}'$ ($\mathcal{J}'$ corresponds to the contribution from the cloused channels) will be self-adjoint. The two body scattering amplitude has a form

$$f = f_0 + \Delta f \tag{4}$$

where f_0 is a scattering amplitude associated with H^0 and

$$\Delta f^{\{\lambda\}} \sim -\mu E^2 \sum_{ij} < \varphi_0^- \mid u_i^{\lambda_i} > \beta_i A_{ij}^{-1} \beta_j < u_j^{\lambda_j} \mid \varphi_0^+ > \tag{5}$$

Here $\varphi_0^\pm$ are the solutions of the Schrödinger equation, involving H^0 with appropriate boundary conditions for scattering. The matrix elements of A are given by

$$A_{ij} = \beta_i \delta_{ij} + z \beta_i \delta_{ij} \beta_j + z^2 \beta_i g_{ij}(z) \beta_j + \beta_i H_{ij}^0 \beta_j \tag{6}$$

where $g(z) = (H^0 - z)^{-1}$. The additional poles of S-matrix being solutions of the following dispersion equation can appear at $E < 0$ and on the non physical sheet as a resonances

$$\det\{A_{ij}\} = 0 \tag{7}$$

The index λ_i corresponds to the dynamical symmetry group of Ψ_M^{i,λ_i}, and Δf can be specified by multiple index $\{\lambda\} = \lambda_1, \lambda_2, \ldots$. In applications to the three body problem OIM may include the specific three body correlation term.

Pauli-Forbidden States and the Levinson Theorem in $\alpha + \alpha$ Scattering

D. Baye and M. Kruglanski

Physique Nucléaire Théorique et Physique Mathématique, CP 229,
Université Libre de Bruxelles, B-1050 Brussels, Belgium

The scattering of two α particles is reinvestigated in a microscopic cluster model with distorted internal wave functions. The wave function of each α particle is represented by a linear combination of the internal parts of $(0s)^4$ harmonic oscillator Slater determinants with different parameters (four in the present case). Removing spurious c.m. components from the scattering wave function requires a double integral transformation [1]. The scattering asymptotic behaviour is obtained with the microscopic R-matrix method [2].

Elastic $\alpha + \alpha$ scattering is studied with both α in their ground state only. Phase shifts for $\ell = 0$ to 4 are displayed in fig.1 for the Minnesota interaction [4]. Whereas each phase shift presents a physical resonance at low energies, other resonances appear at higher energies: 68.5 and 220 MeV for $\ell = 0$ and 74.4 MeV for $\ell = 2$. These Pauli resonances are related to the states forbidden by antisymmetrization, encountered in the two-cluster model when a single oscillator parameter is employed.

The existence of the Pauli resonances explains why simple realistic potentials for $\alpha + \alpha$ scattering satisfy the generalized Levinson theorem [4] $\delta_\ell(0) - \delta_\ell(\infty) = m_\ell \pi$ where m_ℓ is the number of forbidden states. After removing the Pauli resonances, the phase shifts resemble those obtained with a single oscillator parameter. The two-centre model with a single oscillator parameter provides a Levinson theorem which remains physically valid in more realistic models.

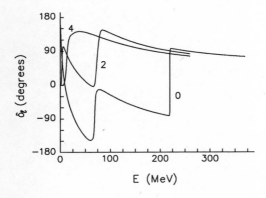

1) M. Hanck, Nucl. Phys. A **439** (1985) 1
2) D. Baye, P.-H. Heenen and M. Libert-Heinemann, Nucl. Phys. A **291** (1977) 230
3) D.R. Thompson, M. LeMere and Y.C. Tang, Nucl. Phys. A **286** (1977) 83
4) P. Swan, Proc. Roy. Soc. A **228** (1955) 10

Springer Series in Nuclear and Particle Physics Clustering Phenomena in Atoms and Nuclei
Editors: M. Brenner · T. Lönnroth · F.B. Malik   © Springer-Verlag Berlin, Heidelberg 1992

Construction of Permutational Symmetry Adapted Non-Spurious States for Multi-Cluster Systems

J. Katriel[1] and A. Novoselsky[2]

[1]Department of Chemistry, Technion, Israel Institute
of Technology, 32000 Haifa, Israel
[2]The Racah Institute of Physics, The Hebrew University
of Jerusalem, 91104 Jerusalem, Israel

An algorithm for the construction of multi-cluster non-spurious harmonic oscillator wavefunctions with arbitrary permutational symmetry is developed. Each cluster is presented in terms of harmonic oscillator wavefunctions, expressed in an appropriate set of Jacobi coordinates. The cluster wavefunctions are coupled into non-spurious multi-cluster states with an overall well-defined angular momentum and permutational symmetry. This is achieved by diagonalizing an appropriate set of single-cycle class operators of the symmetric group involving all the constituent particles. The results are immediately applicable to the study of nuclear cluster models, nuclei as quark clusters, nuclear fusion, fission and multifragmentation, as well as in atomic and molecular physics.

I. Introduction

The treatment of a finite system of identical particles within non-relativistic quantum mechanics starts with the construction of a many-particle basis set with appropriate symmetry properties. We shall deal with spherically symmetrical systems, for which the wavefunctions are characterized by their total angular momentum as well as by their behaviour with respect to the symmetric group. There may be additional symmetries, such as the symplectic symmetry which gives rise to the seniority scheme, which we shall not discuss explicitly; they can always be incorporated in order to take care of whatever degeneracies remain after the classification in terms of the symmetric group and the total angular momentum has been carried out.

An important distinction is the one between the system of electrons in an atom on the one hand and the system of nucleons in a nucleus, on the other hand. In the former case a natural origin exists, namely the atomic nucleus which is much heavier than the system of electrons anchored to it. In the latter case the internal motions should be referred to the center-of-mass (c.m.) of the particles themselves.

An entirely analogous situation occurs in the van-der-Walls clusters of identical rare-gas atoms or in alkali-metal clusters.

One of the technical difficulties that have to be faced in the study of systems of the second kind involves the construction of a basis set in which the spurious c.m. motion is excluded. The common procedures involve the construction of a basis set contaminated by spurious states, which have than to be eliminated. This is usually done by adding an appropriate interaction to the hamiltonian, which pushes up the spurious states in such a way that they do not affect the eventual calculations[1]. However, these states are still present, yielding large, inefficient basis sets.

A more serious difficulty is encountered in the calculation of spectroscopic factors in nuclear reactions[2]-[3]. These involve non-spurious two-cluster wavefunctions. Several methods have been proposed to eliminate two-cluster spurious states, but each one of them is only applicable within a small class of special cases[1].

In the present talk we propose the framework and the systematic notation for the actual treatment of a system consisting of several, not necessarily identical, clusters of indistinguishable particles. We obtain an overall state which is at the same time non-spurious, has a well defined total angular momentum, and is well behaved with respect to the cluster as well as overall permutational symmetry. Some technical details which can be inferred from the treatment of multi-cluster states in arbitrary basis sets[4] and from the treatment of the non-spurious single-cluster[5]-[6] and two-cluster[7] states, are omitted. The only technical ingredients required in the present context, which were not discussed in refs. [4]-[7] are the symmetric-group recoupling transformation and the cluster-interchange transformation. The former is a well-known procedure[8], and the latter will be briefly discussed.

II. The intracluster and intercluster coordinates

We consider a system of N identical particles distributed in k clusters. Let n_i be the number of particles in the i'th cluster and

$$N_i = \sum_{j=1}^{i} n_j \tag{1}$$

the total number of particles in the first i clusters. Obviously, $N_k = N$. We shall denote the set of indices corresponding to the n_i particles in the i'th cluster by $\{1, 2, \ldots, n_i\}^{(i)}$ or by (i). The particle coordinates in the i'th cluster are $\vec{r}_1{}^{(i)}, \vec{r}_2{}^{(i)}, \ldots, \vec{r}_{n_i}{}^{(i)}$. Within each cluster we define a set of Jacobi coordinates

$$\vec{\rho}_{\{j\}^{(i)},[j-1]^{(i)}} = \sqrt{\frac{j-1}{j}} \left[\vec{r}_j{}^{(i)} - \frac{1}{j-1} \sum_{m=1}^{j-1} \vec{r}_m{}^{(i)} \right] \quad i = 1, 2, \ldots, k \;\; j = 2, 3, \ldots, n_i \tag{2}$$

where $[j-1]^{(i)}$ is the set consisting of the first $j-1$ particles in cluster i. This set of $n_i - 1$ internal Jacobi coordinates of the i'th cluster will be denoted collectively by $\vec{\rho}_{(i)}$. Thus,

$$\vec{\rho}_{(i)} = \left\{ \vec{\rho}_{\{2\}^{(i)},[1]^{(i)}}, \ \vec{\rho}_{\{3\}^{(i)},[2]^{(i)}}, \ldots, \ \vec{\rho}_{\{n_i\}^{(i)},[n_i-1]^{(i)}} \right\} \tag{3}$$

Furthermore, we introduce the intercluster Jacobi coordinates

$$\vec{\rho}_{(i),[i-1]} = \sqrt{\frac{n_i \cdot N_{i-1}}{N_i}} \left[\frac{1}{n_i} \sum_{j=1}^{n_i} \vec{r}_j{}^{(i)} - \frac{1}{N_{i-1}} \sum_{j=1}^{i-1} \sum_{l=1}^{n_j} \vec{r}_l{}^{(j)} \right] \tag{4}$$

where $[i-1] = (1) \cup (2) \cup \ldots \cup (i-1)$ is the set of particle indices in the first $i-1$ clusters.

The set of internal coordinates for the first i clusters, including the appropriate intercluster coordinates, will be denoted by $\vec{\rho}_{[i]}$. These coordinates are conveniently defined recursively, starting from

$$\vec{\rho}_{[2]} = \left\{ \vec{\rho}_{(1)}, \vec{\rho}_{(2)}, \vec{\rho}_{(2),(1)} \right\} \tag{5}$$

and noting that

$$\vec{\rho}_{[i]} = \left\{ \vec{\rho}_{[i-1]}, \vec{\rho}_{(i)}, \vec{\rho}_{(i),[i-1]} \right\} \qquad i = 2, 3, \ldots, k \tag{6}$$

Finally, the system of clusters as a whole has a center-of-mass defined by

$$\vec{R} = \sqrt{\frac{1}{N}} \sum_{i=1}^{k} \sum_{j=1}^{n_i} \vec{r}_j{}^{(i)} \tag{7}$$

III. The multi-cluster wavefunctions

The permutational symmetry adaptation of the multi-cluster wavefunction is based on the fact that the centers of the symmetric groups $S_{(i)}$ $i = 1, 2, \ldots, k$ commute with one another as well as with the sequence of symmetric groups $S_{[2]} \subset S_{[3]} \subset \ldots \subset S_{[k]}$.

The recursive construction of the internal wavefunction for a single cluster was presented in refs. [5]-[6]. This wavefunction can be written in the form

$$|Y_{(i)} \vec{\Lambda}_{(i)} \epsilon_{(i)} \alpha_{(i)}; \vec{\rho}_{(i)} > \tag{8}$$

where the quantum numbers specify the Yamanouchi symbol $Y_{(i)}$, the internal resultant angular momentum $\vec{\Lambda}_{(i)}$ and the total internal energy $\epsilon_{(i)}$. The additional label $\alpha_{(i)}$ takes care of any remaining degeneracies.

The various clusters are now coupled to form the overall system. This is carried out recursively, adding one cluster at a time but with all the values of the single cluster and relative energies and angular momenta with which the good quantum numbers are consistent.

179

We start by forming the angular-momentum coupled internal states of the first two clusters

$$| Y_{(1)} Y_{(2)} \left(\vec{\Lambda}_{(1)} \vec{\Lambda}_{(2)} \right) \vec{\mathcal{L}}_{[2]} \, \vec{\lambda}_{(2),(1)} \vec{\Lambda}_{[2]} \, \left(\epsilon_{(1)} \epsilon_{(2)} \epsilon_{(2),(1)} \right) \epsilon_{[2]} \, \left(\alpha_{(1)} \alpha_{(2)} \right) ; \vec{\rho}_{[2]} > \qquad (9)$$

The states are formed by coupling the internal angular momenta of the first and second cluster into $\vec{\mathcal{L}}_{[2]}$, which is further coupled with the angular-momentum $\vec{\lambda}_{(2),[2]}$ of the intercluster relative motion into a two-cluster internal angular-momentum $\vec{\Lambda}_{[2]}$.

The center of $S_{[2]}$ which commutes with the centers of $S_{(1)}$ and $S_{(2)}$, also commutes with the two-cluster harmonic hamiltonian

$$H_{[2]} = \sum_{i=1}^{2} \sum_{j=1}^{n_i} \left[-\frac{\hbar^2}{2m} \left(\vec{p}_j{}^{(i)} \right)^2 + m\omega^2 \left(\vec{r}_j{}^{(i)} \right)^2 \right] \qquad (10)$$

and with the two-cluster total angular momentum

$$\vec{L}_{[2]} = \sum_{i=1}^{2} \sum_{j=1}^{n_i} \vec{\mathcal{L}}_j{}^{(i)} \qquad (11)$$

where $\vec{\mathcal{L}}_j{}^{(i)} = \vec{r}_j{}^{(i)} \times \vec{p}_j{}^{(i)}$. It does not, however, commute with the single-cluster hamiltonians and angular momenta. Consequently, the elements of the center of $S_{[2]}$ transform each angular-momentum-coupled two–cluster internal state into a linear combination of states of the same form, with the same value of $Y_{(1)}$, $Y_{(2)}$, $\vec{\Lambda}_{[2]}$ and $\epsilon_{[2]} = \epsilon_{(1)} + \epsilon_{(2)} + \epsilon_{(2),(1)}$. Diagonalizing the center of $S_{[2]}$ we obtain the set of states

$$| \left(Y_{(1)} Y_{(2)} \right) \Gamma_{[2]} \, \vec{\Lambda}_{[2]} \, \epsilon_{[2]} \, \alpha_{[2]} ; \vec{\rho}_{[2]} > \qquad (12)$$

which, in addition to the good quantum numbers listed above are also labeled by an irrep symbol $\Gamma_{[2]}$ corresponding to $S_{[2]}$

To proceed, we couple the two-cluster state just produced with the internal state of the third cluster to obtain the angular momentum $\vec{\mathcal{L}}_3$, which is than coupled with the angular momentum associated with the motion of the third cluster relative to the first two, into a three-cluster internal angular momentum $\vec{\Lambda}_{[3]}$. Again, to form a subspace which is invariant (closed) under $S_{[3]}$ we take into consideration all the states whose partial energies and angular momenta are consistent with the values of $\epsilon_{[3]}$ and $\vec{\Lambda}_{[3]}$ specified.

Diagonalizing $S_{[3]}$ we obtain a set of states of the form

$$| \left(\left(Y_{(1)} Y_{(2)} \right) \Gamma_{[2]} \, Y_{(3)} \right) \Gamma_{[3]} \, \vec{\Lambda}_{[3]} \, \epsilon_{[3]} \, \alpha_{[3]} ; \vec{\rho}_{[3]} > \qquad (13)$$

Proceeding in the same manner we finally obtain the $k-$cluster internal state

$$| \left(\dots \left(Y_{(1)} Y_{(2)} \right) \Gamma_{[2]} \dots Y_{(k)} \right) \Gamma_{[k]} \, \vec{\Lambda}_{[k]} \, \epsilon_{[k]} \, \alpha_{[k]} ; \vec{\rho}_{[k]} > \qquad (14)$$

The only degree of freedom which is not included in this state is the (spurious) motion of the center-of-mass of the multi-cluster system. It can now be incorporated in a trivial way.

IV. Particle and cluster interchange

The diagonalization of the centers of the consecutive symmetric groups requires the evaluation of matrix elements of certain representative permutations, as explained in ref. [4]. The formation of an auxiliary cluster consisting of the particles affected by the permutation has been proposed as a means for the convenient evaluation of these matrix elements (cf. ref. [7]).

In order to achieve this transformation by means of a series of pairwise rotations, for which the Talmi-Moshinsky transformation [9]-[11] is applicable, the particle to be transformed into the auxiliary cluster has to be in the last (k'th) cluster. To satisfy this requirement we subject the system to a sequence of cluster interchanges.

Consider the interchange of clusters i and $i + 1$. The intercluster coordinates which are affected by this interchange are $\vec{\rho}_{(i),[i-1]}$ and $\vec{\rho}_{(i+1),[i]}$ which are transformed into

$$
\begin{aligned}
\vec{\rho}_{(i),[i-1]\cup(i+1)} &= \vec{\rho}_{(i),[i-1]}\, cos\Theta - \vec{\rho}_{(i+1),[i]}\, sin\Theta \\
\vec{\rho}_{(i+1),[i-1]} &= \vec{\rho}_{(i),[i-1]}\, sin\Theta + \vec{\rho}_{(i+1),[i]}\, cos\Theta
\end{aligned}
\tag{15}
$$

with

$$
sin\Theta = \sqrt{\frac{n_i n_{i+1}}{n_i n_{i+1} + N_{i-1} N_{i+1}}} \qquad (0 \le \Theta \le \frac{\pi}{2})
\tag{16}
$$

Note that the intracluster coordinates are not affected by this intercluster interchange.

V. Conclusions

A computationally efficient procedure for constructing non-spurious h.o. states for multi-cluster systems with arbitrary symmetry is presented. The explicit elimination of the spurious states results in a very significant reduction in the size of the basis employed. This is a very crucial point for calculations involving a very large number of states.

The multi-cluster wavefunctions can be used in several very different physical contexts. They provide a very straightforward means for the evaluation of nuclear spectroscopic factors and enable the study of interacting clusters of identical particles ranging from quarks through nucleons to rare-gas atoms. Among the many applications we find the study of fusion, fission and multifragmentation of clusters of nuclei or atoms particularly exciting. In the atomic case these are crucial steps in the process of nucleation and droplet formation in the gas phase. The coordinate transformations discussed in the present article are equally useful in the study systems of identical particles within classical mechanics, such as in molecular dynamics simulations of the behaviour of gas-phase atomic clusters.

References

[1] R. D. Lawson, *Theory of the Nuclear Shell Model* (Clarendon, Oxford, 1980).

[2] Yu. F. Smirnov and Yu. M. Tchuvil'sky, Phys. Rev. C **15**, 84 (1977).

[3] V. G. Neudatchin, Yu. F. Smirnov and N. F. Golovanova, *Clustering Phenomena and High-Energy Reactions*, in *Advances in Nuclear Physics* Vol. 11, edited by J. W. Negele and E. Vogt, (Plenum, New York, 1979).

[4] J. Katriel and A. Novoselsky, *Multi-cluster Wavefunctions With Arbitrary Permutational Symmetry* , Ann. Phys. (N. Y.) (in press).

[5] A. Novoselsky and J. Katriel, Ann. Phys. (N. Y.) **196**, 135 (1989).

[6] A. Novoselsky and J. Katriel *Non-Spurious Harmonic Oscillator States for Many Body Systems*, in *Proceedings of the VI International Conference on Recent Progress in Many body Theories*, edited by Y. Avishai, (Plenum, New York, 1991).

[7] A. Novoselsky and J. Katriel, *Non-Spurious Two-Cluster Harmonic Oscillator Wavefunctions*, Phys. Rev. Lett. (submitted).

[8] J.-Q. Chen, *Group Representation Theory for Physicists* (World Scientific, Singapore, 1989).

[9] I. Talmi, Helv. Phys. Acta **25**, 185 (1952).

[10] M. Moshinsky, Nucl. Phys. **13**, 104 (1959).

[11] A. Gal, Ann. Phys. (N. Y.) **49**, 341 (1968).

Correlated Basis Approach to Strong Interacting Systems

M. Viviani

INFN and Department of Physics, University of Pisa, I-56100 Pisa, Italy

Summary

The method of correlated wave functions can be applied to study systems interacting via strong forces with satisfactory results. A well known example is the Jastrow ansatz where the wave function for an N–particle system is written as

$$\Psi = \left[\prod_{j>i=1}^{N} f(r_{ij}) \right] \Phi_0 \,, \tag{1}$$

where Φ_0 is a model wave function constructed in terms of single–particle states and the remaining part is the so–called correlation factor; the two–body function $f(r_{ij})$ is chosen variationally.

More flexible wave functions can be constructed by including in the correlation factor effective three–body correlations, state–dependent correlations, etc. However, to obtain a more accurate description, it is convenient to expand the "missing part" of the wave function onto a complete function basis. For example, the wave function given by eq. (1) can be improved by

$$\Psi = \left[\prod_{j>i=1}^{N} f(r_{ij}) \right] \sum_{k=0}^{K} a_k |k\rangle \, \Phi_0 \,, \tag{2}$$

where $|k\rangle$ are the basis states and the coefficients a_k are considered as variational parameters. By increasing the number K of states included in the expansion, more and more accurate wave functions are obtained. However, when the interparticle potential is strong and the correlation factor is taken to be one, the expansion converges slowly and a large number K of states must be taken into account; on the contrary, the convergence of the expansion is appreciably fastened by the presence of the correlations. By increasing K, one has also a criterion to evaluate the quality of the calculated wave function.

One possibility which has been extensively investigated is the expansion in the Hyperspherical Harmonic (HH) basis. With this method[1], it is possible to obtain very accurate wave functions for the ground states of different systems (nuclei, helium atom drops, α–particles clusters, etc.) with $N = 3, 4$ and central interparticle potential. The extention of the method to more "realistic" potentials and to a large number of particles ($N = 8 \div 20$) is at present being investigated.

Reference

 1) S.Rosati, M.Viviani and A.Kievsky, Few–Body Systems **9**, 1 (1990).

Springer Series in Nuclear and Particle Physics   **Clustering Phenomena in Atoms and Nuclei**
Editors: M. Brenner · T. Lönnroth · F.B. Malik   © Springer-Verlag Berlin, Heidelberg 1992

Non-Closed-Shell Nuclear Clusters and the Algebraic Approach

J. Cseh

Institut für Theoretische Physik, Justus-Liebig-Universität,
W-6300 Giessen, Fed. Rep. of Germany
On leave from the Institute of Nuclear Research of the Hungarian
Academy of Sciences, Debrecen, Hungary

Abstract: The $U(3)$ basis of the vibron model connects this phenomenological algebraic approach to the harmonic oscillator cluster model. This connection enables us to study the consequences of the antisymmetrization, and it gives preference to the description of the internal cluster degrees of freedom in terms of the $SU(3)$ shell model. This new form of the model is better suited to several cluster problems, and its relation to the microscopic description becomes transparent.

1. Introduction

In the early 1980's Iachello proposed a group theoretical approach to clusterization of atomic nuclei[1-3]. The relative motion of two clusters is described in this approach by a phenomenological model of the dipole collectivity, called vibron model, which is similar in spirit to the interacting boson model (IBM) of the quadrupole motion[4]. The vibron model was applied also to the rotational–vibrational motion of molecules[2,5], and more recently to the relative motion of quarks within a hadronic particle[6-8]. In nuclear physics the first applications of this model concerned the molecular resonances, which appear in reactions of light heavy–ions[9-11], and to clusterization of heavy nuclei[12-15].

So far no real microscopic foundation of the vibron model is known. Recent applications[16-18] to some well known states of light nuclei seem to suggest, that such a microscopic connection can be built up based on the oscillator cluster model[19,20].

As for the description of internal degrees of freedom in this phenomenological algebraic approach, two possibilities were considered. For clusters of even mass number the quadrupole internal excitations are described in terms of the IBM. Then the coupled model has a $U_C(6) \otimes U_R(4)$ group structure, and it is called nuclear vibron model[13-15]. (Here C stands for cluster, and R for relative motion.) For clusters of odd mass number an interacting fermion model can

Springer Series in Nuclear and Particle Physics **Clustering Phenomena in Atoms and Nuclei**
Editors: M. Brenner · T. Lönnroth · F.B. Malik © Springer-Verlag Berlin, Heidelberg 1992

be applied[21], and then the coupled vibron–fermion model has a $U_C^F(m) \otimes U_R(4)$ group structure, where m is the number of the single–fermion states.

Here I wish to draw attantion to an alternative possibility, namely, to describe the internal cluster degrees of freedom in terms of Elliott's $SU(3)$ shell model[22], in which case the model has an $U_C^{ST}(4) \otimes U_C(3) \otimes U_R(4)$ group structure, where $U_C^{ST}(4)$ is the spin–isospin group. This modified form of the algebraic approach is better suited to several cluster problems, and in addition its relation to the microscopic description becomes fairly transparent.

In finding the connection between the phenomenologic and microscopic descriptions the equivalence between the Hamiltonians of the harmonic oscillator cluster model and harmonic oscillator shell model[19,20] plays a crucial role. Due to this equivalence several consequences of the antisymmetrization can be seen without carrying out the actual calculation[20]. We can apply the same kind of considerations to the vibron model through its $U(3)$ basis.

The vibron model and qualitative effects of the antisymmetrization are discussed in section 2. Among them one finds the preference of the description of the internal cluster degrees of freedom in terms of the $SU(3)$ shell model. In section 3 is shown the model emerging from the coupling of the vibron model and the $SU(3)$ shell model. Some applications are discussed in section 4.

Although the algebraic description can be extended to multi–cluster systems as well, here we consider mainly two–cluster systems.

2. Vibron model and the effects of antisymmetrization

2.1. Vibron model and its harminic oscillator limit

In the vibron model the spectrum is generated by a finite number (N) of bosons, which interact with each other. They can occupy single particle states of angular momenta $l = 0$ (σ bosons) and $l = 1$ (π bosons). The model has a group structure of $U(4)$, and two limiting cases (called dynamical symmetries) which provide us with analytical solutions of the eigenvalue problem, and with complete sets of basis states that can be used in the numerical calculations of the general case. One of them is labelled by the group chain:

$$U(4) \supset U(3) \supset O(3)$$

$$\mid N \; , \; n_\pi \; , \; L \;). \tag{1}$$

Here L is the angular momentum of a state of the N boson system, and n_π is the number of π bosons, i.e. the number of dipole phonons. When considering only linear and quadratic i.e. one– and two–body operators in the Hamiltonian, the energy–eigenvalue corresponding to dynamical symmetry (1) is given by:

$$E = \epsilon + \beta L(L+1) + \gamma n_\pi + \delta n_\pi^2. \qquad (2)$$

This dynamical symmetry gives a description in terms of an anharmonic oscillator. Neglecting the anharmonic (i.e. the last) term in (2) and for $\beta = 0$ we arrive at an harmonic oscillator limit. In the application to some well–known bands in ^{20}Ne, considered as $^{16}O + \alpha$ system, the basis states (1) proved to be more suitable, than the other possible basis[16–18].

2.2. Effects of antisymmetrization

It was shown in the early days of cluster studies, that the Hamiltonians of the harmonic oscillator shell model and the harmonic oscillator cluster model can be transformed into each other via an appropriate coordinate transformation[19]. (In the harmonic oscillator cluster model the clusters are described by harmonic oscillator shell models and their interaction is described by harmonic oscillator potential, each heaving the same size parameter.) Consequently, the energy spectrum of these two models are the same, and the wave function of a state in one of these descriptions can be expressed as a linear combination of those basis states of the other model, which belong to the same energy–eigenvalue. This theorem connects the *properly antisymmetrized* wave functions of the two pictures. Via this harmonic oscillator connection one can find interesting relations between different clusterizations and the shell model configuration[20]. Here we apply such kind of considerations to the vibron model. Some of the consequences are relevant for the reletive motion, and they can be studied also for systems of strucureless clusters, while others have importance only in the presence of non-closed-shell cluster(s).

a) **The microscopic content of π bosons:** In the oscillator basis (1) n_π denotes the number of quanta carried by the relative motion. In the shell model description an oscillator quantum corresponds to a major shell excitation of a nucleon. This shows an important difference between the bosons of the vibron model and the IBM, inasmuch the IBM bosons describe intrashell excitations and they correspond to nucleon pairs, while the π boson of the vibron model describes intershell excitation and corresponds to an oscillator quantum.

b) Basis truncation: In the shell model discription the lowest–lying Pauli allowed states carry a certain number of oscillator quanta, and consequently the same number of quanta have to be present in the cluster model description, too. It gives a limit from the side of small n_π values, known as Wildermuth condition. When the smaller cluster with nucleon number A_1 carries no oscillator quantum due to its internal structure, this condition reads:

$$n_\pi \geq \sum_{i=1}^{A_1} 2n_i + l_i \quad , \tag{3}$$

where n_i and l_i are the node number and angular momentum of the ith nucleon in the shell model description of the united nucleus. If the lighter cluster has some nucleons outside the $0s$ shell, then the number of its internal quanta has to be subtracted. The relative motion with smaller number of oscillator quanta is Pauli forbidden. The findings of a) and b) are discussed with some more detail in Refs. 16,17.

c) Cluster spectroscopic factor: Several non–antisymmetric cluster wave functions may become identical after the antisymmetrization. As a consequence the antisymmetric cluster model wave function may have a normalization factor different from that of the non–antisymmetric wave function. The cluster spectroscopic factor is obtained from a properly antisymmetrized cluster model wave function as the reciprocal square of the normalization factor[23,24], and this may deviate from unity even for pure cluster states. In the phenomenological approach of the vibron model one can take this effect into account by introducing a cluster spectroscopic factor operator[18]. This is obtained, just like any other operator of a physical quantity in this model, as a series expansion in terms of boson operators coupled to the proper tensorial character. Since the spectroscopic factor is an $O(3)$ scalar, this expension is similar to that of the Hamiltonian, and in case the $U(3)$ dynamical symmetry holds it has a closed expression. The relation to the spectroscopic factor of the microscopic cluster model in the harmonic oscillator approximation is discussed in Ref. 18, together with some applications.

d) Internal cluster degrees of freedom: When we apply the $SU(3)$ shell model for the description of internal degrees of freedom then some extra simplicity and flexibility appears, due to the alike nature of the bosons appearing in the two sectors of the model. In this case the buliding blocks are oscillator quanta, and they are related to the shell model excitation quanta of the united nucleus via the harmonic oscillator connection. This gives us the possibilities: i) to define the model space in such a way, that the Pauli principle is included, and ii) to relate the interactions of the phenomenological and microscopical descriptions. These questions are discussed further on in sections 3,4.

e) **Dipole (cluster) and quadrupole degress of freedom:** The $SU(3)$ model gives a description of the quadrupole deformation in terms of shell model configurations. Furthermore, some shell model configurations prove to be identical with some cluster model configurations. The relation of the cluster model and the collective model via the shell model was discussed right after the invention of $SU(3)$ model and harmonic oscillator cluster model[25]. In the cluster picture the space vector of the relative displacement has a dipole character, so in this sense the dipole and quadrupole degrees of freedom of certain nuclear states are identical.

f) **Equivalence of different clusterizations:** Different clusterizations may become equivalent as a rusult of the antisymmetrization. In case there is only one shell model state with certain quantum numbers and states with these quantum numbers appear among the basis states of different clustarization, they have to be identical. E.g. the ground state of ^{16}O can be considered as a spherical shell model state, as well as $^{12}C + \alpha$, $^{8}Be + {}^{8}Be$, or $\alpha + \alpha + \alpha + \alpha$. Similarly the ground state of ^{24}Mg can be seen as $^{12}C + {}^{12}C$, or $^{16}O + {}^{8}Be$, etc.

3. The $U_C^{ST}(4) \otimes U_C(3) \otimes U_R(4)$ model and its $U(3)$ dynamical symmetry

3.1. Model space

In the Elliott model the wave function of a nucleus is characterized by the $U^{ST}(4) \otimes U(3)$ symmetry. Via the requirement of the total antisymmetry the representation of the spin–isospin group is uniquely related to the representation of the orbital symmetry group; namely they have adjoint Young patterns.

There is another difference, worth mentioning here, between this model and the nuclear vibron model and the vibron–fermion model. In those approaches one associates a dynamical group to a cluster, which generates a full spectrum, while here we have only the symmetry group of the cluster. One can involve, of course, also in this description several $SU(3)$ representations, as it is done in the Elliott model. However, usually there is no need for that, since in a typical cluster study we take into account only a few internal cluster excitations.

For the sake of simplicity we consider a system, in which one of the clusters is a $U_C(3)$ and $U_C^{ST}(4)$ scalar, e.g. an alpha–particle. When it is not so the calculations become more involved; instead of considering the representations of just one $U_C(3)$ and $U_C^{ST}(4)$ groups, one has to consider the outer product representations of two groups for clusters C_1 and C_2.

The basis states have the quantum numbers of the group–chain:

$$U_C^{ST}(4) \ \otimes \ U_C(3) \ \otimes \ U_R(4) \ \supset \ U_C^{ST}(4) \otimes U_C(3) \otimes U_R(3) \supset$$

$$|[f_1^C, f_2^C, f_3^C, f_4^C], [n_1^C, n_2^C, n_3^C], [N, 0, 0, 0], \qquad\qquad [n_\pi, 0, 0],$$

$$\supset \ U_C^{ST}(4) \otimes U(3) \supset U_C^S(2) \otimes O(3) \supset U(2) \qquad\qquad (4)$$

$$[n_1, n_2, n_3], \quad s_C, \quad \chi, L, \quad\quad j \ \rangle \ .$$

In order to exclude the Pauli forbidden states we can follow two procedures.

i) Since the set of basis states (4) is identical with that of the harmonic oscillator cluster model, one can apply the results of the microscopic calculations, i.e. take only those states which belong to non–zero eigenvalues of the norm–kernel operator[26]. To solve this problem also the $SU(3)$ basis was extensively used[27]. This procedure would resemble to that of the orthogonal condition model[28], where the calculations are done semi–microscopically in the sense, that on a microscopically defined model space one applies phenomenological interactions.

ii) Another possibility is to require the matching of the $U(3)$ and $U_C^{ST}(4)$ representations of (4) to those of the shell model description of the united nucleus[29]. For the lowest–lying shell model configurations, where the effect of the antisymmetrization is the strongest, this matching can be calculated usually very simply. The representations which are excluded in this way are definitely Pauli forbidden: they are orthogonal to each of the Pauli allowed shell model states; so this condition is a necessery one. However, if a representation is present in the shell model basis, and it can be obtained in more than one way in a specific clusterization when coupling the internal cluster wave functions to the relative motion, $|\Phi_{C_1}\Phi_{C_2}\Phi_R\rangle$, then the matching condition can not select one of them. In other words, it is not a sufficiant condition if we want to define the allowed states belonging to certain internal cluster states $|\Phi_{C_1}\rangle$ and $|\Phi_{C_2}\rangle$. For many cases it may give the result in agreement with the microscopic calculations, but for some other cases it may give extra $U(3)$ representations. Since, however, even these extra states have physical relevance, the model space defined this way also seems to be reasonable.

The center of mass motion can also be separeted in the $U(3)$ basis[30], so the model space can be built up in such a way, that it is free from the spurious excitations.

The treatment of the internal cluster degrees of freedom in this model is based on the L–S coupled shell model. This scheme is known to be good for light nuclei[22], but fail to work in the heavier region. There are indications, however, about the usefulness of a pseudo L–S coupling scheme for heavy nuclei, in which the shell model space is reorganised in pseudo shells of an oscillator with one quantum less per shell[31]. Through this generalisation this model may find some applications related to heavy cluster systems.

3.2. Physical quantities

The operators of physical quantities should be constructed similarly, as in other algebraic models of coupled degreees of freedom. This means series expansion in terms of the generators of the $U_C^{ST}(4)$, $U_C(3)$, and $U_R(4)$ groups. In the algebraic models the special limits, called dynamical symmetries, play an important role, e.g. the eigenvalue problem has analytical solutions for these limits. In the present description a dynamical symmetry holds, when the Hamiltonian can be expressed with the Casimir invariants of the group chain (4). In that case some operators have the same structure, as those of the nuclear vibron model[13] or the vibron–fermion model[21], for the $U_C^{ST}(4)$ scalar, and non–scalar systems, respectively. The reason is that similarly to (4) a $U_C(3)$ (or a $SU_C(3)$) group appears also in the corresponding dynamical symmetries of those models. In the same way, a combined $U(3)$ (or $SU(3)$) group is also present, which accounts for the coupling between different degrees of freedom. The expression of the Casimir operators in terms of the group generators are independent of the realization of the group. Due to this fact the same Casimir invariants appear in the physical operators, independently, whether e.g. $U_C(3)$ is realized by monopole and quadrupole boson operators (nuclear vibron model), by fermion operators (vibron–fermion model), or by dipole boson operators (used here).

The same argument is valid for the $U_C^S(2)$ and $U(2)$ groups of (4) and those of Ref. 21.

When only one and two–body terms are considered, the energy–eigenvalue is obtained as a linear combination of the first and second order Casimir operators of the group chain (4). In case of $U(3)$ the one–body term is ruled out by the requirement of the particle number conservation for the shell model and the vibron model sectors separately. Further simplifications take place when the core has a spin and isospin zero, and only one $U(3)$ representation is important. Then the contribution from the Casimir operators of the groups $U_C^{ST}(4)$, $U_C^S(2)$, $U_C(3)$, $U(2)$ can be summarised in an overall constant ϵ. So, the total energy–eigenvalue is given by:

$$E = \epsilon + \gamma n_\pi + \delta n_\pi^2 + \eta(\lambda^2 + \mu^2 + \lambda\mu + 3\lambda + 3\mu) + \beta L(L+1) . \tag{5}$$

The functional form of S_α is similar to that of the Hamiltonian, so it is also diagonal in the $U(3)$ basis[18]. Specifically, for our case its expectation value is:

$$S_\alpha = \alpha_0 + \alpha_1 n_\pi + \alpha_2 n_\pi^2 + \eta_2(\lambda^2 + \mu^2 + \lambda\mu + 3\lambda + 3\mu) + \beta_2 L(L+1) . \tag{6}$$

Electromagnetic transitions can also be described simply in the dynamical symmetry approach. In Ref. 29 the $B(E2)$ transition rates are expressed in terms of the $SU(3) \supset O(3)$ Wigner coefficients.

3.3. Relation to two–nucleon interactions

The physical quantities of the previous subsection were obtained in a purely phenomenological way, and the parameters of eqs. (5,6) are to be determined from a fit to the experimental data. It is an interesting question, how these phenomenological parameters are related to effective two–nucleon interactions of microscopic studies. Due to the fact that the basis states (4) are in a one–to–one correspondence with those of the harmonic oscillator microscopic cluster model, such a relation can be found easily. We can apply the Otsuka–Arima–Iachello mapping, which was used to relate the IBM and the shell model[32]. In this procedure one obtains the relation of the microscopic and phenomenologic inteactions by equating the matrix elements, which are calculated between the corresponding states. The first application of the next section shows an example for this mapping.

191

4. Applications

4.1. $^{12}C + \alpha$ system

As a first example, I consider here a well–known system of $^{12}C + \alpha$, for which microscopic calculation is also available. The ^{12}C is taken here with a $U(3)$ reresentation: $[4, 4, 0]$.

The model space is defined by the matching between the cluster basis and the shell model basis, as discussed in section 3.1. Since both clusters are $U(3)$ scalars, only the $[4, 4, 4, 4]$ spin–isospin representations have to be considered. The $U(3)$ labels of the cluster basis are obtained as the outer product:

$$[4, 4, 0] \otimes [n_\pi, 0, 0] = \sum_{i=0}^{4} [4 + n_\pi - i, 4, i] . \tag{7}$$

The matching condition selects the representations given in Table I. The $n_\pi = 5$, $(\lambda, \mu) = (1, 0)$ representation is present in the shell model space only with the multiplicity 1, so it has to correspond to the spurious center of mass excitation. For this system the matching condition gives exactly the model space of the microscopic calculation[26].

The energy spectrum corresponding to the eq. (5) is shown in Figure 1 together with the experimental levels[33]. The alpha–particle spectroscopic factors are shown in Figure 2. The parameters, obtained from the respective fits are: $\epsilon = -35.51 MeV$, $\gamma = 11.47 MeV$, $\delta = -0.61 MeV$, $\eta = -0.05 MeV$, $\beta = 0.17 MeV$,

Table I. $SU(3)$ representations for the $^{12}C + \alpha$ system from the matching of the cluster basis and shell model basis

n_π	(λ, μ)				
4	(0, 0)				
5		(2, 1)			
6	(2, 0)	(3, 1)	(4, 2)		
7	(3, 0)	(4, 1)	(5, 2)	(6, 3)	
8	(4, 0)	(5, 1)	(6, 2)	(7, 3)	(8, 4)
9	(5, 0)	(6, 1)	(7, 2)	(8, 3)	(9, 4)
⋮	⋮	⋮	⋮	⋮	⋮

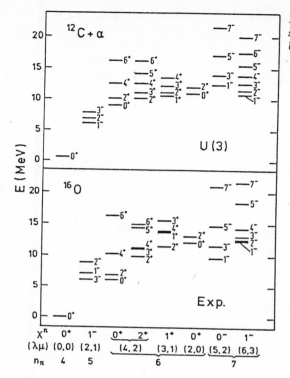

Figure 1. Experimental and model spectra of ^{16}O according to the $U(3)$ dynamical symmetry

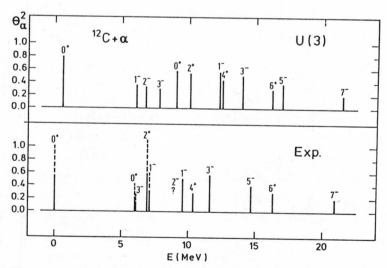

Figure 2. Alpha–particle reduced widths as ratios to the Wigner limit for some ^{16}O states. The solid and dashed lines in the low–energy part correspond to the two different values of Table 16.13 in Ref. 33.

and $\alpha_0 = 6.46$, $\alpha_1 = -1.89$, $\alpha_2 = 0.118$, $\eta_2 = 0.026$, $\beta_2 = -0.0069$. The ratios of the $B(E2)$ values can be obtained parameter–free in the dynamical symmetry approach. Their agreement with the experimental data has a quality similar to those of the energy–spectrum and the spectroscopic factor[29].

As it is discussed in section 3.3. the relation of the phenomenological parameters to microscopic interaction can be obtained from the equation between the corresponding matrix elements. In Ref. 34 the energy–eigenvalue of the Volkov–2 interaction[35] with Majorana exchange mixture $M = 0.6$ is given for the $SU(3)$ basis states. The correspondence of eq. (5) and Table I. of Ref. 34 gives: $\epsilon = -68.8 MeV$, $\gamma = 12.45 MeV$, $\delta = 0.0325 MeV$, $\eta = -1.295 MeV$ (in this case the energy of the ground state of ^{16}O is: $E_0 = -18.3 MeV$). In determining these parameters all the 9 energy–eigenvalue listed in Ref. 34 were taken into account. We can similarly determine the parameters of S_α, which correspond to the eigenvalues of the norm–kernel in the microscopic description[18]. Taking into account the $n_\pi \leq 10$ values of Ref. 26, we obtain: $\alpha_0 = 1.85$, $\alpha_1 = -0.0137$, $\alpha_2 = -0.0063$, $\eta_2 = -0.005$.

As shown in Figs. 1,2 the simple dynamical symmetry approach can reproduce some general properties of the experimental data. Several detailes fail in this description, but it is not surprising because of the extremly simple form of the interactions applied here. Considerable improvement is likely with the use of more realistic interactions and wave functions. The latter ones can be obtained from the numerical diagonalization of the Hamiltonian.

4.2. $^{12}C + ^{16}O$ system

One of the heavy ion reactions which show molecular resonances is the $^{12}C + ^{16}O$. Various theoretical approaches were made to these resonances. Among them there are considerations based on the rotation–vibration picture. Bromley et. al.[10] discussed the distribution of these states from the viewpoint of the dynamical symmetries of the simple vibron model. Their conclusion is that these states are not describable with the $U(3)$ or $O(4)$ dynamical symmetry. In Ref. 36 more resonances were analysed in terms of Morse potential, which corresponds to the $O(4)$ limit of the dipole model, as well as in terms of the anharmonic quadropole oscillator.

As a second example for the application of this new model, I show here the kind of description that can be given within a dynamical symmetry limit when internal degrees of freedom of the ^{12}C are coupled to the relative motion. The ^{12}C is taken again to have the [4, 4, 0] representation. We have addressed two further questions within this analysis: i) Whether or not low–lying cluster bands can be described together with the molecular rsonances within the same approach, ii) Is the quasimolecular basis sufficient to account for the observed density of resonances, when we take only a single oscillator function for the relative motion (for each parity) and the ground–state band of ^{12}C. For the $^{12}C + {}^{12}C$ system, with the same internal degrees of freedom for both clusters, such a basis proved to be satisfactory[37]. The model space is defined, again by the matching between the shell model basis and the cluster model basis. The smallest Pauli allowed quantum numbers are $n_\pi = 16$, $(\lambda, \mu) = (12, 0)$, and we get always one new μ value till $n_\pi = 20$, similar to that in Table I. However, in this case there is no missing label due to the center of mass motion.

In Ref. 38 the authors determined a band structure and (λ, μ) labels for some low–lying levels. From among them our cluster basis accounts for the $(12, 0)$, $(14, 1)$ and $(16, 2)$ bands, with 0, 1 and 2 excitation quanta, respectively. To the molecular resonance we associate $n_\pi = 20$, as the next positive parity relative motion quantum number, and $n_\pi = 21$ for the negative parity levels. When calculating the energy spectrum, shown in Figure 3, we also included a strong rotational–vibrational interaction[17] in the form: $\beta_1 n_\pi L(L + 1)$. The parameters obtained from a fitting procedure are: $\epsilon = -47.10 MeV$, $\gamma = 1.63 MeV$, $\delta = 0.1125 MeV$, $\eta = -0.016 MeV$, $\beta = 0.429 MeV$, $\beta_1 = -0.0179 MeV$.

In order to be consistent with the parametrization of section 4.1, we have applied also the energy–formula (5). In that case the agreement with the experiment was slightly worse, and the values of the parameters are: $\epsilon = -54.15 MeV$, $\gamma = 3.54 MeV$, $\delta = 0.032 MeV$, $\eta = -1.49 MeV$, $\beta = 0.071 MeV$.

In spite of the fact that the low–lying bands put a rather strong constraint both on the quantum numbers, and on the parameters, the agreement between the energies of the molecular resonances and the corresponding model states are reasonable; similar to that of Ref. 36. So, the $U(3)$ dynamical symmetry gives an acceptable approach, if the non–closed–shell structure of the ^{12}C is taken into account. The cluster basis is sufficient to describe the observed resonances

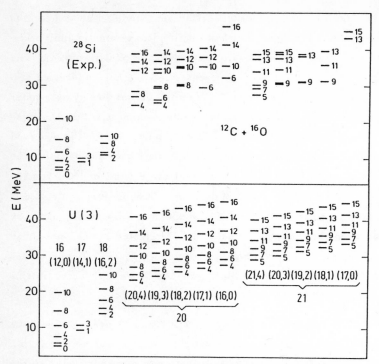

Figure 3. *Experimental and model spectra of* ^{28}Si. *The angular momenta are given next to the states, and all the parities are natural ones. Above or below the bands the* (λ, μ), *and* n_π *quantum numbers are indicated*

with one relative motion quantum number for each parity, when the ^{12}C can change its orientation in space but no further intrinsic exitations are included.

The description of the molecular resonances in terms of the basis (4) is remarkable from the viewpoint of the multipole character of these states. For the model states of Figure 3 the dipole (cluster) and quadrupole degrees of freedom are equivalent, as dicussed in section 2.2.

Dipole and quadrupole models seem to give comparable descriptions of resonances for other heavy ion systems, too[9,11,39,40]. The reason why the experimental data do not distinguish between these two kind of models may turn out to be, that due to the antisymmetrization these degrees of freedom become identical.

It is also worth mentioning, that bands with the same quantum numbers, as those of the model spectrum in Figure 3 are present in other clusterizations, e.g. $^{24}Mg + \alpha$, $^{14}N + {}^{14}N$ as well. For the bands: $(12, 0), (14, 1), (16, 2), (20, 4)$ the overlap between different clusterizations is 100%, because these representations appear in the shell model basis with single multiplicity.

4.3. Clusterization in heavy nuclei

The cluster studies of heavy nuclei are especially interesting in view of the recently observed heavy cluster radioactivity[41]. In the framework of the algebraic model[12-15] so far only alpha–like clusterizations were considered. With the present modification, however, one can apply the group–theoretical approach to other kind of clusters, as well. In Ref. 42 we have considered the $(core + X; \quad X = {}^{4}He, \ {}^{8}Be, \ {}^{12}C, \ {}^{14}C)$ clusterizations of the ground state of ^{224}Ra. Due to the antisymmetrization effects these fragmentations turn out to be equivalent within a resonable approximation for the pseudo–$SU(3)$ scheme of the heavy core and parent nuclei.

5. Conclusions

In this contribution I have proposed an algebraic description for the nuclear cluster states in terms of the vibron model coupled to the $SU(3)$ shell model, for the relative motion and internal cluster degrees of freedom, respetively. The novel feature, from the viewpoint of the phenomenologic approach, is the treatment of the internal cluster excitations. Due to this modification the microscopic connection to the harmonic oscillator cluster model becomes straightforward.

Here I have put much emphasis on the harmonic oscillator limit, because the relation of different approaches are evident there. Nevertheless, our description does not mean an harmonic oscillator approximation. In fact, even in the dynamical symmetry limit it can be largely anharmonic, and by including the $O(4)$ generators[2], the wave function of a state expand to many oscillator shells.

From the microscopic viewpoint the present model can be considered as an approximation, in particular an interacting boson type approximation, of the microscopic cluster model. It is still an open question, how useful this approximation is in relation with cluster degrees of freedom.

Acknowledgments

I gratefully acknowledge the support of the Alexander von Humboldt Foundation, and extend my sincere thanks to Professor W. Scheid for the hospitality of the Institut für Theoretische Physik der Justus–Liebig–Universität. I am also indebted to Prof. R. K. Gupta, Prof. K. Katō, Dr. G. Lévai, and Prof. W. Scheid for the collaboration on different parts of this work.

References

1) F. Iachello, Phys. Rev. **C23**, 2778 (1981).
2) F. Iachello and R. D. Levine, J. Chem. Phys. **77**, 3046 (1982).
3) F. Iachello, Proc. 4th Int. Conf. on Clustering Aspects of Nuclear Structure and Nuclear Reactions, Chester, 1984, eds.: J. S. Lilley and M. A. Nagarajan (Riedel, Dordrecht, 1985) p. 101.
4) F. Iachello and A. Arima, The Interacting Boson Model (Cambridge University Press, Cambridge, England 1987).
5) R. Lemus and A. Frank, Ann. Phys. (N.Y.) **206**, 122 (1991).
6) F. Iachello, Nucl. Phys. **A497**, 23c (1989).
7) P. Halse, Phys. Lett. **253B**, 9 (1991).
8) F. Iachello, N. C. Mukhopadhyay, and L. Zhang, Phys. Lett. **256B**, 295 (1991)
9) K. A. Erb and D. A. Bromley, Phys. Rev. **C23**, 2781 (1981).
10) D. A. Bromley, Proc. 4th Int. Conf. on Clustering Aspects of Nuclear Structure and Nuclear Reactions, Chester, 1984, eds.: J. S. Lilley and M. A. Nagarajan (Riedel, Dordrecht, 1985) p. 1.
11) J. Cseh, Phys. Rev. **C31**, 692 (1985).
12) F. Iachello and A. D. Jackson, Phys. Lett. **108B**, 151 (1982).
13) H. J. Daley and F. Iachello, Ann. Phys. (N.Y.) **167**, 73 (1986).
14) H. J. Daley and B. Barrett, Nucl. Phys. **A449**, 256 (1986).
15) H. J. Daley, Symmetries and Nuclear Structure (Dubrovnik 1986) Nucl. Sci. Conf. Ser. 13 eds.: R. A. Meyer and V. Paar (Harwood Acad. Publ., 1987) p. 359.
16) J. Cseh and G. Lévai, Phys. Rev. **C38**, 972 (1988).
17) J. Cseh, J. Phys. Soc. Jpn. Suppl. **58**, 604 (1989).
18) J. Cseh, G. Lévai, and K. Katō, Phys. Rev. **C43**, 165 (1991).
19) K. Wildermuth and Th. Kanellopoulos, Nucl. Phys. **7**, 150 (1958).
20) K. Wildermuth and Y. C. Tang, A Unified Theory of the Nucleus, (Acad. Press, N. Y., 1977).
21) G. Lévai and J. Cseh, Phys. Rev. C, in press
22) J. P. Elliott, Proc. Roy. Soc. **A245**, 128 and 562 (1958).
    M. Harvey, Adv. Nucl. Phys. **1**, 67 (1968).
23) M. Ichimura, A. Arima, E. C. Halbert, and T. Terasowa, Nucl. Phys. **A204**, 225 (1973).
24) A. Arima, Heavy Ion Collisions. ed.: R. Bock (North-Holland, Amsterdam, 1979) Vol. 1, p. 417.
25) B. F. Bayman and A. Bohr, Nucl. Phys. **9**, 596 (1958/59).
26) H. Horinchi, Prog. Theor. Phys. **51**, 745 (1974).
27) K. T. Hecht, Nucl. Phys. **A238**, 223 (1974).
    K. T. Hecht, E. J. Reske, T. H. Seligman, and W. Zahn, Nucl. Phys. **A356**, 146 (1981).

28) S. Saito, Prog. Theor. Phys. Suppl. **62**, 11 (1977).

29) J. Cseh, to be publisched

30) J. P. Elliott and T. H. K. Skyrme, Proc. Roy. Soc. **A232**, 561 (1955).
D. M. Brink and G. F. Nash, Nucl. Phys. **40**, 608 (1963)
K. T. Hecht, Nucl Phys. **A170**, 34 (1971).

31) J. P. Draayer, Nucl. Phys. **A520**, 259c (1990).

32) A. Arima and F. Iachello, Adv. Nucl. Phys. **13**, 139 (1984).

33) F. Ajzenberg-Selove, Nucl. Phys. **A460**, 1 (1986).

34) Y. Goto and H. Horiuchi, Prog. Theor. Phys. **62**, 662 (1979).

35) A. B. Volkov, Nucl. Phys. **74**, 33 (1965).

36) U. Abbondanno, S. Datta, N. Cindro, Z. Basrak, and G. Vannini, J. Phys. **G15**, 1845 (1989).

37) K. T. Hecht, H. M. Hoffmann, and W. Zahn, Phys. Lett. **103B**, 92 (1981).

38) R. K. Sheline, S. Kubono, K. Morita, and M. H. Tanaka, Phys. Lett. **119B**, 263 (1982).

39) N. Cindro and W. Greiner, J. Phys. **G9**, L175 (1983).

40) J. Cseh and J. Suhonen, Phys. Rev. **C33**, 1553 (1986).

41) H. J. Rose and G. A. Jones, Nature (London) **307**, 245 (1984).

42) J. Cseh, R. K. Gupta, and W. Scheid, contribution to this conference.

Microscopic Cluster Models for Nuclear Astrophysics

D. Baye

Physique Nucléaire Théorique et Physique Mathématique, CP 229,
Université Libre de Bruxelles, B-1050 Brussels, Belgium

The determination of nuclear reactions rates requires the knowledge of cross sections for radiative capture and transfer reactions at very low energies. Our group is developing a number of cluster models aimed at providing such cross sections in a microscopic way without direct use of experimental data [1]. In microscopic models, Pauli antisymmetrization and the conservation of angular momentum and parity are taken exactly into account. These models are essentially free of parameters.

One or several cluster configurations are selected on physical grounds to describe the initial and final states of the system as realistically as possible. The calculation proceeds in three steps. (i) Matrix elements of the Hamiltonian and of other operators are calculated with simple basis functions, related to Slater determinants. (ii) Bound-state and scattering wave functions are then determined in this basis with the help of the microscopic R-matrix method. (iii) Radiative-capture cross sections are calculated from matrix elements of multipole operators.

The two-cluster model is based on the two-centre harmonic-oscillator model. If the harmonic oscillator parameters of both clusters are equal, the basis functions are Slater determinants and the calculation of matrix elements is systematic. This model has been applied successfully to a number of reactions [1], and in particular to the $^{13}N(p,\gamma)^{14}O$ reaction [2] which has recently been studied with a radioactive ^{13}N beam. The three-centre harmonic oscillator model allows a description of collisions in which one of the colliding nuclei presents a two-cluster structure. This model provides useful results about the $^{8}Be(\alpha,\gamma)^{12}C$, $^{7}Be(p,\gamma)^{8}B$ and $^{19}Ne(p,\gamma)^{20}Na$ reactions [1-3]. An extension to four clusters is in progress [4].

The equal-parameter assumption has the drawback of overestimating the coupling between different configurations. Its validity at astrophysical energies will be evaluated with a generalization of the two-cluster model involving distorted internal wave functions for the clusters. A double integral transformation is then necessary to remove spurious c.m. components [5]. The model has been applied to $\alpha + \alpha$ scattering [6]. Its application to other systems is under investigation.

1) D. Baye and P. Descouvemont, Proc. Fifth Int. Conf. Clustering aspects in nuclear and subnuclear systems, Kyoto 1988, J. Phys. Soc. Jpn suppl. **58** (1989) 103
2) P. Descouvemont and D. Baye, Nucl. Phys. A **500** (1989) 155
3) P. Descouvemont and D. Baye, Nucl. Phys. A **517** (1990) 143
4) P. Descouvemont, Phys. Rev. C, in press
5) M. Hanck, Nucl. Phys. A **439** (1985) 1
6) D. Baye and M. Kruglanski, Contribution to this Conference

Springer Series in Nuclear and Particle Physics **Clustering Phenomena in Atoms and Nuclei**
Editors: M. Brenner · T. Lönnroth · F.B. Malik   © Springer-Verlag Berlin, Heidelberg 1992

The Visualization of the Energetic Structure of Nuclear Molecules

G. Mouze, A. Rocaboy, and C. Ythier

Faculté des Sciences, F-06034 Nice, France

An attempt is made to introduce chemical concepts in the description of nuclear molecules and nuclear clusters. For instance, the binding energy of a nuclear molecule is given by the sum of the binding energies of the various internal bonds. The corresponding mass formula is shown to be the mass-data-based extension, including the various neutron-proton interactions, of Talmi's quadratic equation.

As an example of application of this mass equation, the energetic structures of the neutron-halo nucleus ^{11}Be and of ^{12}C, belonging to the ^{4}He valence shells, and of ^{23}Na and of ^{24}Mg, belonging to the ^{16}O valence shells, are visualized and discussed.

There are specific reactions leading to the formation of diclusteric and polyclusteric molecules; in some cases the reverse reaction can be shown to exist. Various important phenomena are related to rearrangement reactions within such molecules, or to their dissociation: $\alpha-$ and cluster–emission, and fission.

An attempt is also made to detect the factors favouring the formation of clusters. Among these factors are the presence of anti-bonding nucleons or nucleon pairs, the attractive, eventually enhanced, neutron-proton interaction, and magic Z, N or A numbers. Lack of information concerning the nuclear entropies unfortunately still limits the application of chemical concepts to nuclear molecules.

Springer Series in Nuclear and Particle Physics   **Clustering Phenomena in Atoms and Nuclei**
Editors: M. Brenner · T. Lönnroth · F.B. Malik   © Springer-Verlag Berlin, Heidelberg 1992

Nuclear Structure Effects in Elastic and Inelastic Scattering of Light Ions

K. Shitikova and A. Katashov

Moscow State University, Institute of Nuclear Physics,
119899 Moscow, Russia

Elastic and inelastic scattering of light ions with the excitation of the monopole states has been analyzed with the respect to the information on the radial shape of the nuclear matter distributions of the target nuclei.

The possible manifestation of the nuclear structure in these processes is being investigated.

1. INTRODUCTION

Some time ago in [1] has been proposed a microscopic approach for description the reactions with the heavy ions.

In this approach the heavy-ions interaction potential is constructed with the use of the nuclear densities calculated in the microscopic approaches in the double-folding model and in the energy-density formalism.

The angular distributions for elastic and inelastic scattering of ions were calculated in the coupled channel model.

It was assumed that the forms of the real and imaginary potentials were the same, so that

$$U=U_R(1+i\alpha) \tag{1}$$

The factor α in the elastic channel was found by fitting theoretical differential cross sections to experimental data. The criterion of fitting is the usual one, namely, we minimazed the value:

$$\chi^2=\sum_i \frac{[\ \sigma_{exp}(\theta_i)-\sigma_{theor}(\theta_i)\ ]^2}{[\ \Delta\sigma_{exp}(\theta_i)]^2} \tag{2}$$

Springer Series in Nuclear and Particle Physics   **Clustering Phenomena in Atoms and Nuclei**
Editors: M. Brenner · T. Lönnroth · F.B. Malik   © Springer-Verlag Berlin, Heidelberg 1992

The possible manifestations of the nuclear structure in these processes is being investigated.

We present here our investigation in this approach for light ions. The analysis shows that the specific shapes of the nucleon density distributions affect slightly the elastic scattering cross sections. The inelastic scattering of light ions with the excitation of the monopole states is very sensitive to the nuclear structure of the target nuclei.

2. HYPERSPHERICAL FUNCTIONS METHOD AND MONOPOLE EXCITATIONS OF LIGHT NUCLEI.

The hyperspherical functions method[2] involves a collective variable (hyperradius ρ), which is related to the mean-square radius of the nucleus $\rho^2 = A <r^2>$, i.e. to the mean nuclear density. The excitations of this degree of freedom correspond to the monopole oscillation of the nucleus as a whole, i,e. the density is a dynamic variable. So, in this method such nuclear properties as radial density distribution and rms-radius are functions of the excitation energy of the nucleus and the increase in the nuclear size of the excited states is automatically accounted for in the hyperspherical functions method.

The method expresses the nuclear wave functions by an expansion in hyperspherical harmonics:

$$\psi = \chi_{K\gamma}(\rho) Y_{K\gamma}(\theta_\iota) \tag{3}$$

which are eigenfunctions of the angular part of the Laplacian

$$\Delta_{\Omega_n} Y_{K\gamma}(\theta_i) = -K(K+n-2) Y_{K\gamma}(\theta_i) \tag{4}$$

The hyperspherical coordinates (angles θ_i and radius ρ) are related to the normalized Jacobi coordinates. The diagonal (i=j) and transition (i j) densities are given by

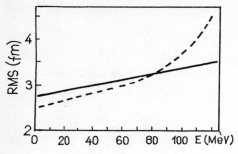

Fig.1. Dependence of the rms-radii on excitation energy for the nucleus ^{16}O.

$$\rho_{ij}^{m}(r) = \frac{16}{\pi} \frac{\Gamma(\frac{5A-11}{2})}{\Gamma(\frac{5A-14}{2})} \int\limits_{r}^{\infty} \frac{(\rho^2-r^2)^{\frac{5A-16}{2}}}{\rho^{5A-13}} \chi_{\iota}(\rho)\chi_{j}(\rho)d\rho +$$

$$\tag{5}$$

$$+ \frac{8(A-4)}{3\pi^{1/2}} \frac{\Gamma(\frac{5A-11}{2})}{\Gamma(\frac{5A-16}{2})} \int\limits_{r}^{\infty} \frac{r^2(\rho^2-r^2)^{\frac{5A-18}{2}}}{\rho^{5A-13}} \chi_{\iota}(\rho)\chi_{j}(\rho)d\rho$$

Fig.1 shows the dependence of the rms-radius on excitation energy in the hyperspherical functions method for the ^{16}O nuclear in comparison with that of finite temperature Hartree-Fock-method.[3] The fact is that our calculation leads to a dependence of the rms-radius on excitation energy so that the nuclear size goes to infinity when the excitation energy of the nucleus reaches its binding energy, while the mean-field theory predicts a finite value. It is necessary to stress that in such type of calculation all the energy of the nucleus is concentrated on one degree of freedom, especially the nuclear radius, while in the Tomas-Fermi model, which introduces temperature into nuclear physics, and in the Hartree-Fock approximation the excitation energy is distributed over a many single-particle degree of freedom. Consequently, one has to expect a stronger energy dependence of the nuclear size than it is given by the mean-field theory.

The above discussion may be illustrated by the calculation of the amplitude of the reduced width of the virtual decay of the ^{16}O nucleus into the ^{12}C nucleus and an α-particle for the nuclear

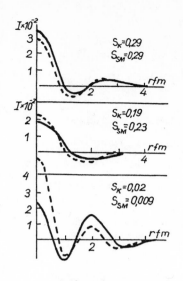

Fig.2. The amplitude of the reduced virtual decay of ^{16}O into ^{12}C plus α-particle in the hyperspherical functions method (solid curves) and with the oscillator potential (dashed lines).

(a) Calculation results for the ground states of ^{16}O, ^{12}C and ^{4}He.

(b) Monopole excitation of ^{12}C.

(c) Monopole excitation of ^{16}O.

states $J^{\pi}=0^{+}$, LST=000 and the relative orbital angular momentum $\Lambda=0$. In this case, the reduced width for α-decay is of the form:

$$J(r)=\Sigma\ <AK[f]LST\ A_1K_1[f_1]L_1S_1T_1;\Lambda,4K_2[f_2]L_2S_2T_2\{K\}>\ x$$

$$x\ <N_1K_1N-N_1K\ NK><n\Lambda\ N_2K_2\ N-N_1K>C_A^{N}C_{A_1}^{N_1}C_4^{N_2}R_{n\Lambda}(r/r_0)$$

(6)

where r is the distance between the centers of mass of particles $A_1=A-4$ and $A_2=4$; the C-are coefficients of the expansion of the K-harmonics functions in the wave functions of the oscillator potential.

The spectroscopic factor for an α-cluster was also estimated from

$$S_{\alpha}=\frac{A!}{4!(A-4)!}\int\left[J(r)\right]^2r^2dr$$

(7)

The results are displayed in Fig2.

According to the calculations the monopole-excited levels 0^{+} of the ^{12}C and ^{16}O nuclei are in the 21 and 26 MeV regions respectively. The results for the ground states of the nuclei differ but little in the oscillator and hyperspherical approaches. This difference is, however, significant for the monopole-excited states.

3. ELASTIC AND INELASTIC SCATTERING OF LIGHT IONS.

In the approach [1] we are calculated elastic [4,5] and inelastic [6] scattering of light ions with the excitation of the monopole states.

Two forms of the nucleon density distributions are studied: the density resulting from the hyperspherical functions method (GSF) and the cluster model densities (CL) calculated with the 3α-resonating group method by Kamimura which are parameterized in the form:

$$\rho_{ij}^{\mu}(\lambda) = \sum_{s=1}^{N} c_{ij}^{\lambda}(s)(r/r_s)^{\lambda} \exp(-r^2/r_s^2) \tag{8}$$

The real part of the optical potential is generated equation:

$$U_R(R) = \int dr_1 dr_2 \rho_1^{m}(r_1)\rho_2^{m}(r_2)V(r_{12}=R+r_2-r_1) \tag{9}$$

with the M3Y(density independent) and DDM3Y (density depen-dent) energy dependent effective nucleon-nucleon interactions:

$$V(E,\rho^{m},s) = g(E,s)f(E,\rho^{m}) \tag{10}$$

where

$$\rho(E,\rho^{m}) = C(E)[1+\alpha(E)e^{-\beta(E)\rho^{m}}] \tag{11}$$

and $g(E,s)$ is the original M3Y interaction whose spin and isospin-independent part is given in MeV by

$$g(E,s) = [7999e^{-4s}/4s - 2134e^{-2.5s}/2.5s] + J(E)\delta(s) \tag{12}$$

Here E is the bombarding energy per nucleon, $s=r$ is the internucleon separation, and ρ^m is the density of nuclear matter in which the interacting nucleons are embedded. We take $\rho^m = \rho_1^{m}(r_1) + \rho_2^{m}(r_2)$ for a nucleon at r_1 in nucleus 1 interacting with a nucleon at r_2 in nucleus 2 (sudden approximation) since mainly the nuclear surface is involved. The term $J(E)$ represents the effect of knock-on exchange between the interacting nucleons with

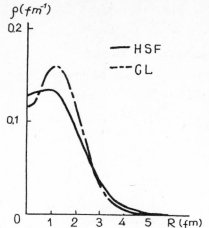

Fig.3.Two forms of the
nuclear density distri-
butions of ^{12}C resulting
from theoretical calcula-
tions GSF-(5) and CL-(8).

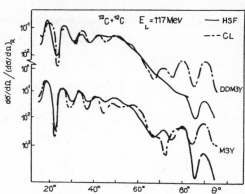

Fig.4. Experimental and
theoretical differential
cross section at T_L=117MeV
^{12}C+^{12}C elastic scattering.

$$J(E)=-276(1-0.005E) \quad (MeV.fm^3) \qquad (13)$$

When calculating the cross sections on the basis of the folding
model it is generally found that the interaction strength U_r has
to be renormalized by an overall factor N (around 1) in order to
fit the experimental data. Values of N=1 indicate success of the
model while any deviation of N from unity implies deficiencies of
the model calculations.

The angular distributions for elastic and inelastic scattering of
ions were calculated in the coupled channel model.
The calculation results are shown in Fig.3-5.
Fig.3 shows two forms of the nuclear density distributions of ^{12}C
resulting from theoretical calculations GSF- (5) and CL-(8).
Fig.4 displays the experimental and theoretical differential cross
sections at T_L=117MeV ^{12}C+^{12}C elastic scattering.

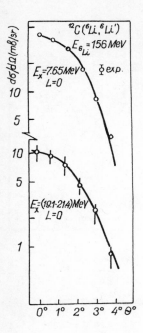

Fig.5.Experimental and theoretical angular distributions of inelastic nucleus-nucleus scattering at incident energy T_L=156MeV leading to the excitation at 7.65 MeV and 20.3 MeV of the monopole states in ^{12}C.

Fig.5 shows the experimental and theoretical angular distribution of inelastic nucleus-nucleus scattering for the system ^{6}Li +^{12}C at the incident energy T_L=156MeV leading to the excitation of the monopole states in ^{12}C. From the analyses of the results presented in Fig.3-5 one can conclude that there is a general agreement between calculated and experimental cross sections.Though we have to admit that details of the experimental cross sections(for which a better determination of the oscillation pattern by additional data points in the forward hemisphere would have been help full)are less well preference for the cluster model density distribution,in particular through the fact that the renormalization factor N approaches best the unity in that case. As to the inelastic scattering of ^{6}Li+^{12}C it,s shows strong dependency of the nature of the monopole excited states. As a result we are using α-cluster density distribution for the description of the monopole excited states at 7.65 MeV excitation energy and we are using the"breathing mode" nuclear density of ^{12}C for description monopole excited states at 20.3 MeV excitation energy in ^{12}C.

4.CONCLUSION.

Results of our calculations indicates the validity of further application of microscopical approach [1]to study the elastic and inelastic ions scattering with the excitation of the monopole states in the light ions. The possible manifestations of the nuclear structure in these processes is being investigated.

REFERENCES.

1. K.V. Shitikova Part. Nuclei.16, 824 (1985)

2.K.V.Shitikova Nucl. Phys.A331,365 (1979)

3.K.V.Shitikova,P.Rozmei.GSI-89-20,Preprint,Fabruary.1989.

4.S.Galachmatova,E.Romanovsky,A.Shirokova,K.Shitikova,H.Gils, H.Rebel.Preprint.KfK.4584,Mai 1989.

5.V.Garistov,S.Galachmatova,E.Romanovsky,A.Shirokova,K.Shitikova. To be publshed in Izv.ANSSR.ser.fiz.

6.S.Galachmatova,E.Romanovsky, K.Shitikova.To be published in Izv. ANSSR ser. fiz.

Part III

Cluster Radioactivity
(f-Decay) and Fragmentation

From Nuclear Molecules to Cluster Radioactivities

W. Greiner

Institut für Theoretische Physik der J.W. Goethe-Universität,
W-6000 Frankfurt 11, Fed. Rep. of Germany

1. Introduction

Nuclear molecules have been first experimentally discovered by Bromley et al. [1] in the scattering of ^{12}C on ^{12}C, as resonant phenomena. Other systems where they could be observed since then include $^{17}O + ^{12}C$, $^{16}O + ^{16}O$, $^{28}Si + ^{28}Si$, $^{24}Mg + ^{24}Mg$, etc [2]. Resonances in the scattering excitation functions typically exhibit gross structure widths ($\Gamma_{CM} \sim 2$ - 4 MeV and superimposed intermediate structure widths ($\Gamma_{CM} \sim 0.5$ MeV). The correlation of structures between different exit channels have been successfully explained by the double resonance mechanism [3] based on two centers shell modell (TCSM) [4,5].

The principle behind an atomic molecule is well known. There are usually two or more centers (the nuclei) around which electrons are orbiting and thus binding them. An atomic quasimolecule is a short-living intermediate system, formed in nuclear scattering, in which electrons feel - during the collision - the binding of both atomic nuclei. Similarly, in a nuclear molecule the outermost bound nucleons are orbiting around both nuclear cores, binding them together for times $\tau(mol)$ longer than the collision time $\tau(coll)$. Depending on the ratio of these times, a whole variety of more or less pronounced intermediate phenomena do occur: from molecules of virtual and quasibound type to longer living intermediate, nearly compound systems.

The analogy to the atomic case should be stressed. There the potential minimum between the atoms comes from certain two-center orbitals with lower energy, which expresses the binding effect due to valence electrons. Similarly, in the nuclear case the two-center shell structure of the various valence nucleons causes the additional binding. The nuclear molecules are, however, much more complex than the atomic ones. Because of the strong force many more nucleons participate in the molecular interaction, depending on the overlap and fragmentation of the two subsystems. This leads to various types of nuclear molecules. Also there are many more closed and open channels and collective degrees of freedom, which damp the molecular states more than in atomic physics. At higher energies compression effects become important.

Springer Series in Nuclear and Particle Physics **Clustering Phenomena in Atoms and Nuclei**
Editors: M. Brenner · T. Lönnroth · F.B. Malik   © Springer-Verlag Berlin, Heidelberg 1992

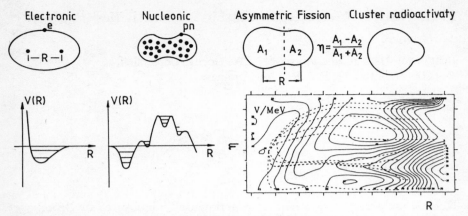

Fig. 1. Electronic and nucleonic molecules. Shape isomers are also nuclear molecules. Shapes and potential energy surfaces for asymmetric fission and cluster radioactivity indicate intermediate molecular configurations. The radial R and mass asymmetry η coordinates in the fragmentation theory are illustrated. The last diagram indicates the focussing collective flow leading to asymmetric or superasymmetric fission or even to cluster decay. The valley at $\eta = 0.8$ corresponds to cluster radioactivity and at $\eta = 0.2$ to fission. Only the half of the figure (for $\eta > 0$) was plotted; the other half is symmetric.

Even the fission isomers or the recently discovered shape isomers (superdeformed or hyperdeformed shapes produced at very high spins) as well as the nuclear systems exhibiting cluster radioactivities, might be considered to be nuclear molecules. At least there are quasimolecular stages in a fission and cluster decay process. For such objects it is essential that the potential energy has a certain "pocket", i. e. one or more minima, as a function of the two-center distance R, or as a function of the mass asymmetry coordinate $\eta = (A_1 - A_2)/(A_1 + A_2)$. In these minima the system is captured for some time during their relative motion (see Fig. 1), or the collective flow is channeled (focussed) during the separation process.

In order to explain the ground state deformations of many nuclei, the asymmetric distribution of fission fragments, fission isomers, intermediate structure resonances in fission cross-sections, etc, it is important to take into account both the collective and single-particle aspects of nucleonic motion, or, in other words, to add a shell and pairing correction to the liquid-drop energy [6] when the potential barrier is calculated. Otherwise, within the liquid-drop model, the nuclear shapes are always highly symmetric : spherical in the ground state, axially and reflection symmetric at the saddle point of the interaction potential. By using the TCSM or describing the single-particle states, one can follow the shell structure all the way from the original nucleus, over the potential barriers, up to the final stage of individual well separated fragments. Within fragmentation theory [7,8] all kinds

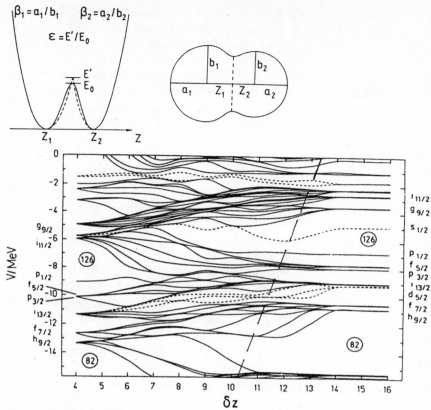

Fig. 2. Two-center shell model potential along the z-axis, nuclear shape and level diagram for superasymmetric fission of ^{232}U with the light fragment ^{24}Ne. The levels leading to final states of the light fragment are plotted with dashed-lines. The asymptotic shells are reaching for inwards, practically close or up to the the saddle point. The approximate position of the saddle as a function of the two-center distance $\delta z = z_1 - z_2$ is indicated by a straight line.

of fission fragment mass distributions experimentally observed (symmetric, asymmetric, superasymmetric) have been explained. Typical shapes are shown in Fig. 1. Superasymmetric fission and bimodal fission (experimentally discovered by Hulet et al.) [18,19] are particular forms of the influence of shell structure on the dynamics of low energy collective motion. The basic information about these phenomena can be found in Figs. 2 and 5 - 8. The cold fission phenomenon is intimately connected with the high energy mode in bimodal fission and also with cluster radioactivity dicussed further below. In fact, extrapolating from asymmetric to superasymmetric fission and further to extreme asymmetric break-up of a nucleus we are lead to cluster radioactivity; in particular to the inside that there is no difference between fission and radioactivity.

Cluster radioactivity has been predicted in a classical paper by Sandulescu, Poenaru and Greiner [9] . Among other examples given in this paper, it was shown that ^{14}C should be the most probable cluster to be emitted from [222]Ra and [224]Ra. Four years later, Rose and Jones [10] detected ^{14}C radioactivity of [223]Ra. Since then, other types of emissions (Ne, Mg, Si) have been identified. Theories and experiments have been recently reviewed [11-13]. Cluster radioactivities to excited states of the daughter and of the cluster, predicted by Martin Greiner and Werner Scheid [14] have also been observed as a fine structure [15].

Other experimental facts prove the presence of strong shell effects in cold rearrangement processes. Cold fusion with one of the reaction partner [208]Pb was of practical importance for the synthesis of the heaviest elements [16]. This has been suggested rather early [17]. [132]Sn and its neighbours plays a key role in cold and bimodal fission phenomena [18,19].

A pocket due to shell effects is present in the potential barrier of superheavy nuclei for which liquid drop barrier vanishes. It was estimated that magic numbers for protons and neutrons higher than 82; 126, are 114; 184, respectively, hence [298]114 is supposed to be double magic. Unfortunately no experimental evidence was obtained up to now in spite of a long search in nature and many attempts to produce superheavies by nuclear reactions.

From the rich variety of interconnected phenomena we have mentioned only few; some others will be presented in the following sections.

2. The Two-Center Shell Model

A single-particle shell model asymptotically correct for fission and heavy-ion reactions was developed [4,5] on the basis of the two-center oscillator, previously used in molecular physics. Besides the usual kinetic energy [20] $\mathbf{p}^2/2m$, a shape dependent potential V_{shp}, a spin-orbit coupling term V_{ls} and a l^2-correction term V_{l2} are present in the Hamiltonian, $\hat{H}$ of the TCSM. It is convenient to use a cylindrical system of coordinates (z, ρ, φ) due to the assumed axial symmetry of the problem. One has:

$$\hat{H} = \mathbf{p}^2/2m + V_{shp}(\rho, z; R, \eta, \varepsilon, \beta_1, \beta_2) + V_{ls}(\mathbf{r}, \mathbf{p}, \mathbf{s}; \kappa_i) + V_{l2}(\mathbf{r}, \mathbf{p}; \kappa_i, \mu_i) \quad (1)$$

where $\mathbf{r}, \mathbf{p}, \mathbf{s}$ are the coordinate, momentum and spin of a single nucleon of mass m, $R = z_2 - z_1$ is the separation distance between centers, $\eta = (V_1 - V_2)/(V_1 + V_2)$ is the mass asymmetry parameter, V_i $(i = 1, 2)$ are the volumes of the fragments, $\varepsilon = E'/E_o$ is the neck parameter, $\beta_i = a_i/b_i$ are the deformations, κ_i and μ_i are mass-dependent quantities defined in the Nilsson model (which is obtained for $R = 0$ within TCSM). E_o is the actual barrier height and $E' = m\omega_{z_i}^2 z_i^2/2$ is the barrier of the two-center oscillator. For $R = 0$ and $R \to \infty$ one nucleus and two fragments, respectively, are described.

The Hamiltonian is diagonalized in a basis of eigenfunctions analytically determined. As a result, the proton and neutron energy level diagrams for a given set of deformation parameters are obtained. An example of neutron levels for the split of ^{232}U in ^{24}Ne and ^{208}Pb is presented in Fig. 2. In this case a folded-Yukawa potential has been used for a nuclear shape parametrization obtained by smoothly joining two spheres with a third surface generated by rotating a circle of radius $R_3 = 1/c_3$ around the symmetry axis. A small radius $R_3 = 1$ fm has been chosen in the above diagram.

3. Nuclear Molecules

In peripheral heavy ion reactions, only the nucleons located near the surface of each participant are moving on molecular orbits. The lower shells nuclei are forming the nuclear "cores" of the molecule. Molecular orbits are formed and can be observed when the ratio of the collision time to nuclear period (orbiting time) of the valence nucleon, roughly given by $\sqrt{\epsilon_F A_{pr}/E_{lb}}$, is high enough. Here ϵ_F is Fermi energy, and E_{lb}/A_{pr} is the bombarding energy per nucleon. For example in a collision of ^{17}O with ^{12}C at energies near the Coulomb barrier, this ratio is about 3.5, and consequently the effect may be viewed. Resonances are observed in a narrow window of the excitation energy versus angular momentum plane - the *molecular window*.

Signatures of the formation of molecular orbits are observable in reaction channels involving enhanced nuclear transitions of nucleons between molecular levels (elastic and inelastic transfer, or inelastic excitation of the valence nucleons), due to avoided crossing of molecular single nucleon levels - the *nuclear Landau - Zener effect*. According to a theorem of Neumann and Wigner, levels with the same symmetry cannot cross. This avoided level crossing is the point of nearest approach of two levels with the same projection of angular momentum Ω, on the internuclear axis. The nuclear Landau - Zener effect [21] consists in a large enhancement of transition probability between the two levels at pseudocrossing (avoided crossing).

For a constant relative velocity v of two nuclei, the transition probability P from the lower adiabatic state to the higher one (see Fig. 3) is

$$P_{1\rightarrow 2} = 2P_{LZ}(1 - P_{LZ}); P_{LZ} = exp(-2\pi G)$$

$$G = |H_{12}|^2 / [\hbar v | \frac{d}{dR}(\varepsilon_1 - \varepsilon_2) |] \tag{2}$$

where $2H_{12}$ is the closest distance between the adiabatic eigenvalues E_1, E_2 at $R = R_C$, and ε_i are the corresponding diabatic (crossing) levels given by

$$E_{1,2} = \frac{1}{2}(\varepsilon_1 + \varepsilon_2)^{1/2} \mp [\frac{1}{4}(\varepsilon_1 - \varepsilon_2)^2 + |H_{12}|^2]^{1/2} \tag{3}$$

217

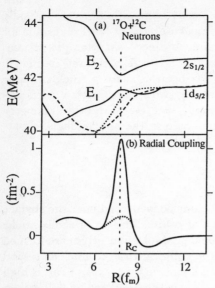

Fig. 3. (a) TCSM neutron level diagram for $^{17}O + {}^{12}C$ system. Only the levels used in calculations are shown. The full, dashed, and dotted lines corresponds to $\Omega = 1/2$, $3/2$, and $5/2$, respectively. (b) The radial coupling matrix element for the transition from the ground state to the first excited state of ^{17}O by promoting the valence neutron from the $1d$ to the $2s$ level. Dotted curve was obtained by supressing the peak at the avoided level crossing.

As a function of incident energy, $P_{1\rightarrow 2}$ steeply rises to 0.5 and then drops to zero.

An example of avoided level crossing is shown in Fig. 3 at $R = R_C$, where two $\mid \Omega \mid = 1/2$ levels, are approaching the single particle states $1d_{5/2}$ and $2s_{1/2}$ of ^{17}O for $R \rightarrow \infty$. Assuming the valence neutron on $1d_{5/2}$ level in the ground state of ^{17}O and the $2s_{1/2}$ level in its first excited state, signatures of nuclear molecular resoances have been predicted [21] in the inelastic excitation function of $^{17}O + {}^{12}C$ reaction. The valence neutron of ^{17}O is strongly excited from ground state to the $2s_{1/2}$ excited state $(1/2^+)$ at $E^* = 0.871$ MeV.

Two excitation mechanisms between molecular single-particle levels are effective radial and rotational couplings. The former acts between states with the same Ω, and is largest (see the lower part of Fig. 3) at the point of an avoided level crossing (Landau - Zener transition). The later (Coriolis coupling) arises from the rotation of the body fixed coordinate system. It causes transitions between states with $\Omega \pm 1$, and is the largest at the crossing point of levels. The TCSM level diagrams (as in Fig. 3) are used to predict enhanced transitions and to determine the most important couplings in specific reactions.

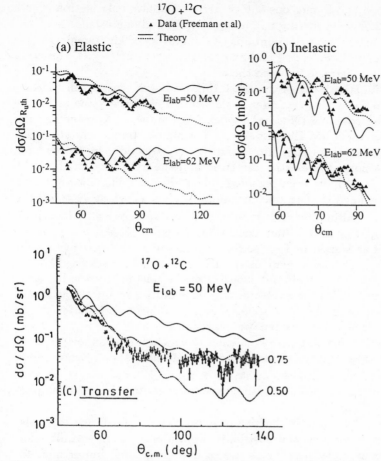

Fig. 4. Differential cross sections for: (a) the elastic scattering of ^{17}O on ^{12}C; (b) inelastic excitation of the first $1/2^+$ state of ^{17}O at the incident energies 50 and 62 MeV. The solid and dotted lines are calculations in correspondence with similar curves from Fig. 3b. (c) The calculated transfer yield is found to be very sensitive to the energy splitting of the avoided crossing. The figure demonstrates the influence of a variable intensity of the radial coupling.

The oscillatory behaviour of the differential cross sections for elastic scattering of ^{17}O on ^{12}C and inelastic excitation of the first $1/2^+$ state of ^{17}O, at incident energies of 50 and 62 MeV, experimentally determined by the Strasbourg group [22] (see Fig. 4) is an evidence of nuclear molecular phenomena. Complete quantum - mechanical molecular reaction calculations for this system have been performed within *molecular particle-core model*. In this model, first introduced in 1972 [23] and improved successively [24-27] the outermost bound nucleons are eigenstates of the TCSM. The mean field is generated by all nucleons and contains adiabatic polarisation effects between nuclei. By considering two unequal cores and one valence neutron, the Hamiltonian of the model contains two operators of kinetic energy (for valence neutron and two cores) and an optical potential $(U + iW)$ function of the separation distance between the cores.

The wave function ψ_{IM} of the scattering problem , written in the laboratory system for a given angular momentum I, has the same structure as in the strong coupling (Nilsson) model, but here $\Phi_{\alpha j \Omega}$ is eigenfunction of the asymmetric TCSM Hamiltonian. The coupled channel equations for the radial wavefunctions $R_{\alpha lI}$ are obtained by projection. In the coupled channel calculations, the elastic scattering and the excitation of the $1/2^+$ state at 0.871 MeV have been included. The radial coupling shown in Fig. 3, has a strong narrow peak due to Landau- Zener effect at $R = 7.6$ fm. This peak is located just behind the barrier of the real optical potential U. Rotational couplings also induce transitions between elastic and inelastic channels. As it is shown in Fig. 4, the experiment is in good agreement with the theory. The agreement of the theory with the backward rise of the neutron–transfer crossection demonstrates the existence of the Landau–Zehner–effect and hence the level–crossing. This in turn is proof that we are dealing with nuclear molecules, because level crossings as a function of the two–center distance are characteristic molecular effects.

We believe that fusion reaction calculations including channels with enhanced single-particle transitions due to the nuclear Landau-Zener effect, could clarify the role of molecular states in the fusion process. A possible connection of nuclear molecules with cluster radioactivity has been emphasized [28].

A pocket in the potential barrier of giant systems (U-U for example) is a good indication about the possibility to produce experimentally giant atomic or nuclear molecules. These are of extremely high importance for testing the electrodynamics of strong fields, predicting spontaneous production of electron-positron pairs. Let us also mention the analogy of these nuclear molecules with atomic cluster molecules, as first discussed by R. Schmidt (these proceedings). For these do also exist internuclear potentials with boundary pockets. Figure 5 illustrates a molecular dynamics–calculation for a $(Na)_9$–$(Na)_9$–cluster collision clearly visualizing the formation of a $(Na)_9$–$(Na)_9$–molecule. In figure 6 the inter–cluster potential is sketched.

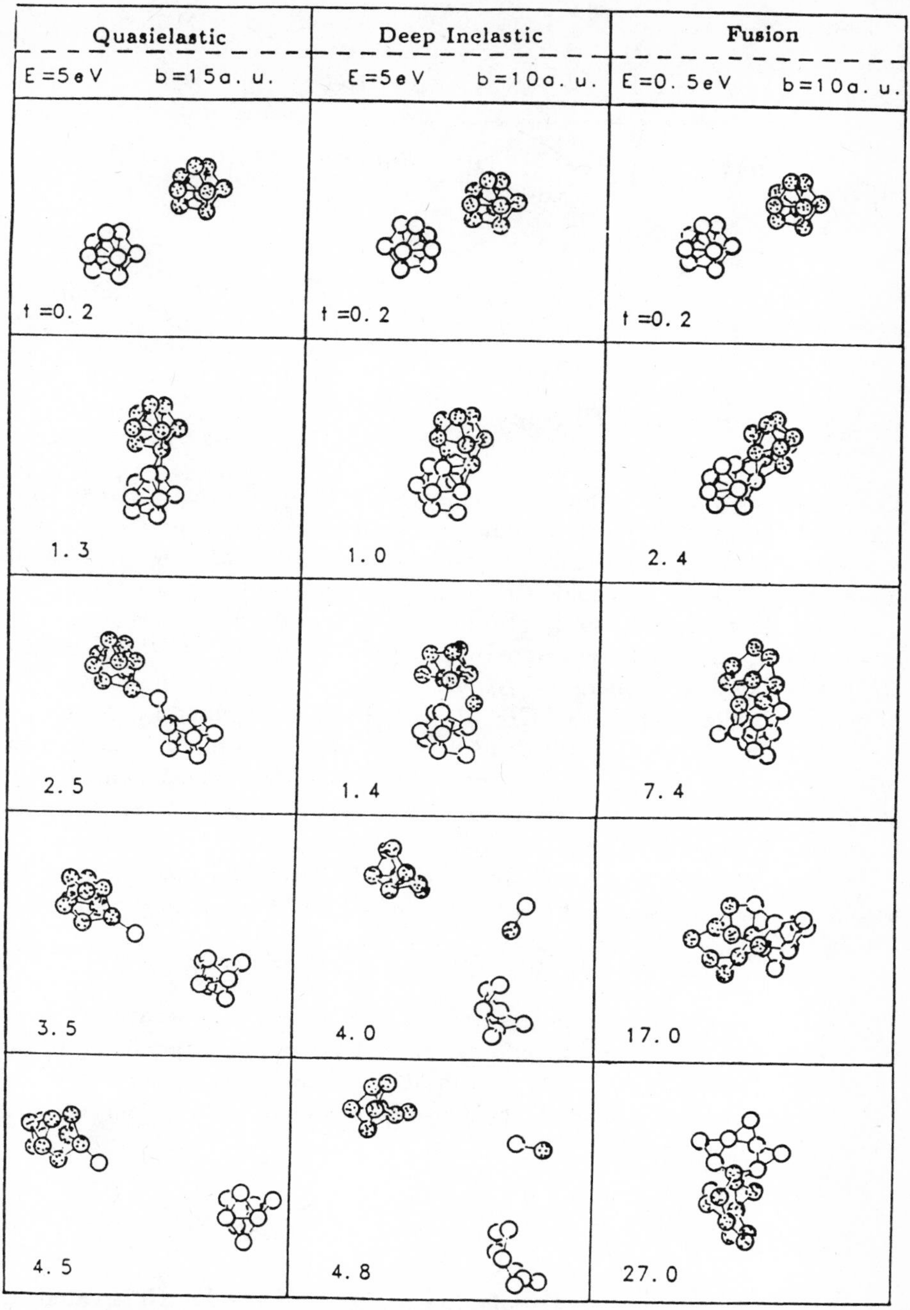

Quasielastic	Deep Inelastic	Fusion
E = 5 eV b = 15 a. u.	E = 5 eV b = 10 a. u.	E = 0.5 eV b = 10 a. u.
t = 0.2	t = 0.2	t = 0.2
1.3	1.0	2.4
2.5	1.4	7.4
3.5	4.0	17.0
4.5	4.8	27.0

Fig. 5. The formation of a $(Na)_9 - (Na)_9$ atomic cluster molecule within a molecular dynamic simulation.

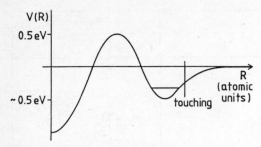

Fig. 6. Typical form of the inter–cluster–potential for $(Na)_9$–$(Na)_9$ according to R. Schmidt.

4. Fragmentation Theory

The theory of fragmentation is a consistent method allowing to treat two-body and many-body breakup channels in fission, fusion, and heavy ion scattering. Within this theory we get a unified point of view on fission, superasymmetric fission, cold- and bimodal- fission and cluster decay.

The main collective coordinates have been already defined in section 2. By assuming the adiabatic approximation, for each pair of (R, η) coordinates, the liquid drop energy $V_{LDM}(R, \eta, \varepsilon, \beta_1, , \beta_2)$ can be minimised with respect to $\varepsilon, \beta_1, \beta_2$, leading to $E_{LDM}(R, \eta)$. Then the shell and pairing corrections $\delta U(R, \eta)$ are added. In this way the time consuming procedure of calculating single-particle levels is applied only once for a given combination (R, η). Also the collective mass tensor $B_{ik}(R, \eta)$ can be calculated by using the states of the TCSM supplemented with pairing forces.

At high excitations the shell effects could be completely washed out. To account for this physical behavior, a Gaussian factor is introduced :

$$V(R, \eta, \Theta) = E_{LDM}(R, \eta) + \delta U(R, \eta) \exp(-\Theta^2/\Theta_o^2), \qquad (4)$$

where at a given excitation energy, E^*, the nuclear temperature equals $[\Theta = (10E^*/A)^{1/2}.]$ One can choose $\Theta_o = 1.5$ MeV in order to obtain a vanishing contribution of shell effects at excitation energies of the order of 60 MeV. The wave function $\Phi_k(R, \eta, \Theta)$ is a solution of a Schroedinger equation in η:

$$(-\frac{\hbar^2}{2\sqrt{B_{\eta\eta}}} \frac{\partial}{\partial \eta} \frac{1}{\sqrt{B_{\eta\eta}}} \frac{\partial}{\partial \eta} + V(R, \eta, \Theta))\Phi_k(R, \eta, \Theta) = E_k \Phi_k(R, \eta, \Theta). \qquad (5)$$

Here R is kept fixed and only the η-degree of freedom is treated in order to simplify the procedure and the understanding. This can be justified as long as fission is an adiabatic process. The fission fragment mass yield normalized to 200% is given by :

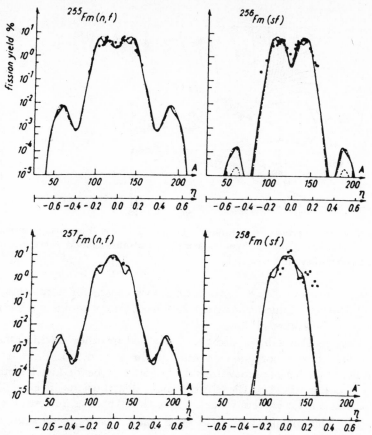

Fig. 7. Mass yields for various Fm isotopes calculated by using different inertia : $B_{\eta\eta}$ (full curve); average B (chain curve), and smoothed B (dashed curve). Experimental data are denoted by points. The observation of superasymmetric fission with one of the clusters being ^{78}Ni is a formidable task.

$$Y(A_2) = |\Psi(R,\eta,\Theta)|^2 \sqrt{B_{\eta\eta}}(400/A) \qquad (6)$$

where $B_{\eta\eta}$ is a diagonal component of the inertia (effective mass) tensor, and:

$$|\Psi(R,\eta,\Theta)|^2 = \sum_k |\Phi(R,\eta,\Theta)|^2 \exp(-E_k/\Theta). \qquad (7)$$

The example of transition from asymmetric to symmetric fission in Fm isotopes (see Fig. 7) illustrates again the dominant influence of shell effects [29] on fission fragment mass asymmetry. This transition was confirmed by experiment [30]. Moreover, bimodal fission has been observed in this region of nuclei [19]. A typical potential energy surface for a nucleus undergoing

223

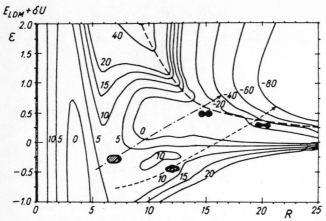

Fig. 8. Potential energy surface for symmetric fission of ^{256}Fm The dashed curve with arrow is the elongated shapes fission path. The dot-and-dashed curve is the compact shapes (cold) fission path.

bimodal fission is shown in Fig. 8, containing the two paths for normal and cold fission [45]. Rather quantitative investigations of this phenomenon have been given by Sobichewski et al. [33].

The side peaks in Fig. 7 have also been obtained in other calculations based on fragmentation theory, as for example in U and No. First experimental indications for superasymmetric fission can be deduced from Gönnenwein's experiments [18]. Clearly the mass distributions in Eq. (6) yield also contributions for extreme break-ups of the fissioning nucleus, i. e. for $\eta \sim 0.8 - 1.0$. This means that nuclei emit small clusters, even though with small probability.

We are thus lead to the idea of cluster radioactivity and, furthermore, to the inside that there is no difference between (cold) fission and a radioactive process. To further illustrate this idea, let us look at the groundstate wavefunction and its mass distribution for a two-center potential , as shown in Fig. 7. Obviously

$$P(\eta) = \int \Psi_o^*(R, \eta)\Psi_o(R, \eta)d^3R \tag{8}$$

can be interpreted as the cluster preformation probability in a nucleus of mass A within the interval $\eta, \eta + d\eta$ of cluster formation. In order to describe cluster radioactivity one has to extend the theory to very asymmetric shapes.

The ATCSM contains odd multipole moments due to the use of the mass asymmetry coordinate η. The octupole moment is given by :

$$Q_{30}(R, \eta) = \sqrt{\frac{7}{4\pi}} \int Y_{30}^*(\theta, 0)r^3 \rho_e(\mathbf{r}, R, \eta)d^3r \tag{9}$$

where ρ_e is the charge density resulting for this isolated fragmentation.

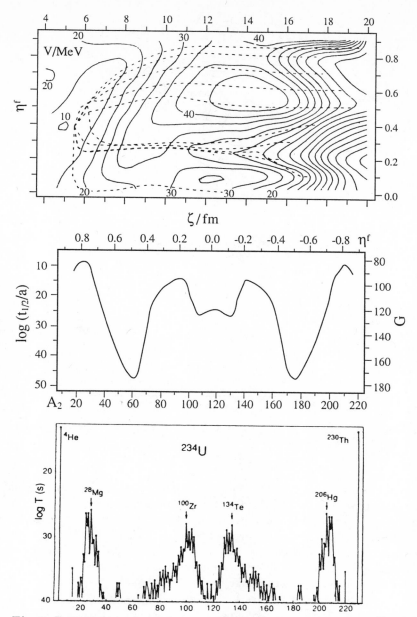

Fig. 9. Potential energy surfaces versus mass asymmetry and separation distance between centers for ^{232}U at a fixed neck coordinate $c_3 = 0.2/$fm. The dashed lines are fission paths leading to a given final mass asymmetry. The corresponding half-life (in years) along these paths are plotted in the middle part of the figure. It is comparable with the lifetime (in seconds) spectrum of ^{234}U calculated within ASAFM, given in the lower part of the figure. The upper two pictures have been calculated by D. Schnabel and H. Klein.

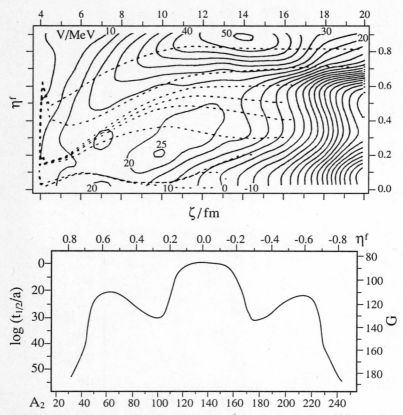

Fig. 10. The same as the preceding figure for the nucleus $^{270}110$. Both diagrams have been obtained by D. Schnabel and H. Klein.

Another interesting feature of the ground state wave function follows from the quantity $P(\eta) = \int | \Psi_o(R, \eta)|^2 dR$. As mentioned above, this is the preformation probability for the cluster configuration characterized by mass asymmetry η. With this interpretation we can see that the observable octupole moment is not a result of one particular configuration of the type shown in Fig. 1, but a superposition of all clusters, weighted by the corresponding preformation probability.

On the potential energy surface [32] given in Figs.9 and 10, the fission paths leading to various final asymmetries are shown with dashed lines. They are determined by minimizing [31] the WKB-integral. Shell corrections based on the energy levels of the kind shown in Fig. 2 have been added to the Yukawa-plus-exponential potential energy extended for fragments with different charge densities [34]. The inertia tensor has been computed by using Werner-Wheeler approximation within hydrodynamical model. A fixed neck radius has been assumed. The corresponding halflife obtained along each path in the upper part of Fig. 9 is plotted in the middle part, showing that according to these calculations, Ne radioactivity of ^{232}U has a larger probability than the spontaneous fission of the same nucleus. It is interesting to note the similarity with the time spectrum of the ^{234}U nucleus calculated within ASAFM. Finally we mention that fragmentation potentials can also be calculated for metallic clusters, as done by Rüdiger Schmidt (see contribution to this conference) which must be seen in analogy to the nuclear ones for U fission and Ca fission respectivly.

5. Cluster Decay Modes

Spontaneous cluster emission is allowed if the released energy $Q = M - (M_e + M_d)$ is a positive quantity. In this equation M, M_e, M_d are the atomic masses of the parent, the emitted cluster and the daughter nuclei, respectively, expressed in units of energy.

The most important observable is the parent nucleus life-time, T relative to this disintegration mode, and the corresponding branching ratio with respect to α-decay, $b = T_\alpha / T$. If T is low enough and b is sufficiently high, the phenomenon may be detected experimentally, and the kinetic energy of the emitted cluster $E_k = Q A_d / A$ is also measured. Up to now the longest measured lifetime is of the order of 10^{26} seconds and the lowest branching ratio is almost 10^{-17} !

The difficult task of the theory is to study the dynamics of the process. Three fission models and one cluster preformation model have been used in our papers of 1980 to predict the new decay modes (see [11,12] and the references therein). All are using the shape parametrization of TCSM. In order to be able to take into consideration the large number of combinations parent - emitted cluster (at least of the order of 10^5), we developed since 1980, the analytical superasymmetric fission model (ASAFM) with which we made the first predictions of nuclear lifetimes. In 1984 , before any other model was developed, we published the first estimates of the half-lives and branching ratios relative to α decay for more than 150 decay modes, including all cases experimentally confirmed up to now on ^{14}C, $^{24-26}$Ne, ^{28,30}Mg, and ^{32}Si radioactivities. A comprehensive table was produced by performing calculations within that model. Subsequently, the numerical predictions of ASAFM have been improved by taking better account of the pairing effect in the correction energy [35]. Cold fission fragments [36,37] were also considered in a new version of the tables [38]. The above mentioned systematics was further extended in the region of heavier clusters with mass numbers $A_e > 24$ [39].

Recently, the half-life estimations within ASAFM have been updated and the region of parent nuclei expanded far from stability and toward superheavy elements using the 1988 mass tables as input data for the Q-value [40] calculation.

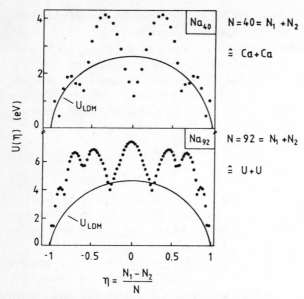

Fig. 11. Fragmentation potentials for atomic clusters according to R. Schmidt. They must be seen in analogy to the fragmentation potentials of U (asymmetric fission) and Ca (symmetric fission) respectivly. (Dotted Kohn-Sham (LDA) - jellium approx., solid Liquid Drop), $U = E(N_1) + E(N_2) - E(N)$.

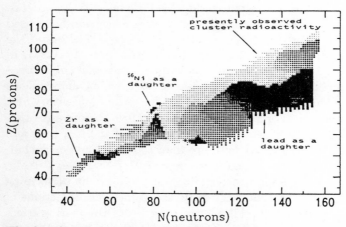

Fig. 12. Chart of nuclides with the regions of the most probable cluster emissions ($Z_e = 4 - 28$). Light points are clusters with small number of protons; dark points - with large Z.

The region of cluster emitters (see Fig. 12) extends well beyond that of α-emitters. Light clusters are preferentially emitted from neutron-deficient nuclei and the heavy ones from neutron-rich parents. Nevertheless, from practical point of view, the above mentioned conditions to observe cluster radioactivities are fulfilled in a few smaller areas.

The half-life of a nucleus (A, Z) against the split into a cluster (A_e, Z_e) and a daughter (A_d, Z_d), is calculated with analytical relationships derived from

$$T = [(h \ln 2)/(2E_v)] exp(\frac{2}{\hbar} \int_{R_a}^{R_b} \{2\mu[(E(R) - E_{cor}) - Q]\}^{(1/2)} dR \qquad (10)$$

where $\mu = m A_e A_d / A$ is the reduced mass, m is the nucleon mass, and $E(R)$ is the interaction energy of the two fragments separated by the distance R between centers. It is known that the fission barrier heights are too large within the liquid drop model. E_{cor} is a correction energy allowing to get a more realistic, lower and thinner barrier, and to introduce shell and pairing effects. R_a and R_b are the turning points of the WKB integral, h is the Planck constant, and E_v is the zero point vibration energy. For practical reasons (to reduce the number of fitting parameters) we took $E_v = E_{cor}$ though it is evident that, owing to the exponential dependence, any small variation of E_{cor} induces a large change of T, and thus plays a more important role compared to the preexponential factor variation due to E_v. By confusion, this correction was misquoted [41] as violating the energy conservation. In fact one can see easily that the Q-value is not changed; it only affects the height and the width of fission barrier in a way similar with the shell correction method.

The unified approach of three groups of decay modes (cold fission, cluster radioactivities and α-decay) within ASAFM is best illustrated on the example of ^{234}U nucleus (see the lower part of Fig. 9), for which all these processes have been measured. For α-decay, Ne- and Mg- radioactivities the experimental half-lives are in good agreement with our calculations (in fact the predicted lifetimes for cluster radioactivities have been used as a guide to perform the experiment). Spontaneous cold fission of this nucleus was not measured, but the experiments on induced cold fission are showing that ^{100}Zr is indeed the most probable light fragment.

An extensive study of the fission dynamics over a wide range of mass asymmetry has been performed [42-44], by replacing the reduced mass with the Werner-Wheeler inertia tensor and the liquid-drop model energy with Yukawa-plus-exponential potential extended to fragments with different charge densities [34].

In order to calculate the inertia tensor for very asymmetric fission, we had demonstrated the importance of taking into account a correction term, B_{ij}^c, due to the center of mass motion. We got [42] analytical expressions of the inertia for two parametrizations of nuclear shapes, obtained by intersecting two spheres of radii R_1, R_2 and volumes V_1, V_2 : "cluster-like" (with $R_2 =$ constant) and more compact ($V_2 =$ constant). Pik-Pichak [41] claimed that the later is the best parametrization for studying cluster radioactivities as fission processes. In fact he made some errors. By ignoring the above mentioned contribution of the center of mass motion he obtained wrong values for the inertia. As we have shown [42] in case of α-decay, the ratio of wrong to correct value of inertia may be as high as 30/4. A comparison of action integral K along the fission paths just mentioned shows that cluster-like shapes are more suitable (action integral lower) when the mass number of the emitted cluster is less than 34, giving additional support for the two-center shape parametrization within ASAFM.

From the same dynamical calculations [42] we concluded that the choice of the shape coordinate has no consequence on the value of action integral, determining the observable half life. In conclusion the separation distance of geometrical centers R is as good as the distance between mass centers of the fragments, again in contrast to what was claimed in Ref. [41]. A three-dimensional parametrization [45] has been used [43] to study the influence of a smooth neck on the fission dynamics. We performed calculations for α-decay, ^{28}Mg radioactivity and cold fission (with the light fragment ^{100}Zr) of ^{234}U . For every mass asymmetry, the optimum fission path in the plane of two variables: separation distance R, and the neck radius R_3 was determined by minimization of action integral. The corresponding shapes, along the fission path and in the absence of the smooth neck, are compared in Fig.13. One can see that the neck influence is stronger when the mass asymmetry parameter is small; for a very large mass asymmetry the parametrization of two intersected spheres is a good approximation of the optimum fission path.

Some authors are trying to show [46] that fission models are in conflicting relations with preformation cluster models. They have invoked even naive arguments, as we have mentioned [47]. We have compared the two kinds of models and our predictions with experimental data [48] pointing out that there was no method available which would allow preformation probabilites or quantum penetrabilities to be measured and thereby check the validity of the two kinds of theories. Recently [47] we gave a new interpretation of the cluster preformation probability within a fission model, as the penetrability of the prescission part of the barrier. On this basis we have shown that preformation cluster models are equivalent with fission models and we introduced one universal curve for each kind of cluster radioactivity. A number of successful experiments have been performed since 1984 (see the review [13], and [52]). They are compared with our early predictions and with those of 1986, made after including an even-odd effect in Fig.14. One can see the good agreement between the calculated and experimentally determined half-lives in a range of 15 orders of magnitude. The strong shell effect predicted by the theory has been confirmed.

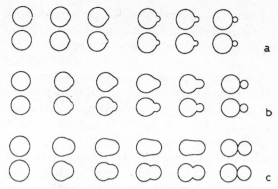

Fig. 13. Neck influence on fission dynamics of ^{234}U in a wide range of mass asymmetry. The nuclear shape along optimum fission path are compared with two intersected spheres with the same separation distance between centers for α-decay (a), ^{28}Mg radioactivity (b), and cold fission with light fragment ^{100}Zr (c).

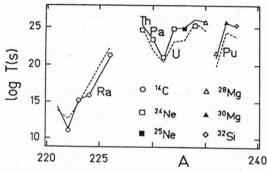

Fig. 14. Comparison of our early life-time predictions (dashed line) and the calculations of 1986 including an even-odd effect (full lines) with experimental points for cluster radioactivities confirmed up to now.

Summary

A unifying point of view of various nuclear phenomena may be best achieved on the basis of two center shell model, allowing to describe di-nuclear systems and the dynamical evolution from one initial nucleus to two final nuclei, or vice versa.

Nuclear molecules are very general phenomena. Light and intermediate nuclear molecule signatures have been already detected as resonances in the scattering cross sections, as various kinds of shape isomers (fission isomers, superdeformed and hyperdeformed nuclei at very high spin). In a sens even the nuclear systems exhibiting cluster radioactivities may be viewed as nuclear molecules.

Many important physical phenomena are expected to be seen in very heavy dinuclear systems (U-U, U-Cm, etc). They could be produced by using the existing heavy ion accelerators. Potential energy pockets supporting giant nuclear molecules have been predicted in various theoretical approaches [49] [50]. Such giant nuclear molecules are of fundamental importance for the observation of the decay of the vacuum in supercritical fields [51]. Moreover, these giant molecules seem to be the most superdeformed nuclei one can imagine. They deserve to be studied in their own right and may lead to exciting new nuclear structure in the years ahead.

Cluster radioactivity is a well established phenomenon. Conceptually it broadened our horizon by telling us that radioactivity is not limited to the three classical modes namely, α-decay, β-decay, and γ-decay. Furthermore, fission and especially cold fission and cluster radioactivity are the same phenomena.

References

1. D. A. Bromley, J. A.Kuehner, and E. Almquist, Phys. Rev. Lett. **4** (1960) 65.
2. N. Cindro, W. Greiner, and R. Caplar, eds., *Frontiers of Heavy Ion Physics* (World Scientific, Singapore), 1987.
3. W. Scheid, W.Greiner,and R. Lemmer, Phys. Rev. Lett. **25** (1970) 176.
4. P. Holzer, U. Mosel, and W. Greiner, Nucl. Phys. **A 138** (1969) 241.
5. J. Maruhn, and W. Greiner, Z. Phys. **251** (1972) 431.
6. M. Brack, J. Damgaard, A. Jensen, H. C. Pauli, V. M. Strutinsky, and C. Y. Wong, Rev. Mod. Phys. **44** (1972) 320.
7. H. J. Fink, J. Maruhn, W. Scheid, and W. Greiner, Z. Phys. **268** (1974) 321.
8. J. A. Maruhn, W. Greiner and W. Scheid, *in Heavy Ion Collisions* edited by R. Bock (North Holland, Amsterdam), Vol. 2, 1980, p. 399.
9. A. Sandulescu, D. N. Poenaru, and W. Greiner, Sov. J. Part. Nucl. **11**(1980) 528.
10. H. J.Rose and G. A. Jones, Nature **307** (1984) 247.
11. W. Greiner, M. Ivaşcu, D. N. Poenaru, and A. Sandulescu, *in Treatise on Heavy Ion Science* edited by D. A. Bromley (Plenum, New York), Vol. 8, 1989, p. 641.
12. D. N. Poenaru, M. Ivaşcu, and W. Greiner, *in Particle Emission from Nuclei* edited by D. N. Poenaru and M. Ivaşcu (CRC, Boca Raton, Florida), Vol. III, 1989, p. 203.
13. P. B. Price, Annu. Rev. Nucl. Part. Sci. **39** (1989) 19.
14. M. Greiner and W.Scheid, J. Phys. G. **12** (1986) L 285.
15. L. Brillard, A.G. Elayi, E.Hourani, M. Hussonnois, J.F. Le Du, L.H. Rosier, and L. Stab, C.R. Acad. Sci. **309** (1989) 1105.
16. G. Münzenberg, Rep. Prog. Phys. **51** (1988) 57.
17. A. Sandulescu, R. K. Gupta, W. Scheid, and W. Greiner, Phys. Lett. **60 B** (1976) 225.

18. F. Gönnenwein, B. Börsig and H. Löffler, in *Proc. Internat. Symp. on Collective Dynamics*, Bad Honnef, edited by P. David (World Sci., Singapore), 1986, p. 29.

19. E. K. Hulet, J. Wild, R. Dougan, R. Lougheed, J. Landrum, A. Dougan, M. Schädel, R. hahn, P. Baisden, C. Henderson, R. Dupzyk, K. Sümmerer, and G. R. Bethune, Phys. Rev. Lett. **56** (1986) 313.

20. J. M. Eisenberg and W. Greiner, *Nuclear Theory*, Vol. 1, (North Holland, Amsterdam), 3d edition, 1987.

21. J. Y. Park, W. Greiner, and W. Scheid, Phys. Rev. **C 21** (1980) 958.

22. R. M. Freeman, C. Beck, F. Haas, A. Morsad, and N. Cindro, Phys. Rev. **C 33** (1986) 1275.

23. J. Y. Park, W. Scheid, and W. Greiner, Phys. Rev. **C 6** (1972) 1565.

24. J. Y. Park, W. Scheid, and W. Greiner, Phys. Rev. **C 20** (1979) 1988.

25. G. Terlecki, W. Scheid, H. J. Fink, and W. Greiner, Phys. Rev. **C 18** (1978) 265.

26. R. Könnecke, W. Greiner, and W. Scheid, Phys. Rev. Lett. **51** (1983) 366.

27. A. Thiel, W. Greiner, J. Y. Park, and W. Scheid, Phys. Rev. **C 36** (1987) 647.

28. N. Cindro and M. Bozin, Phys. Rev. **C 39** (1989) 1665.

29. H. J. Lustig, J. A. Maruhn, and W. Greiner, J. Phys. G. **6** (1980) L 25.

30. D. C. Hoffman and L. P. Somerville *in Particle Emission from Nuclei* edited by D. N. Poenaru and M. Ivaşcu (CRC, Boca Raton, Florida), Vol. III, 1989, p. 1.

31. R. Herrman, J.A.Maruhn, and W.Greiner, J. Phys. G. **12** (1986) L 285.

32. D. Schnabel and H. Klein, private communication.

33. S. Cwiok, P. Rozmej, A. Sobichewski, and Z. Patyk, Nucl. Phys. **A 491** (1989) 281.

34. D. N. Poenaru, M. Ivaşcu, and D. Mazilu, Comp. Phys. Communications, **19** (1980) 205.

35. D. N. Poenaru, W. Greiner, M. Ivaşcu, D. Mazilu, and I. H. Plonski, Z. Phys. **A 325** (1986) 435.

36. D. N. Poenaru, M. Ivaşcu, and W. Greiner, Nucl. Tracks, **12** (1986) 313.

37. D. N. Poenaru, J. A. Maruhn, W. Greiner, M. Ivaşcu, D. Mazilu, and R. Gherghescu, Z. Phys., **A 328** (1987) 309.

38. D. N. Poenaru, M. Ivaşcu, D. Mazilu, R. Gherghescu, K. Depta, and W. Greiner, Central Institute of Physics, Bucharest, Report NP-54-86, 1986.

39. D. N. Poenaru, M. Ivaşcu, D. Mazilu, I. Ivaşcu, E. Hourani, and W. Greiner, in *Developments in Nuclear Cluster Dynamics* edited by K. Akaishi et al. (World Scientific, Singapore, 1989), p.76.

40. D. N. Poenaru, D. Schnabel, W. Greiner, D. Mazilu, R. Gherghescu, Atomic Data and Nuclear Data Tables, in print.

41. G. A. Pik-Pichak, Sov. J. Nucl. Phys. **44** (1987) 923.

42. D. N. Poenaru,J. A. Maruhn, W. Greiner, M. Ivaşcu, D. Mazilu, and I. Ivaşcu, Z. Phys. **A 333** (1989) 291.

43. D. N. Poenaru, M. Ivaşcu, I. Ivaşcu, M. Mirea, W. Greiner, K. Depta, and W. Renner, in *50 Years with Nuclear Fission* (American Nuclear Society, Lagrange Park) 1989, p 617.

44. D. N. Poenaru, M. Mirea, W. Greiner, I. Căta, and D. Mazilu, Modern Physics Letters A **5** (1990) 2101.

45. K. Depta, R. Herrmann, J. A. Maruhn, and W. Greiner, in *Dynamics of Collective Phenomena* edited by P. David (World Scientific, Singapore) 1987, p. 29.

46. B. G. Novatsky and A.A.Ogloblin, Vestnik AN SSSR, 1988, p. 81.

47. D. N. Poenaru and W. Greiner, J. Phys. G, in print.

48. D. N. Poenaru, W. Greiner, and M. Ivaşcu, Nucl. Phys., **A 502** (1989) 59c.

49. M. Seiwert, W. Greiner, and W. T. Pinkston, J. Phys. G., **11** (1985) L-21.

50. J. F. Berger, J. D. Anderson, P. Bonche, and M. S. Weiss, Phys. Rev., **41** (1990) R2483.

51. W. Greiner, B. Müller, and J. Rafelski, *Quantum Electrodynamics of Strong Fields* (Springer Verlag, Berlin), 1985.

52. A. A. Ogloblin et al., Phys. Lett. **B 235** (1990) 35.

Theories of Cluster Radioactivities

D.N. Poenaru* and W. Greiner

Institut für Theoretische Physik der J.W. Goethe-Universität,
W-6000 Frankfurt 11, Fed. Rep. of Germany
*Permanent Address: Institute of Atomic Physics, P.O. Box MG-6,
RO-76900 Bucharest, Romania

Fission- (FM) and preformation cluster- (PCM) models are reviewed, pointing out the differences between the adopted nuclear radii, zero point vibration energies, potential barriers, and the methods of calculating the halflives. FM are equivalent with PCM because cluster preformation probability is the penetrability of the prescission part of the barrier. A universal curve for each type of cluster radioactivity is derived. The analytical superasymmetric FM shape parametrization is the optimum fission path for clusters with mass numbers smaller than 34. The influence of a smooth neck (in a two dimensional fission dynamics) is important for symmetric and lowasymmetric fission.

1 Introduction

The unified approach of cluster radioactivities(CR), cold fission (CF) and α decay (αD) had been reviewed recently [1] [2]. A system exhibiting such kind of processes may be viewed as a nuclear molecule [3].

From the mass asymmetry point of view, CR are intermediate phenomena between CF and αD. Hence, it was quite normal to extend to this new field either fission theories or traditional methods used to study αD, reformulated correspondingly, in order to fit the new conditions.

In our papers of 1980 (for detailed references see [1] [2]) we predicted CR by using both fission models (fragmentation theory (FT) [4] based on asymmetric two-center shell model (ATCSM); three variants of numerical- (NSAFM) and one analytical- (ASAFM) superasymmetric fission model) and the Gamow penetrabilities [5] [6]. Side peaks have been obtained in the fission fragment mass distributions calculated with FT. On the basis of NSAFM computations, we concluded that α-decay can be considered a fission process. A semiempirical relationship based on fission theory gave the best agreement with experimental partial half-lives of almost 400 α-emitters. In this way, CR are making a bridge between extremely asymmetric αD and almost symmetric CF. Wrong conclusions have been drawn in other papers published before 1980, including speculations since 1924 concerning the abundance of some gases (like N and Ne) in uranium ores.

Springer Series in Nuclear and Particle Physics **Clustering Phenomena in Atoms and Nuclei**
Editors: M. Brenner · T. Lönnroth · F.B. Malik   © Springer-Verlag Berlin, Heidelberg 1992

The ^{14}C radioactivity of ^{222}Ra and ^{224}Ra was one of the examples we gave (with A. Sandulescu) in our paper of 1980. Four years later, the experiment performed by Rose and Jones [7] on ^{14}C spontaneous emission from ^{223}Ra, agreeing with the conclusions of our penetrability calculations, triggered a great excitement among the experimentalists and theorists. Solid state track detectors and the superconducting solenoidal spectrometer SOLENO, played a key role in obtaining further experimental results (see the review papers [8] [9] and the references therein). The fine structure in ^{14}C radioactivity of ^{223}Ra [10] has also been discovered [11] with SOLENO.

The measured quantity is the half-life, T, or the branching ratio with respect to α-decay, $b = T_\alpha/T$. In 1983 we have used ASAFM, allowing to find rapidly the lifetime, in a systematic search for new decay modes, beginning with He isotopes; ^5He radioactivity was predicted. In 1984 we continued this search for heavier and heavier clusters. Before any other model was published, we have estimated the half-lives for more than 150 decay modes, including all cases experimentally confirmed until now on ^{14}C, $^{24-26}Ne$, $^{28,30}Mg$, and ^{32}Si radioactivities. The numerical predictions have been improved by taking better account of the pairing effect [12]. In a new table [13], cold fission as cluster emission [14] [15] has been included. The unifying point of view [1] [2] has been stressed [16] [17]. The systematics was extended in the region of heavier emitted clusters (mass numbers $A_e > 24$) [18], and of parent nuclei far from stability and superheavies [19].

Theoretical works of other groups are published since 1985 (see [20-34] and the references therein). We have performed an extensive study of the one-dimensional and two-dimensional fission dynamics over a wide range of mass asymmetry [35 - 38]. In these calculations, we have used the same Yukawa-plus-exponential potential extended to fragments with different charge densities [39], as in 1979 (when we have shown that αD is a superasymmetric fission process), and a smoothed neck parametrization of the nuclear shape [40].

One can divide the theories in two main categories : fission- (FM) and pre-formation cluster- (PCM) models. P. B. Price [8] did a quantitative comparison between various theories and experiments. Good agreement with measurements have been obtained within almost all models.

We also have compared FM and PCM [41], and we proved the equivalence of these models [42] by interpreting the penetrability of the inner part of the barrier, within fission theory, as a cluster preformation probability.

The essential difference between a fission model and a preformed cluster model is the shape of the potential barrier : it has both an inner (prescission) and outer (postscission) part in a fission model, but only an outer one in a PCM. This difference is also reflected in the expression of the decay costant :

$$\lambda_f = \nu_f P_f \; ; \; \lambda_p = \nu_p S P_p \tag{1}$$

where the subscripts f and p denote fission and preformation, respectively, ν is the frequency of assaults on the barrier , P is a potential barrier penetrability, and S is the preformation probability. In a PCM the width $\Gamma = \hbar\lambda$ is sometime also expressed by $\Gamma = \gamma^2 P$, hence the reduced width $\gamma^2 = \hbar\nu S = E_\nu S/\pi$, where E_ν is the zero-point vibration energy.

The decay of a parent nucleus (AZ) by cluster emission leading to the emitted cluster $(A_e Z_e)$ and a daughter $(A_d Z_d)$, is calculated within ASAFM by the above equation, where

$$P_f = \exp(-K); \quad K = \frac{2}{\hbar} \int_{R_a}^{R_b} \{2B[(E(R) - E_{cor}) - Q]\}^{(1/2)} dR \qquad (2)$$

K is the WKB action integral, B is nuclear inertia approximated by the reduced mass $\mu = mA_e A_d/A$, m is the nucleon mass, and $E(R)$ is the interaction energy of the two fragments separated by the distance R between centers; it vanishes for $R \to \infty$. The correction energy E_{cor} allows to obtain realistic barriers; it is well known that they are too high in the Myers-Swiatecki liquid-drop model (LDM). It has no consequence on the Q-value; the energy conservation is not violated as it was argued [21] by misunderstanding. In fact this is a phenomenological shell and pairing correction, combined with the LDM bias mentioned above. R_a and R_b are the turning points of the WKB integral, and h is the Planck constant. In the following we shall analyze different prescriptions adopted for nuclear radii, zero point vibration energies, potential barriers, nuclear inertia, halflife calculations, and we shall discuss fission dynamics.

2 Nuclear Radii

For practical purposes it is useful to assume a nuclear volume proportional to the total number of nucleons. In fact this is a good approximation, and we get a radius

$$R = r_o A^{1/3} \qquad (3)$$

Various constants r_o have been used in different models (see Fig. 1). The LDM value adopted by ASAFM and by PP[2] is $r_o = 1.2249$ fm. In the Yukawa-plus-exponential model (Y+EM), used by SK, $r_o = 1.16$ fm. Unusually small values $r_o = 1.04$ fm and $r_o = 0.928$ fm, are needed to get a very high barrier height, compensating a preformation factor chosen equal to unity, by BM and PR, respectively.

By taking into account the surface difuseness, W. Myers had defined three types of radii: central (C); equivalent sharp (R), and equivalent rms (Q_r). For a Fermi distribution of the nuclear charge density and a mass number $A \geq 20$, one has :

$$C \simeq R(1 - d^2); \quad Q_r \simeq R(1 + 2.5d^2) \qquad (4)$$

where $d = b/R$ and $b = 1$ fm.

---

[2]For notations see the legend of Fig. 1 and 2.

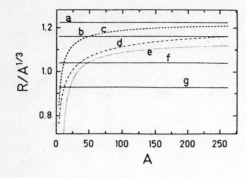

Figure 1: Variation of nuclear radii of different models with the nuclear mass number. (a) ASAFM and Pik - Pichak (PP); (b) Shanmugam - Kamalaharan (SK); (c) Blendowske - Walliser (BW); (d) Shi - Swiatecki (SS) and Gupta et al. (GM); (e) de Carvalho et al. (CMT); (f) Buck and Merchant (BM); (g) Price (PR).

From an effective sharp radius

$$R = 1.28A^{1/3} - 0.76 + 0.8A^{-1/3} \tag{5}$$

SS obtained a central radius $C = R - 1/R$, which was used in calculations. The expression

$$R = 1.233A^{1/3} - 0.978A^{-1/3} \tag{6}$$

has been considered by BW. The CMT choice

$$Q_r = 1.15A^{1/3} + 1.8A^{-1/3} - 1.2A^{-1} \tag{7}$$

leads to $R = Q_r/2 + \sqrt{(Q_r/2)^2 - 2.5}$, hence it has no real solution for small A, when $Q_r < 10$.

3 Frequency of Assaults

In eq. (2) the penetrability depends exponentially on the action integral. Consequently any small variation of a quantity which influences the potential barrier, like nuclear radius, or correction energy, induces a large transformation of the decay constant λ or of the halflife $T = \ln 2/\lambda$. A quantum linear harmonic oscillator has a "zero-point" energy $E_v = h\nu/2 = \hbar\omega/2$; its energy can never be smaller than this quantity. The zero-point energy for vibrations leading to fission is given by the same relationship, and the frequeny ν is very likely the same as the number of assaults on the barrier per second, from the PCM. For practical reasons (to reduce the number of fitting parameters) we took $E_v = E_{cor}$ though it is evident that, owing to the exponential dependence, a small variation of E_{cor} leads to a large change of T, and thus plays a more important role compared to the preexponential factor variation due to E_v. The values of $\log \nu(s^{-1}) = 20.684 + \log E_v(MeV)$ of other models are close to those of ASAFM (see Fig.2). $\nu(s^{-1}) = 4.836 \cdot 10^{20} E_v(MeV)$. Within NSAFM variant using Y+EM, we have obtained [18] :

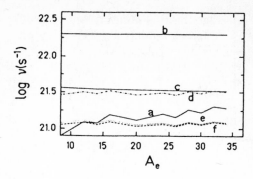

Figure 2: Decimal logarithm of the assault frequencies of different models versus emitted cluster mass number for even-even parents and clusters. (a) ASAFM; (b) SS: 22.30; (c) BW; (d) GM; (e) CMT; (f) SK. PR : 26.63; Barranco et al. (BBB): 21.00 for cluster decay and 20.30 for fission.

$$E_v = \hbar\sqrt{2E_b/\mu}/[2(R_t - R_i)]; \; E_b = E(R_t) - Q \tag{8}$$

where $R_t = R_1 + R_2$ and $R_i = R_o - R_2$ are the separation distances at the touching- and initial- point, respectively. Other equations are :

$$\nu = 6.9175 \cdot 10^{21}\sqrt{Q/\mu_A}/(C_1 + C_2) \; ; \; CMT \tag{9}$$

$$E_v = 14.306\sqrt{Q/\mu_A}/(C_1 + C_2) \; ; \; SK \tag{10}$$

$$\nu = v/R_o = 1.383 \cdot 10^{22}\sqrt{E_2/\mu_A}/R_o; \; E_2 = QA_1/A \; ; \; GM \tag{11}$$

$$\nu = v/(2R_a) = 3.45875 \cdot 10^{22}\sqrt{A/A_2}/R_a; \; \mu v^2/2 = 25A_2 \, MeV \; ; \; BW \tag{12}$$

Some authors (SS, PR, BBB) are keeping constant values for ν. This is a good approximation, because as shown in Fig. 2 the variations with A_e are small. For odd-mass parents SS and PR gave lower numbers as compared to the even-even case : 20.30; 25.04, respectively. The very large value of PR is an indication that the high barrier height of this model overcompensates the missing low preformation factor. The corresponding zero-point vibration energies, proportional with ν, are lying in the range 1 - 8 MeV for the majority of the models. They are larger (41.35 MeV and 889 GeV) for even-even nuclei, in case of SS and PR, respectively.

4   Potential Barriers

According to the classification given in the introduction, we denote by FM all the models in which the finite size of the fragments is taken into consideration when the potential barrier is calculated. Consequently, for example the models ASAFM, SK, SS, and BBB (see Fig.3), belong to this category. By considering the finite radius R_2 of the light fragment, the potential barrier in a fission model has an important inner part, coming from the early stage of the deformation process, when the fragments are not separated (overlapping

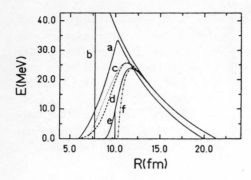

Figure 3: Potential barriers of different models for ^{14}C decay of ^{222}Ra. (a) ASAFM ; (b) PR; (c) SK; (d) SS; (e) BBB; (f) BW.

or prescission region), from the initial $R_i = R_0 - R_2$ to the touching point (or scission) configuration R_t. For heavier clusters the relative width of this inner part is broader. In numerical computations (see for example [39] used by NSAFM) both strong- and Coulomb- interactions contribute to the prescission part. The parameters of the Y+EM have been fitted not only to the elastic scattering data (like in case of PCM), determining the external tail (when the fragments are well separated), but also to the mass values, fission- and fusion-barriers, which are properties of the prescission regions of the potential energy surfaces (PES).

Nuclear structure may be also taken into account via Strutinsky type of shell- and pairing- corrections. In this way, for odd systems the barrier increases due to "specialization energy" and the fine structure of cluster radioactivity (discussed by [43] [44]) can be explained within FM. The specialization energy arises from the conservation of spin and parity of the odd particle during the fission process. With an increase of deformation (distance between fragment centers in the TCSM) the odd nucleon may not be transfered on a low energy at a level crossing if in this way it cannot conserve the spin and parity. The corresponding barrier becomes higher and wider, compared to that of the even-even neighbour. There would also be an increase of the nuclear inertia. Up to now there are technical difficulties to apply Strutinsky method at very large mass asymmetry, hence we could only use phenomenological shell and pairing corrections.

In a PCM (for example BW and PR in Fig. 3) the potential barrier is mainly of Coulomb nature and extends from the "channel radius" $R_t = R_1 + R_2$ up to the outer turning point $R_b = Z_1 Z_2 e^2 / Q$ if spherical shapes of nuclei are assumed. Usually the finite cluster extension is not taken into account. As above, we are using the subscript 2 or e for the emitted cluster and 1 or d for the daughter. The Y+EM potential barrier (SK) is very close to the proximity potential barrier (SS). Both BW and BBB potentials are semiempirical, suggested in heavy ion scattering processes.

5 Nuclear Halflives

The nuclear partial half-life, T, for CR is estimated in the framework of ASAFM with a smooth variation of correction energy versus mass number of the emitted cluster (see the new tables [19]). One has an analytical relationship derived from

$$T = [(\hbar \ln 2)/(2E_v)]exp(K) \tag{13}$$

and Eq. (2). The two parts of the action integral K, corresponding to the overlapping and separated fragments, respectively, are given explicitly in Ref.[1, 2]. Examples of results of a systematic search for new decay modes in the whole nuclear chart, and of unified approach of cluster radioactivities, α-decay and cold fission, are given in Ref. [3].

A comparison of experimental results with our early predictions and with those of 1986 [13] is given in Fig. 4. One can see the good agreement between the calculated and experimentally determined half-lives in a range of 15 orders of magnitude, though at the beginning we gave optimistic numbers for $A_e \geq 24$. Similar agreement of theory with experiment has been obtained within other models (see for example [8]). Nevertheless one should stress that many of them have been developed after most of existing experimental work had been done.

The strong shell effect predicted by the theory has been confirmed. It is evident in Fig. 4 from the variation of the half-life for ^{14}C radioactivity of Ra

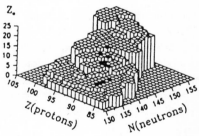

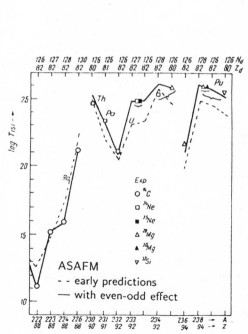

Figure 4: Comparison of the ASAFM early predictions (dashed lines) and calculations of 1986 with even-odd effect taken into account (full lines) with experimental points. In experiments with ^{233}U and ^{238}Pu , ^{24}Ne from ^{25}Ne and ^{28}Mg from ^{30}Mg are not resolved. The atomic number of the most probable cluster is plotted on the right side. Eleven confirmations are marked with a black top.

isotopes with the neutron number, N_d of the daughter and from the variation with both Z_d and N_d for ^{24}Ne radioactivity. The even-odd effect is also very clearly seen for ^{14}C radioactivity of Ra isotopes.

In almost all models the penetrability is calculated by the semiclassical Wentzel - Kramers - Brillouin (WKB) method. SS, PP and SK fission models proceed in a similar manner. There are two exceptions . Fermi's Golden Rule for the cluster transition rates is used by BBB to compute the decay constant :

$$\lambda = \frac{2\pi}{\hbar} \mid \langle \psi_E \mid \psi_0(\xi_n) \rangle \mid^2 v^2 \mid \psi_0(\xi_{n-1}) \mid^2 \frac{dN}{dE} \qquad (14)$$

where the initial state is described by the eigenfunction $\psi_0(\xi_n)$ of the Hamiltonian at the touching configuration and ψ_E is a continuum wavefunction of the fragments treated as particles. The matrix element connecting these two states is $v = -\Delta^2/(4G)$ with Δ the pairing gap and G the pairing force coupling constant. Here $\mid \psi_0(\xi_{n-1}) \mid^2$ is the probability that the system deforms into a configuration just one step in ξ previous to ξ_n. The generalized deformation variable ξ along the fission path is zero at the parent ground state and unity at the touching point; $dN/(dE)$ is the density of final states. The inertial mass

$$B = \hbar^2 \frac{2G}{\Delta_n^2 + \Delta_p^2} \left(\frac{dn}{d\xi} \right) \qquad (15)$$

depends on the pairing effects.

Another exception is the fission model GM which has a preformation cluster probability, like a PCM. This S is calculated within fragmentation theory, by solving a Schroedinger equation with the mass asymmetry $\eta = (A_1 - A_2)/A$ as a variable (see [3]). The expression giving the mass yield is taken as preformation probability at the touching point, where one has an excitation $E(R_t)$. The penetrability is a product of three terms : WKB transmission of the barrier at excitation energy $E(R_t)$; deexcitation $W_i = \exp(-b(E(R_t) - Q))$ with $b = 0$, like [10], and another WKB integral after deexcitation up to the external turning point R_b at the final energy Q.

In a typical PCM model (BW), the spectroscopic factor, expressing the probability of finding the open channel structure (the antisymmetrized (ϕ_d plus ϕ_e) structure) in the parent nucleus, is given by the expectation value of a projection operator $\hat{P}$ onto this structure :

$$S = \langle \phi \mid \hat{P} \mid \phi \rangle \qquad (16)$$

where ϕ, ϕ_e, ϕ_d are the many-body wave functions of the parent, emitted cluster and daughter, respectively. S is related to the overlaps of the states of ϕ_e with the last A_e states in ϕ. Calculations have been performed for emission of ^{4}He, ^{12,14}C and ^{16}O. A law of linear decrease of $\log S$ has been obtained by using these calculated values and a fit to experimental data. It is difficult to understand a choice by BM and PR of a spectroscopic factor equal to unity in a PCM.

6 Preformation as Barrier Penetrability

In our fission models we have expressed P_f as a product of two probabilities: $P_f = P_{ov}P_s = \exp[-(K_{ov} + K_s)]$. P_{ov} corresponds to the inner part of the barrier (overlapping or prescission stage) and is calculated by changing R_b with R_t in Eq. (2). P_s is the penetrability in the region of separated fragments. The Eqs. (1) look now

$$\lambda_f = \nu_f P_{ov}P_s \; ; \; \lambda_p = \nu_p SP_p \tag{17}$$

One can say [42] that *in a fission model the preformation probability is the penetrability of the inner part of the barrier* :

$$P_{ov} = \exp(-K_{ov}) = S_f \tag{18}$$

This semiclassical method of finding the spectroscopic factor could be also generalized to lighter systems and other kinds of processes.

A linear variation of $\log S$ with A_e has been found by BW within PCM, by fitting the experimental data, as we mentioned above, but it was not clear up to which value of A_e this law remains still valid. We have performed within ASAFM, a calculation of the corresponding quantity $\log P_{ov}$ and found also a straight line up to A_e around 32, but for heavier clusters there is a bending (limitation) due to the fact that fission barriers are no longer increasing.

The equation of *universal curves*, obtained by taking ν constant and $\log P_{ov}$ proportional to A_e for even-even combinations is

$$\log T = -\log P_s - 22.169 + 0.598(A_e - 1) \tag{19}$$

In the plane $(\log T, -\log P_s)$, it represents one straight line for each cluster radioactivity. For separated fragments

$$-\log P_s = 0.22873(\mu_A Z_1 Z_2 R_b)^{1/2}[\arccos \sqrt{r} - \sqrt{r(1-r)}] \tag{20}$$

where $r = R_t/R_b$, $R_t = 1.2249(A_1^{1/3} + A_2^{1/3})$, $R_b = 1.43998 Z_1 Z_2/Q$, and $\mu_A = A_1 A_2/A$ is the reduce mass number.

The slope of any universal curve equals unity, hence it can be easily drawn. The vertical distance between two universal curves corresponding to A_{e1} and A_{e2} is $0.598(A_{e2} - A_{e1})$.

The advantage of such a simple relationship (see also [27]) is evident. Nevertheless one should be aware of its limitations. A closer estimation of the deviations from experimental results shows that the shell effects not taken into account in ν, are very clear present, leading to a minimum around the magic number of neutrons of the daughter nucleus. We have found deviations from experimental data of α - decay halflives of the universal curve within $\pm$ one order of magnitude. Consequently, it is important to consider the variation of ν with A_1, Z_1, A_2, Z_2, as we have shown in Eq. (8).

7 Fission Dynamics

The inertia tensor, $\{B_{ij}\}$, depends on the arbitrarily chosen set of shape coordinates $\{q_i\}$ $(i = 1, 2, ..i, ..j, ..n)$. The Werner-Wheeler approximation to irrotational flow motion within hydrodynamical model can be used in a wide range of mass asymmetry, and allows to obtain analytical results for two parametrizations : the ASAFM "cluster-like"- (R_2 = constant) and PP more compact- (fragment volumes = constant) shapes.

In view of some confusions and errors [21] one has to stress how important is to take into account a correction term, B_{ij}^c, due to the center of mass motion, or to place the origin of coordinates in this center (see [35]).

$$B_{ij}^c = -(\pi^2 \sigma / V) \int_{z_{min}}^{z_{max}} \rho_s^2 X_i dz \int_{z_{min}}^{z_{max}} \rho_s^2 X_j dz \qquad (21)$$

where $\rho_s = \rho_s(z)$ is the nuclear surface equation in cylindrical coordinates, with z_{min}, z_{max} intercepts on z axis. Functions X_i are calculated as a sum of two terms for the left (l) and right (r) side of the shape. V is the volume of the system, assumed to be conserved, $\sigma = 3m/(4\pi r_o^3)$ is the mass density, m is the nucleon mass, and $r_o = 1.16$ fm within Y+EM.

$$X_{il} = -\rho_s^{-2} \frac{\partial}{\partial q_i} \int_{z_{min}}^{z} \rho_s^2 dz, \qquad X_{ir} = \rho_s^{-2} \frac{\partial}{\partial q_i} \int_{z}^{z_{max}} \rho_s^2 dz, \qquad (22)$$

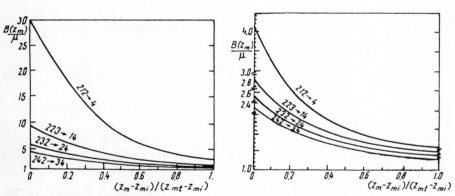

Figure 5: Comparison of wrong (left) and correct (right) values of inertia for compact shapes with the origin of coordinate system in the separation plane. Only the mass numbers are indicated, because B is independent on atomic numbers.

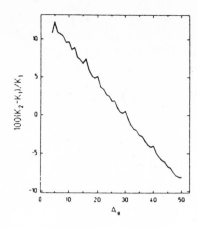

Figure 6: Relative difference of the action integrals along cluster-like - (K_1) and more compact- (K_2) shapes fission paths for cluster radioactivities.

By changing the shape coordinate from R to z_m (the distance between mass centers of the fragments), the inertia becomes :

$$B(z_m) = B(R) \left(\frac{dR}{dz_m} \right)^2 \tag{23}$$

For cluster-like shapes, both $B(R)$ and $B(z_m)$ are increasing functions of the respective variable. On the contrary, $B(z_m)$ decreases but $B(R)$ increases for the more compact shapes. This may be taken as a justification for the empirical law used by Fiset and Nix to study almost symmetrical fission. The parametrization of $V_2 =$ constant is not suitable for CR with $A_2 < 34$, hence a better choice for the law of variation of $B(z_m)$ in SK would be an increasing function.

When the motion of the center of mass is not taken into account (like in PP), the inertia are much higher (see Fig. 5) than the corrected values. One, arbitrarily chosen, independent shape coordinate (separation of mass centers z_m) can not be better than the other (separation of geometrical centers, R) describing the same shapes, as it was argued by PP, because the action integral is independent of this choice. This property follows immediately by substituting $B(R)$ with $B(z_m)$ in K given by Eq. (23). In fact if the final result - the halflife would depend on the deformation coordinate, all calculations would be nonsens.

In a next step, we have introduced a second shape degree of freedom - the neck curvature. The asymmetric two-center shape [40] is obtained by smoothly joining two spheres of radii R_1 and R_2 with a third surface generated by rotating around the symmetry axis a circle of radius $R_3 = S$.

Calculations have been performed for a fixed mass asymmetry, hence the (symmetric) inertia tensor had three components B_{RR}, B_{SS} and B_{RS}.

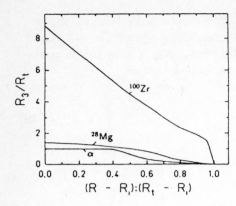

Figure 7: Optimum fission paths for α-decay, ^{28}Mg radioactivity and cold fission with the light fragment ^{100}Zr.

The multi-dimensional penetration problem through the potential barrier is reduced to a single-dimensional one by defining an inertia scalar, $B(s)$, along a path of equation, given parametrically : $q_i = q_i(s)$ $(i = 1, 2, ...n)$. One has :

$$B = B_{RR} + 2B_{RS}(dS/dR) + B_{SS}(dS/dR)^2 \qquad (24)$$

The optimum fission path is determined by minimizing the integral K_{ov}. For separated fragments, the neck is no more present.

We performed calculations for α-decay, ^{28}Mg radioactivity and cold fission (with the light fragment ^{100}Zr) of ^{234}U - the first nucleus for which all three groups of decay modes have been experimentally detected. In the absence of the smoothed neck, for two intersected spheres with cusp, $K_{ov} = 7.1\%$, 35.2%, and 62.5% from the total value of K .

In the presence of the smoothed neck, the optimum fission trajectories are plotted in Fig. 7. One can see an almost linear decrease of $R_3(R)$ for cold fission, and a kind of plateau at much lower values for small distances between centers in case of α - decay and ^{28}Mg radioactivity.

As a result of the minimization, K_{ov}/K has been reduced to $(K_{ov}/K)_{opt} = 5.9\%$, 32.9%, and 55%, respectively.

In conclusion, the neck influence is stronger when the mass asymmetry parameter is small; for a very large mass asymmetry the parametrization of two intersected spheres is a good approximation of the optimum fission path.

Alternatively [45], we are using a method based on the solution of a differential equation:

$$DS'' + D_3 S'^3 + D_2 S'^2 + D_1 S' + D_0 = 0 \qquad (25)$$

where coefficients D are expressions containing partial derivatives with respect to S and R of inertia tensor components and of energy.

Acknowledgments

This work is supported by the Bundesministerium für Forschung und Technologie, Bonn, the Institute of Atomic Physics, Bucharest, and the German-Romanian agreement for research and technology directed by the Stabsabteilung Internationale Beziehungen of the KfK Karlsruhe.

References

[1] W. Greiner, M. Ivaşcu, D. N. Poenaru, and A. Sandulescu, *in Treatise on Heavy Ion Science* edited by D. A. Bromley (Plenum, New York), Vol. 8, 1989, p. 641.

[2] D. N. Poenaru, M. Ivaşcu, and W. Greiner, *in Particle Emission from Nuclei* edited by D. N. Poenaru and M. Ivaşcu (CRC, Boca Raton, Florida), Vol. III, 1989, p. 203.

[3] W. Greiner, invited talk at the present Conference.

[4] J. A. Maruhn, W. Greiner and W. Scheid, *in Heavy Ion Collisions* edited by R. Bock (North Holland, Amsterdam), Vol. 2, 1980, p. 399.

[5] G. Gamow, Z. Phys. **51** (1928) 204.

[6] E. H. Condon and R. W. Gurney, Nature **122** (1928) 439.

[7] H. J. Rose and G. A. Jones, Nature **307** (1984) 247.

[8] P. B. Price and S.W.Barwick, in *Particle Emission from Nuclei* edited by D. N. Poenaru and M. Ivaşcu (CRC, Boca Raton, Florida), Vol. II, 1989, p. 206; P. B. Price, Annu. Rev. Nucl. Part. Sci. **39** (1989) 19; invited talk at the present Conference.

[9] E. Hourani and M. Hussonnois, in *Particle Emission from Nuclei* edited by D. N. Poenaru and M. Ivaşcu (CRC, Boca Raton, Florida), Vol. II, 1989, p. 171; E. Hourani, M. Hussonnois, and D. N. Poenaru, Ann. Phys. (Paris) **14** (1989) 311.

[10] M. Greiner and W.Scheid, J. Phys. G. **12** (1986) L 285; M. Greiner, W. Scheid, and V. Oberacker, J. Phys. G. **14** (1988) 589.

[11] L. Brillard, A.G. Elayi, E. Hourani, M. Hussonnois, J.F. Le Du, L.H. Rosier, and L. Stab, C.R. Acad. Sci. **309** (1989) 1105.

[12] D. N. Poenaru, W. Greiner, M. Ivaşcu, D. Mazilu, and I. H. Plonski, Z. Phys. A **325** (1986) 435.

[13] D. N. Poenaru, M. Ivaşcu, D. Mazilu, R. Gherghescu, K. Depta, and W. Greiner, Central Institute of Physics, Bucharest, Report NP-54-86, 1986.

[14] D. N. Poenaru, M. Ivaşcu, and W. Greiner, Nucl. Tracks, 12 (1986) 313.

[15] D. N. Poenaru, J. A. Maruhn, W. Greiner, M. Ivaşcu, D. Mazilu, and R. Gherghescu, Z. Phys., A 328 (1987) 309.

[16] R. Herrman, J.A.Maruhn, and W.Greiner, J. Phys. G. 12 (1986) L 285.

[17] R. Herrman, K. Depta, D. Schnabel, H. Klein, W. Renner, D. N. Poenaru, A. Săndulescu, J. A. Maruhn, and W. Greiner, in Proc. Symp. on *Physics and Chemistry of Fission*, Gaussig, 1988, in print.

[18] D. N. Poenaru, M. Ivaşcu, D. Mazilu, I. Ivaşcu, E. Hourani, and W. Greiner, in *Developments in Nuclear Cluster Dynamics* edited by K. Akaishi et al. (World Scientific, Singapore, 1989), p.76.

[19] D. N. Poenaru, D. Schnabel, W. Greiner, D. Mazilu, R. Gherghescu, Atomic Data and Nuclear Data Tables, in print.

[20] Y. J. Shi and W. J. Swiatecki, Nucl. Phys. A 464 (1987) 205.

[21] G. A. Pik-Pichak, Sov. J. Nucl. Phys. 44 (1987) 923.

[22] F. Barranco, R. A. Broglia, and G. F. Bertsch, Phys. Rev. Lett. 60 (1988) 507; R. Bonetti, E. Fioretto, C. Migliorino, A. Pasinetti, F. Barranco, E. Vigezzi, and R. A. Broglia, Phys. Lett. B 241 (1990) 179.

[23] S. S. Malik and R. K. Gupta, Phys. Rev. C 39 (1989) 1992.

[24] G. Shanmugam and B. Kamalaharan, Phys. Rev. C 41 (1990) 1184.

[25] S. Landowne and C. H. Dasso, Phys. Rev. C 33 (1986) 387.

[26] M. Iriondo, D. Jerrestam, and R. J. Liotta, Nucl. Phys. A 454 (1986) 252.

[27] R. Blendowske and H. Walliser, Phys. Rev. Lett. 61 (1988) 1930; R. Blendowske, T. Fliessbach, and H. Walliser, unpublished, 1991.

[28] V. A. Rubchenya, V. P. Eysmont, and S. G. Yavshits, Izv. A. N. SSSR, ser. Fiz. 50 (1986) 1017.

[29] S. G. Kadmensky, S. D. Kurgalin, V. I. Furman, and Yu. M. Tchuvilsky, Yad. Fiz. 51 (1990) 50.

[30] I. Rotter, J. Phys. G. 15 (1989) 251.

[31] M. Ivaşcu and I. Silişteanu, Nucl. Phys. A 485 (1988) 93.

[32] H. G. de Carvalho, J. B. Martins, and O. A. P. Tavares, Phys. Rev. C 34 (1986) 2261.

[33] N. Cindro and M. Bozin, Phys. Rev. C 39 (1989) 1665.

[34] B. Buck and A. C. Merchant, Phys. Rev. C **39** (1989) 2097.

[35] D. N. Poenaru,J. A. Maruhn, W. Greiner, M. Ivaşcu, D. Mazilu, and I. Ivaşcu, Z. Phys. A **333** (1989) 291.

[36] D. N. Poenaru, M. Ivaşcu, I. Ivaşcu, M. Mirea, W. Greiner, K. Depta, and W. Renner, in *50 Years with Nuclear Fission* (American Nuclear Society, Lagrange Park) 1989, p 617.

[37] D. N. Poenaru, M. Mirea, W. Greiner, I. Căta, and D. Mazilu, Modern Physics Letters A **5** (1990) 2101.

[38] K. Depta, J. A. Maruhn, Hou-Ji Wang, A. Sandulescu, and W. Greiner, Int. J. Mod. Phys. **5** (1990) 3901.

[39] D. N. Poenaru, M. Ivaşcu, and D. Mazilu, Comp. Phys. Communications, **19** (1980) 205.

[40] K. Depta, R. Herrmann, J. A. Maruhn, and W. Greiner, in *Dynamics of Collective Phenomena* edited by P. David (World Scientific, Singapore) 1987, p. 29.

[41] D. N. Poenaru, W. Greiner, and M. Ivaşcu, Nucl. Phys., A **502** (1989) 59c.

[42] D. N. Poenaru and W. Greiner, J. Phys. G, in print.

[43] M. Hussonnois, J. F. Le Du, L. Brillard, and G. Ardisson, Phys. Rev. C **42** (1990) R495.

[44] R. K. Sheline and I. Ragnarsson, Phys. Rev. C **43** (1991) 1476.

[45] M. Mirea, D. N. Poenaru and W. Greiner, to be published.

Mechanism of the Spontaneous and Induced Two-Body Fragmentation of Heavy Nuclei

V.A. Rubchenya and S.G. Yavshits

V.G. Khlopin Radium Institute, 194022 St. Petersburg, Russia

1. Introduction

The study of nuclear binary decay is one of the main methods in nuclear physics to investigate the large-scale nuclear collective motion, the role of the shell structure and cluster effects. The fission and the α—decay are two well-known extreme modes of binary decay of heavy nuclei. The mass distribution of the light fission fragment is cut off near $A_f \leq 80$. Then there is the wide unexplored region of the intermediate fragment masses $4 \leq A_f \leq 80$. As the α—decay is connected with the role of the ^{4}He shell structure, and the fission process is influenced by the strong stability of clusters near ^{132}Sn, we can expect that the binary decays in the intermediate fragment mass region are governed by the shells N,Z = 8, 20, 28, 50, 82 and 126. The gap between the two extreme limits began to fill in the 1970s when the intensive investigations of the fragmentation mechanism in in reactions with high-energy beams was started, see for example [1,2]. The discovery of the spontaneous emission of ^{14}C by ^{223}Ra [3-5] showed the shortcomings in the theory of two-body fragmentation phenomena. The known features of this phenomenon give us the possibility to treat it as a special kind of decay from ground and excited states which is often named "cluster radioactivity". In our early work [6] we proposed the name "f-decay" to emphazise the similarity of this process both for the ground and the excited states.

At present the spontaneous fragmentation has been found for a number of nuclei from Ra to Pu. The common feature for all identified f-decays is the crucial role of the doubly-magic cluster ^{208}Pb. It is necessary to note also that the probability of spontaneous fragmentation decreases quite rapidly with the increase of the parameter Z^2/A in contrast to that of the spontaneous fission [8]. A recent observation of the fine structure of the ^{14}C yields emitted from ^{223}Ra [9] indicates the influence of the single-particle effects of the new radioactivity.

The spontaneous f-decay is a very rare process and its experimental investigation is a very hard task. The fragmentation from the highly excited nuclear state occurs with fairly large cross sections ($\sigma \approx 1$ mb), and a large amount of experimental data has been obtained for such reactions [10]. One of the most important experimental results is the power law of the fragment yield, $Y \propto A^{-\tau}$, see [11]. The value of the parameter τ is 2-4 for different

Springer Series in Nuclear and Particle Physics Clustering Phenomena in Atoms and Nuclei
Editors: M. Brenner · T. Lönnroth · F.B. Malik   © Springer-Verlag Berlin, Heidelberg 1992

beam energies and particles. Such a dependence may be associated with the multifragmentation process of nuclear matter in the critical state at the last stage of the compression-expansion process. However, in order to have reliable conclusions about the multifragmentation presence it is necessary to pick out the contribution of two-body fragmentation.

The f-decay phenomenon is placed between α—decay and spontaneous fission and gives a new impulse to the study of nuclear collective motion. The explanation of spontaneous fragment emission has been proposed in a number of works which one can arrange in three groups.

First, there are a number of works where spontaneous f-decay is treated as a super asymmetrical fission process [12-18]. In these works the different nuclear shapes or potential barriers along the path in the space of collective variables leading to the strong asymmetrical fission have been used. In refs. [12-14] using a simple parametrization of two crossed spheres the fission half-lives are calculated using an analytical expression that corresponds to the tunneling through a potential barrier represented by a simple parametrization. An unphysically large value of the zero-point vibration energy that violates the energy conservation for the fragments has been introduced in such an approach. In the work [15] the parametrization of two crossed spheroids is used but the shell effects in potential energy and mass parameter are not taken into consideration. A proximity potential in the potential barrier calculations near the tangent point has been used by Y. Shi and W. Swiatecki [16-18]. In the nuclear overlapping region the potential energy is approximated by a smooth polynomial approximation. A similar method is used in ref. [19] where the potential energy in the overlapping region is approximated by a cubic parabola.

In the second group a number of works treat the spontaneous emission of intermediate mass fragments as a kind of α—decay [20-25]. The main difficulty in this approach is connected with the microscopic calculation of spectroscopic factors for massive fragments. The calculation for the known ^{14}C decays provide a good description, but for the emission of heavier fragments this approach predicts very small probability.

Thirdly, an intermediate approach between the two previous ones has been proposed in our early works [6,7,26]. In our model it is assumed that in the process of zero-point vibrations the nucleus reaches some decay configuration which strongly overlaps with a tangent fragment configuration and then the fragment moves through the coulomb barries like in the case of α—decay. The fragment formation probability is then given by the square of the collective ground-state wave function at the decay configuration. A similar idea is used in [27] emphasizing the role of the pairing residual interaction in the transition from one Hartree-Fock state to another. The main problem here is to obtain the corresponding Schrödinger equation, and the calculation of the appropriate collective parameters. Later in [26] we have generalized this treatment to include the induced high-energy f-decay case.

In this recent report the improved model of the f-decay is formulated and new results both for spontaneous and induced two-body fragmentation are presented.

2. Spontaneous Fragmentation

The many-particle nuclear wave function contains the mixtures of different channel functions. The channel-decay widths for different fragments are defined in a similar fashion as in the case of the one-body theory of the α—decay, by a probability of fragment formation on the nuclear surface, γ_f, and a penetrability potential P_f via the expression

$$\Gamma_f = \hbar \frac{v_f}{2R} \cdot \gamma_f \cdot P_f, \tag{1}$$

where v_f is the velocity of the fragment cluster (f-cluster), and R is the nuclear radius. The collective effects are not included in this expression. The effects of shell structure and residual interactions are taken into account in the calculation of γ_f. In the case of α—decay in the Ra region the probability is $\gamma_f \sim 1$, but the ^{14}C emission experiments give a value of $\gamma_f \sim 10^{-5}$. The calculations of spectroscopic factors in the framework of the α—decay theory [20,21] show a rapid decrease of the cluster formation probability in the ground state. The probabilities γ_f are reduced by 7 orders of magnitude for the formation of ^{8}Be as compared to ^{20}Ne clusters in heavy nuclei. It is apparent that with the increase of the cluster mass the role of collective properties must increase also due to deeper rearrangements of a large number of nucleons that must take place between the initial ground state of the parent nucleus and the two two ground states of the daughter fragments.

The probability of f-cluster formation at the equilibrium deformations is small but with the increase of the deformation the value of γ_f can rapidly increase. It is difficult to extend the microscopic calculations of the preformation probability γ_f to large deformations. We shall assume that at some nuclear shape to be named 'decay configuration' the value of preformation is close to unity. It seems that the decay configuration must strongly overlap with the touching configuration of fragment and daughter nuclei. In the case of the decay from the ground state, the nucleus deforms to reach the decay configuration in the process of zero-point shape oscillations. Then the decay width Γ_f is defined by two factors: the first one is the probability to reach the decay configuration N_f, and the second is the potential barrier penetrability P_f, that is

$$\Gamma_f(A_f, Z_f, A, Z) = \hbar N_f P_f(Q_f). \tag{2}$$

Here Q_f is the decay energy release in the case of emission of a fragment (A_f, Z_f) by the nucleus (A, Z).

The shape of the nucleus is described by a set of collective variables which includes the parameters of quadrupole, octupole and hexadecapole deformations. The ground state is characterized by the equilibrium values of the deformation parameters $\{\alpha^0\}$ which can be found within the framework of the Strutinsky shell correction method [28], or can be taken from experiments. The decay configuration is defined by the deformation parameter set $\{\alpha^f\}$. The touching configuration of the daughter and fragment nuclei is described by the set $\{\alpha^t\}$ consisting of the mass and charge asymmetries, nuclear deformations and some other parameters. In order to determine the connection between the parameters $\{\alpha^f\}$ and a given set $\{\alpha^t\}$ we use the moments method proposed in our previous work [29] to study the fragment configuration in fission. The method is based on the condition of strong overlapping of the density distribution function of the nucleons before and after the nuclear division. The parameters of the decay configuration $\{\alpha^f\}$ can be found from the solution of the system of equations for lower density distribution moments which are equal for both configurations:

$$I_{lm}(\{\alpha^f\}) = I_{lm}(\{\alpha^t\}), \quad l, m = 0, 1, 2, \ldots \qquad (3)$$

In cylindrical coordinates these moments are, for axially symmetric shapes, given by the integral

$$I_{lm} = 2\pi \int dz z^l \int dr r^{m+1} \rho(r, z),$$

where $\rho(r, z)$ is a density distribution function. These equations have to be solved by requiring the equality of potential deformation energy for the decay configuration $\{\alpha^f\}$ and the interaction energy between the fragments if the touching configuration $\{\alpha^t\}$, i.e.

$$V_{def}(\{\alpha^f\}) = V_{int}(\{\alpha^f\}) - Q_f. \qquad (5)$$

In actual numerical calculations it is supposed that the two fragments are spherical and only the Coulomb interaction energy is taken into consideration.

To calculate the collective wave function it is in general necessary to have a potential energy surface $V_{def}(\{\alpha\})$ and a mass parameter tensor. In order to simplify the calculations, we shall use the one-dimensional Schrödinger equation with a variable mass

$$-\frac{\hbar^2}{2\sqrt{M(x)}} d/dx \frac{1}{\sqrt{M(x)}} \frac{d\Psi(x)}{dx} + V(x)\Psi(x) = E \cdot \Psi(x). \qquad (6)$$

This equation describes the evolution of the system along the path in the space of deformation parameters which lead to the decay configuration point $\{\alpha^f\}$. After point x_f, which corresponds to the decay configuration, the variable x transforms to the relative distance between the centers of mass of the two nuclei R, and the mass parameter becomes equal to the reduced mass. In the ground state at $x = x_0$, the mass parameter $M(x)$ can be estimated from the expression for the frequency of collective oscillations $\omega_0 = \sqrt{C_2/M(x_0)}$, where C_2 is the stiffness coefficient at the equilibrium point.

The formation probability N_f is given by the square of the ground-state wave function at the decay configuration $x = x_f$

$$N_f(A, Z, A_f, Z_f) = \frac{\omega_0}{2\pi} |\Psi_0(x_f)|^2 |\frac{dx}{dA_f}|_{x=x_f}. \tag{7}$$

Here ω_0 and Ψ_0 are the frequency and the wave function, respectively, of the ground state. The Schrödinger equation (6) with a variable mass has been solved by diagonalizing the hamiltonian using an oscillator basis. In this case the problem is reduced to the generalized problem of the eigenvalues of the wave function $\phi = M^{-1/4}\Psi$. For the mass parameter transformation at the point $x = x_f$ we use the expression $M(x_f) = \mu(\frac{dR}{dx})^2_{x=x_f}$, where μ is the resduced mass. In the calculations we hae used a linear approximation for the mass parameter dependence on the collective coordinate

$$M(x) = M_0, \qquad x \leq x_0 \tag{8}$$
$$= M_0(1 + ax), \quad x_0 < X < x_f.$$

In the equilibrium state the mass parameter depends on the odd-evenness of the mass and charge number due to the pairing effects. The calculation shows that in the case of an even-even nucleus the value of M_0 exceeds the value of $M(x_f)$ by a factor of about 3. We have used the following approximate relation for M_0

$$M_0 = b \cdot M(x_f) \cdot (1 + p \cdot \delta_p),$$

where $\delta_p = 0$ for even A and $=1$ for odd A, and the constant $b = 3.1$. The coefficient p is due to the pairing interaction and has been chosen equal to unity $(p \sim 1)$.

The calculated values of the stiffness coefficient in the ground state C_2 and the mass parameter define the frequency of the oscillator $\hbar\omega_0 = \sqrt{C_2/M_0}$. The dependence of the potential energy at $x < x_f$ is approximated by a third-order polynomial

$$V_{def}(x) = 1/2C_2x^2 + C_3x^3. \tag{10}$$

In our calculations we used the parametrization of the shapes of nuclei proposed by V. Pashkevich [30] where the Cassini ovaloids in a lemniscate coordinate system are used as basic figures. The set $\{\alpha^f\}$ has been constrained by four variables, namely $\epsilon, \alpha_1, \alpha_3$ and α_4. Here ϵ is the lemniscate

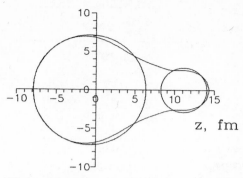

r, fm

z, fm

Fig. 1. Shapes of nuclei at the decay point for the decay $^{223}\text{Ra} \Rightarrow {}^{14}\text{C} + {}^{209}\text{Pb}$.

parameter which is connected with the quadrupole deformation, α_1 and α_3 define the octupole type of deformation, and α_4 is the hexadecapole deformation parameter. Because the f-decay of heavy nuclei is determined by the influence of the nucleon shells N=126 and Z=82, the daughter nuclei are considered to be spherical. Then the set $\{\alpha^t\}$ includes three variables: R_f and R_d are the fragment and daughter nucleus radii, and R is the distance between the centers of the nuclei at the decay configuration. The density distribution function is chosen to be the Fermi function with the radius $R_0 = 1.2 \cdot A^{1/3}$ and the diffuseness parameter a=0.54 fm.

To find the decay configuration parameters $\{\alpha^t\}$ the system of equations (4) has been solved together with equation (5). The result of the solution presented in figure 1 is obtained for the ^{223}Ra decay configuration $\{\alpha^f\}$ described by the parameters $\epsilon^f = 0.60, \alpha_1^f = -0.25, \alpha_3^f = 0.25$ and $\alpha_4^f = 0.02$. The calculated values of deformation parameters in the ^{223}Ra ground state are $\epsilon^0 = 0.12, \alpha_1^0 = -0.15, \alpha_3^0 = 0.10$ and $\alpha_4^0 = 0.02$. In the Ra - Th region the neutron-deficient isotopes have a stable octupole defomation and the f-decay probabilities will be enhanced.

The experimental and theoretical values of the f-decay half-lives are presented in Table 1. The following qualitative conclusions may be drawn from the analysis of the experimental data. The f-decay probability of the nuclei with odd mass number is lower (if we take into account the Q_f variation) than in the case of even neighbouring nuclei. Compare, for example, the decays of ^{222}Ra and ^{223}Ra. This effect may be explained by the influence of the pairing correlations on the mass parameter. For an even nucleus the mass parameter is lower, and the ground state of equation (6) is higher, and therefore the wave function in the subbarrier region is damping more slowly. One can also conclude that in the case of an odd-odd nucleus the f-decay is very weak and its observation would be very improbable. The parity effects are fairly clearly manifested in the Ra region. In the U - Am

255

Table 1. Experimental and theoretical values[a] of f-decay half-lives.

Decay	$T_{1/2}^{theor.}$, years	$T_{1/2}^{exp.}$, years
$^{221}Ra \Rightarrow {}^{14}C + {}^{207}Pb$	$3.69 \cdot 10^7$	$> 7.4 \cdot 10^6$
$^{222}Ra \Rightarrow {}^{14}C + {}^{208}Pb$	$4.57 \cdot 10^3$	$3.9 \cdot 10^3$
$^{223}Ra \Rightarrow {}^{14}C + {}^{209}Pb$	$7.96 \cdot 10^3$	$4.9 \cdot 10^7$
$^{224}Ra \Rightarrow {}^{14}C + {}^{210}Pb$	$9.27 \cdot 10^8$	$2.3 \cdot 10^8$
$^{225}Ac \Rightarrow {}^{14}C + {}^{211}Bi$	$1.02 \cdot 10^{19}$	$> 2.5 \cdot 10^{18}$
$^{226}Ra \Rightarrow {}^{14}C + {}^{212}Pb$	$5.05 \cdot 10^{13}$	$5.0 \cdot 10^{13}$
$^{230}Th \Rightarrow Ne + Hg$	$1.86 \cdot 10^{17}$ (^{24}Ne)	$1.3 \cdot 10^{17}$
$^{232}Th \Rightarrow Ne + Hg$	$5.32 \cdot 10^{20}$ (^{26}Ne)	$> 3.0 \cdot 10^{20}$
$^{231}Pa \Rightarrow Ne + Tl$	$5.60 \cdot 10^{15}$ (^{24}Ne)	$8.6 \cdot 10^{15}$
$^{232}U \Rightarrow Ne + Pb$	$1.76 \cdot 10^{13}$ (^{24}Ne)	$3.4 \cdot 10^{13}$
$^{233}U \Rightarrow Ne + Pb$	$2.81 \cdot 10^{17}$ (^{25}Ne)	$3.0 \cdot 10^{17}$
$^{234}U \Rightarrow Ne + Pb$	$4.97 \cdot 10^{17}$ (^{26}Ne) $(4.62 \cdot 10^{17})$	$5.6 \cdot 10^{17}$
$^{234}U \Rightarrow Mg + Hg$	$1.34 \cdot 10^{15}$ (^{28}Mg) $(2.18 \cdot 10^{18})$	$1.8 \cdot 10^{18}$
$^{235}U \Rightarrow Ne + Pb$	$4.27 \cdot 10^{22}$ (^{26}Ne)	$> 1.4 \cdot 10^{20}$
$^{235}U \Rightarrow Mg + Hg$	$5.39 \cdot 10^{20}$ (^{29}Mg)	$> 9.0 \cdot 10^{20}$
$^{236}U \Rightarrow Ne + Pb$	$8.38 \cdot 10^{23}$ (^{26}Ne)	$> 6.0 \cdot 10^{18}$
$^{236}U \Rightarrow Mg + Hg$	$2.20 \cdot 10^{19}$ (^{30}Mg)	$> 6.0 \cdot 10^{18}$
$^{237}Np \Rightarrow Mg + Tl$	$4.96 \cdot 10^{20}$ (^{30}Mg)	$> 5.0 \cdot 10^{19}$
$^{236}Pu \Rightarrow Mg + Pb$	$5.60 \cdot 10^{13}$ (^{28}Mg)	$\sim 1.5 \cdot 10^{14}$
$^{238}Pu \Rightarrow Mg + Pb$	$3.02 \cdot 10^{17}$ (^{30}Mg) $(6.34 \cdot 10^{17})$	$1.5 \cdot 10^{18}$
$^{238}Pu \Rightarrow Si + Hg$	$7.20 \cdot 10^{15}$ (^{32}Si) $(1.24 \cdot 10^{22})$	$6.5 \cdot 10^{17}$
$^{241}Am \Rightarrow Si + Tl$	$1.49 \cdot 10^{21}$ (^{34}Si)	$> 9.0 \cdot 10^{16}$

[a] The values are from ref. [8], except for $^{235,236}U$ which are from [34], and ^{236}Pu which is from [35].

region the effect of the pairing correlations on the mass parameter is weak and the value of the parameter p decreases from unity in the Ra region to the value $p = 0.2$. It is also interesting to note the observed effect of the f-decay branching for ^{234}U and ^{238}Pu.

The Q_f-value for the decay $^{234}U \Rightarrow {}^{28}Mg + {}^{206}Hg$ is 74.13 MeV while in the case $^{234}U \Rightarrow {}^{26}Ne + {}^{208}Pb$ it is lower, Q_f=59.47 MeV. The coulomb barrier penetrability for the latter decay branch is about four orders of magnitude lower, but the experimental half-life is longer for this branch. Because the fragments are almost of the same kind in both branches, the collective parameters of the decay configurations are about the same in both cases. Consequently, in order to describe the f-decay probability branching ration, it is necessary to take into account the isovector mode of the collective oscillations, i.e. the oscillations of the charge density within the nuclear volume. It is well known [31] that the relaxation time of charge variables is less than the typical time of deformation processes. So, one can suppose that in each moment of time along the deformation path the equilibrium of charge density is established. The situation is very similar to the nuclear fission where the distribution of fragments in an isobaric chain is defined at the scission point, and its dispersion is about 0.4 for fragments after neutron emission [32]. In f-decay branching ratio calculations the formula (7) must be augmented by a charge factor W, defined as

$$W(A_f, Z_f) = |\Psi_0^{iv}|^2 = \frac{1}{\sqrt{2\pi}\sigma_z} \cdot exp[-\frac{(Z - Z_f)^2}{2\sigma_z^2}], \qquad (11)$$

where the most probable charge state $\bar{Z}_f = \bar{\rho}_z A$ and $\bar{\rho}_z$ is determined either by the charge density of the parent nucleus or that of the daughter nucleus if $Z - Z_f = 82$. In the case of the ^{234}U decay to A_f=28 the value of $Z_f = 11.0085$ and of $\sigma_f = 0.25$, the decrease of decay probability along this branch is about three orders of magnitude. So the contribution of the zero-point oscillations along the isovector mode provides an explanation of the observed branching ratio of this f-decay.

In order to understand the fine structure observed in the ^{14}C spectra of the ^{223}Ra decay [9] it is necessary to calculate more accurately the dynamic parameters of the deformation energy for those orbits where the odd neutron may be placed. In the ground state the nucleus ^{223}Ra has a stable octupole deformation, and the odd neutron occupies an orbital with $\Omega = 3/2$, while the odd neutron in ^{209}Pb occupies the low-lying orbits with $I^\pi = 9/2^+, 11/2^+$ and $15/2^-$. The structure of the parent nuclear state with $\Omega = 3/2$ in a decay configuration is favourable with respect to the transition of the odd nucleon to the spherical orbit $I^\pi = 11/2^+$ in accordance with experiment [9]. In the region of heavy nuclei where the f-decay has been observed the Ra-Ac region stands out, because the stiffness coefficient of the octupole deformation is less here than compared to its value for stiff nuclei relevant for higher f-decay half-lives.

3. Two-Body Fragmentation of Highly Excited Nuclei

The partial width of binary decay from highly excited states may be obtained by a generalization of the above model as follows

$$\Gamma_f(A, Z, E^\star, A_f, Z_f, E_f) = \hbar N_f^T(E^\star)P(E_f)\rho(E^\star + Q_f - E_f)dE_f. \qquad (12)$$

Here N_f^T is the probability to reach the decay configuration for the excited

257

nucleus, $P(E_f)$ is the barrier penetrability at the fragment kinetic energy E_f and ρ is the level density of the fragment system

$$\rho(E^{\star}+Q_f-E_f) = \int dU \rho_d(A-A_f, Z-Z_f, U)\rho_f(A_f, Z_f, E^{\star}+Q_f-E_f-U),$$

(13)

where ρ_d and ρ_f are the level densitiies of the daughter and fragment nuclei, repectively.

The calculation of the factor N_f^T in the case of highly excited nuclei is a hard problem in general. In this case it is necessary to solve the Schrödinger equation (6) for the collective variables taking into consideration the influence of the thermal effects on the collective and liquid-drop parameters. The decay configuration is being reached as a result ot the thermal fluctuation and in accordance with expression (7) we can write

$$N_f^T = \frac{\omega_T}{2\pi\hbar} [\sum_n e^{-\hbar\omega_n/T} |\Psi_n|^2] \cdot |\frac{dx}{dA},$$

(14)

where ω_n and Ψ_n are the eigenvalues and eigenstates of equation (6) with the parameters corresponding to the excited nucleus, where $\omega_T \approx T$ is the effective thermal frequency. We have performed calculations to estimate the f-decay probabilities within the framework of the liquid drop model. Such estimates show that the probability N_f^T depends strongly on the mass parameter, the stiffness coefficient and the location of the decay point x_f. For the mass parameter we used the approximations of (8) and (9) without pairing corrections. Due to the spherical shape of the excited nuclei in their equlibrium state the value of x_f increases for this case. It is also necessary to take into account the temperature dependence of the liquid drop parameters for which qualitative estimations only can be obtained. In [33] some approximate dependence of the liquid drop parameters on the temperature have been obtained based on temperature dependent Hartree-Fock calculations. These calculations show that the stiffness coefficient decreases with increasing temperature. The decrease of the stiffness coefficient leads to an increase of x_f. As a result of this the formation probability N_f^T decreases with increasing temperature. The level density factor strongly grows with the nuclear temperature. Therefore the induced f-decay width increases strongly with the excitation energy. The dependences of N_f^T and Γ_f as a function of the excitation energy for the decay $^{235}U \Rightarrow ^{26}Ne + ^{209}Pb$ are shown in figure 2.

In these calculations the following temperature dependence of x_f is used

$$x_f = x_f(0) + c_x \cdot T^2.$$

(15)

For the decay $^{235}U \Rightarrow ^{26}Ne + ^{209}Pb$ the values of $x_f(0) = 0.75$ and $c_x = 0.016$ have been used. The dependence of the mass parameter M_0 on the increasing temperature is supposed to be weak. The level density is

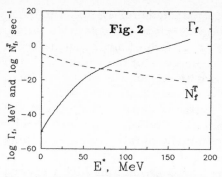 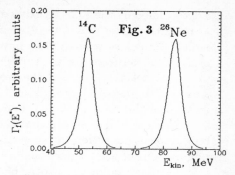

Fig. 2. The probability N^T and the decay width Γ as a function of excitation energy for the decay $^{235}\text{U} \Rightarrow {}^{26}\text{Ne} + {}^{209}\text{Pb}$.

Fig. 3. The kinetic energy spectra of ^{14}C and ^{26}Ne in the decay of ^{235}U at $E^\star = 100$ MeV.

calculated in terms of a Fermi-gas model with the level density parameter a=A/10. The penetration probability through the barrier, $P(E_f)$, has been calculated using the approximation relevant to the parabolic barrier. At an excitation energy of $E^\star \geq 200$ MeV the fragmentation width becomes very large.

The kinetic energy spectra of the ^{14}C and ^{26}Ne emitted by ^{235}U at $E^\star = 100$ MeV are shown in figure 3. The shape of the spectra near the maximum is close to a Gaussian-like shape, which is similar to the fission-fragment spectra [38].

The ratio of partial widths $\Gamma_f(^{26}\text{Ne})/\Gamma_f(^{14}\text{C})$ for the ^{234}U fragmentation is equal to $9.74 \cdot 10^{-8}\,\text{MeV}/3.02 \cdot 10^{-7}\,\text{MeV} = 0.32$. The parameter τ in the power law for the fragment yields $Y_f \propto A_f^{-\tau}$, equals 1.84. This parameter strongly depends on the values of the collective parameters, and its detailed behaviour may be analyzed only after the investigation of the temperature dependence of the collective parameters and the level density function. Our calculation shows that the two-body fragmentation becomes a strongly competing channel at $T \geq 3$ MeV.

At excitation energies $E^\star < 50$ MeV the shell corrections should be taken into account and one can expect that the most favourable situation for fragment emission will be at $E^\star \approx 20$ MeV, because the equilibrium nuclear shape is still deformed but the stiffness coefficient is not as large as in the ground state.

4. Conclusions

In the present report we discussed a new model of the spontaneous emission of fragments by heavy nuclei. The probabilities of the spontaneous f-decay are determined by the collective dynamical parameters of the nuclei and they have a high value for nuclei which are relatively soft to octupole deformation. The calculations of probabilities have been performed for the spontaneous f-decay of nuclei from Ra to Am. The comparison of our results with the experimental data shows that the proposed model can account for the main features of the experimental data. The odd-even effect in the spontaneous f-decay probabilities can be explained by the odd-even differences in the mass parameters. It is shown that the branching ratios can be explained by taking into consideration the isovector zero-point vibrations at the touching configurations of the fragments.

The expression for the widths of fragment emission from highly excited nuclei has been proposed as a generalization of the spontaneous emission width. The probabilities to reach the decay configuration are defined by the dynamical collective parameters and their dependence on the excitation energy. The fragment kinetic energy spectra have a Gaussian-like shape. The emission width increases rapidly at $T \geq 3$ MeV where the transition to a multifragmentation mechanism may be taking place.

References

1. Hufner, J.: Phys. Rep. **125** 129 (1985)
2. Proc. of the Symp. on Central Coll. and Fragmentation Processes, Denver, Nucl. Phys. A427 (1987)
3. Rose, H.J., Jones, J.A.: Nature **307** 245 (1984)
4. Alexandrov, D.V., Belyatsky, A.F., Gluhov, Yu.A., Nicolsky, F.Yu., Novatsky, B.V., Stepanov, D.N.: JETP Lett. **40** 909 (1984)
5. Gales, S., Hourani, E., Hussonnois, M., Shapira, H.P., Stab, I., Vergnes, M.: Phys. Rev. Lett. **53** 759 (1984)
6. Rubchenya, V.A., Eismont, V.P., Yavshits, S.G.: Izv. AN SSSR, ser. fiz. **50** 1017 (1986)
7. Rubchenya, V.A., Yavshits, S.G.: Proc. of the XVI Int. Symp. on Nucl. Phys, Gaussig, GDR (1986), Report ZfK-610 138 (1987)
8. Price, P.B.: Proc. of the Int. Conf. on Fifty Years in Nuclear Fission, Berlin (1989). Nucl. Phys. **A502** 41c (1989)
9. Brillard, L., Elayi, A.G., Hourani, E., Hussonnois, M., Le Du, J.F., Rosier, L.H., Stab, L.: Proc. of Int. School-Seminar on Heavy Ion Physics, Dubna 333 (1990)
10. Klotz-Engman, G., et al: Phys. Lett. **B187** 245 (1987)
11. Finn, J.E., et al.: Phys. Rev. Lett. **49** 1321 (1982)
12. Sandulescu, A., Lustig, H.J., Hahn, J., Greiner, W.: J. Phys. G: Nucl. Phys. **4** L279 (1978)
13. Poenaru, D.N., Ivascu, M., Sandulescu, A., Greiner, W.: Phys. Rev. C **32** 572 (1985)
14. Sandulescu, A.: J. Phys. G: Nucl. Phys. **15** 529 (1989)
15. Pic-Pichak, G.A.: Sov. J. Nucl. Phys. **44** 1421 (1986)
16. Shi Yi-Jin, Swiatecki, W.J.: Phys. Rev. Lett. **54** 300 (1985)

17. Shi Yi-Jin, Swiatecki, W.J.: Nucl. Phys. **A438** 450 (1985)
18. Shi Yi-Jin, Swiatecki, W.J.: Nucl. Phys. **A464** 205 (1987)
19. Shanmugam, G., Kamalaharan, B.: Phys. Rev. **C38** 1377 (1988)
20. Kadmensky, S.G., Furman, V.I., Tchuvilsky, Yu.M.: JINR Communications, Dubna **R4**-85-368 (1985)
21. Kadmensky, S.G., et al.: Izv. AN SSSR, ser. fiz. **50** 1776 (1986)
22. Iriondo, M., Jerrestam, D., Liotta, R.J.: Nucl. Phys. **A454** 252 (1986)
23. Blendowske, R., Fliessbach, T., Walliser, H.: Nucl. Phys. **A464** 75 (1987)
24. Ivascu, M., Silistenau, I.: Nucl. Phys. **A485** 93 (1988)
25. Buck, B., Merchant, A.C.: J. Phys. G: Nucl. Part. Phys. **15** 615 (1989)
26. Rubchenya, V.A., Eismont, V.P., Yavshits, S.G.: Proc. of Int. School-Seminal on Heavy Ion Physics, Dubna, p. 200 (1986)
27. Barranco, F., Broglia, R.A., Bertch, G.F.: Phys. Rev. Lett. **60** 507 (1988)
28. Strutinsky, V.M.: Nucl. Phys. **A122** 1 (1968)
29. Rubchenya, V.A., Yavshits, S.G.: Yadernaya Fizika **40** 649 (1984)
30. Pashkevich, V.V.: Nucl. Phys. **A169** 275 (1971)
31. Weidenmüller, H.A.: Prog. on Part. and Nucl. Phys. **3** 42 (1980)
32. Vandenbosch, V., Huizenga, J.R.: Nuclear Fission (New York, Academic Press, 1973)
33. Bartel, J., Quentin, Ph., Brack, M., Guet, C., Håkansson, H.B.: Nucl. Phys. **A386** 79 (1982)
34. Tretyakova, S.P., et al.: Preprint JINR, Dubna, **E7**-88-803 (1988)
35. Ogloblin, A.A., et al.: Kratkie soobsch. OIYAI **2(35)**-89, Dubna, 43 (1989)

Cluster Radioactivity and Nuclear Structure-Clusters as Solitons

A. Săndulescu, A. Ludu, and W. Greiner

Institut für Theoretische Physik der J.W. Goethe-Universität,
W-6000 Frankfurt 11, Fed. Rep. of Germany

Abstract: Based on liquid drop model, by including nonlinear terms in the hydrodynamical equations we have shown that stable solitons could exist on the surface of the most rigid double magic spherical nucleus ^{208}Pb. The shell effects related with the soliton lead to a new minimum in the total potential energy choosen to be degenerated in energy with the ground state minimum. A new coexistence model consisting of the usual shell model and a cluster-like model described by a soliton moving on the nuclear surface is obtained. The corresponding amplitudes describe excellently the experimental spectroscopic factors for alpha and cluster decays.

I. Introduction

The experimental discovery of the cluster decays [1], i.e. spontaneous emission of carbon, neon, magnesium and silicon from heavy nuclei, indicates a large enhancement of such clusters on the nuclear surface. Alpha decay was the first example for a cluster enhancement, the shell model underestimating the preformation factor by at least two orders of magnitude. This suggests that the external nucleons join together to form the emitted cluster leaving the residual nucleus unpolarized, i.e. to new shapes which could not be described by multipole expansion of the nuclear surface. In a phenomenological description we have to find some arguments of collective type which may justify the existence of new exotic shapes. Probably such shapes have not been considered up to now due to the fact that the many body correlations of this type have not been included in the present microscopic calculations.

In the present paper, by introducing nonlinear terms in the hydrodynamical equations, we have shown that stable solitons could exist on the surface of a sphere. This is based on the hypothesis that the outside nucleons do not polarize the double magic core ^{208}Pb. The total potential energy consists of four terms : surface ,centrifugal, Coulomb and shell energies. The last term was introduced phenomenologically proportional with the overlap between the

Springer Series in Nuclear and Particle Physics   **Clustering Phenomena in Atoms and Nuclei**
Editors: M. Brenner · T. Lönnroth · F.B. Malik   © Springer-Verlag Berlin, Heidelberg 1992

emitted cluster, considered as a sphere, and solitons with different sizes and shapes. Evidently this term leads to a new minimum in the total potential energy. Due to the fact that alpha and cluster decays are spontaneous we choose this minimum to be degenerate in energy with the ground state minimum.

This description leads to a new coexistence model : the usual shell model and a cluster-like model described by a soliton moving on a sphere. Due to the large barrier between the two minima the amplitude of the cluster-like state is much smaller than the usual ground state. The ratios of the square of the two amplitudes give the preformation probabilities of the corresponding clusters at the nuclear surface. The comparison with the present existing experimental data shows an exccelent agreement.

We conclude that a new large amplitude collective motion is existing in nuclei. We have shown that, in addition to the usual shapes described by multipole expansions, the introduction of nonlinear terms in the description of nuclear surface leads to the existence of the solitons as a new cluster-like component added to the shell model. This is not in conflict with our previous conclusion that cluster decays including alpha emission could be regarded as a superasymmetric fission process with compact shapes [2]. Now we are stressing the fact that such decays start already from very compact shapes like solitons moving on the nuclear surface.

II. Hydrodynamical Equations

Let us consider a small perturbation propagating on the surface of a sphere of radius R. For an axial symmetry, with the symmetry axis in the direction of the perturbation, the problem reduces to a small perturbation propagating on a circle with the shape $r = R + \eta(\theta, t)$. We make three assumptions : first that the amplitude of the perturbation η_0 is small compared with the radius R so that we can introduce a small parameter $\xi = (r - R)/R \ll 1$, second that the core is unperturbed up to the radius $r = R - h$ with $h \ll R$, and third that we have the case of an ideal, incompressible and irrotational fluid which leads to a field of velocities $\mathbf{V} = \nabla \Phi$ given by a scalar potential Φ satisfying the Laplace equation $\Delta \Phi = 0$.

In order to describe the dynamics of such a perturbation we have to find the equations for shapes and velocities. Writing Φ as a power series in the small parameter ξ :

$$\Phi(r, \theta, t) = \sum_{n \geq 0} \xi^n f_n(\theta, t) \tag{1}$$

we get from the Laplace equation an infinite system of recurence relations for the functions f_n. It is easy to show that the radial $v = \Phi_r$ and tangential $u = \Phi_\theta / r$ velocities depend only on two independent functions $f_{0,\theta}(\theta, t)$ and $f_1(\theta, t)$ denoted in the following by $g(\theta, t)$ and $j(\theta, t)$ respectively. In second order in ξ we have :

$$u = \frac{1}{R}[g + \xi(-g + j_\theta) + \frac{1}{2}\xi^2(2g - g_{\theta\theta} - 3j_\theta)] \tag{2a}$$

$$v = \frac{1}{R}[j + \xi(-j - g_\theta) + \xi^2(j - \frac{1}{2}j_{\theta\theta} - \frac{3}{2}g_\theta)] \tag{2b}$$

The first equation is given by the time derivative of the radial coordinate on the surface Σ :

$$v\Big|_\Sigma = \frac{dr}{dt}\Big|_\Sigma = \eta_t + \frac{1}{r}u\eta_\theta \tag{3}$$

By imposing the condition that $v = 0$ at $r = R - h$ we obtain in the first order in η, a relation between j and g_θ :

$$j = -\frac{h}{R}g_\theta \tag{4}$$

The second equation is given by Euler equation :

$$\rho_m\left(\frac{\partial \mathbf{V}}{\partial t} + (\mathbf{V} \cdot \nabla)\mathbf{V}\right) = -\nabla P + \nabla \Phi_C \tag{5}$$

where ρ_m is the constant mass density, P the pressure and $\nabla \Phi_C$ the external coulomb force density given by the electrostatic potential Φ_C .

The pressure at the surface, in the first order in η, is given by :

$$P\Big|_\Sigma = \frac{\sigma}{R}\left[1 - \frac{1}{R}(\eta + \eta_{\theta\theta})\right] \tag{6}$$

where σ is the surface tension.

The electrostatic potential Φ_C is obtained by solving the Poisson equation with the charge distribution :

$$\rho_{el}(r, \theta, t) = (\rho_{sph} - \rho_{sol})H(1 - \frac{r}{R}) + \rho_{sol}H(1 - \frac{r}{R(1 + \frac{\eta}{R})}) \tag{7}$$

where $H(x)$ are Heaviside distributions. Here we assumed that the sphere of radius R is uniformly charged with the density ρ_{sph} corresponding to the residual nucleus and the disturbance is, also, uniformly charged with ρ_{sol}, corresponding to the emitted cluster.

Like previously, writing the solution as an expansion in $\xi^n f_n$ with $n \geq -2$ we obtain a system of five recurrence relations for f_0, f_1, f_2, f_3 and f_{-3} which leads to the following expression of the electrostatic potential at the surface Σ :

$$\Phi_C\Big|_\Sigma = \frac{\rho_{sph} R^2}{3\epsilon_0}\left(1 - \frac{\eta}{R}\right) - \frac{\rho_{sol}\eta^2}{2\epsilon_0} \tag{8}$$

where by ϵ_0 we denote the vacuum dielectric permitivity.

Now, from the Euler equation written in a gradient form we obtain the following equation for the scalar potential Φ at the surface Σ :

$$\Phi_t + \frac{1}{2}(\nabla\Phi)^2 + \frac{1}{\rho_m}P + \frac{1}{\rho_m}\Phi_C = N(t) \tag{9}$$

where $N(t)$ is a constant of integration depending only on time. Using the expressions of u, v, P, Φ_C and the relation (4) we obtain, from the above equation, in second order in η, the following equation :

$$g_t + \frac{1}{R^2}gg_\theta - \frac{\sigma}{\rho_m R^2}(\eta_\theta + \eta_{\theta\theta\theta}) - \frac{\rho_{sol}\rho_{sph}R}{3\epsilon_0\rho_m}\eta_\theta + \frac{\rho_{sol}}{\rho_m\epsilon_0}\left(\rho_{sol} - \frac{\rho_{sph}}{3}\right)\eta\eta_\theta = 0 \tag{10}$$

The eq.(3), by using rel.(4), gives in the second order in η :

$$R\eta_t + \frac{h}{R}g_\theta + \frac{1}{R}(\eta g)_\theta + \frac{h}{R^2}\eta g_\theta = 0 \tag{11}$$

In order to solve the system of eqs.(10) and (11) for η and g we make the transformation

$$g = \chi\eta + \psi(\theta, t), \tag{12}$$

where χ is an arbitrary real parameter and $\psi(\theta, t)$ an arbitrary function. We have :

$$k\chi\eta_t - R\eta_t - 2\frac{\chi}{R}\eta\eta_\theta + k\psi_\theta - \frac{1}{R}(\eta\psi)_\theta = 0 \tag{13}$$

$$\chi\eta_t + \psi_t + \frac{\chi^2}{R^2}\eta\eta_\theta + \frac{\chi}{R^2}(\eta\psi)_\theta + \frac{1}{R^2}\psi\psi_\theta + \alpha\eta\eta_\theta + \beta\eta_\theta - \gamma(\eta_\theta + \eta_{\theta\theta\theta}) = 0 \tag{14}$$

with the constants $\alpha = \frac{\rho_{sol}}{\rho_m\epsilon_0}\left(\frac{\rho_{sph}}{3} - \rho_{sol}\right)$, $\beta = -\frac{1}{\rho_m R}\left(\frac{\sigma}{R} + \frac{R^2\rho_{sph}\rho_{sol}}{3\epsilon_0}\right)$, $\gamma = \frac{\sigma}{\rho_m R^2}$ and $k = \frac{h}{R}$.

Further we use the functional transformation :

$$\psi(\theta, t) = \frac{k\chi\eta - \frac{1}{R}\chi\eta^2 - R\int \eta_t d\theta + \psi^{(1)}}{k - \frac{1}{R}\eta} \tag{15}$$

which satisfies eq.(13). Choosing the arbitrary function $\psi^{(1)}$ so that :

$$\psi_t^{(1)} - \frac{\beta}{\chi}\psi_\theta^{(1)} + \frac{1}{R^2}\psi^{(1)}\psi_\theta^{(1)} + \left(\frac{\chi}{R^2} + \frac{\beta}{hR\chi}\right)(\eta\psi^{(1)})_\theta = 0 \tag{16}$$

265

we obtain from eq.(14) a Korteweg-de Vries equation for η :

$$\mathcal{A}\eta_t + \mathcal{B}\eta\eta_\theta + \mathcal{C}\eta_{\theta\theta\theta} = 0 \tag{17}$$

with the coeficients :

$$\mathcal{A} = \chi + \frac{R^2\beta}{h\chi} \quad ; \quad \mathcal{B} = \frac{\chi^2}{R^2} + \alpha + \frac{2\beta}{h} \quad ; \quad \mathcal{C} = -\gamma \tag{18}$$

This equation has a soliton solution, [3], [4] :

$$\eta = \eta_0 sech^2\left(\frac{\theta - Vt}{L}\right) \tag{19}$$

characterized by the half-width $L = (12\mathcal{C}/\mathcal{B}\eta_0)^{\frac{1}{2}}$:

$$L = \left\{ \frac{\eta_0\rho_m R^2}{12\sigma}\left[\frac{2}{h\rho_m R}\left(\frac{\sigma}{R} + \frac{\rho_{sol}\rho_{sph}R^2}{3\epsilon_0} \right) + \frac{\rho_{sol}}{\rho_m\epsilon_0}\left(\rho_{sol} - \frac{\rho_{sph}}{3} \right) - \frac{\chi^2}{R^2} \right] \right\}^{-1/2} \tag{20}$$

and the velocity $V = \mathcal{B}\eta_0/3\mathcal{A}$:

$$V = \frac{\eta_0\chi}{3R^2}\left[\frac{\chi^2 - \frac{R^2\rho_{sol}}{\rho_m\epsilon_0}\left(\rho_{sol} - \frac{\rho_{sph}}{3}\right) - \frac{2R}{h\rho_m}\left(\frac{\sigma}{R} + \frac{R^2\rho_{sol}\rho_{sph}}{3\epsilon_0}\right)}{\chi^2 - \frac{R}{h\rho_m}\left(\frac{\sigma}{R} + \frac{R^2\rho_{sph}\rho_{sol}}{3\epsilon_0}\right)} \right] \tag{21}$$

The cluster volume V_{cl} is given by :

$$V_{cl} = \frac{2\pi R^3}{3}\int_0^\pi \left(1 + \frac{\eta}{R}\right)^3 sin(\theta)d\theta - \frac{4\pi R^3}{3} \tag{22}$$

In terms of our parameters the amplitude of the soliton η_0, the surface thickness h, parameters which are small compared with the core radius R, and χ which plays the role of the coupling constant between the soliton shape η and its dynamics u we have :

$$L(\eta_0, h, \tilde{\chi}) = \eta_0^{-1/2}\left[\mathcal{N}_1\frac{(\lambda^2 - \frac{\lambda}{3})Z^2}{A^{4/3}} + \left(\frac{1}{6} + \mathcal{N}_2\frac{\lambda Z^2}{A}\right)\frac{1}{h} - \tilde{\chi} \right]^{-1/2} \tag{23}$$

$$V(\eta_0, h, \tilde{\chi}) = \eta_0\frac{\tilde{\chi}V_0}{A^{2/3}}\left[\frac{\tilde{\chi}^2 - \mathcal{N}_1\frac{(\lambda^2 - \frac{\lambda}{3})Z^2}{A^{4/3}} - \frac{1}{2h}\left(\frac{1}{6} + \mathcal{N}_2\frac{\lambda Z^2}{A}\right)}{\tilde{\chi}^2 - \frac{1}{4h}\left(\frac{1}{6} + \mathcal{N}_2\frac{\lambda Z^2}{A}\right)} \right] \tag{24}$$

where :

$$\mathcal{N}_1 = \frac{3e^2}{64\pi^2\sigma\epsilon_0 r_0^4} \quad ; \quad \mathcal{N}_2 = \frac{e^2}{32\pi^2\sigma\epsilon_0 r_0^3} \quad ; \quad \mathcal{N}_3 = \frac{\rho_m}{12\sigma} \tag{25}$$

$$\tilde{\chi} = \sqrt{\mathcal{N}_3}\chi \quad ; \quad \lambda = \frac{\rho_{sol}}{\rho_{sph}} \quad ; \quad R = r_0 A^{1/3}$$

with $r_0 = 1.3 fm$ and $V_0 = \frac{2}{r_0^2}\sqrt{\frac{\sigma}{3\rho_m}}.$

We should like to stress that all three terms of K-dV eq.(17) must have the same order of magnitude. Consequently considering that the corresponding angle θ is given by $\theta = V\tau$ we have the conditions :

$$VA \sim B\eta_0 \sim \frac{C}{V^2\tau^2} \tag{26}$$

Now we have to impose some physical restrictions on K-dV eq.(17) and its solutions (19). First, in order to satisfy the condition of quasiperiodicity and to have a well defined soliton, we must satisfy the geometrical condition that the angular half width L must be smaller than π, $L < \pi$. This implies a lower band for η_0^{min}, in addition to the condition $\eta_0/R_0 \leq 0.1$ which gives an upper bound for η_0^{max}. Considering also that $h/R \leq 0.1$ we obtained for actinide nuclei an upper bound for $\tilde{\chi}_{max} \simeq 2 \cdot 10^7$ meter$^{-1/2}$. Also we have to satisfy the kinematic condition that the soliton half width L must be of the same order as the product of the velocity V and the time scale of the process. The above physical conditions are automatically satisfied if rels.(26) are satisfied. In order to avoid the relativistic effects, we choose the denominator of rel.(24) to be positive and different from zero which implies a lower bound for $\tilde{\chi}_{min} \simeq 10^7$ meter$^{-1/2}$. Consequently the values of our parameters η_0, h and $\tilde{\chi}$ are very much restricted by the above physical conditions.

Finally we should like to stress that, due to the fact that V is proportional with η_0, the general behaviour of solitons, i.e. the velocity is increasing with its amplitude is satisfied.

III. Potential Energy

The total potential energy E_p, around the soliton shape, can be written as the sum of the surface E_S, the centrifugal E_{cf}, the Coulomb E_C and the shell E_{shell} energies. The surface energy is given by :

$$E_S = \sigma \int_\Sigma dS = 2\pi R^2 \sigma \int_0^\pi \left[1 + \frac{2\eta}{R} + \frac{1}{R^2} \left(\eta^2 + \frac{1}{2}\eta_\theta^2 \right) \right] sin\theta d\theta \tag{27}$$

In order to avoid the explicit expressions of tangential and radial velocities, i.e. to solve eq.(16), we use for E_{cf} the approximate expression :

$$E_{cf} = \rho_m V_{cl} r_{cm}^2 V^2 \tag{28}$$

where V is the soliton angular velocity, eq.(24), and r_{cm} is the soliton center of mass which in second order in $\frac{\eta_0}{R}$ is given by :

$$r_{CM} = R \left(1 + \frac{\eta_0}{3R} + \frac{\eta_0^2}{15R^2} \right) \tag{29}$$

267

The Coulomb energy E_C can be written as the sum of the proper energy of the sphere :

$$E_C^{sph} \simeq \frac{4\pi}{15\epsilon_0} \rho_{sph}^2 R^5, \tag{30}$$

the Coulomb energy of the soliton in the fifth order in $\frac{\eta_0}{R}$ is

$$E_C^{sol} = \frac{\pi}{120\epsilon_0} \rho_{sol}^2 \eta_0^5 \tag{31}$$

and the Coulomb interaction energy between them reads

$$E_C^{inter} = \frac{\pi}{6\epsilon_0} \rho_{sol} \rho_{sph} R^3 \int_0^\pi (2R\eta + \eta^2) sin\theta d\theta \tag{32}$$

In the following, we introduce a phenomenological shell model correction E_{shell} proportional with the overlap integral between the soliton and a cluster of given volume :

$$E_{shell} = -U_0 I \tag{33}$$

In this way we avoid the microscopic evaluation of the soliton self energy due to the many body correlations inside the soliton. We chose U_0 in such a way that this minimum is degenerate in energy with the ground state minimum.

We would like to stress that the conditions

$$\frac{\partial E_p}{\partial \eta_0} = 0 \qquad ; \qquad \frac{\partial E_p}{\partial \tilde\chi} = 0 \tag{34}$$

are also satisfied which determine, for a given h, the parameters η_0^{min} and $\tilde\chi^{min}$.
Evidently the above expressions are not valid around the ground state minimum. Consequently we introduce the normal modes :

$$\eta(\theta, t) = \eta_0 \sum_{l \geq 0} \alpha_l(t) P_l(cos\theta) \tag{35}$$

with :

$$\alpha_l(t) = \int_{-\pi}^\pi P_l(cos\theta) \eta(\theta, t) d\theta \tag{36}$$

This leads to the following expression of the potential energy in terms of normal modes :

$$E_p^{n.m.} = E_p(0) + \frac{1}{2} \sum_{l \geq 0} C_l \alpha_l^2 \tag{37}$$

with :

$$C_l = \frac{l-1}{4\pi} \left[E_s(0)(l+2) - 10E_C(0)\frac{1}{2l+1} \right] \tag{38}$$

In order to have an unique expression for the total potential energy valid for any value of the soliton amplitude η_0, $E_p^{n.m.}$ must be smoothly connected with E_p. A small number of terms in expansion (35) is obtained by choosing the solitons shape with an amplitude η_0 as small as possible and a half width L as close as possible to π.

In this way we dynamically describe the formation of the cluster-like component, as a soliton moving on a sphere, from the ground shell model state. Evidently the two dimensional problem in the coupling constant $\tilde{\chi}$ and the soliton amplitude η_0 or $\epsilon = \frac{\eta_0}{R_0}$ where R_0 is the radius of the parent nucleus could be reduced to an one dimensional problem by choosing the path along the valley in the potential.

IV. Spectroscopic Factors

The spectroscopic factors, as ratio of the experimental decay (λ_{exp}) over the theoretical one-body decay constant (λ_G) were first introduced in alpha decay in order to separate the nuclear structure effects from the main factor due to the product between the frequency of the collisions with the walls and the barrier penetrability (λ_{Gamow}). Similar spectroscopic factors have been introduced for cluster decays. Consequently it was concluded that the cluster decays are more similar with alpha decay than with a superasymmetric fission.

It is usually believe that the spectroscopic factors could be computed in the frame of the shell model without many-body correlations. For alpha decay the relative ratios are well reproduced, the absolute values being underestimated at least by two orders of magnitude. One accepted possible explanation of absolute values for alpha and cluster decay constants is the correct antysymmetrisation of the channel wave function [5]. Evidently for clusters larger than ^{14}C these calculation are not possible.

In the present paper we consider that without the introduction of the many body calculations in the many body correlations in the shell model is not possible to explain such a large enhancement of clusters on the nuclear surface. This is done by considering the solitons as nonlinear solutions of the hydrodynamic equations. The new shapes are grouping together the external nucleons to forme the emitted cluster. In this way we have a new coexistence model consisting of the usual shell model and a cluster-like model (soliton).

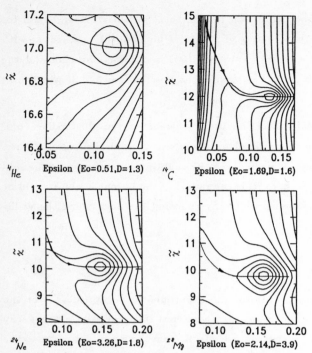

Fig. 1. The potential energy surfaces E_p for ^{4}He, ^{14}C, ^{24}Ne and ^{28}Mg - decays with the residual nucleus ^{208}Pb in the plane $\tilde{\chi}$, the coupling constant between the soliton speed and its shape and $\epsilon = \eta_0/R_0$, the ratio between the soliton amplitude η_0 and the radius of the parent nucleus R_0. Also the paths which define the barriers between the ground and soliton minima are indicated. By E_0 we denote the value of the lowest potential contour line and by D the energy interval between two contour lines.

The difference between the fission and cluster decays consists in the fact that the fission starts by the deformation of the whole nucleus leading to two very alongated fragments and that the cluster decays starts from the soliton component leading to two very compact fragments.

In our collective model the spectroscopic factors are given by the ratios of the wave amplitude in the two wells, evaluated with the help of barrier penetrability between the two minima. The barrier was chosen along the path in the plain $(\epsilon, \tilde{\chi})$ coresponding to the minimum potential energy. With this procedure we fix-up in an unique way the barrier from the soliton minimum up to the smallest $\epsilon(\eta_0)$ and biggest $\tilde{\chi}$ (L close to π). The rest of the barrier is determined by joining smoothly the barrier at this point with the ground state minimum normalized to zero. Along this path the mass of the emitted cluster is constant.

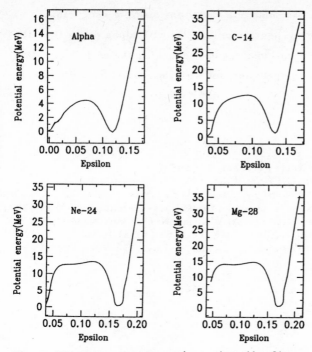

Fig. 2. The soliton barriers $B^{A_{cl}}$ for ^{4}He, ^{14}C, ^{24}Ne, and ^{28}Mg - decays with the residual nucleus ^{208}Pb along the paths shown in Fig.1 The spectroscopic factors, defined as ratio of the amplitude in the two wells, are evaluated by WKB - method between the two minima using the mass of the emitted cluster.

In Figs. 1 and 2 for $k = 1.6 \times 10^{-3} \times A_{Cl}^{\frac{11}{6}}$ we give the potential energy surfaces and the soliton barriers B$^{A_{Cl}}$ for ^{4}He, ^{14}C, ^{24}Ne and ^{28}Mg decays with the daughter nucleus ^{208}Pb. The obtained spectroscopic factors S(A_{Cl}) together with the empirical ones given by the equation : $S_{exp} = (6.3 \times 10^{-3})^{\frac{A_{Cl}-1}{3}}$, [6], are presented in the following table :

$A_{Cluster}$	^{4}He	^{14}C	^{24}Ne	^{28}Mg
S_{exp}	6.3×10^{-3}	2.91×10^{-10}	1.34×10^{-17}	1.56×10^{-20}
S_{th}(soliton)	6.4×10^{-3}	3.25×10^{-10}	7.8×10^{-17}	1.12×10^{-19}

We stress again that our spectroscopic factors are calculated for each decay and not extrapolated from known values. More than that our spectroscopic factors depend also on the cluster charges and not only on cluster masses as the present empirical systematics [6]. No fittings of the data are made. Further

271

studies are necessary to obtain some final conclusions. We consider that is to early to consider cluster decays much more similar with alpha decay than the fission. Anyhow such a comparison must be made with the cold fission data [7] where the fragments are compact and not with the normal fission with very alongated fragments.

References

1) P.B. Price, Ann. Rev. Nucl. Part. Sci. **39** (1989) 19
2) A. Sandulescu, J. Phys. G : Nucl. Past. Phys. **15** (1989) 529
3) E. Lamb, "Theory of Solitons", Academic Press, New-York,1985
4) S. Novikov et al., "Theory of Solitons", Consultants Bureau, New-York,1986
5) R. Blendowske, T. Fliessbach and A. Walliser, Nucl. Phys. **A464** (1987) 75
6) R. Blendowske, H. Walliser, Phys. Rev. Lett. **61** (1988) 1930.
7) A. Sandulescu, A. Florescu and W. Greiner, J. Phys. G : Nucl. Part. Phys. **15** (1989) 1815.

Cluster Radioactivity

P.B. Price

Physics Department, University of California, Berkeley, CA 94720, USA

Following the discovery of the emission of ^{14}C by ^{223}Ra, many other actinide nuclei have been found to decay by emission of neutron-rich clusters, ranging from ^{14}C to ^{32}Si. Experimentally determined branching ratios relative to alpha emission range from 10^{-16} to 10^{-9}, and partial lifetimes range from 10^{11} to 10^{28} sec. Even-A nuclides have higher cluster emission rates, typically by a factor $\sim 10^2$, than their odd-A neighbors. Both quantal (cluster) and macroscopic (fission) models can account for most observed lifetimes to within a factor three. Two experiments illustrate the importance of nuclear structure in controlling the quantitative rates of cluster emission: (1) For ^{223}Ra $\rightarrow$ ^{14}C + ^{209}Pb, decay to an excited level of ^{209}Pb occurs 80% of the time in preference to the ground state. (2) In the cluster radioactivity of ^{233}U, the observed emission rate of Mg relative to that of Ne is an order of magnitude lower than expected without consideration of Nilsson levels.

Introduction

At this conference, it is appropriate to ask whether there are any similarities between atomic and nuclear clusters. Metallic clusters produced in an oven and studied by a quadrupole mass analyzer in flight have been found to exhibit abundance peaks at certain masses.[1] From these masses is inferred the existence of closed electronic shells derived by assuming that, by analogy with the shell model of nuclear physics, the valence electrons move in elliptical orbits throughout the cluster rather than around individual atoms. In the case of nuclear clusters such as ^{14}C emitted in the two-body radioactive decay of heavy nuclei, the role of shell structure is seen only if one looks at the masses of both the light and the heavy daughter. In Fig. 1, which shows the binding energy per nucleon for light nuclei, arrows indicate the positions of the lighter of the two daughters emitted in cluster radioactivities. Instead of a preference for the most tightly bound light daughter, the isotope of a species actually observed is the most neutron-rich one that is still relatively tightly bound. It is obvious why this is so when we realize that cluster radioactivity is the two-body decay of a heavy nucleus by barrier penetration, the rate of which is governed largely by the Q value,

$$Q = M(A,Z) - M(A_1,Z_1) - M(A_2,Z_2)$$

which enters into the expression for the Gamow penetrability,

$$P = \exp(-2G); \qquad G = \int \sqrt{2\mu(r) \, (V(r) - Q)} \, dr$$

Springer Series in Nuclear and Particle Physics **Clustering Phenomena in Atoms and Nuclei**
Editors: M. Brenner · T. Lönnroth · F.B. Malik   © Springer-Verlag Berlin, Heidelberg 1992

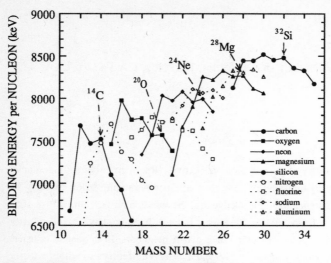

Figure 1. Binding energy per nucleon for light nuclei. Most common clusters are marked.

Here the subscripts 1 and 2 refer to the light and heavy daughter, and $\mu(r)$ is the reduced mass. The decay rate, which is proportional to P, is maximized when the barrier height, V - Q, is as small as possible. For given values of Z_1 and Z_2 this occurs when *both* of the emitted fragments are as tightly bound as possible. Alpha decay is a special case of cluster radioactivity in which the barrier height is small because Z_1 is small (making V small) and $M(A_1,Z_1)$ is small (making Q large). In cluster radioactivity, V - Q is smallest if the light daughter is neutron-rich and the heavy daughter is close to the doubly magic nuclide ^{208}Pb. In order for a parent nuclide lying near the stability line to decay into a heavy cluster close to ^{208}Pb, the light cluster must be neutron-rich.

Cluster radioactivity with $Z_1 > 2$ is extremely rare relative to alpha decay and was not discovered until 1984, when Rose and Jones[2] and Aleksandrov et al.[3] reported the two-body decay ^{223}Ra $\rightarrow$ ^{14}C + ^{209}Pb with branching ratio relative to alpha decay B(C/α) = 8.5 x 10^{-10}. In all subsequently discovered cases, the branching ratios have been even smaller, ranging

Table 1. Number of Radioactive Decay Modes Observed to Lead to Clusters

light cluster	no. of cases		heavy cluster	no. of cases
^{14}C	4		^{208}Pb	7
^{20}O	1		^{209}Pb	2
^{24}Ne	5		^{210}Pb	3
^{25}Ne	1		^{212}Pb	1
^{26}Ne	1		^{206}Hg	3
^{28}Mg	3		^{207}Tl	1
^{30}Mg	1			
^{32}Si	1			

down to 10^{-16}. For this reason experimental progress has been slow. Table 1 lists the number of cases for which various species have been detected (or suspected to have been detected, when isotopic resolution has not been possible).

Ref. 4 gives a thorough discussion of experimental techniques and of the status of the field in 1989. Ref. 5 gives a detailed discussion of the various theoretical approaches to the description of cluster radioactivity.

Systematics of Decay Rates and Branching Ratios

The branching ratio for emission of ^{14}C from ^{223}Ra is large enough that it was practicable to use a semiconductor detector telescope in contact with the radioactive source to recognize and reject alpha particle pileups[2,3]. Carbon-14 emission by ^{226}Ra was discovered by Hourani et al.[6], who used an electronic detector at the focal plane of a magnetic spectrometer to select only the ^{14}C ions. For all of the other positive detections, the emitted light cluster, with kinetic energy of order 2 to 2.5 MeV per nucleon, has been identified with track-recording plastic or glass detectors.[4] Because of the short range of the emitted clusters, the source must be no thicker than ~1 mg/cm^2, which makes difficult the search for cluster radioactivities in nuclides with low specific activity, such as ^{232}Th and ^{237}Np. Most of the searches for cluster radioactivities have been guided by reference to the tables of Poenaru and co-workers[7], who have calculated partial half-lives and branching ratios relative to alpha decay for nuclides with $47 \leq Z \leq 106$ and half-lives longer than 10^{-6} sec, light daughters with $2 \leq Z \leq 26$, and predicted cluster decay half-lives shorter than 10^{50} sec.

Track-recording plastic and glass detectors have a number of features that make them the detector of choice for most cluster studies:

1. One can choose a track-recorder that is insensitive to all ions with Z less than Z_1. In their study of ^{14}C emission from isotopes of Ra, Fr, and Ac, the Berkeley group used polycarbonate film,[8,9] and the Milano-Berkeley collaboration is using BP-1 phosphate glass in an ongoing search. Both of these detectors are insensitive to ions lighter than carbon. To study Ne and Mg emission, the detectors of choice are either polyethylene terephthalate[10] or PSK-50 phosphate glass[11]. To study Si emission, which has been seen only at very low branching ratios, one must discriminate against the accumulation of short tracks due to recoils of atoms struck by alpha particles impinging on the detector. Here LG-750 phosphate glass is used because it is insensitive to Ne and lighter ions but can record Mg as well as Si (ref. 12).

2. Track-recording solids can collect events with an efficiency that is independent of rate, and they are immune to backgrounds other than heavy charged particles such as fission fragments. There is thus no problem such as the pileup of alpha particle pulses during a narrow time interval.

3. They can be tailored in size and shape to fit into the collector of a facility such as ISOLDE (where the ^{14}C radioactivity of ^{222}Ra and ^{224}Ra was discovered[8]) or into large planar or hemispherical arrays.

To identify a charged particle with a track-recording solid one immerses the solid in a selective reagent that etches the entire surface at a uniform rate v_G and etches along the trajectory of the particle at a higher rate v_T, which is a strongly increasing function of ionization rate. The competition between the two rates leads to a conical etchpit of half-angle θ = arcsin (v_G/v_T). Both Z and range (and thus energy) can be determined from microscopic measurement of the size and shape of the etchpit. See ref. 4 for details of the method.

In some applications the track-etching method of particle identification can provide information on mass as well as charge. However, in the case of clusters, which have only 2 to 2.5 MeV per nucleon of kinetic energy, the resolution is not sufficient. Using magnetic spectrometry, two groups were able to show, for cluster radioactivity of ^{223}Ra, that the emitted clusters were carbon ions with A = 14. For all other cases studied, the branching ratio for cluster emission was too low to be measured by magnetic spectrometry, and the cluster masses had to be inferred from calculated rates.

Table 2 summarizes the results of searches for cluster radioactivities. Column 2 gives the kinetic energies for the light cluster, assuming ground state to ground state decays. The last two columns give the measured values of partial half-life and branching ratio relative to alpha decay. Where limits are given, they refer to 90% confidence level. The other six columns give half-lives predicted (or postdicted) by various authors. Values calculated by other authors can be found in a recent review[4]. Beginning with the column labeled "Poe.", the calculations are discussed in refs. 7, 8, 13, 14, 15, and 16. Most of the models have been fine-tuned with two or more adjustable parameters after seeing results of some of the measurements. (Usually one parameter takes into account the fact that odd-A nuclei emit clusters at a rate hindered by a factor ~10^2 relative to the rate for even-even nuclides.) It is impressive that the standard deviation in log T, for several models, is less than 0.4, which corresponds to a factor 2.5 in half-life T.

Figure 2 presents the results in the form of Geiger-Nuttall plots, named after Rutherford's colleagues who found that log T is approximately a linear function of $1/\sqrt{Q}$. One sees that the hindrance factor for odd-A is larger for cluster decay than for alpha-decay.

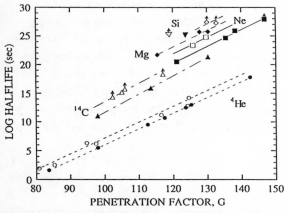

Figure 2. Dependence of partial half-life on penetration factor for various cluster radioactivities. Solid points refer to even-even parent nuclides; open points refer to odd-A nuclides.

Table 2. Comparison of Calculated and Measured Values of log T (half-life, sec)

Decay Mode	E_k (MeV)	Poe.	SqW	S-S	B-M	B-W	I-S	Measured log T (sec)	-log B
^{221}Fr $\rightarrow ^{14}$C	29.28	14.3	15.2	16.0	14.0	15.5	18.7	>15.77	>13.3
^{221}Ra $\rightarrow ^{14}$C	30.34	14.2	14.1	14.8	>12.4	14.2	17.6	>14.35	>12.9
^{222}Ra $\rightarrow ^{14}$C	30.97	11.1	11.2	11.6	11.4	11.8	10.5	11.0±.06	9.4±.06
^{223}Ra $\rightarrow ^{14}$C	29.85	15.1	15.0	15.7	15.3	15.1	14.9	15.2±.05	9.2±.05
^{224}Ra $\rightarrow ^{14}$C	28.63	15.9	16.0	16.8	16.1	16.2	15.3	15.9±.12	10.4±.12
^{225}Ac $\rightarrow ^{14}$C	28.57	17.8	18.7	19.7	18.8	18.6	22.1	>18.34	>12.4
^{226}Ra $\rightarrow ^{14}$C	26.46	20.9	21.0	22.2	21.0	21.1	21.7	21.3±.2	10.6±.2
^{231}Pa $\rightarrow ^{23}$F	46.68	25.9	26.0	25.5	--.-	26.8	<u>24.4</u>	>25.4	>13.4
^{230}Th $\rightarrow ^{24}$Ne	51.75	25.2	24.8	24.9	24.7	24.8	25.2	24.6±.07	12.3±.07
^{232}Th $\rightarrow ^{26}$Ne	49.70	30.2	29.1	28.4	28.7	29.3	<u>26.7</u>	>27.9	>10.3
^{231}Pa $\rightarrow ^{24}$Ne	54.14	23.3	23.7	23.5	21.6	23.4	23.9	23.4±.08	11.4±.08
^{232}U $\rightarrow ^{24}$Ne	55.86	20.8	20.7	20.0	20.9	20.8	19.8	20.5±.03	11.1±.03
^{233}U $\rightarrow ^{24}$Ne	54.27	25.2	24.9	24.8	23.7	25.4	24.4	24.8±.06	12.1±.06
$\rightarrow ^{25}$Ne	54.32	25.7	25.1	24.4	--.-	26.0	24.2		
^{234}U $\rightarrow ^{24}$Ne	52.81	26.1	25.8	25.7	25.5	25.6	25.8	25.9±.2	13.0±.2
$\rightarrow ^{26}$Ne	52.87	27.0	26.2	25.0	25.9	26.4	26.1		
^{235}U $\rightarrow ^{24}$Ne	51.50	29.9	29.7	30.1	--.-	29.9	--.-	>27.4	>11.1
$\rightarrow ^{25}$Ne	51.68	30.6	29.7	29.6	--.-	28.0	--.-		
^{233}U $\rightarrow ^{28}$Mg	65.32	27.4	26.9	27.5	--.-	28.0	<u>23.9</u>	>27.8	>15.1
^{234}U $\rightarrow ^{28}$Mg	65.26	25.9	25.4	25.7	25.4	25.4	25.7	25.7±.2	12.8±.2
^{237}Np $\rightarrow ^{30}$Mg	65.52	28.3	28.3	27.7	>27.3	29.9	<u>25.6</u>	>27.4	>13.6
^{236}Pu $\rightarrow ^{28}$Mg	70.22	21.1	21.2	20.5	21.5	22.0	--.-	21.7±.3	13.7±.3
^{238}Pu $\rightarrow ^{30}$Mg	67.00	26.2	25.9	24.3	25.6	25.8	26.2	25.7±.25	16.3±.25
$\rightarrow ^{28}$Mg	67.32	26.2	25.5	--.-	25.7	26.9	24.9		
^{238}Pu $\rightarrow ^{32}$Si	78.95	26.1	25.7	--.-	25.8	25.7	26.3	25.3±.16	15.9±.16
^{241}Am $\rightarrow ^{34}$Si	80.60	25.8	26.5	26.2	25.3	28.8	<u>23.8</u>	>25.3	>15.1
^{242}Cm $\rightarrow ^{34}$Si	82.88	<u>23.5</u>	<u>23.4</u>	<u>22.6</u>	<u>--.-</u>	<u>24.1</u>	<u>--.--</u>	>21.5	>14.4
σ_{logT} *		0.36	0.27	0.77	0.62	0.33	0.46		

* Evaluated only for positive detections. Underlined values are discordant by more than 10x from experimental limits and would worsen the value of σ_{logT} if future experiments lead to positive detection. All upper limits are 90% confidence level.

Following an exposure of 690 days to ^{236}Pu produced in the cyclotron at the Kurchatov Institute, Ogloblin et al.[17] detected two decays to energetic Mg ions, which pins down the partial half-life to within a factor 2. This result appears in Table 2.

In order to search for various modes of cluster radioactivity and to determine the spontaneous fission decay constant, Bonetti et al.[18,19] obtained a source of ^{232}U of certified activity and carried out two separate investigations. In the first[18], they set a stringent upper limit, $\lambda_{SF} < 3.2 \times 10^{-24}$ sec^{-1}, a factor 10^2 lower than had been reported several decades earlier (ref. 20). This result confirmed a conjecture[21] that the particles emitted by ^{232}U and attributed to spontaneous fission in ref. 20 were actually due to Ne decays. In ref. 18, Bonetti and co-workers also set an upper limit of 0.006 on the branching ratio for Mg emission relative to Ne emission, and showed that the Ne emission rate reported by Barwick et al.[21] was too low by a factor 4.6. The most likely explanation for the discrepancy was that the authors of ref. 21 purchased the ^{232}U from a commercial firm which measured its activity incorrectly. In ref. 19, Bonetti et al. made an independent determination of the Ne emission rate of ^{232}U which confirmed their earlier result. As a result of their work, we now know that ^{232}U is the most copious emitter of Ne clusters. One consequence of the factor 4.6 increase in the Ne emission rate of ^{232}U is that the Ne emission rate of ^{234}U must be revised downward by a factor 4.8. The reason is that the contribution to Ne emission from the 0.002% (by mass) ^{232}U in the ^{234}U source[12,22] had previously been underestimated. Bonetti et al. also pointed out that the revised Ne emission rate of ^{234}U requires that the Ne emission rate of ^{235}U reported in ref. 23 also be revised.

In an experiment aimed at measuring the relative emission rates of Mg and Ne from ^{233}U, the Berkeley-Livermore-Milano collaboration[24] collected tracks of more than 1000 clusters emitted from 0.76 g of ^{233}U during a 582 day exposure. They found a surprisingly low branching ratio for Mg/Ne -- less than 0.0018 (90% confidence level) -- which is inconsistent with existing models unless modified to take into account the Nilsson states of the parent and daughter nuclei. We will later discuss the implications of this result.

An experiment in progress by the Milano group[25] shows that, as predicted by various models, ^{228}Th does indeed decay by ^{20}O cluster radioactivity. To date, four ^{20}O events have been detected, for a branching ratio tentatively given as ~6 x 10^{-14} relative to alpha decay. This is the first example of ^{20}O cluster emission.

Spectroscopy

With their recent experiment using the Orsay magnetic spectrometer, Brillard et al.[26] have advanced the study of cluster radioactivity from the exploratory phase -- the determination of lifetimes and branching ratios for various cluster types and parent nuclei -- to the spectroscopic phase, where the role of nuclear structure is becoming apparent. By improving the energy resolution of their spectrometer, they were able to show that only 15±3% of the ^{14}C decays left the daughter ^{209}Pb in the 9/2$^+$ ground state, that 81±6% of the decays went to the 11/2$^+$ first excited state at 0.779 MeV, and that the remainder went to the next two excited states, 15/2⁻ at 1.423 MeV or 5/2⁺ at 1.567 MeV. The inhibited decay of the odd-A nuclide ^{223}Ra seems to be due to some kind of selection rule that forbids decay to the ^{209}Pb ground state. Recalling that the hindrance to alpha decay of certain odd-A nuclei is usually attributed to the necessity of of an unpaired neutron to change its single-particle wave-function, Hussonnois et al.[27] offered a similar explanation for the preference for decay to an excited level of ^{209}Pb in ^{14}C emission from ^{223}Ra. Unfortunately, the magnetic spectrometric technique seems to be limited to one or two favorable nuclides for which the cluster emission rate is particularly high.

Fission vs Cluster Models

Two apparently quite different classes of models can account equally well for the data on partial half-lives and branching ratios of all the cluster radioactivities studied before about 1990. The macroscopic or fission models depict the decaying parent as deforming from a spheroidal shape through a shape barrier into two separated spheroidal daughters. The decay rate is given simply by

$$\lambda_f = \nu_f \, P_f$$

where ν_f is the frequency of oscillations in the fission mode and P_f is the penetrability. As shown in Fig. 3, in the fission model the potential barrier has an inner or prescission part and an outer or postscission part, and is thicker than the barrier used in the second class of models, called the microscopic or cluster models, which have only a postscission barrier (see Fig. 3), but include a cluster preformation factor. The inner part, which is missing in cluster models, comes from the early stages of fission in which the two fragments have finite spatial extensions when the deformation energy is calculated. In one such fission model[13] the barrier includes both a phenomenological "proximity potential" and a Coulomb potential. In the versions presented in the literature, fission models contain parameters that take into account on average a hindrance against decay of odd-A nuclides relative to even-even nuclides, but do not explicitly include the role of nuclear structure except as it affects the Q value.

The microscopic or cluster models explicitly include three factors -- the cluster preformation probability at the nuclear surface, S_c, the frequency of assaults on the barrier, ν_c, and the penetrability through the external Coulomb barrier, P_c.

$$\lambda_c = \nu_c \, S_c \, P_c$$

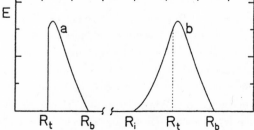

Figure 3. Potential barrier shapes typical of (a) a preformed cluster model and (b) a fission model.

Generally the finite radius of the cluster is not taken into account. In principle, S_c should be calculated quantum mechanically and should include the Nilsson states of the parent and two daughters. In practice it is too difficult to calculate S_c exactly and one resorts to empirical fits to clusters for which S_c can be calculated. In one such model[15] a plausible phenomenological expression was used

$$S = S_\alpha{}^{(A-1)/3}$$

in which A is the mass number of the light cluster and S_α is assigned the value 0.0063 for even-even parent nuclei and the value 0.0032 for odd-A parent nuclei. In this form, S is related to the probability of finding A nucleons together in the surface of the parent nucleus, and S_α is the value of S for preformation of an alpha particle.

Poenaru and Greiner[28] have argued that the two classes of models can be reconciled if one expresses P_f as the product of two penetrabilities, P_{inner} P_{outer}, and equates the preformation probability to the penetrability of the inner part of the barrier:

$$P_{inner} = S_c.$$

Formally, this procedure explains all the data before 1990 if the frequencies ν_c and ν_f are the same.

Nuclear Structure Information from Cluster Radioactivity

The two surprising results -- that ^{223}Ra undergoes ^{14}C emission primarily to the 11/2+ state of ^{209}Pb and that the Mg/Ne branching ratio in the cluster radioactivity of ^{233}U is an order of magnitude smaller than expected in cluster models to date -- require for their interpretation that the quantum states of the parent and daugher nuclei be taken into account. It has been known for decades that alpha decays of odd-A nuclides are strongly hindered when the odd nucleon has to change its quantum state[29]. Analogous reasoning has been applied to the hindered ^{14}C decay of ^{223}Ra (refs. 30-32) and to the hindrance against Mg emission relative to Ne emission of ^{233}U (ref. 24). According to ref. 31, the ground state of the deformed nuclide ^{223}Ra is mainly built from the 3/2+ state emerging from the i 11/2 orbital, a result suggested by the Nilsson orbitals in Fig. 4. The authors of ref. 31 have calculated that the squared overlap between the $\Omega = 3/2$ Fermi level orbitals for N = 135 and deformation parameters appropriate for ^{223}Ra and the $\Omega = 3/2$ spherical orbitals for ^{209}Pb is by far greatest for i 11/2, which explains the preference for ^{14}C decay to the i 11/2 state of ^{209}Pb.

In the case of ^{24}Ne emission by ^{233}U, the spherical ground state of daughter ^{209}Pb is g 9/2. A distortion of spherical symmetry into spheroidal symmetry would break the fivefold degeneracy of the g 9/2 state: the Nilsson configuration of one of the resulting states is 5/2 [633], which is the same configuration as the ground state of the deformed ^{233}U nucleus. The transition for which the overlap integral between the parent and daughter nuclei is maximized is ground state ^{233}U to ground state ^{209}Pb. In contrast, for ^{28}Mg emission of ^{233}U, the ground state of the daughter ^{205}Hg is the neutron p 1/2 hole state, which is unrelated to the ground state of ^{233}U. In fact, decays to the first five states of ^{205}Hg all involve a change of parity.

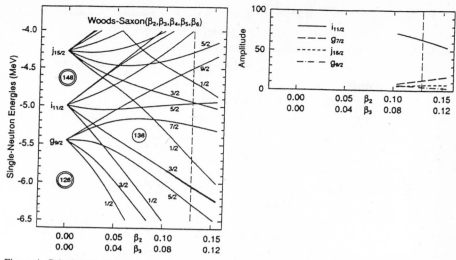

Figure 4. Calculation of shell-model composition of the $\Omega = 3/2$ g. s. orbital of ^{223}Ra (ref. 31). At left, neutron orbitals are drawn where dashed line corresponds to deformation of ^{223}Ra. All five deformation parameters increase linearly from left to right. At right, the squared overlap betwen the $\Omega = 3/2$ Fermi level orbitals for N = 135 and the most important $\Omega = 3/2$ spherical orbitals is shown.

Conclusions

1. By looking at the distribution of both daughter nuclei, one sees that the heavier daughter, always ^{208}Pb or a nuclide one or two nucleons away from it, plays a role as important in cluster radioactivity as does the ^{4}He cluster in alpha radioactivity.

2. The results on cluster decay of odd-A nuclides discussed in the previous section suggest that there is a closer quantitative connection between cluster radioactivity and alpha decay than between cluster radioactivity and fission.

3. Both cluster models and fission models, fine-tuned with adjustable parameters, fit most of the measured partial half-lives to within a factor three. Larger deviations from the predictions of the models point to a role of nuclear structure analogous to that seen in alpha decay of odd-A nuclides. Additional examples of the role of nuclear structure are now being sought among such odd-A nuclides as ^{221}Fr, ^{221}Ra, ^{225}Ac, ^{231}Pa, and ^{237}Np, which tend to show larger deviations from the predictions of the models than the even-even nuclides.

This work was supported in part by the U. S. Department of Energy at Lawrence Berkeley Laboratory.

References

1. W. D. Knight et al., Phys. Rev. Lett. 52, 2141 (1984).

2. H. J. Rose and G. A. Jones, Nature 307, 245 (1984).

3. D. V. Aleksandrov et al., JETP Lett. 40, 909 (1984).

4. P. B. Price, Ann. Rev. Nucl. Part. Sci. 39, 19 (1989).

5. Yu. S. Zamyatnin, V. L. Mikheev, S. P. Tretyakova, V. I. Furman, S. G. Kadmenskii, and Yu. M. Chuvilskii, Sov. J. Part. Nucl. 21, 231 (1990).

6. E. Hourani et al., Phys. Lett. 160B, 375 (1985).

7. D. N. Poenaru et al., At. Data Nucl. Data Tables 34, 423 (1986) and subsequent unpublished revisions.

8. P. B. Price et al., Phys. Rev. Lett. 54, 297 (1985).

9. S. W. Barwick et al., Phys. Rev. C 34, 362 (1986).

10. See, for example, S. P. Tretyakova et al., Z. Phys. A 333, 349 (1989).

11. See, for example, Shicheng Wang et al., Phys. Rev. C 36, 2717 (1987).

12. See, for example, Shicheng Wang et al., Phys. Rev. C 39, 1647 (1989).

13. Y.-J. Shi and W. J. Swiatecki, Nucl. Phys. A438, 450 (1985); Nucl. Phys. A464, 205 (1987); and unpublished tables.

14. B. Buck and A. C. Merchant, Phys. Rev. C 39, 2097 (1989); J. Phys. G: Nucl. Part. Phys. 16, L85 (1990); and unpublished tables.

15. R. Blendowske and H. Walliser, Phys. Rev. Lett. 61, 1930 (1988); unpublished tables.

16. M. Ivascu and I. Silisteanu, Nucl. Phys. A485, 93 (1988); J. Phys. G: Nucl. Part. Phys. 15, 1405 (1989); and unpublished tables.

17. A. A. Ogloblin et al., Phys. Lett. 235, 35 (1990).

18. R. Bonetti et al., Phys. Lett. 241, 179 (1990).

19. R. Bonetti et al., submitted to Phys. Rev. C (1991).

20. A. H. Jaffey and A. Hirsch, unpublished result, quoted in R. Vandenbosch and J. R. Huizenga, Nuclear Fission (Academic Press, New York, 1973).

21. S. W. Barwick, P. B. Price, and J. D. Stevenson, Phys. Rev. C 31, 1984 (1985).

22. K. J. Moody et al., Phys. Rev. C 39, 2445 (1989).

23. S. P. Tretyakova et al., Z. Phys. A 333, 349 (1989).

24. P. B. Price et al., Phys. Rev. C 43, 1781 (1991).

25. R. Bonetti et al., Istituto di Fisica Generale Applicata, Milano, unpublished results.

26. L. Brillard et al., C. R. Acad. Sci. Paris 39, ser. 2, 1105 (1989).

27. M. Hussonnois et al., J. Phys. G: Nucl. Part. Phys. 16, L77 (1990).

28. D. N. Poenaru and W. Greiner, University of Frankfurt preprint UFTP 253/1990.

29. J. O. Rasmussen, Ark. Fys. 7, 185 (1953).

30. M. Hussonnois et al., Phys. Rev. C 42, R495 (1990).

31. R. K. Sheline and I. Ragnarsson, Phys. Rev. C 43, 1476 (1991).

32. B. Buck, A. C. Merchant, and S. M. Perez, Nucl. Phys. A512, 483 (1990).

Observation of Nuclear Clusters in the Spontaneous Decay of Heavy Clusters

S.P. Tretyakova[1], A.A. Ogloblin[2], V.L. Mikheev[1], and Yu.S. Zamyatnin[1]

[1]Joint Institute for Nuclear Research, Dubna, P.O. Box 79,
10100 Moscow, Russia
[2]I.V. Kurchatov Institute of Atomic Energy, 123182 Moscow, Russia

Abstract: The technique used in the research of the cluster-decay studies from the ground state, along with some results, are presented for atomic numbers Z=90 to 95. Preliminary results of the experiments searching for cluster decay from the excited nuclear states formed by irradiation of ^{233}U with thermal neutrons and with 22 MeV bremsstrahlung are also presented

1. Introduction

At present the radioactive decay involving the emission of monoenergetic heavy ions of ^{14}C, ^{24}Ne, ^{28}Mg and ^{32}Si has been detected experimentally in 11 nuclides with atomic numbers ranging from 88 to 94. In addition, the upper limits for the probability of this kind of decay have been established for 9 other nuclides [1].

Theoretical approaches to the description of this phenomenon consider it as a process similar to $\alpha-$decay, but with the emission of a heavy cluster, or treat it as a strongly asymmetric fission. There have been attempts to create unified models that include the features of the models of $\alpha-$decay as well as the conventional spontaneous fission. One can hope that the study of cluster decay will help to reveal the interdependence between the processes of spontaneous transmutation of heavy nuclei into lighter ones, caused by $\alpha-$particle, cluster and fission fragment emission. In this respect semi-empirical systematics of decay probabilities can be of use.

Presently the experimental investigations of cluster decay are developed in three directions:

i) accumulation of more precise definition of data on cluster decay probabilities from ground states of a wider range of nuclei;

ii) study of decay into the excited state of daughter nuclei. The most interesting results are obtained by the French group that has studied the emission of ^{14}C in the decay of ^{223}Ra [2], and

iii) cluster emission study from the excited states of nuclei. The region of excitation energies below the maxima of potential barriers relevant for a particular decay mode is, of course, highly interesting. Experiments performed up to now only allow to establish an upper limit of this effect [3-5].

Springer Series in Nuclear and Particle Physics   **Clustering Phenomena in Atoms and Nuclei**
Editors: M. Brenner · T. Lönnroth · F.B. Malik   © Springer-Verlag Berlin, Heidelberg 1992

2. Cluster Decay Investigations from Ground States

The wide use of heavy element isotopes, accelerators of heavy ions with a large variety of accelerated particles necessary for calibration, and great registration experience of different particles by solid state track detectors have given good possibilities for cluster radioactivity research in Dubna. Practically all measurements of cluster radioactivity are carried out by track registration in polyethyleneterephtalate (PETF), sensitive to particles with $Z \geq 6$. The detectors are exposed in the air in a geometry of nearly 2π.

For obtaining the visible (under a microscope) optical image of a particle track the detectors are etched by a 20% NaOH solution at a temperature of 60° C in several stages (2-4) at intervals of 0.5 to 1 hour for the purpose of determining the etch rate V_T at several points of the track. After each stage the search and measurement of the parameters of the track are carried out by scanning under an optical microscope. The track detection is done in the angular range $15 - 70°$ with respect to the detector plane. This yields a cluster detection efficiency equal to 0.33 of 4π.

The scattering of α−particles and fission fragment in light nuclei are incorporated in the source and detector materials produce the secondary background due to recoil nuclei with $Z \geq 6$. In order not to exceed the number of recoil nuclei that makes cluster observation difficult, the time of exposure is limited for each detector by the admissible α−particle flux of $\leq 10^{12}$ α−particles per cm^2. In this connection a two-layer detector was specially designed for studies of the cluster decay of heavy nuclei with short α−decay and spontaneous-fission half-lives. The upper layer (relative to the ion source) has been manufactured from polycarbonate and the lower one from polyethyleneterephtalate. The initial thickness of the upper layer was chosen so that its residual thickness in the given operational regime is exceeded by 3-5 μm and hence the range of recoils from the elastic scattering perpendicular to the detector could not reach the lower PETF layer. After the termination of the chemical treatment the upper layer was removed by dissolving it in dichloroethane.

The two-layer detector technique has permitted the studies of the decay of heavy nuclei at α−particle fluxes of $10^{14} - 10^{15}$ and with fluxes of spontaneous-fission fragments of $\sim 10^{-6}$ per cm^2.

To identify the particle producing the track the dependence of the etching selectivity V_T/V_M upon the value of the residual particle range R is used. Here V_T is the velocity of etching along the track in the detector material, and V_M is the etching velocity outside the damaged zone produced by the detected particle. The ratio V_T/V_M is the difference between the length of the fully etched track and that of its part etched at the instant of terminating a given etchin step. Measurements of track during the multistage etching have the advantage that one and the same track can be used repeatedly for identifying the particle which has produced it.

In order to study the effect of a high α—particle flux on the detection properties of the PETF the detectors were calibrated using 10-30 MeV/amu ^{12}C, ^{16}O, ^{20}Ne, ^{26}Mg, ^{27}Al and ^{28}Si ion beams from the U-300 cyclotron of JINR. It was established that the ratio V_T/V_M is affected by the magnitude of the integrated α—particle flux to which the detector material was exposed when placed in air on a radioactive sample. This dependence can be approximated in our measurements by the expression

$$V_T/V_M = a(dE/dx)b, \tag{1}$$

where $a = 0.0063$ for $\Phi_\alpha > 10^9$ cm^{-2}, $b = 1.83 + 0.81(\Phi_\alpha \cdot 10^{-9})^{0.044}$, dE/dx is the specific ionization in units of MeV·cm^2/mg, and Φ_α is the number of α—particles that have passed through 1 cm^2 of detector. This expression allows us to predict the change of detector properties as a function of Φ_α.

Two kinds of neon and magnesium are observed in the decay of ^{234}U nuclei [6]. The calibrations and measurements for ^{234}U are given in Figure 1. For reliable visual identification a difference between the etching rates for these ion tracks are used. The photomicrographs of the neon and magnesium tracks (cf. Figure 1b) and those of the ^{234}U fission fragments demonstrate a difference between the track sizes for the same etching time.

Our technique of processing solid-state track detectors in the regions of atomic numbers Z$\leq$ 15 and mass numbers A$\leq$ 30 provides resolutions of ΔZ$\sim$ $\pm$0.15 and ΔA$\sim$ $\pm$1, respectively.

The mass of clusters with a known charge is found from the Q value of the reaction. In the case of full track etching the cluster range R in the detector and its energy E are measured. The ion energy is determined from the total ion range with a relative probability ΔE/E$\sim$ 3$-$5%. The results of investigation of the cluster decay of ground state nuclei in our experiments are given in table 1.

Appreciable correction to the experimental data on radioactive characteristics of ^{234}U and ^{235}U are made to ^{24}Ne decay of $1.2 \cdot 10^{-4}$ % admixture of ^{232}U in our uranium samples. The data on cluster decay probability of ^{232}U are taken from paper [7]. However, prof. Bonetty with colleagues [8] have established that the ^{24}Ne decay probability from ^{232}U is 4.5 times larger than it was shown in the work [7]. Taking into account this fact the decay probability of ^{24}Ne from ^{234}U in table 1 has to be decreased by a factor of 2.1. As a consequence the ^{24}Mg decay from ^{234}U becomes more probable than the ^{24}Ne decay, in accordance with the theoretical predictions [9]. Unfortunately, the effect that can be related to cluster decay of ^{235}U in our experiments [6], and taking into account the data of [8], are within the error limits of the measurement.

During the irradiation time of the detectors some ^{232}U is produced as a result of the α—decay of ^{236}Pu. Estimates show that about two Ne tracks might be produced [12]. They are not observed because the detector during

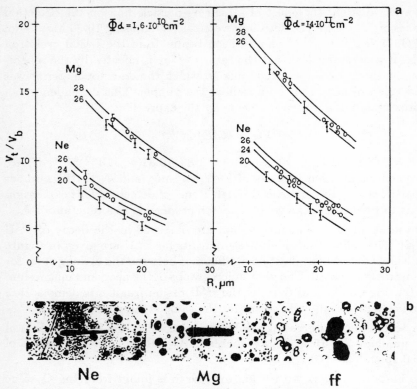

Fig. 1. a. The etching selectivity V_T/V_M as a function of the residual range. The curves are drawn according to (1). The calibration data on ^{20}Ne and ^{26}Mg are indicated by solid circles. The results of processing some of the tracks revealed in studies of the ^{234}U decay for two expositions are shown by open circles. **b.** Profiles on Ne, Mg and fission fragment tracks with the same magnification of 1200× after etching during 3.5 h. The circular tracks of different diameters are due to recoil nuclei knocked out by alpha particles.

the etching time (1.5 hours) is adjusted to the optimal detection of Mg tracks. Under that condition the tracks of Ne, having less specific energy loss than Mg, have not been fully etched and are excluded from registration in the optical scanning [6].

The systematics of experimental data are very useful for their analysis and different extrapolations. Figure 2, see ref. [12], presents the systematics of relative cluster decay probabilities which are analogous to the Geiger-Nuttal systematics for α−decay. The barrier penetrability P in figure 2 was calculated using the relation

$$P = exp\Big\{-2/\hbar \int_{R_N}^{Z_1 Z_2 e^2/Q} [2\frac{A_1 A_2}{A_1 + A_2} \cdot u \cdot \frac{Z_1 Z_2 e^2}{r} - Q]^{1/2} dr\Big\},$$

Table 1. The results of investigations of the cluster decay of heavy nuclei.

Isotope	Cluster	Isotope weight (mg)	Exposure time (days)	N_{cl} tracks	$(\lambda_{cl}/\lambda_\alpha)$ $\times 10^{-12}$	$(T_{1/2})^\star_{cl}$ (years)
^{230}Th	^{24}Ne	210	64	165	0.56	$1.3 \cdot 10^{17}$
^{231}Pa	^{24}Ne	7+10	138	252	3.8	$8.6 \cdot 10^{15}$
^{233}U	^{24,25}Ne	75	28	16	0.75	$2.2 \cdot 10^{17}$
^{234}Th	^{24,25}Ne	11.5+18.5	233+420	7+24	0.39	$6.3 \cdot 10^{17}$
	^{28}Mg			3+13	0.23	$1.1 \cdot 10^{18}$
^{235}U	^{25,26}Ne	1200	233	0	< 5.0	$> 1.4 \cdot 10^{20}$
	^{28}Mg				< 0.79	$> 9 \cdot 10^{20}$
^{236}U	^{26}Ne	25.3	233	0	< 4.0	$> 6 \cdot 10^{18}$
^{237}Np	^{30}Mg	320	122	0	< 0.04	$> 5.4 \cdot 10^{19}$
^{236}Pu	^{28}Mg	$3.4 \cdot 10^{-4}$	689	2	0.02	$1.5 \cdot 10^{14}$
^{241}Am	^{34}Si	3.7	30	0	> 0.005	$> 9 \cdot 10^{16}$

* Partial half-life for the cluster emission

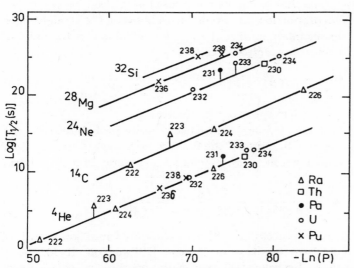

Fig. 2. Dependence of the common logarithm of the partial half-life on the natural logarithm of the Coulomb potential barrier penetrability P for ^{4}He, ^{14}C, ^{24}Ne, ^{28}Mg, ^{32}Si emission.

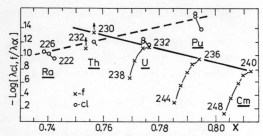

Fig. 3. Dependence of ratios of spontaneous fission and cluster decay probabilitites to the $\alpha-$decay probability on the fissility parameter X.

where A_1, A_2 and Z_1, Z_2 are the mass and atomic numbers of the cluster and the daughter nucleus, respectively, Q is the mass difference between the initial and final nuclei, $\hbar$ is the Planck constant, u is the unit mass, and e is the elementary charge. The value of R_N is taken to be equal to $1.44(A_1^{1/3} + A_2^{1/3}) \cdot 10^{-13}$ cm.

The distances between the straight lines for the decay by the emission of different clusters are proportional to their mass number A. This fact agrees well with the hypothesis that the mechanism of cluster decay is similar to that of $\alpha-$decay. The proportionality of distances between the straight lines in figure 2 to the cluster mass number A could be explained on the basis of the equation for the cluster spectroscopic factor:

$$S(A) = [S(\alpha)]^{(A-1)/3}, \tag{2}$$

where $S(\alpha)$ is the $\alpha-$particle spectroscopic factor.

The ratio of spontaneous fission and cluster decay probabilities to the $\alpha-$decay probabilities as a function of the fissility parameter X, determined according to [11], are shown in figure 3. The fissility parameter X is given by

$$X = \frac{Z^2/A}{50.833\{1 - 1.7826[(N-Z)/A]^2\}}.$$

The latest data on cluster decay of ^{236}Pu, investigated in a joint work of Dubna and the Kurchatov Institute in Moscow [12] and on cluster decay of ^{238}Pu, performed at Berkeley [13], are of importance for this kind of systematics.

We have presented in figure 3 even-even nuclei only, in order to exclude odd-even hindrance factors. Double points (circles) for the ^{234}U and ^{236}Pu decay indicate the emission by each nuclide of clusters of two types. The relative probability of spontaneous fission increases with the increasing of the fissility parameter, and the relative probability of cluster decay diminishes. On one hand, it could be interpreted as an evidence for the difference

288

between the processes of spontaneous fission and cluster decay. But, on the other hand, it could be assumed that, starting from fissility parameter $\simeq 0.76$ and cluster mass number $\simeq 24$, the transition to the spontaneous fission valley on the energy surface for heavier clusters becomes more probable, instead of escaping from the initial nuclei along the cluster valley.

Presently, collaborating with Kurchatov Institute in Moscow, we are studying the cluster decay of ^{242}Cm with emission of ^{34}Si in order to obtain the additional data of competition of spontaneous fission and cluster decay.

3. Cluster Decay From Excited Nuclear States

Theoretical estimations of cluster decay probabilities of heavy nuclei deal with, firstly, the cluster formation probabilities from nucleons and, secondly, the exit potential barrier penetrabilitites. The value of excitation energy is very essential for both these factors.

The cluster emission probabilities from the excited nuclei increases sharply in comparison to those from the ground states. Unfortunately, the competitive processes, namely α—decay and fission, become much more probable also. Figure 4 shows the excitation energy dependence of the partial half-lives of ^{234}U emitting ^{4}He and ^{24}Ne nuclei. The calculations have been performed by means of a "Decay" program kindly given to us by dr. R. Blendowski [10]. We have summed the Q value for a given decay chan-

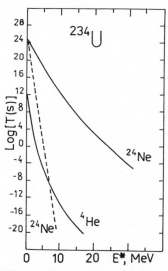

Fig. 4. Excitation energy dependence of the partial half-lives of ^{234}U emitting ^{4}He and ^{24}Ne nuclei according to [10] (full lines). The dashed lines shows the partial half-life of ^{234}U emitting ^{24}Ne nuclei as a function of the nuclear zero-point vibration [15].

nel from the ground state and the values of excitation energy. In this way of excitation energy realization the ratio of ^{24}Ne and ^{4}He emission probabilities become less than those from the ground state. However, there are experimental data on cluster emission from highly excited nuclear systems, for instance Sm, Ag + ^{40}Ar at 336 MeV, which provides evidence for the decrease of the emission barrier with excitation energy. The decrease of the emission barrier height is larger for the heavier clusters. This is because a dynamical deformation sets in due to the transformation of excitation energy to vibrational degree of freedom. The dashed line in figure 4 shows the partial lifetime for the emission of ^{24}Ne from ^{234}U as a function of energy. The dashed line is obtained by adding to the energy the correction due to zero-point vibration in accordance with the theoretical estimations [15]. Correcting for the transformation of excitation energy to vibrational energy one can hope to obtain the yield of ^{24}Ne clusters about $10^{-7} - 10^{-9}$ relative to the yield for binary fission.

In the work of [5], performed at the Grenoble reactor LOHENGRIN spectrometer, the limit of 10^{-9} is established for ^{24}Ne cluster emission relative to the binary fission yields in the case of thermal neutron capture. It seems to us that we can obtain 1-2 orders lower limit due to some modifications of the solid state track detector technique. It is essential that solid state track detectors can provide the same sensitivity for the desirable range of mass and atomic numbers of detected particles.

The main difficulties are due to the background of large fission fragments and scattering of $\alpha-$particles and fission fragments by light nuclei present in the source and detector material, which produce a secondary background due to recoil nuclei with $Z \geq 6$. We hope that our technique will be able to overcome this background problem.

We use a track detector with a thickness equal to the cluster range. We etch it only from side, opposite to particle entrance surface. In this case one can reveal cluster tracks without revealing the fission fragment tracks because the ranges of the latter ones are two times less than the former.

The group of dr. V. Rubchenya at the Leningrad Radium Institute has performed trajectory calculations for the ternary fission of ^{234}U. In accordance with these calculations the ranges of Ne and Mg nuclei emitted from ^{234}U in ternary fission are two times less than in the case of two-body cluster decay. Consequently, their range will be in the fission fragment range region.

The track detection is carried out in the angular range $\pm 30°$ with respect to the normal to the detector plane. This yields a cluster detection efficiency equal to 13% of 2π. The main problems are due to the recoil of C, N and O fission fragments and the possibility of extremely large deviations of the ranges of fission fragment from the mean values. The reliable identification of Ne and Mg tracks is very important. We hope to do it by the careful choice os solid state track detector materials and variation of the chemical etching regime [6].

We have used as a thermal neutron source the microtron MT-25 at JINR, Dubna, with ^{233}U bremsstrahlung target placed in the center of the graphite-moderated cube with dimensions $1.2 \times 1.2 \times 1.2$ m^3. The electron beam with intensity 16μA and 22 MeV energy has provided a flux of thermal neutrons of $\sim 10^8$ cm^{-2}s^{-1}. The area of the ^{233}U target was 7×9 cm^2, and the thickness 0.3 mg/cm^2. In the case of the tungsten bremsstrahlung target and without graphite moderator we have the possibility to irradiate our ^{233}U target by $\gamma-$rays of 22 MeV end point energy and intensity $7 \cdot 10^{12}$ cm^{-2}s^{-1}.

The products of uranium decay are recorded by PETF track detectors having a thickness of 31 μm which equal the range of Ne and Mg nuclei emitted in binary fission from the ground state of ^{234}U. The detectors are calibrated using 52 and 43 MeV ^{20}Ne ion beams from the U-400 cyclotron of JINR, Dubna. The face side of the detectors after irradiation but before etching is protected by special covering to prevent etching.

The numbers of tracks going through the whole detector thickness is about $\sim 10^7$ relative to the fission fragment number in our first experiments with neutrons and about $\sim 10^8$ in experiments with $\gamma-$rays. The fission fragment track density is up to $\sim 10^9$ 1/cm^2. Due to that the tracks have a distinctive halo at the face surface of the detector because of partial etching of fission fragment tracks intersecting the cluster track. This halo makes the searching of similar events easier in detector.

Hard background conditions do not allow us to identify definitely that part of the registered tracks which is due to decay from the two-body excited state of ^{234}U. We continue the data analysis and the improvement of our methods.

4. Conclusions

The solid state nuclear track technique appears to be a very effective method for investigations of cluster decay from the nuclear ground state. We hope that its improvement will allow to obtain the cluster decay data in the excitation energy range below emission barrier.

The authors express their appreciatio to Academician G.N. Flerov[†] and Professors Yu.Ts. Oganession, A. Sandulescu, Yu.P. Popov and V. Rubchenya for useful discussions. We also thank L.V. Jolos, K.I. Merkina, I.V. Ivanova, E.I. Kurenkova, E.A. Petrova and N.M. Macarie for the processing and scanning of the detectors.

[†] Deceased

References

1. Zamyatin, Yu.S., Mikheev, L.V., Tretyakova, S.P., Furman, Yu.M.: Fiz. Elem. Chastitz At. Yadra (USSR) —bf 21 537 (1990)
2. Hussonois, M., Le Du, J.F., Brillard, L., Ardison, G.: Phys. Rev. C **42** 495 (1990)
3. Oganessian, Yu.Ts., Gerbisch, Sh., Lobanov, Yu.V., Korotkin, Yu.S., Hussonois, M.: JINR Rapid Comm. (Dubna) **12-85** 30 (1985)
4. Gangrski, Yu.P., Khristov, Kh.G., Vasko, V.M.: Yad. Fiz. (USSR) **44** 294 (1986)
5. Börsig, B., Geltenbort, P., Gonennwein, F., Löffler, H.: Proc. Int. Conf. on "Fifty Years of Nuclear Fission", 3-7 April, 1989, Berlin, contrbuted papers HMI-B464 10 (1989)
6. Tretyakova, S.P., Zamyatin, Yu.S., Kvantsev, V.N., Korotkin, Yu.S., Mikheev, V.L., Timofeev, G.v.: Z. Phys. **A333** 349 (1989)
7. Barwik, S.W., Price, P.B., Stevenson, J.D.: Phys. Rev. C **31** 1984 (1985)
8. Bonetty, R., Fioretto, E., Migliorano, H.C., Pasinetty, A., Barranco, F., Vigezzi, E., Broglia, R.A.: Phys. Lett. **241B** 179 (1990)
9. Poenaru, D.N. Greiner, W., Depta, K., Ivascu, M., Mazilu, D., Sandulescu, A.: At. Data Nucl. Data Tables **34** 423 (1986)
10. Blendowski, R., Walliser, H.: Phys. Rev. Lett. **61** 1930 (1988)
11. Myers, W.D., Swiatecki, W.J.: Nucl. Phys. **81** 1 (1986)
12. Ogloblin, A.A., Venikov, N.I., Lisin, S.K., Pirozhkov, S.M., Pchelin, V.A., Rodionov, Yu.F., Semochkin, V.M., Shabrov, V.A., Shvetsov, I.K., Shubko, V.M., Tretyakova, S.P., Mikheev, V.L.: Phys. Lett. **B235** 35 (1990)
13. Wang, S., Showden-Ifft, D., Price, P.B., Moody, K.J., Hulet, E.K.: Phys. Rev. C **39** 1647 (1989)
14. Vaz, L.C., Logan, D., Alexander, J.M., Duek, E., Gueirea, D., Kowalski, L., Rivet, V.F., Zisman, M.S.: Z. Phys. **A311** 89 (1983)
15. Depta, K., Greiner, W., Maruhn, J.A., Hou-Ji Wang, Sandulescu, A., Herrman, R.: Preprint GSI-89-60 (Darmstadt, 1989)

Evidence of Spontaneous Emission of Oxygen Clusters from ^{228}Th

R. Bonetti, C. Chiesa, A. Guglielmetti, and C. Migliorino

Istituto Nazionale di Fisica Nucleare and
Istituto di Fisica Generale Applicata, via Celoria 16,
I-20133 Milano, Italy

Since 1984, many clusters of nucleons (^{14}C; $^{24/25/26}$Ne; $^{28/30}$Mg, $^{32/34}$Si) have been observed to be spontaneously emitted from heavy nuclei.
Fission models describe the mechanism of this so called "exotic decay" as the deformation of parent nucleus followed by the separation of the two fragments, while cluster models depict it as the preformation of the cluster inside the parent nucleus, followed by emission without any dynamic deformation.
All of the accessible cases are alpha emitters, with branching ratio with respect to alpha decay ranging from 10^{-9} to 10^{-17}.

We decided to search for the decay ^{228}Th $\rightarrow$ ^{20}O + ^{208}Pb since it is a favourable case, leading to the bimagic ^{208}Pb, and since the oxygen cluster has never been observed to be spontaneously emitted.
Theoretical predictions for the branching ratio relative to alpha decay are:
1×10^{-14} (fission model of Poenaru, Ivascu, Greiner) and 4×10^{-14} (cluster model of Blendowske, Fliessbach, Walliser). The phosphate glass track detector BP1 has demonstrated to be the most suitable one for this experiment because it is insensitive to alpha particle (like most of track detectors), and it can withstand about 2×10^{12} α/cm^2 without loosing sensitivity. Besides that, with a careful choice of etching conditions, it can reveal oxygen tracks and reject carbon tracks, in our case 1000 time more probable due to the presence, in the ^{228}Th radioactive serie, of ^{224}Ra, a well known carbon emitter.

We exposed a BP1 hemisphere under vacuum to a (1.84 ± 0.09) mCi initial activity ^{228}Th source , covered with a 2 μm aluminum foil absorber in order to stop the recoil nuclei, for 134 days. After having etched the glasses in HBF4, 50%, 65 °C for 40 hours, we manually scanned glasses. In order to improve the optical contrast of the oxygen tracks in respect to background, the irradiated surface was covered with a glicerine layer during scanning.

The analysis, at present limited to 19% of the total exposed surface, gives four tracks, whose sensitivity S lies on the calibration line of oxygen ions. (See Figure).
The resulting branching ratio is (6.8 ± 3.4) x 10^{-14}, in good agreement with the microscopic cluster model of Blendowske, Fliessbach and Walliser.

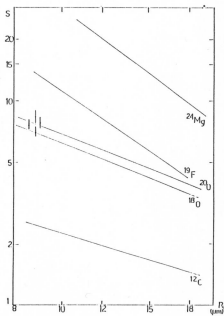

New Points of View on the Symmetric Fission of ^{258}Fm

G. Mouze and C. Ythier

Faculté des Sciences, F-06034 Nice, France

An attempt is made to interpret the symmetric spontaneous fission of the heavy actinides, e.g. of ^{258}Fm, according to the general scheme developed by the authors [1] in the case of asymmetric fission. This attempt leads to observations suggesting that this kind of symmetric fission occurs for nuclei in which the initial formation of a cluster within the valence shells of the fissioning nucleus yields an energy higher than the threshold for formation of a pion. It is also suggested that this mode of fission could be related to a phase change of nuclear matter.

Reference

1.   Mouse, G. and Ythier, C: Nuovo Cimento **103A** 617 (1990)

Springer Series in Nuclear and Particle Physics   **Clustering Phenomena in Atoms and Nuclei**
Editors: M. Brenner · T. Lönnroth · F.B. Malik   © Springer-Verlag Berlin, Heidelberg 1992

Shell Model Approach to the Description of Cluster Radioactivity of Heavy Nuclei

W.I. Furman[1], S.G. Kadmensky[2], and Yu.M. Tchuvil'sky[3]

[1]Joint Institute for Nuclear Research, Dubna, P.O. Box 79,
  10100 Moscow, Russia
[2]Voronezh State University, Voronezh, Russia
[3] Nuclear Physics Institute, Moscow State University,
  Moscow, Russia

A microscopic version of the theory of cluster decay is developed. The proposed model is based on the concept of a preformation of a cluster due to nonadiabatic quantum fluctuations and its tunneling through a potential barrier. The main point of the model is to derive the formalism and to calculate quantitative characteristics of the fluctuations, i.e. the spectroscopic factors of investigated clusters in parent nuclei using the shell model approach.

The root-mean square deviation of the logarithms of theoretical values of cluster emission half-life times from experimental ones is 0.7. It is rather satisfactory in the absence of fitting parameters in the proposed scheme. So our approach allows the explanation of the scale of experimental absolute probabilities of cluster decay within the accuracy of the theoretical method including the odd-even effects and the fine structure. This makes it possible to conclude that the existing experimental data on heavy cluster emission as a whole give good evidence in favour of the domination of the mechanism similar to that of α-decay in comparison with the adiabatic formation of clusters due to nuclear surface vibrations.

Table. A comparison of the shell model W_x^{sh} and phenomenological W_x^{exp} cluster spectroscopic factors

X	W_x^{sh}	W_x^{exp} (parent nucl.)	X	W_x^{sh}	W_x^{exp} (parent nucl.)
^{8}Be	6.6×10^{-7}	–	^{28}Mg	1.5×10^{-21}	2.0×10^{-22} (^{234}U)
^{14}C	5.9×10^{-11}	2.0×10^{-10} – mean value for even Ra-isotopes			6.1×10^{-22} (^{236}U)
					4.6×10^{-21} (^{238}Pu)
^{24}Ne	7.0×10^{-19}	7.1×10^{-19} (^{232}U)	^{32}Si	5.1×10^{-25}	2.1×10^{-24} (^{238}Pu)
		7.8×10^{-18} (^{234}U)	^{40}Ca	6.0×10^{-33}	–
^{26}Ne	2.8×10^{-21}	–	^{48}Ca	10^{-42}	–

Springer Series in Nuclear and Particle Physics   Clustering Phenomena in Atoms and Nuclei
Editors: M. Brenner · T. Lönnroth · F.B. Malik   © Springer-Verlag Berlin, Heidelberg 1992

Exotic Cluster Decays and Reinforcing and Switching of Shell Gaps in Nuclei*

R.K. Gupta[1,2], W. Scheid[3], and W. Greiner[1]

[1]Institut für Theoretische Physik der J.W. Goethe-Universität,
  W-6000 Frankfurt, Fed. Rep. of Germany
[2]Physics Department, Panjab University, Chandigarh-160014, India
[3]Institut für Theoretische Physik, Justus-Liebig-Universität,
  W-6300 Giessen, Fed. Rep. of Germany

The measured large ground-state deformations for $Z(N)=38,40$ nuclei call for a natural breaking of spherical shell closures at $Z=N=40$. This leads to an instability of these nuclei against exotic cluster decays. The recent calculation of Puri et al.[1] shows that ^{80}Zr, though stable against alpha-decay, is metastable (having positive Q-values) with respect to many heavier clusters (with $A_2 \gtrsim 16$). The calculated half-life times for ^{24}Mg and ^{28}Si decays of ^{80}Zr are of the same order as predicted for many cluster decays of heavy deformed "stable" nuclei[2,3]. In the following, however, we show that the neighbouring $^{76,78}Sr$ nuclei with $Z=38$, $N=38,40$, which are even more deformed and called superdeformed, are rather more stable against exotic cluster decays than ^{80}Zr.

For metastable systems ($Q>0$), the half-life time $T_{1/2}$ or decay constant λ is

$$\lambda = \ln2/T_{1/2} = P_0 \nu P. \qquad (1)$$

Here, P_0 is the cluster preformation probability in the ground state, P the tunneling probability of the confining nuclear interaction barrier and ν the assault frequency given as the number of attempts per second. We use here the model of Malik and Gupta[1,2,4] where in the decoupled approximation the cluster preformation probability P_0 is calculated as the quantum mechanical fragmentation probability depending on the mass-asymmetry coordinate $\eta=(A_1-A_2/(A_1+A_2)$ at a fixed relative separation R and charge asymmetry coordinate $\eta_Z=(Z_1-Z_2)/(Z_1+Z_2)$. The fragmentation probability in the ground state is obtained by solving the stationary Schrödinger equation in η:

$$[-\frac{\hbar^2}{2\sqrt{B_{\eta\eta}}}\frac{\partial}{\partial\eta}\frac{1}{\sqrt{B_{\eta\eta}}}\frac{\partial}{\partial\eta}+V(\eta,\eta_Z,R)]\psi(\eta) = E_{g.s.}\psi(\eta). \qquad (2)$$

with

$$P_0(A_2)= |\psi(A_2)|^2 \sqrt{B_{\eta\eta}(A_2)}\ \frac{2}{A}. \qquad (3)$$

Using the two touching spheres approximation ($R=R_1+R_2=R_t(\eta)$), the collective potential in (2) is defined as

$$V(\eta,\eta_Z,R)=-\sum_{i=1}^{2} B_i(A_i,Z_i)+ \frac{Z_1Z_2e^2}{R} +V_P \qquad (4)$$

with charge numbers Z_i (and hence η_Z) fixed by minimizing the sum of the two binding energies and the Coulomb interaction in the η_Z coordinate. The additional proximity potential is taken from Blocki et al.[5]. The mass parameters $B_{\eta\eta}(\eta)$ in (2) are calculated with a classical model[6].

For the relative motion, Malik and Gupta[4] assumed a tunneling probability P in WKB approximation and solved it analytically for each cluster decay. For detailed expressions, we refer the reader to the paper of Malik and Gupta[4]. The assault or escape frequency ν is calculated by the relation

---

* Work supported in part by Deutsche Forschungsgemeinschaft (DFG).

Springer Series in Nuclear and Particle Physics   **Clustering Phenomena in Atoms and Nuclei**
Editors: M. Brenner · T. Lönnroth · F.B. Malik   © Springer-Verlag Berlin, Heidelberg 1992

$$\nu = \frac{v}{R_0} = \left(\frac{2Q}{mA_2}\right)^{1/2}/R_0 \qquad (5)$$

where mA_2 is the mass of the emitted cluster.

Table I gives the calculated Q-values, the preformation probabilities P_0, the WKB penetrabilities P, the dacay constants λ and $\log_{10} T_{1/2}$ values for ^{78}Sr and ^{80}Zr nuclei. Comparing the results for the two nuclei we notice that for all the possible cluster-decays the λ-values in the case of ^{78}Sr are very much smaller. Even for the most probable ^{24}Mg cluster decay, $T_{1/2} > 10^{100}$s for ^{78}Sr nucleus, as compared to $\sim 10^{50}$s for ^{80}Zr. A closer look at the numbers in Table I shows that both the preformation probabilities P_0 and the penetrabilities P are much smaller in the case of ^{78}Sr. This happens for P_0 because of the relatively shallow minima in the potential energy surface and for P because of the Q-values being relatively small. We have listed the decays with $T_{1/2} > 10^{100}$s as stable, since the measured cluster-decay half-life times are $\lesssim 10^{25}$s.

Such an unexpected stability of the $^{78}_{38}$Sr nucleus against cluster decays must apparently be due to stable deformed shapes of a doubly magic deformed nucleus at Z=N=38 rather than at Z=N=40 (Ref.7). This can be understood in terms of what Hamilton et al.[8] call the reinforcing and switching of shell gaps in nuclei. These authors have observed that Z (or N) = 40 is not a spherical magic number until and unless it is reinforced by another spherical magic N (or Z) number. In the absence of such a reinforcement, there is a switch in importance of Z (or N)=40 spherical to Z(or N)=38 deformed shell. Here, we observe a new reinforcement shell gap effect of Z=38 deformed shell on N=38 deformed shell by studying the exotic cluster emission life times. This shell gap makes the ^{76,78}Sr nuclei far more stable than $^{80}_{40}$Zr$_{40}$ and other neighbouring nuclei like $^{72}_{36}$Kr$_{36}$ and $^{84}_{42}$Mo$_{42}$.

Concluding, we have shown that strong nuclear structure effects can be found by exotic cluster decay studies.

Table I: Calculated quantities characteristic of possible cluster decays.

Parent nucleus	Cluster nucleus	Q-value (MeV)	Preformation probability P_0	Penetrability P	Decay constant λ	Half-life times $\log_{10} T_{1/2}$
$^{78}_{38}$Sr	^{16}O	2.89	4.06×10^{-28}	9.97×10^{-132}	4.87×10^{-138}	stable
	^{20}Ne	4.25	2.09×10^{-34}	7.13×10^{-128}	1.94×10^{-140}	stable
	^{24}Mg	7.16	5.55×10^{-37}	1.87×10^{-98}	1.60×10^{-113}	stable
	^{28}Si	8.73	3.26×10^{-39}	5.33×10^{-94}	2.74×10^{-111}	stable
$^{80}_{40}$Zr	^{16}O	5.10	4.34×10^{-21}	3.66×10^{-89}	2.52×10^{-88}	87.44
	^{20}Ne	7.17	4.96×10^{-26}	2.58×10^{-88}	2.15×10^{-92}	91.51
	^{24}Mg	13.76	5.23×10^{-22}	1.71×10^{-54}	1.90×10^{-54}	53.56
	^{28}Si	15.75	9.38×10^{-24}	8.92×10^{-55}	1.76×10^{-56}	55.59

References:

1. R.K. Puri, S.S. Malik and R.K. Gupta, Europhys.Lett. 9 (1989) 767.
2. S.S. Malik et al. Pramana-J.Phys. 32 (1989) 419.
3. D.N. Poenaru et al., At.Data Nucl.Data Tables 34 (1986) 423.
4. S.S. Malik and R.K. Gupta, Phys.Rev. C39 (1989) 1992.
5. J. Blocki et al., Ann.Phys. (N.Y.) 105 (1977) 427.
6. H. Kröger and W. Scheid, J.Phys.G:Nucl.Phys. 6 (1980) L85.
7. R. Bengtsson, P. Möller, J.R. Nix and J.-Y. Zhang, Phys.Scr. 29 (1984) 402.
8. J.H. Hamilton et al., J.Phys.G:Nucl.Phys. 10 (1984) L87.

Alpha Widths in Deformed Nuclei:
A Microscopic Approach

A. Insolia[1], *R.J. Liotta*[2], *and D.S. Delion*[3]

[1]Department of Physics, University of Catania and INFN,
 I-95129 Catania, Italy
[2]Manne Siegbahn Institute, S-104 05 Stockholm, Sweden
[3]Institute of Atomic Physics, Bucharest Magurele POB MG-2, Romania

We have studied the alpha decay of deformed nuclei by assuming that the decay proceeds in two steps. First the four nucleons that eventually constitute the alpha particle are clustered at some point close to the nuclear surface. From here the alpha particle thus formed penetrates the Coulomb barrier. This second step (penetration of the already formed α-particle through the Coulomb barrier) has been carried out by using the WKB approximation.

The alpha formation amplitude[1] is given by

$$F_L(\vec{R}) = \int d\xi_\alpha d\xi_A [\phi_\alpha(\xi_\alpha)\phi_A(\xi_A)Y_L(\hat{R})]^*_{J_B M_B} \phi_B(\xi_A, \vec{r}_1, \vec{r}_2, \vec{r}_3, \vec{r}_4) \tag{1}$$

where ξ indicates internal coordinates, $B(A)$ labels the mother (daughter) nucleus and $\vec{r}_i$ is the coordinate of the nucleon i measured from the center of the nucleus B. We write the wave function of the mother and daughter nucleus whitin the BCS approximation, considering only axially symmetric nuclei. The potential that defines our single-particle representation $\{\varphi_\Omega\}$ has a axially simmetric deformed Woods-Saxon plus spin-orbit form[2].

Within the BCS approximation the formation amplitude then becomes

$$F_0(\vec{R}) = \sum_{N_\alpha L_\alpha} W_{N_\alpha L_\alpha} \times \phi_{N_\alpha L_\alpha}(\vec{R}) \tag{2}$$

where α labels the quantum numbers of the α-particle and W - coefficients can be expressed in terms of the trasformation coefficients from the individual nucleon coordinates to center of mass and relative coordinates (including occupation amplitudes from BCS calculation and the corresponding coefficients of the expansion in the spherical basis). The calculations have been performed including up to N=18 h. o. major shells. Coherence properties of the α formation amplitude are found to play a crucial role in fixing the absolute value of the calculated widths. Both quadrupole and octupole deformations have been considered. It has been found that the effect of the octupole deformation is to increase the formation probability (and the corresponding alpha decay widths) by 30%. Applications of the proposed method are presented in ^{222,224,226}Ra isotopes , ^{218,220,222}Rn isotopes and ^{232}Th. In the considered cases, the calculated widths agree with the experimental ones within a factor of about 2.

1. A. Insolia, P. Curutchet, R. J. Liotta and D. S. Delion, Phys. Rev. C43 (1991)

2. S. Cwiok, J. Dudek, W. Nazarewicz, J. Skalski and T. Werner, Comp. Phys. Comm. 46(1987)379

Springer Series in Nuclear and Particle Physics   **Clustering Phenomena in Atoms and Nuclei**
Editors: M. Brenner · T. Lönnroth · F.B. Malik   © Springer-Verlag Berlin, Heidelberg 1992

Clusters in Nuclear Fission

V. Rubchenya and S. Yavshits

V.G. Khlopin Radium Institute, 197022 St. Petersburg, Russia

The study of nuclear fission gives us an unique possibility to investigate the clusterization effects in heavy nuclei. The special role in the formation of clusters in fission plays the late stage of fission process near the scission point. On the early stage the fission mechanism is governed by the hydrodynamical and shell properties of collective motion along the fission path because the cluster substructures are washed out due to internucleon interaction. In the clusterization region near the scission point the initial mass distribution is defined by double magic nuclei ^{132}Sn, mass ditribution is rather narrow and is centered around A ~132. To obtain the final (observed) mass distribution it is necessary to take into account the nucleon exchange between fragments. The process of nucleon exchange which corresponds in cluster approach to the antysymmetrization of the total wave function taken in a two-cluster representation leads to the increasing of initial mass asymmetry and width of distribution. This results are due to the high stability of double magic cluster ^{132}Sn. The difference between chemical potentials of fragments violates the chemical equilibrium in the system of separated nuclei and cause the mass flow from the light fragment to a complimentary double magic one in order to equilibrate the chemical potentials over the whole system [1].

Wide mass spectrum of light clusters is observed in the fission accompanied by the light charged particle emission or ternary fission. Ternary fission may be treated as a formation of the three-cluster system near the scission point where the ternary nuclear system is a two fragments with the light particle situated between them. In our work [2] the formation of the three-cluster configuration has been considered as a result of two ruptures in the neck region. The configuration of the unbroken nucleus here the same as in the case of a binary fission. So, the binary and ternary fission may be treated as a two or three cluster decay of one and the same fissioning system. The formation of the mass and charge distribution of light clusters in the ternary fission is connected with the nucleon exchange effects similar to the binary fission again [2].

References

1. Rubchenya V.A., Yavshits S.G.: Sov. J. of Nucl. Phys., **40**, 649(1984)

2. Rubchenya V.A., Yavshits S.G.: Z. Phys. A, **A239**, 217(1988)

Springer Series in Nuclear and Particle Physics   **Clustering Phenomena in Atoms and Nuclei**
Editors: M. Brenner · T. Lönnroth · F.B. Malik   © Springer-Verlag Berlin, Heidelberg 1992

Excitation and Fission in Alkali-Atom Clusters

C. Bréchignac, Ph. Cahuzac, F. Carlier, J. Leygnier, and A. Sarfati

Laboratoire Aimé Cotton, Bât. 505, F-91405 Orsay, France

The study of the electronic structure of clusters is subjected
to various experimental and theoretical approches. Currently
used are the evolution with size of ionization potentials,
static polarizabilities, collective excitation and stability
against an excess of charge or internal energy. Results are
interpreted in terms of simple models as the classical metallic
drop, successefully applied to alkali-atom clusters, or with
quantum mechanical models which reproduce in more details the
variations with size. Here we present new results obtained for
alkali-atom clusters in two different domains. First we have
analyzed in details the fission of doubly charged potassium
clusters around the critical size of stability. Second we have
studied the response of singly ionized potassium clusters K_n^+
to an external electromagnetic field, in an extended domain of
size.

Assymmetric fission in doubly charged potassium clusters

Neutral clusters are produced during the adiabatic expansion
of the neat metallic vapor and ionized by a pulsed N_2 laser at
a photon energy $h\nu = 3.67$ eV. charged species are accelerated
within a multiplate system. Mass spectrometry is ensured by a
tantem time-of-flight device. In the first drift tube a couple
of deflecting electrostatic plates isolates ionized clusters
of a given size. Before entering the second time-of-flight,
the selected ion packets are decelerated by a retarding
potential. For metastable species which decompose in the first
time-of-flight, this retarding field separates in time parent
and fragments, according to their charge-to-size ratio.
Doubly charged species formed directly in the ionizing region
are visible in the mass spectrum, i.e. without size selection,
up to n=21. They cannot be observed below this limit, the
critical size of stability n_c [1,2]. We have studied the

Springer Series in Nuclear and Particle Physics   **Clustering Phenomena in Atoms and Nuclei**
Editors: M. Brenner · T. Lönnroth · F.B. Malik   © Springer-Verlag Berlin, Heidelberg 1992

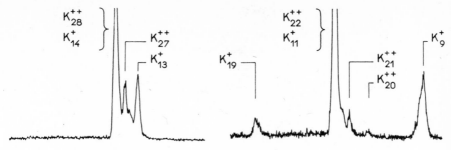

Fig. 1 Typical fragmentation spectra

unimolecular dissociation around this limit, from n=20 to n=28.
The assymetric chararter of the process (2) is well explained
in the frame of the metallic drop model [3]. Calculations show
that the energy change for process (2) is the sum of two terms.
The surface term due to the energy change for the extra surface
created during the dissociation. It leads to a totally
assymetric fission. The secondterm corresponds to the coulomb
effects. It would lead to a symetric fission. For clusters the
first term dominates and assymetric fission is predicted.
Taking into account the shell structure of the delocalised
valence electrons for alkali-atom clusters [4,5] the net result
is in favor of the ejection of a singly charged trimer around
n_c, in agreement with our observations [3]. From the classical
drop model, it is possible to evaluate the energy barrier wich
has to be surpassed to promote coulombic fission. Around n_c
this energy is close to the energy change for the evaporation
process (1). This explains the competition between fission and
evaporation [2,3].

For the largest size of the parent, fragmentation spectra show
only doubly charged framents corresponding to a dissociation
via neutral monomer evaporation :
$$K_n^{2+} \longrightarrow K_{n-1}^{2+} + K \tag{1}$$
Decreasing the size, unimolecular spectra show evaporation
products in competition with singly charged fragments attributed
to fission process :
$$K_n^{2+} \longrightarrow K_{n-p}^+ + K_p^+ \tag{2}$$
The fission is assymetric with preferred channels for p=3, a
stable two-electron species, and p=5 as well (Fig.1).

Collective excitation in singly ionized potassium clusters

For this study, the set-up is modified as follows. The size selected ionized clusters are excited by a second pulsed laser source, conveniently delayed against the ionizing laser. This excitation takes place in between a multiplate decelerating-accelerating system located in front of the second time-of-flight. The photoexcitation of the ion packets is then monitored by a fast photoevaporation process. Ionized fragments are mass-separated during their propagation in the second time-of-flight.

For small cluster sizes, we have measured the absorption cross-section σa as follows. At low laser fluence, under one photon absorption conditions, σa is related to the fragmentation rate by the equation :

$$\frac{I_F}{I_P + I_F} = A\left[1 - \exp(-\sigma_a \phi \delta t)\right]$$

Where I_F and I_p are the ion peak intensities of fragments and parent, ϕ is the photon flux of the excitation laser, δt its pulse duration and A is a geometrical factor. For the closed-shell clusters K_9^+ and K_{21}^+ the spectral profile of a is well interpreted by a resonance curve [6] (Fig.2).

This agrees with the classical Mie theory, and with the spherical symmetry predicted for the valence electron cloud of these 8 and 20 electron clusters [7,8]. However the maximum of the resonance at 1.93 eV for K_9^+ and 1.98 eV for K_{21}^+ is red shifted against the classical Mie (bulk) value 2.31 eV. This is understood as a spill out effect of the valence electrons which are distributed within a volume slightly larger than the ionic core volume [9]. The RPA calculations and related methods offer a usefull comparison with the experiment. Calculated values are close but still larger than the experimental ones [10,11].

For large clusters the above procedure is no longer valid. The evaporative photodissociation process may accur after multiphoton absorption only. Moreover the fragments cannot be discriminated from the parent at low laser fluence.

302

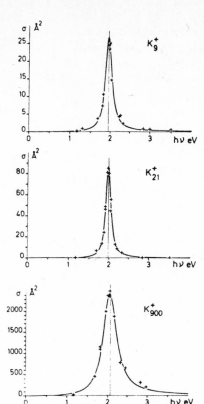

Fig.2 : absorption cross-section profiles

To extend our measurements toward large sizes we have used another experimental procedure, based on the analysis of the fragment intensity distribution, observed at relatively high laser fluence, and published in details elsewhere [12]. Absorption cross-section profiles are abtained in the large size domain up to n=900. The profiles are analogous to those of K_{21}^+. The new interesting feature is that the resonances are peaked at a value which is still far from the bulk value. For instance 2.05 eV is found for K_{900}^+ (Fig.2). Further investigations are necessary in order to interpret this too slow evolution toward the bulk.

References

[1] T.P.Martin, J.Chem. Phys. **81**, 4426 (1984)

[2] C.Bréchignac, Ph.Cahuzac, F.Carlier, M.de Frutos, Phys. Rev. Lett. **64**, 2893 (1990)

[3] C.Bréchignac, Ph.Cahuzac, F.Carlier, J.Leygnier, A.Sarfati Phys. Rev. B, to be published, and references therein.

[4] W.E.Ekardt, Phys. Rev. B **29**, 1558 (1984)

[5] W.D.Knight, K.Clemenger, W.A. de Heer, W.A. Saunders, M.Y.Chou, M.L.Cohen, Phys. Rev. Lett. **52**, 2141 (1984)

[6] C.Bréchignac, Ph.Cahuzac, F.Carlier, J.Leygnier, Chem. Phys. Lett. **164**, 433 (1989)

[7] W.A. de Heer, K.Selby, V.Kresin, J.Masui, M.Vollmer, A.Chatelain and W.D.Knight, Phys. Rev. Lett. **59**, 1805 (1987)

[8] W.A. de Heer, W.D.Knight, M.Y.Chou, M.L.Cohen, Solid State Phys. **40**, 93 (1987)

[9] S.Pollack, C.R.Chris Wang, M.M.Kappes, J.Chem. Phys. **94**, 2496 (1991)

[10] V.Kresin, Phys. Rev. B $\underline{40}$, 12507 (1989-II)

[11] C.Yannouleas, J.M.Pacheco, R.A.Broglia, Phys. Rev. B $\underline{41}$, 6088 (1990-II)

[12] C.Bréchignac, Ph.Cahuzac, N.Kebaili, J.Leygnier, A.Sarfati, To be published.

Fragmentation of Doubly Charged Alkali-Metal Clusters

M. Barranco[1], *J.A. Alonso*[2], *F. Garcias*[3,*], *and J.M. López*[2]

[1]Departament ECM, Facultat de Física, E-08028 Barcelona, Spain
[2]Departamento de Física Teórica y Física Atómica,
  Molecular y Nuclear, Universidad de Valladolid, E-47071 Valladolid, Spain
[3]Institut des Sciences Nucléaires, F-38026 Grenoble, France
*Permanent Address: Departament de Física,
  Universitat de les Illes Balears, E-07071 Palma de Mallorca, Spain

Abstract: We have studied the competition between neutral atom evaporation and asymmetric fission as fragmentation channels for doubly charged alkali-metal clusters. We have used an Extended Thomas-Fermi (ETF) method and the jellium model to compute the evaporation energy and the completely asymmetric fission barrier height in the case of Na_N^{2+}. Preliminary calculations are also presented for the symmetric fission channel.

Since the experimental discovery that the electrostatic repulsion in isolated multiply charged metal clusters X_N^{q+} may lead them to fragment into smaller aggregates (Coulomb explosion) [1], the question of what is the critical size N_c beyond which multiply charged clusters can be observed has attracted a considerable theoretical and experimental interest.

A pure energy criterium would lead to the conclusion that N_c (which depends on the chemical species X and on the charge q) is the size below which the sum of the ground state (g.s.) energies of the fragments is lower than the g.s. energy of the parent cluster. By considering only this energy balance, different dissociation channels can be easily studied and it is possible to provide the correct trends for neutral and singly charged aggregates. However, multiply charged clusters with N lesser than predicted N_c values have unexpectedly been observed [2, 3], thus suggesting that some of these clusters may be stabilized against fission by large barriers, which cannot be reached by an energy criterium.

The experimental situation has been substantially clarified by a series of experiments carried out on excited doubly charged alkaline [4, 5] and gold [6] clusters. These authors have noticed the competition between neutral atom evaporation and fission, here defined as the decay into two charged fragments. Atom evaporation is dominant for large clusters but fission competes with it when N decreases, X_N^{2+} being undetectable below the critical value N_c.

The explanation proposed in ref. [5] is that an excited Na_N^{2+} (or K_N^{2+} in ref. [4]) cluster with large N preferently evaporates a neutral atom because in this size range the fission barrier is larger than the binding energy of the neutral monomer. Consequently, when X_N^{2+} clusters are formed by atom evaporation from hot clusters of higher masses, N_c is the size below which the fission barrier becomes lower than the binding energy of the monomer. Smaller doubly ionized clusters dissociate into two charged fragments. On the other hand, if the X_N^{2+} clusters are formed from cold neutral aggregates by two-step ionization, metastable doubly charged clusters can exist below the critical size defined above. Therefore, the size of the smallest observable doubly ionized aggregate N_c strongly depends on the formation process of charged clusters [5].

In a recent publication [7] we have carried out a calculation of the fission barrier of Na_N^{2+} that has allowed us to estimate N_c and to propose a mechanism for neutral monomer evaporation which is different from the one of ref. [5]. It is the aim of this contribution to review the results we have obtained in ref. [7] and to present some preliminary results on symmetric fission. The symmetric fragmentation of charged alkali-metal clusters has been recently addressed in ref. [8] within the Liquid Drop Model, and in ref. [9] adding to it shell effects.

Let us consider the completely asymmetric fission of a doubly charged sodium cluster:

$$Na_N^{2+} \quad \rightarrow \quad Na_{N-1}^+ \quad + \quad Na^+ \quad . \tag{1}$$

As indicated before, this is a barrier-controlled process.

The heat of reaction (1) is defined as:

$$\Delta H_f = E(Na_{N-1}^+) + E(Na^+) - E(Na_N^{2+}) \quad . \tag{2}$$

ΔH_f turns out to be negative for small N and positive otherwise. We want to compare the probability for reaction (1) with that for the evaporation of a neutral monomer:

$$Na_N^{2+} \quad \rightarrow \quad Na_{N-1}^{2+} \quad + \quad Na \quad . \tag{3}$$

This evaporation is endothermic, the heat of the reaction ΔH_e being positive:

$$\Delta H_e = E(Na_{N-1}^{2+}) + E(Na) - E(Na_N^{2+}) > 0 \quad . \tag{4}$$

It means that evaporation is not a spontaneous process if Na_N^{2+} is in its g.s., but may occur for a highly excited parent cluster. A schematic representation of the competition between processes (1) and (3) is presented in fig. 1. The fission barrier height is indicated as F_m. It is lower than ΔH_e in the left panel and larger than it in the right panel.

The heats ΔH_f and ΔH_e have been obtained from a spherical ETF calculation within the jellium model. The details about the method of solving the corresponding Euler-Lagrange equation can be found in [10]. The electron energy density functional we have employed is described in [7].

To obtain the fission barrier we have used a deformed, fully selfconsistent ETF model. We have considered as initial configuration of the fissioning system that of a deformed cluster with $N-2$ electrons moving in the mean-field created by two tangent jellium spheres corresponding to cluster sizes

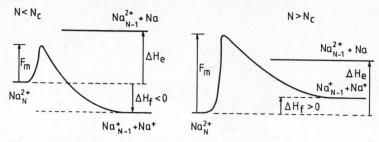

Fig. 1. Schematic representation of the competition between the fission and evaporation reactions.

$N - 1$ and 1, respectively. The other configurations along the dissociation path to the final state $Na_{N-1}^+ + Na^+$ have been obtained by increasing the separation between the two jellium spheres representing the emerging fragments. For a given separation, the electron density is obtained by solving the corresponding Euler-Lagrange equation which is now a partial differential equation. Technical details concerning the method used to solve it have been given in [7] and references therein.

We have also studied the symmetric fission of doubly charged Na_N^{2+} clusters:

$$Na_N^{2+} \quad \rightarrow \quad 2 \ Na_{N/2}^+ \quad . \tag{5}$$

In experiments, symmetrical division has not been identified as an important decay channel [6], but from the theoretical point of view it is interesting to compare the relative probabilities of symmetric (5) and completely asymmetric fission (1). As in the previous case, the heat of reaction (5) is defined as:

$$\Delta H_f \ = \ 2 \ E(Na_{N/2}^+) \ - \ E(Na_N^{2+}) \quad . \tag{6}$$

We have applied the deformed ETF model as before but modelling now the fissioning cluster by two identical jellium spheres made of $N/2$ atoms each.

Figure 2 is a three-dimensional plot of the electron density for the symmetric fission of Na_{40}^{2+} at separations (measured by the distance between the sharp jellium surfaces) D = 0, 2, 4 and 6 atomic units (a.u.) (we have used a.u. throughout, i.e. $\hbar = m = e^2 = 1$, length unit: $a_0 = 0.53$ Å, energy unit: 1 Hartree = 27.2 eV).

Equidensity lines associated to these configurations are shown in figure 3. From outside to inside, the lines correspond to densities n = 0.5, 1, 2, 3 and 4×10^{-3} (twice) a.u. A similar figure for the completely asymmetric fission of Na_{27}^{2+} can be found in [7]. These figures show an appreciable polarization in the electronic density which has a sizeable effect on the height and shape of the fission barrier at small distances.

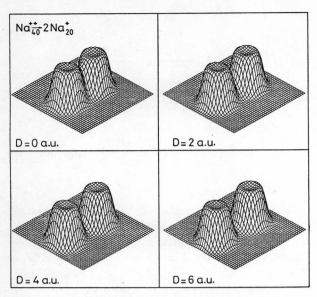

Fig. 2. Electron densities corresponding to the symmetric fission of Na_{40}^{2+} at selected separations D.

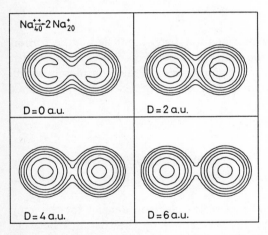

Fig. 3. Equidensity lines associated to the fission configurations shown in figure 2.

Figure 4 shows the ETF fission barriers (solid lines) of the processes $Na_{40}^{2+} \rightarrow Na_{39}^{+} + Na^{+}$ (lower curve) and $Na_{40}^{2+} \rightarrow 2\ Na_{20}^{+}$ (upper curve) as a function of the distance between the centers of the two fragments. Both barriers are measured from the g.s. energy of Na_{40}^{2+}. The ETF barriers tend to the classical Coulomb barriers (dashed lines) at large distances. One can see from this figure that the fission barrier for symmetric fragmentation

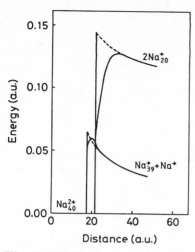

Fig. 4. ETF fission barriers (solid lines) of symmetric and completely asymmetric fission of Na_{40}^{2+}. The dashed lines correspond to the classical Coulomb barriers.

is much higher than the corresponding to the asymmetric division, thus indicating that in this case, the asymmetric dissociation is the most probable fission channel.

A systematic ETF study of the completely asymmetric fission of Na_N^{2+} clusters [7] has shown that, within this model, the fission barrier vanishes for $N_c^* = 9$. For this size and below, a Na_N^{2+} cluster spontaneously fissions. The comparison of ΔH_e and F_m shows that $\Delta H_e > F_m$ for small N, and $\Delta H_e < F_m$ for large N, in agreement with the analysis of [5] and [6]. These quantities become equal at $N_c \simeq 41$, to be compared with $N_c = 27$ as measured by Bréchignac *et al* [5].

Our results suggest the following interpretation of the critical numbers for the Coulomb explosion of doubly charged clusters: We can define an N_c such that fission is the preferred decay mode of excited clusters for $N < N_c$ because $F_m < \Delta H_e$. For $N \geq N_c$, evaporation of a neutral atom competes with fission. It is important to realize that in both processes the fission barrier does exist and must be overcome during the dissociation. Only after passing the barrier and when the system is undergoing fragmentation, the state $X_{N-1}^{2+} + X$ suddenly appears as an available channel, which becomes more and more probable when N increases. We refer the interested reader to [7] for a detailed discussion of this point. The zero-value of F_m determines the other critical number N_c^* below which the fission of X_N^{2+} is a spontaneous process. For $N > N_c^*$, small doubly charged clusters are experimentally observable when the formation mechanism leaves them with an excitation energy below the fission barrier height. Consequently, due to the energy excess of hot X_N^{2+} clusters, it is not surprising that the experimental critical numbers N_c be smaller than the theoretical ones.

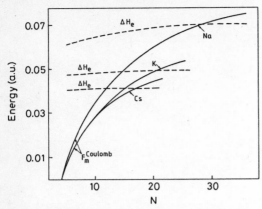

Fig. 5. Coulomb fission barrier $F_m^{Coulomb}$ and evaporation heat ΔH_e vs cluster size for Na_N^{2+}, K_N^{2+} and Cs_N^{2+} clusters.

Figure 4 shows that at large separations the fission barrier is purely coulombic, the difference between the ETF and the Coulomb barriers being sizeable only at small distances. We can thus ask ourselves how far can one go estimating N_c from the simple fission barrier obtained from the repulsion of two point-like charged fragments. To this end, we have approximated the fission barrier for the completely asymmetric fragmentation as:

$$F_m^{Coulomb} = \frac{1}{d_t} + \Delta H_f \quad , \tag{7}$$

where $d_t = r_s \left[1 + (N-1)^{1/3} \right]$ is the distance between the centers of the tangent jellium spheres and ΔH_f is defined in eq. (2).

Figure 5 shows the N_c values obtained for Na, K and Cs using this crude model, which are 29, 20 and 17, respectively. They are in excellent agreement with the experimental results $N_c = 27$ (Na_N^{2+}) [5], 19 (K_N^{2+}) [4] and 19 (Cs_N^{2+}) [11]. Thus, a simple description of the fission barrier height neglecting the spill-out of the electron density seems to give correctly the magnitude of N_c for alkali-metal clusters. Whether or not this is a fortuitous agreement has to be checked against full ETF calculations. We are at present carrying out a systematic ETF study for alkaline clusters and different decay channels (for instance, the emission of a dimer [5] or a trimer [6]). The extension of the study using a deformed Kohn-Sham method which accounts for shell structure effects is also currently under way.

We thank Mario Centelles for his valuable help with figures 2 and 3. This work has been supported by the DGICYT (Grants PB-89-0332 and PB-89-0352-CO2-01), and by the CIRIT. F.G. is also greatful to the DGICYT for a Postdoctoral Fellowship.

References

1. K. Sattler, J. Mühlbach, O. Echt, P. Pfau, and E. Recknagel, Phys. Rev. Lett. **47** (1981) 160
2. C. Bréchignac, M. Broyer, Ph. Cahuzac, G. Delacretaz, P. Labastie, and L. Wöste, Chem. Phys. Lett. **118** (1985) 174
3. W. Schulze, B. Winter, and I. Goldenfeld, Phys. Rev. B **38** (1988) 12937
4. C. Bréchignac, Ph. Cahuzac, F. Carlier, and J. Leygnier, Phys. Rev. Lett. **63** (1989) 1368
5. C. Bréchignac, Ph. Cahuzac, F. Carlier, and M. de Frutos, Phys. Rev. Lett. **64** (1990) 2893
6. W. A. Saunders, Phys. Rev. Lett. **64** (1990) 3046
7. F. Garcias, J. A. Alonso, J. M. López, and M. Barranco, Phys. Rev. B **43** (1991) 9459
8. E. Lipparini and A. Vitturi, Z. Phys. D **17** (1990) 57
9. M. Nakamura, Y. Ishii, A. Tamura and S. Sugano, Phys. Rev. A **42** (1990) 2267
10. Ll. Serra, F. Garcias, M. Barranco, J. Navarro, L. C. Balbás, A. Rubio, and A. Mañanes, J. Phys.: Condens. Matter **1** (1989) 10391
11. T. P. Martin, J. Chem. Phys. **81** (1984) 4426

Effects of Electronic and Atomic Shell Configurations on Fragmentations of Metal Clusters

Y. Ishii

Department of Material Science, Himeji Institute of Technology,
Kamigouri-cho, Akou-gun, Himeji 678-12, Japan

Alkali metal clusters show distinct shell structures quite similar to those in nuclear physics. If valence electrons are assumed to be confined in a spherical cavity, the system is energetically favored at the shell closing configuration. According to the local density functional approximation (LDA) calculation for a spherical jellium model as well as a perturbative treatment of effects of the ionic potentials, *valence electrons in alkali metal clusters move effectively in a uniform and smooth field because the ionic potentials are well screened by valence electrons.* This is the physical reason why the simple shell model works well for alkali metal clusters.

We might think two types of quantum effects on the cohesion of alkali metal clusters; the exchange-correlation effect and quantization of the electron kinetic energy. The former produces most parts of the cohesive energy because the (classical) electrostatic contribution to the total energy exactly vanishes for a bulk jellium. However, we expect that this part as well as the small electrostatic part for a finite system shows no significant size dependence if the charge density is a smooth function of the system size. The quantization of the single-electron orbitals, on the other hand, contributes to the shell oscillation of the cohesive energy of alkali metal clusters. Theory of shell corrections, which has been originally applied to the nuclear fission problem, provides us the way to treat the smooth part and oscillatory parts of the total energy separately in a phenomenological way.

We study fragmentation of multiply-charged alkali metal clusters consisting of several tens of atoms by using the theory of shell corrections. The potential energy surface of fragmentation process is calculated for various sizes of the clusters. It is found that the electronic shell configuration is quite important in discussing the stability and the fragmentation process of multiply-charged clusters. Competing effects on the stability of metal clusters due to the electronic and atomic shell configurations are also studied within a simple model.

Springer Series in Nuclear and Particle Physics   **Clustering Phenomena in Atoms and Nuclei**
Editors: M. Brenner · T. Lönnroth · F.B. Malik   © Springer-Verlag Berlin, Heidelberg 1992

Spontaneous Decay of Ionized Atomic Clusters: Statistical and Non-Statistical Channels

T.D. Märk, M. Foltin, and P. Scheier

Institut für Ionenphysik, Leopold Franzens Universität, Technikerstr. 25, A-6020 Innsbruck, Austria

1. Introduction

If the energy of an electron (or photon) beam colliding with gas phase clusters is greater than a critical value termed appearance energy, some of these neutral clusters will be ionized. The abundance and the variety of the cluster ions produced from a specific neutral precursor P_n depends on geometric, electronic and energetic properties of the neutral and ionized clusters. Whereas inelastic interaction of electrons with single atoms results only in changes of the electronic configuration, interaction of electrons with van der Waals (vdW) clusters consisting of <u>atomic constituents</u> involves - besides electronic excitation - (i) changes in the nuclear motion (vibrational and rotational excitation), (ii) multiple collisions of the incoming electron, and (iii) subsequent intermolecular reactions within the cluster. All of this leads to the production of hot ions, i.e. the deposition of excess energy into various degrees of freedom. Energy flow between different degrees of freedom leads to spontaneous decay reactions of these hot ions on time scales ranging from a few vibrational oscillations for <u>prompt</u> dissociations up to several 100 μs for <u>metastable</u> dissociations. Depending on the energy storage mode and concomitant decay mechanisms metastable decay reactions can be divided into <u>statistical</u> and <u>non-statistical</u> decay channels.

It is clear that these prompt and delayed dissociations will lead to a strong modification of the original neutral cluster distribution. Curiously enough, despite this fact, in a number of earlier studies cluster ion mass spectra were related on a one to one basis to neutral cluster distributions. It was only 10 years ago that it was noted [1] that "one has to account for the unimolecular dissociation when using impact ionization plus mass spectrometry to probe (quantitatively) neutral vdW cluster beams". Today, it is widely accepted [2-5] that abundance fluctuations (anomalies, magic numbers) in mass spectra of vdW clusters are due to variations in the ionization efficiency and the properties (structure) of the ensuing ions.

Here we will summarize today's knowledge on (i) electron impact ionization of atomic vdW clusters, in particular the various excitation and ionization mechanism, and (ii) spontaneous decay reactions of excited atomic cluster ions, including monomer evaporation (vibrational predissociation), metastable decay series, and excimer induced fissioning. For more details on these subjects see also Ref. [2-7].

Springer Series in Nuclear and Particle Physics **Clustering Phenomena in Atoms and Nuclei**
Editors: M. Brenner · T. Lönnroth · F.B. Malik © Springer-Verlag Berlin, Heidelberg 1992

2. Ionization of atomic vdW clusters

Inelastic interaction between an electron and a cluster may proceed through a variety of different channels, including simple one step ionization reactions (direct ionization) and also two step processes where the initial electron interaction leads to an intermediate state followed by some sort of autoionization reaction [8]. The primary ionization event will be followed by further intermolecular reactions, i.e. isomerization, solvation, and other ion molecule reactions. As an example, we shall briefly consider the sequence of events for the overall reaction

$$Ar_2 + e \longrightarrow Ar_2^+ + 2 e \tag{1a}$$

$$\longrightarrow Ar^+ + Ar + 2e \tag{1b}$$

Close to the <u>adiabatic</u> ionization threshold of Ar_2 electron impact can only lead to the production of a (molecular [9]) Rydberg state Ar_2^{**} due to the poor Franck Condon overlap between Ar_2^+ and Ar_2 in their ground states (Fig. 1). A subsequent associative ionization process, i.e. coupling of the Rydberg electron and the nuclear motion, leads to the production of Ar_2^+ in its stable ground state. If the density of Rydberg states close to the adiabatic ionization threshold is high, little vibrational energy will be deposited in the dimer ion via this two step ionization process.

At higher electron energies (above the <u>adiabatic</u> ionization threshold) this threshold mechanism will be replaced by that of direct ionization (vertical transition in Fig. 1). In this case vibrational and/or electronic excess energy will be deposited in the ion. Depending on the storage mode, the final outcome is either a vibrationally excited dimer ion or in case of a repulsive state the production of a fragment ion Ar^+ (dissociative ionization process). Moreover, as the stabilizing entity in a rare gas cluster ion appears to be Ar_2^+ or Ar_3^+ [10-12], similar ionization events will take place in larger clusters, resulting - in general - in extensive heating (via vibrational excitation transfer from the intramolecular ion mode to the weakly bound van der Waals modes) of the cluster, via the reaction sequence

$$Ar_n + e \longrightarrow Ar_2^+(v') \cdot Ar_{n-2} + 2e$$
$$\downarrow \tag{2}$$
$$Ar_2^+(v'') \cdot Ar_{n-2}(v) \text{ with } v'' < v'$$

Besides ionization, inelastic electron interaction is expected also to result in the formation of electronically excited states, the lowest states being the ^3P_o and $^3P_2$ metastable and $^3P_1$ and $^1P_1$ resonant states (see Fig. 1). Extensive information regarding exciton trapping in liquids and solids indicate that two of these excited states (excitons) become subsequently trapped by self-localization into the bound excimer states $^1\Sigma_u^+$ and $^3\Sigma_u^+$ (see Fig. 1) [13]. Using electron energy loss spectroscopy Ding and coworkers were recently able to identify the excitations of surface and bulk atoms in Kr clusters corresponding to surface and bulk exciton states in solid Kr [14]. An ultrafast vibrational energy flow [15] from these

314

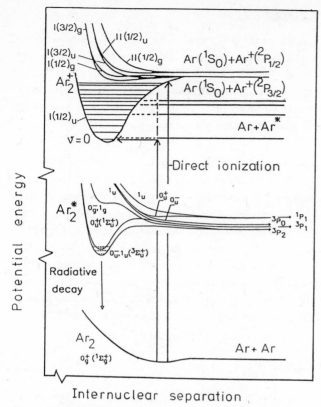

Fig. 1 Schematic view of the Ar_2, Ar_2^* and Ar_2^+ potential energy curves and possible electron impact excitation pathways

vibrational excited excimer states $R_2^*(v)$ into the cluster leads again to extensive heating of the cluster entity. Moreover, radiative decay to the ground state may lead to additional energy disposal.

Furthermore, besides these single collision events, electron (and photon) interaction with a cluster may also lead to multiple collisional cascades via interaction of the secondary electrons (photoelectrons) with cluster constitutents. This yields (i) multiply charged cluster ions consisting of several singly charged states within the clusters [16-18] or - as proved quite recently - (ii) excited cluster ions consisting of a singly charged and an electronically excited state at different sites within the cluster [19].

Energy degradation and relaxation following the primary ionization event as outlined above leads to various types of predissociation processes. These spontaneous decay reactions, their properties and relationship to nuclear phenomena will be discussed in the next three papragraphs.

3. Statistical (vibrational) predissociation

The most common spontaneous decay reaction in the metastable time regime for cluster ions is monomer evaporation. Electronic predissociation and barrier penetration has been named as possible mechanism for small cluster ions [20-22]. A typical example (see Fig. 1) is the metastable decay of Ar_2^+ ($II(1/2)_u$) via Ar_2^+ ($I(1/2)_g$) into $Ar^+(^2P_{3/2})$ + Ar with a lifetime of 91 μs [20,21]. A particular variant of barrier penetration is the tunneling through a centrifugal barrier (rotational predissociation), which has been proposed to account for the slow decay of small Ar cluster ions [3,20,22].

Conversely, vibrational predissociation is thought to be the dominant metastable dissociation mechanism for large (n > 10-20) cluster ions. If a polyatomic cluster ion is complex enough, the random motion of an activated ion on its potential hypersurface will be complicated enough to increase its lifetime into the metastable time regime. This process has to be treated theoretically in the framework of statistical theories (RRKM, QET [23]), where the unimolecular rate k (and other properties such as the release of translational kinetic energy T) are assumed to depend only on the internal energy E^* of the activated ion. The primary excitation process is assumed to have no influence on the values of k or T, that is, k is (i) independent of the ionization mode and (ii) slow relative to the rate of redistribution of the initial vibrational and electronic excitation energy over all degrees of freedom [8,23].

A metastable dissociation reaction of an excited cluster ion P_n^{+*} via

$$P_n^{+*} \xrightarrow{\ k\ } P_{n-1}^+ + P + T \tag{3}$$

may be characterized by the rate k (which is the inverse of the mean lifetime) and the kinetic energy T released in the decay channel. Reaction (3) constitutes the simplest case of ions being excited to a specific energy E^* and decaying via one reaction channel. Cluster ions produced by electron (or photon) impact ionization of a neutral cluster beam normally comprise, however, a broad range of energies due to the broad range of energies deposited into the ions by the primary ionization process (see above). Parent ions with different energies will have different decay rates and produced fragment ions will receive different internal energies. Moreover, in many cases, P_n^{+*} may decay by competing reactions and the produced daughter ion P_{n-1}^+ may not be stable and decay again by further decomposition reactions. This situation makes analysis of experimental data very difficult and usually only averaged or relative values for the rates (and kinetic energy release) may be obtained leading to an apparent time dependence of the rate k (see Fig. 2).

Considering the kinetics of monomer evaporation within the frame of QET and using the concept of an evaporative ensemble, Klots [24] predicted the time dependence of an evaporating parent population, isolated and normalized to unity at a time t_o, to be given by a non-exponential decay function

$$[P_n^{+*}](t) = 1 - \frac{C}{\gamma^2} \ln \left[\frac{t}{t_o + (t-t_o) \exp(-\gamma^2/C)} \right] \tag{4}$$

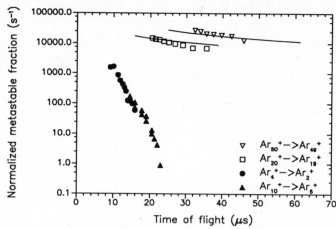

Fig. 2. Apparent decay rates (metastable fractions divided by the flighttime through the field free region) as a function of flighttime from the ion source to the mid point of the field free region for the metastable monomer evaporation reactions $Ar_{50}^+ \rightarrow Ar_{49}^+ + Ar$ ($\triangledown$), $Ar_{20}^+ \rightarrow Ar_{19}^+ + Ar$ ($\square$) and for the excimer driven fission reactions $Ar_{10}^+ \rightarrow Ar_5^+ + 5Ar$ ($\blacktriangle$) and $Ar_4^+ \rightarrow Ar_2^+ + 2Ar$ ($\bullet$). Also shown (full lines) for comparison are the predicted dependences for the monomer evaporation using equ. (4).

where C is the heat capacity of the cluster (in units of the Boltzmann constant k_B) and γ is the modified Gspann parameter, defined by

$$\gamma = \frac{E_{vap}}{k_B \cdot \sqrt{TT^*}} \qquad (5)$$

It contains the energy of evaporation E_{vap} and a geometric means of the before-and-after temperatures T and T^*, respectively. The Gspann parameter is very nearly independent of the size and composition of the cluster and on a typical laboratory time scale of tens of μs equal to about (23.5 ± 1.5). The single unknown parameter in the Klots formula is the heat capacity. Choosing plausible values for C there is very good agreement (at least for larger clusters) with the experimental findings [28] (see Fig. 2).

Moreover, equ. (4) may be used to predict the dependence of the rate k on cluster size n. Again, there is good agreement [24] in the general trend between existing experimental data [25-28] and the predicted curve for atomic cluster ions such as Na, Cu, Xe and Ar (see Fig. 3). It is interesting to note that for certain cluster sizes the experimental values deviate from the predicted curve beyond quoted error bars (Fig. 3). In most cases these anomalous small or large rates (metastable fractions) coincide in a mirror-like fashion with enhanced or depleted ion abundances in the ordinary mass spectrum (Fig. 4). The reason for these anomalies are additional structural stabilities ("magic numbers") not included in the continuum based model of Klots. In view of the qualitative similarity between nuclear and vdW interaction, it should be mentioned that, in the nuclear case the packing responsible for

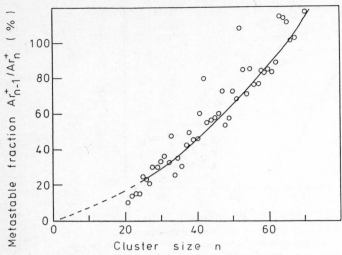

Fig. 3 Comparison of measured and predicted (equ. (4)) metastable decay fractions (product ions produced in the field free region divided by the parent ions arriving at the detector) of argon cluster ions (with $C = 3.9\ k_B$ and $\gamma = 25$).

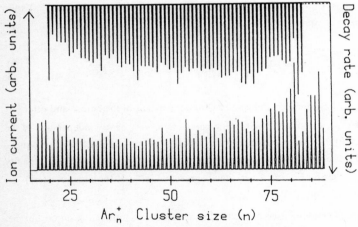

Fig. 4 Ordinary mass spectrum for argon cluster ions ($20 \leq n \leq 83$) and apparent metastable decay rates as a function of cluster size.

the magic number effects is in momentum space, whereas in the cluster case the structure of atomic clusters is likely to be controlled by the direct overlap of repulsive atomic cores. According to calculations and experimental evidence (see Fig. 4) especially stable atomic clusters are obtained for n = 13, 55, 147, 309 etc. atoms, their corresponding structure being icosahedral.

It is interesting to note that Engelbert [29] proposed a modified RRK-QET method to determine the binding energy of a monomer within a cluster ion using measured rates k and concomitant average kinetic energy releases T. Engelbert applied this treatment successfully to data of Ar_n^+ and Brechignac et al. to Na^+ [25].

4. Metastable decay series

Considering also thermochemical aspects Klots [24] predicted that due to the influence of the surface energy sequential evaporations of monomers should be a more likely cooling process for excited cluster ions than single step splitting off of larger fragments. Such sequential evaporations have been assumed in molecular dynamics calculations [30] to occur after ionization of neutral vdW clusters in the ps time regime. Moreover, multiple evaporations (i.e. the loss of more than one monomer) in the metastable time regime (µs) have been observed by numerous authors. However, only recently it became possible to determine the true nature of this process. Using both field free regions of a double focussing sector field mass spectrometer as independent observational windows we were able to demonstrate recently that certain cluster ions P_n^{+*} (with P = Ar [31] and N_2 [32]) decay by sequential decay series (and not single step fissioning), i.e.

$$P_n^{+*} \longrightarrow P_{n-1}^{+*} \longrightarrow P_{n-2}^{+*} .... P^+ \qquad (6)$$

evaporating a single monomer in each of these successive decay steps. Whereas in case of radioactive decay series the various decay rates are constants, in case of these cluster ions individual apparent decay rates are depending on time and on the parent ion due to the fact that each ion may exist (in an ion ensemble probed) in a variety of energy states.

5. Excimer-induced fission

Recently, we have discovered a rather unusual metastable fragmentation channel occurring in argon cluster ions [19,33]. In contrast to the well known case of single monomer evaporation due to vibrational predissociation, in the new decay reaction the number of ejected Ar monomers rises from 2 for Ar_4^+ up to 10 for Ar_{30}^+ [19,33]. After studying the dependence of the metastable fractions on (i) the electron energy, (ii) parent cluster size and the number of ejected monomers, and (iii) the time interval between ion formation and dissociation, we concluded that a metastable, electronically excited excimer Ar_2^* $(^3\Sigma_u^+)$ localized inside the cluster ion could be responsible for the observed unusual decay pattern. The radiative decay of this excimer leads to repulsion of Ar atoms in the ground state Ar_2 $(^1\Sigma_g)$ and to subsequent disintegration of the cluster (see Fig. 1). In order to further investigate this phenomenon we have recently extended our studies to metastable fragmentation of neon cluster ions [34], in particular to the following reaction

$$Ne_4^{+*} \longrightarrow Ne_2^+ + 2\,Ne \qquad (7)$$

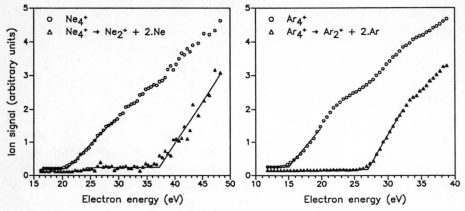

Fig. 5. Ion current as a function of the electron energy for the parent ions Ne_4^+ and Ar_4^+ (o) and for the fragment ions Ne_2^+ and Ar_2^+ (Δ) produced by the metastable decay reaction $Ne_4^+ \rightarrow Ne_2^+ + 2Ne$ and $Ar_4^+ \rightarrow Ar_2^+ + 2Ar$, respectively. Full lines represent linear extrapolations of experimental data in the onset regions

Figure 5 shows the dependence of the fragment ion current resulting from the metastable decay reaction (7) on the electron energy. For comparison, the electron energy dependence of the Ne_4^+ parent ion current is also shown. Linear extrapolation in the onset region shows, that the appearance energy (AE) of the metastable decay (7) is 37.2 ± 0.8 eV, which is approximately 16.4 eV more than the AE of the parent ion Ne_4^+ (in this measurement AE $(Ne_4^+) = 20.8 \pm 0.3$ eV).

The first electronically excited state of the Ne^+ ion is the $(2s2p^6)$ $^2S_{1/2}$ state, lying 26.9 eV above the ground ionic state. The lowest excited states of the neutral Ne are the $(2p^53s)$ $^3P_2$, $^3P_0$ metastable states and the $^3P_1$, $^1P_1$ resonant states, lying 16.62 to 16.85 eV above the ground state. Taking into consideration the measured AE of the metastable fragmentations (7), we have to conclude therefore that electronically excited states of neutral Ne are involved in the fragmentation process (7), populated most likely by either the scattered or ejected electron after a successful ionization process (e + Ne $\rightarrow$ Ne^+ + 2e) inside the same cluster (see above).

Fluorescence studies in liquid and solid neon indicate [13] that after electron or photon impact excitation and subsequent multi-step fast nonradiative transitions, the excimer dimer Ne_2^+ in the 0_u^- and 1_u $(^3\Sigma_u^+)$ or 0_u^+ $(^1\Sigma_u^+)$ bonding state is formed (see Fig. 1). The second excited 0_u^+ state (correlated with the $^3P_1$ atomic level) is radiatively coupled to the 0_g^+ $(^1\Sigma_g^+)$ ground electronic state and therefore it is short living (lifetime t~3ns). The first excited, nearly degenerate 0_u^- and 1_u $(^3\Sigma_u^+)$ states, are metastable, but its radiative lifetime is considerably reduced with respect to the corresponding atomic state $^3P_2$ and lies in the microsecond region. Appearance energy measurements allow us to propose, that the metastable Ne_2^* $(^3\Sigma_u^+)$ excimer is also present in Ne_4^{+*} formed in our experiment. It's radiative decay followed by repulsion of Ne atoms in the ground Ne_2 $(^1\Sigma_g^+)$ state leads to fragmentation of the cluster ion in times of order of microseconds after ionization and excitation. In argon cluster ions we observed for this unusual decay reaction a single exponential decrease of the metastable fraction with increasing time interval between formation and fragmentation of those ions (see Fig. 2). This is evidence for a definitive, unique fragmentation rate. The corresponding fragmentation lifetime deduced was appr. 1.5 μs [19], agreeing well with the radiative liftime of the Ar_2^* ($^3\Sigma_u^+$) excimer in solid and liquid argon, respectively [13] and in argon clusters [35]. Contrary to the argon case, in neon we can observe the metastable decay (7) not only in the first field free region but also in the second field free region, i.e. appr. 17 μs after the ionization. This indicates a much longer radiative lifetime for Ne_2^* ($^3\Sigma_u^+$). This longer lifetime is consistent with the fluorescence measurements, which indicate radiative lifetimes between 5 μs and 11 μs in solids and the gas phase, respectively [36].

In conclusion, this work gives clear evidence that the unusual metastable fragmentation of rare gas cluster ions is initiated by radiative decay of an excimer dimer being present on the surface (trapped surface exciton) or in the bulk (trapped bulk exciton) of the cluster ion. Recently, Hertel and coworkers [37] reported the direct observation of such an excitation process during photoionization of argon clusters using synchroton radiation and threshold photoelectron photoion coincidence TOF analysis. They observed two distinct maxima of the metastable TPEPICO spectrum interpreting the second peak as being due to the excitation of the $n = 2$ exciton, just 1.5 eV above the $n = 1$ state. The existence of this fission process provides an example of a system where statistical energy distribution does not occur upon initial excitation. This mode selective excitation of the excimer in the cluster ion via multiple collisions constitutes a beautiful example for the violation of vibrational energy equipartitioning in a large finite system due to the existence of an "isolated electronic state" [38].

Acknowledgements

Work partially supported by the Österreichischer Fonds zur Förderung der Wissenschaftlichen Forschung and by the Bundesministerium für Wissenschaft und Forschung, Wien.

References

1. K. Stephan, T.D. Märk: Chem. Phys. Lett. 90 51 (1982)

2. T.D. Märk: Int. J. Mass Spectrom. Ion Proc., 79 (1987) 1; Z. Phys. D12, 263 (1989)

3. A.J. Stace: In: Mass Spectrometry (M.E. Rose, Ed.) Royal Chemistry Specialist Report, London (1987) Vol. 9 pp. 96-121

4. O. Echt: In: Elemental and Molecular Clusters (G. Benedek, T.P. Martin, G. Pacchioni, Eds.) Springer, Berlin (1988) pp. 263-284

5. T.D. Märk, O. Echt; In: Clusters of Atoms and Molecules (H. Haberland, Ed.) Springer, Heidelberg (1991) in print

6. T.D. Märk, A.W. Castleman: Adv. Atom. Mol. Phys. $\underline{20}$ 65 (1985)

7. R.G. Keesee, A.W. Castleman, T.D. Märk: In: Swarm Studies and Inelastic Electron Molecule Collisions (L.C. Pitchford, B.V. Mc Koy, A. Chutjian, S. Trajmar, Eds.) Springer, New York (1987) pp 351-366

8. T.D. Märk: In: Electron-Molecule Interactions and their Applications (L.G. Christophorou, Ed.) Vol. 1, Academic Press, Orlando (1984) pp. 251-334

9. P. Dehmer: J. Chem. Phys., $\underline{76}$ 1263 (1982)

10. T.D. Märk: Europhys. Conf. Abstr. $\underline{6D}$ 29 (1982)

11. H. Haberland: Surf. Science $\underline{156}$ 305 (1985)

12. H.U. Böhmer, S.D. Peyerimhoff: Z. Phys. $\underline{D11}$ 239 (1989)

13. E. Morikawa, R. Reininger, P. Gürtler, V. Saile, P. Laporte: J. Chem. Phys. $\underline{91}$ 1469 (1989)

14. A. Burose, C. Becker, A. Ding: Z. Phys. D, in print (1991)

15. D. Scharf, J. Jortner, U. Landman: J. Chem. Phys. $\underline{88}$ 4273 (1988)

16. P. Scheier, T.D. Märk: Chem. Phys. Lett. $\underline{136}$ 423 (1987)

17. M. Lezius, T.D. Märk: Chem. Phys. Lett $\underline{155}$ 496 (1989)

18. O. Echt, T.D. Märk: In: Clusters of Atoms and Molecules (H. Haberland, Ed.) Springer, Heidelberg (1991) in print

19. M. Foltin, G. Walder, A.W. Castleman, T.D. Märk: J. Chem. Phys. $\underline{94}$ 810 (1991)

20. K. Stephan, A. Stamatovic, T.D. Märk: Phys. Rev. $\underline{A28}$ 3105 (1983); P. Scheier, A. Stamatovic, T.D. Märk; J. Chem. Phys. $\underline{89}$ 295 (1989)

21. K. Norwood, J.H. Guo, C.Y. Ng: J. Chem. Phys. $\underline{90}$ 2995 (1989)

22. E.E. Ferguson, C.R. Albertoni, R. Kuhn, Z.Y. Chen, R.G. Keesee, A.W. Castleman: J. Chem. Phys. $\underline{88}$ 6335 (1988)

23. K. Levsen: Fundamental aspects of organic mass spectrometry, Verlag Chemie, Weinheim (1978)

24. C.E. Klots: J. Phys. Chem. $\underline{92}$ 5864 (1988); Int. J. Mass Spectrom. Ion Proc. $\underline{100}$ 457 (1990)

25. C. Brechignac, P. Cahuzac, J. Leygnier, J. Weiner: J. Chem.Phys. $\underline{90}$ 1492 (1989)

26. W. Begemann, K. Meiwes-Broer, H.O. Lutz: Phys. Rev. Lett. $\underline{56}$ 2248 (2986)

27. D. Kreisle, O. Echt, M. Knapp, E. Recknagel: Phys. Rev. $\underline{A33}$ 768 (1986)

28. P. Scheier, T.D. Märk: Int. J. Mass Spectrom. Ion Proc. $\underline{102}$ 19 (1990)

29. P.C. Engelking: J. Chem. Phys. $\underline{87}$ 936 (1987)

30. J.M. Soler, J.J. Saenz, N. Garcia, O. Echt: Chem. Phys. Lett. $\underline{109}$ 71 (1984)

31. P. Scheier, T.D. Märk: Phys. Rev. Lett $\underline{59}$ 1813 (1987)

32. P. Scheier, T.D. Märk: Chem. Phys. Lett. $\underline{148}$ 393 (1988)

33. M. Foltin, G. Walder, S. Mohr, P. Scheier, A.W. Castleman, T.D. Märk: Z. Phys. $\underline{D}$ in print (1991)

34. M. Foltin, T.D. Märk: Chem. Phys. Lett. in print (1991)

35. M. Joppien, F. Grotelüschen, T. Kloiber, M. Lengen, T. Möller, J. Wörmer, G. Zimmerer, J. Keto, M. Kykta, M.C. Castex: J. Lumin. in print (1991)

36. B. Schneider, J.S. Cohen: J. Chem. Phys. $\underline{61}$ 3240 (1974); R. Gaethke, P. Gürtler, R. Kink, E. Roick, G. Zimmerer: phys. stat. sol. $\underline{b124}$ 335 (1984)

37. H. Steger, J. de Vriens, W. Kamke, I.V. Hertel: Z. Phys. $\underline{D}$ in print (1991)

38. C. Lifshitz: J. Phys. Chem. $\underline{87}$ 2304 (1983)

Part IV

Clustering
in Light Nuclei

Alpha-Nucleus Interaction Beyond Woods-Saxon

M. Brenner

Department of Physics, Åbo Akademi, Turku, Finland

Abstract: The existence of alpha cluster states in light nuclei is related to the alpha-nucleus interaction. The Woods-Saxon form factor of the local optical potential does not generally give a good agreement with α-scattering experiments. This is especially true for the scattering by light nuclei at low α particle energies, where the Anomalous Large Angle Scattering (ALAS) is considerable. Different potentials for the alpha-nucleus interaction are reviewed. The resonant character of the excitation function and the fitting of angular distributions with squared Legendre polynomials in the backward direction at resonance are discussed. The angular momentum of the alpha particles scattered at strong resonances equals the spins of the resonances. These spin assignments are in agreement with transfer reaction data. The resonant states are considered to be fragments of quasi-molecular or alpha cluster states.

1 Introduction

Alpha cluster states have been demonstrated in many light nuclei of which ^{44}Ti is the best known. The cluster model implies that the alpha particle as a four-nucleon cluster of two neutrons and two protons exists at the nucleus long enough to form quasi-molecular states. For the discussion a part of a level scheme of Ohkubo et al.[1] is shown in figure 1.

The calculated levels form rotational bands. Below the zero line a ground state band with spin parities up to 6^+ is seen. The experimental data to the left are quite well reproduced by this band of bound states. As for the unbound states above the zero line the agreement between theory and experiment is not as good.

The interaction between the α-particle and a nucleus is usually described with a local potential. The most widely used is the Woods-Saxon (WS) potential. Theoretical results like those in figure 1 can be derived by considering a particle moving in the WS-potential. The question whether or not the WS- potential describes the alpha-nucleus interaction in a proper way should be answered by comparing theory with experiment.

If the potential is not satisfactory we have to look for effects beyond the Woods-Saxon potential. We discuss this question first in a traditional way, i.e. from the point of view of alpha-nucleus scattering. The quasi-molecular states observed in alpha scattering and transfer reactions will be considered last.

Springer Series in Nuclear and Particle Physics   **Clustering Phenomena in Atoms and Nuclei**
Editors: M. Brenner · T. Lönnroth · F.B. Malik   © Springer-Verlag Berlin, Heidelberg 1992

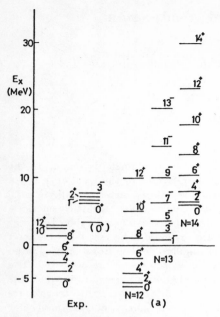

Fig. 1. Experimental ^{44}Ti levels and alpha cluster levels calculated by Ohkubo et al.[1].

2. Woods-Saxon Potentials and Elastic Alpha Scattering

The most correct model for the alpha nucleus interaction involves a non-local potential. When the alpha particle approaches a target nucleus at a distance within the strong interaction radius, pairs of one nucleon in the target and one in the projectile will interact. We have to sum over all pairs. In a homogenized description we integrate over the product of the mass densities of the two interacting clusters and the nucleon-nucleon interaction $U(r + r_c - r_\alpha)$ to get the double-folded potential. The distance between the nucleons in an interacting pair is $r + r_c - r_\alpha$ where r is the distance between the center of masses of the interacting clusters and r_c and r_α are the internal coordinates of the nucleons in the cluster.

The alpha particle is a cluster of great stability. It has no low excited state to populate during the interaction with the nucleus. We may, therefore, consider the interaction between the alpha particle and the individual nucleons in the target nucleus $u(r + r_c)$. The alpha-nucleus interaction potential follows then from an integration over the target nucleus. This single-folded potential has essentially the radial dependence which is known as the Woods-Saxon (WS) form factor

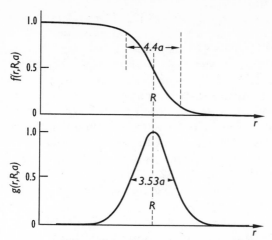

Fig. 2. Volume and surface type optical potentials derived from the WS form factor. $R = rA^{1/3}$

$$f(x) = (1 + e^{x_k})^{-1}$$

$$x_k = \frac{r - r_k A^{1/3}}{a_k.}$$

The interaction may result in either the elastic scattering of the alpha particle or its absorption. The optical potential

$$V(r) = -Vf(x_R) - iWf(x_i) + V_c(r)$$

allows for a parametrisation which by calculation reproduces more or less the experimental cross-sections. The variable x_k has the index $k = R$ for the real part and $k = i$ for the imaginary part. Sometimes a surface potential $g(x_i)$ is used instead of or in addition to the imaginary volume potential. It is obtained by a differentiation of the WS form factor. The volume type and surface type potentials are shown schematically in fig. 2. The meaning of the diffuseness parameter a_k follows from the figure.

The third term of the optical potential is the Coulomb potential of the two homogenously charged clusters. Accordingly we usually have a six parameter optical potential to describe the alpha-nucleus interaction when using the WS form factor. The parameters for the real and the imaginary part of the potential are the potential depths V and W, the radial parameters r_R and r_i and the diffuseness parameters a_R and a_i.

The angular distributions which come out from solving the Schroedinger equation using the WS potential can be fitted to experimental distributions. The best fits yield sets of these parameters. Systematics of the real potential depth V may result in global potentials, which can be used to predict cross-sections within large energy and nuclear mass regions. There are, however, ambiguities in chosing different sets of parameters [2]. The real potential depth and the corresponding radial parameter give good agreement when the product VR^n remains constant. The exponent has roughly the value 2. In addition to this continuous ambiguity there is a discrete ambiguity, which is the result of the different numbers of nodes of the radial wave function inside the nucleus. Good fits of angular distributions have been reported frequently when realistic WS form factors have been used. Systematics of parameters from experiments at 26 MeV alpha energy and targets between A = 50 and 93 eg exhibits a linear dependence of the real potential depth on mass number [3]. Good fits to alpha scattering data for ^{73}Ge are e.g. obtained with the set V = 64.5, 78.5, 104.5, 132.8, 163.9 and 198.1 MeV, the radial parameter $r_R = 1.6$ fm and the diffuseness parameter a_R near 0.5 fm.

Various methods have been suggested to evaluate one of the depths to be the most relevant or most physical to describe the interaction between the projectile and the target. One method is to extrapolate from high energies, where, due to the rainbow effect, the angular distribution has an exponential form and the discrete ambiguity is assumed not to exist [4]. This is by no means a safe method. Although many theoretical relations have been derived for the energy dependence of the real potential depth [5, 6] they fail to describe experimental observations in the low energy region. Moreover, evidence for discrete ambiguity in the rainbow scattering region has recently been reported by Dem'yanova et al. [7].

The low energy region is of special interest in alpha nucleus interaction as the formation of alpha cluster states are more likely to be observed at low excitation. At low energies the ALAS [8] will be more pronounced for many light targets. The studies of α-scattering by ^{16}O [9] ^{40}Ca [20] should be referred to. Others are ^{12}C, ^{20}Ne , ^{24}Mg, ^{28}Si and ^{32}S, which can be characterized as many alpha particle nuclei. The compound systems are resp. ^{16}O, ^{24}Mg, ^{28}Si, ^{32}S, ^{36}Ar and ^{44}Ti, the second and last being the most studied. But light nuclei with other mass numbers exhibit ALAS e.g. ^{22}Ne [9], ^{39}K, ^{42}Ca and ^{44}Ca [8]. Many authors seem to agree on the origin of the ALAS phenomenon considering it to be the result of weak absorption [10]. It is, however, still worth noting as a peculiarity of interest in the description of the alpha nucleus interaction. Before going into detail other potentials should be discussed.

3. Potentials Derived from Fundamental Principles

The properties of nuclear matter are reflected in the energy density formalism [11] for two interacting nuclei. In the sudden approximation their densities are summed in the overlapping region to give the energy of the two body system. Its dependence on the distance between the centers of masses gives in an easy way the potential. The shallow potential, which results from this procedure has a repulsive part related to the Pauli principle. A similar potential was first applied to ^{16}O - ^{16}O scattering and it was considered to generate quasi-molecular states [12, 13].The interaction between light nuclei is preferably described by a shallow potential [14]. There are reasons to believe that it can be applied at low energies better than a deep potential. Angular distributions of 14 to 28 MeV alpha particles elastically scattered by ^{28}Si have been successfully reproduced using a shallow potential (depth $\sim$ 15 MeV), which was based on the energy density formalism and the sudden approximation [15]. This potential could be modified to predict angular distributions for nuclei of different Z and A. In this sense the shallow potential derived by energy density theory can be considered global. The Two Centre Shell Model, TCSM, should be mentioned as a more detailed alternative to the energy density approach. Its use in the study of nucleus-nucleus interaction has been presented by Greiner at this conference.

A more general method than the last mentioned is the Resonating Group Method, RGM. The fundamental effective nucleon-nucleon interaction is the starting point in this method. [1]. Its application to the interaction between two groups has been simplified by the use of the Orthogonality Condition Model, OCM [18]. These two models involve the full expansion of a non-local potential from the pair-wise interaction of the nucleons in two-body collisions. Spectra of molecular states as shown in figure 1 have been derived in this way. The approach has been useful for the understanding of the properties of local potentials. A procedure to deduce from the non-local potential a local potential similar to the phenomenological one has been described by Horiuchi [19].

An essential property of the RGM follows from the Pauli principle. It does not allow particles of an incoming projectile to populate states already occupied by the nucleons in the target. This rule is summarized in the Wildermuth condition $N \geq 2n + L$, where N indicates the allowed harmonic oscillator state, n is the quantum number of relative motion and L the orbital angular momentum. The forbidden or redundant states have N < 2n + L . Rotational bands formed of particles with N = 12 eg are shown in figure 1. Antisymmetrisation effects have an influence on the radial and diffuseness parameters of a WS-type potential [23]. Such considerations have brought into the foreground the use of the squared Woods-Saxon potential, $(WS)^2$. This as well as the non-local potential derived from RGM give good fits to angular distributions in elastic scattering and can be used successfully to deduce rotational bands of α-cluster states. The Woods-Saxon squared potential is similar to the double-folded potential. It has been recommended by several authors because of its global character [20]. The above brief review of potentials for alpha nucleus interaction and their illustration by a few examples does not, however, give a realistic view of the situation for the time being. A look at the experiments will give additional information on our topic.

4. Resonance Phenomena in Elastic Scattering

Collision experiments indicate that there is a complicated nuclear structure in the region where the rotational bands are expected to be observed. At excitation energies over 10 MeV the density of compound states is high. The statistical character of nuclear interaction is the cause of the Ericsson fluctuations [21], which are seen when the cross-section is plotted as a function of the incident energy. This so called excitation function has, therefore, an unregular smooth pattern of peaks and valleys. The peaks observed in elastic scattering by many nuclei have, however, non-statistical character. At low incident energy and light targets the observation of sharp peaks in the excitation function is related to enhanced large angle scattering. Quasi-molecular alpha cluster states and their rotational bands are expected to be found among nuclei which are formed by the non-statistical, resonance-like alpha particle scattering. Measurements of excitation functions with good energy resolution is thus essential in the search for alpha cluster states in scattering experiments.

Van de Graaffs have a high energy resolution. They have mostly been used in excitation function measurements up to 12 MeV. Higher incident energies are needed to cover excitation energies in the broad energy region where the rotational bands occur. Figure 3 shows an example of a simple arrangement comprising a cyclotron, analysing magnet and a scattering chamber for measuring excitation functions and angular distributions of particle scattering and reactions. Alpha particles from 3 to 18 MeV can be used in the experiments. The energy resolution is 0.1 energy in small steps ($\simeq$ 50 keV) takes less than 5 min. An excitation function for alpha particle elastic scattering by silicon is shown in figure 4.

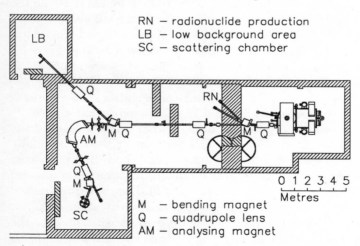

Fig. 3. Layout of the cyclotron and beam transport system at the Åbo Akademi cyclotron laboratory

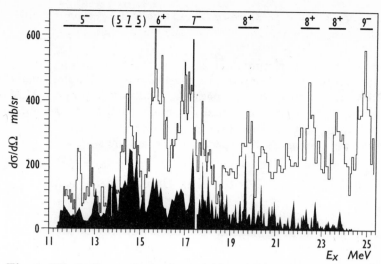

Fig. 4. The dark area shows the excitation function for elastic scattering of alpha particles by ^{28}Si ($\Theta = 173°$) and the curve above the excitation function is the α-^{16}O relative energy spectrum for the transfer reaction $^{28}Si(^6Li, d)^{32}S(\alpha)^{28}Si$. E_x is the excitation energy of ^{32}S. The spin assignments are from the transfer reaction.

The sharp peaks are due to resonant states in ^{32}S at excitation energies between 11 and 25 MeV. The region below 17.2 MeV was measured with a Van de Graaff [22] and the region above with the cyclotron [26]. In the latter case one detector was kept in the backward direction close to 180°, where the Legendre polynomials of all l-values have maximum. Measurements at 180° are most useful as they give an estimate of the total cross-section for resonant scattering. When ALAS occurs, peaks are strong at this angle.

In figure 4 no groups of peaks or other kinds of gross-structure are seen. As the target spin is zero it is possible to determine the spin of the resonant states on condition that the resonant scattering dominates in the backward direction. The angular distribution at large angles can then be fitted by a Legendre polynomial $P_l(\cos \Theta)$. The l is the angular momentum of the spinless system before the formation of the resonant state. It equals, therefore, the spin of the resonant state. The spin can be measured in this way only at the strongest peaks of the excitation function. The potential scattering brought about by the WS or some other potential dominates the pattern of the distribution in the forward direction. The division of the angular distribution into a forward diffraction model part, an intermediate around 90° and another diffraction-like (glory scattering) part in the backward direction has been proposed by Bobrowska et al.[24]. It implies that the cross-section shall be analysed in terms of an interference between the potential and the resonant scattering. The interference is dominating near 90°. In fact the

333

angular distribution off the strong resonances, where the resonant contribution is small, is very different compared to the on-resonance distribution [25].

When we recall the ideal picture in figure 1, we find that the experimental observations exhibit some new phenomena. The unbound alpha cluster states should appear in the excitation function as peaks with approx. 1 MeV spacing . The reduced width should equal the single particle Wigner limit $\gamma_W = \frac{\hbar^2}{\mu_\alpha R^2}$, where μ_α = the reduced mass of the alpha particle . Instead we see many narrow resonances at a spacing less than 100 keV.

5. Fragmentated Quasi-Molecular States and Alpha Scattering

A resonant amplitude due to the narrow resonance state can be introduced into the scattering matrix. This is to encroach upon the simple principle of the particle in a potential . Frekers et al. have successfully generated quasi-molecular resonances in low-energy α-^{40}Ca scattering by modifying the scattering matrix [26]. With a tandem accelerator the energy range from 4.4 to 9.12 MeV was covered in 4 keV steps. Resonances corresponding to the mixed band seen in figure 1 where identified. The spins ranged from 0 to 4 and the parities were natural. The widths varied betwen 40 and 550 keV. Referring to the multilevel R-matrix formulation the authors considered the states as intermediate states or doorway resonances, which decay out into the continuum with a partial width $\Gamma^\uparrow$ and to the compound nucleus levels with $\Gamma^\downarrow$.

The excitation function measured by Frekers et al. clearly shows that the intermediate resonances are split up into fragments, which are seen as narrow peaks. The fragmentation or splitting of single particle states is a well-known phenomenon in nuclear spectroscopy. The fragmentation of well separated states is observed in the fission spectrum of actinide nuclei. It is in that case brought about by the double-humped barrier (see figure 7 of ref. 28).

Transfer reactions are known to pick out single particle states. The transfer of an alpha particle from a ^{16}O projectile to a ^{12}C target leads to broad states in ^{16}O, which decay back to ^{12}C by the emission of an alpha particle. When alpha particles are measured in coincidence with the projectile residue ^{20}Ne, broad peaks in the α - ^{16}O relative energy spectrum are observed [29]. The width of the peaks and there spacing, typically a few tenths of a MeV, are close to the expected value for single particle states. The spin of the states can be determined from angular correlation measurements of coincident ^{20}Ne and alpha particles. The spins increase with the energy of excitation, which is a general feature of rotational bands.

A systematical study of states with α-cluster nature, has been aimed at by the study of the (^{6}Li,d) reaction at the I.V. Kurchatov Atomic Energy

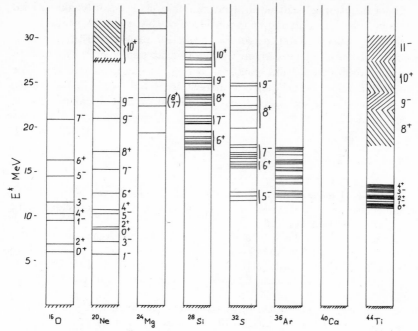

Fig. 5. Levels and groups of levels corresponding to states with α-cluster nature [32]

Institute in Moscow. Levels of mixed parity bands have been observed in ^{16}O [30], ^{20}Ne, ^{28}Si [31] and ^{32}S [27]. This extends the knowledge about the existence of bands similar to the band of Frekers et al. [26] in ^{44}Ti (figure 1). The available information is summarized in figure 5.

The levels in ^{28}Si exhibit groups with spin 6^+, 7^-, 8^+, 9^- and 10^+. The spins of the groups were obtained from α-deuteron angular correlation measurements [31]. Similar bands of fragmented states of cluster nature were observed in ^{32}S. The figure 4 shows peaks corresponding to these states. The spin assignments are shown above. Peaks appear in the excitations function of elastic alpha scattering at the same excitation in ^{32}S. The angular distribution of the alpha particles yielded spin assignments in agreement with the results of the α-deuteron measurement [25].

6. Summary

The formation of α-cluster states affect the alpha-nucleus interaction at low energies ($E_\alpha \leq 20$ MeV). Optical model calculations using a WS potential, (WS)^2 or other simple local potentials do not therefore give appropriate fits to elastic scattering angular distributions at energies where resonances due to the cluster states contribute to the yield. The resonances are resolved in the excitation function of scattered alpha particles as narrow peaks. These peaks give evidence for a strong fragmentation of the α-cluster states.

The states are observed in the spectra of reaction products from α- transfer reactions. There is an agreement in the spin at peaks in these spectra and at corresponding excitation energy in the alpha elastic scattering. The spins give evidence for the existence of mixed parity bands of cluster states, which are not predicted by calculations from local or non -local potentials (c.f. figure 1 and 5).

References

1. S. Okubo, K. Umehara, K. Hiraoka: Proc. Int. Symp. Devel. Nucl. Cluster Dynamics, ed. by Y. Akaishi, K. Katō, H. Noto, S. Okabe (World Sci., Singapore, 1989) pp 114–126

2. G. R. Satchler: Direct Nuclear Reactions: Direct Nuclear Reactions (Clarendon Press, Oxford, 1983) pp 499–505

3. J. B. A. England, S. Baird, D. H. Newton, T. Picazo, E. C. Pollacco, G. J. Pyle, P. M. Rolph, J. Alabau, E. Casal, A. Garcia: Nucl. Phys. A **388** 573 (1982)

4. D. A. Goldberg, S. M. Smith: Phys. Rev. Lett. **29** 500 (1972)

5. S. K. Gupta, K. H. N. Murthy: Z. Phys. A **307** 187 (1982)

6. D. F. Jackson, R. C. Johnson: Phys. Lett. **49** B 249 (1974)

7. A. S. Dem'yanova, A. A. Ogloblin, S. N. Ershov, F. A. Gareev, R. S. Kurmanov, E. F. Svinareva, S. A. Goncharov, V. V. Adodin, N. Burtebaev, J. M. Bang, J. S. Vaagen: Phys. Scripta T **32** 89 (1990)

8. G. R. Gruhn, N. S: Wall: Nucl. **81** 161 (1966)

9. A.A. Cowley, J.C. van Staden, S.J. Mills, P.M. Crouje, G. Heymann and G.F. Burdzik: Nucl. Phys. A **301** 429 (1978)

10. F. Michel, G. Reidermeister and S. Ohkubo: Phys. Rev C **37** 292 (1988)

11. K.A. Brueckner, J.R. Buchler and M.M. Kelly: Phys. Rev. **173** 944 (1968)

12. B. Block and F.B. Malik: Phys. Rev. Lett. **19** 239 (1967)

13. R.H. Siemssen, J.V. Maker, A. Weidinger and D.A. Bromley: Phys. Rev. Lett. **19** 369 (1967)

14. L. McFadden and G.R. Satchler: Nucl. Phys. **84** 177 (1966)

15. P. Manngård, M. Brenner, M.M. Alam, J. Rechstein and F.B. Malik: Nucl. Phys. A **504** 130 (1989)

16. K. Pruess and W. Greiner: Phys. Lett. **33** B 197 (1970)

17. J.A. Wheeler: Phys. Rev. **52** 1033, 1107 (1937)

18. S. Saito: Prog. Theor. Phys. **41** 705 (1969)

19. H. Horiuchi, Prog. Theor. Phys. **69** 886 (1983)

20. Th. Delbar, Gh. Gregoire, G. Paic, R. Ceuleneer, F. Michel, R. Vander-Poorten, A. Budzanowski, H. Dabrowski, L. Freindl, K. Grotowski, S. Micek, R. Planeta, A. Strzalkowski and K.A. Eberhard: Phys. Rev. C **18** 1237 (1978)

21. T. Ericson: Ann. Phys. **23** 390 (1963)

22. J.J. Lawrie, A.A. Cowley, D.M. Whittal, S.J. Mills and W.R. Mc Murray: Z. Phys. A **325** 175 (1986)

23. Z. Majka, A. Budzanowski, K. Grotowski and A. Strzalkowski: Phys. Rev. **18** 114 (1978)

24. A. Bobrowska, A. Budzanowski, K. Grotowski, L. Jarczyk, S. Micek, H. Niewodniczanski, A. Strzalkowski and Z. Wrobel: Nucl. Phys. **A126** 361 (1969)

25. P. Manngård, M. Brenner, K.-M. Källman, T. Lönnroth, M. Nygård, V.Z. Goldberg, M.S. Golovkov and V.P. Rudakov: These proceedings

26. D. Frekers, R. Sauto and K. Langouke: Nucl. Phys. A **394** 189 (1983)

27. K.P. Artemov, M. S. Golovkov, V.Z. Goldberg, V.V. Pankratov, M. Brenner, P. Mannård, K.-M. Källman, T. Lönnroth: Sov. J. Nucl. Phys. (in print)

28. S. Bjørnholm and J.E. Lynn: Rev. Mod. Phys. **52** 725 (1980)

29. S.C. Allcock, W.D.M. Rae, P.R. Keeling, A.E. Smith, B.R. Fulton and D.W. Banes: Phys. Lett. B **201** 201 (1988)

30. K.P. Artemov, V.Z. Goldberg. I.P. Petrov, V.P. Rudakov, I.N. Serikov and V.A. Timofeev: Phys. Lett. B **37** 61 (1971)

31. K.P. Artemov, M.S. Golovkov, V.Z. Goldberg, V.I. Dukhanov, I.B. Mazurov, V.V. Pankratov, V.V. Paramonov, V.P. Rudakov, I.I. Serikov, V.A. Solovjev and V.A. Timofejev: Jadernaja Fisika 51 1220 (1990)

32. K.P. Artemov et al.: I.V. Kurchatov Atomic Energy Institute, private communication

Local Potential Description of Elastic Scattering and of Cluster Structure in Light Nuclei

F. Michel[1] *and G. Reidemeister*[2]

[1]Université de Mons-Hainaut, Faculté des Sciences,
 Avenue Maistriau, 19, B-7000 Mons, Belgium
[2]Physique Nucléaire Théorique et Physique Mathématique,
 Université Libre de Bruxelles, CP 229, B-1050 Brussels, Belgium

The aim of this talk is to survey the progress made these last ten years in the understanding of the anomalous large angle scattering seen in low energy α-particle elastic scattering from some light nuclei. Emphasis will be laid on the relation of the deduced optical potentials to the present microscopic understanding of the nucleus-nucleus interaction, and on how these potentials can be used to clarify the α-cluster structure of the unified system. We will also briefly comment on the coexistence of deep and shallow potential descriptions of elastic scattering, the energy behaviour of the optical potential at low energy, the possible parity dependence of the interaction, and the effects of the deformation of the core. Finally we will present an example of the extension of these ideas to light heavy ion systems.

1. Historical background. ALAS

At the beginning of this century, Geiger and Marsden[1] carried out one of the first nuclear scattering experiments by shooting α-particles emitted by a natural radioactive source at a thin metallic foil. The energy and angular behaviours of the scattering cross section were found to be in complete agreement with the Rutherford formula (which had been derived from classical mechanics assuming that most of the mass of the atom is concentrated in a positively charged nucleus). In these experiments the maximum attainable energy was limited to a few MeV; the distance of closest approach for 5 MeV α-particles impinging on, e.g., a lead nucleus is about 50 fm, that is, considerably larger than the nuclear diameter.

The distance of closest approach decreases for higher energies and/or lighter targets. An estimate of the size of the nucleus can be inferred from the energy where the Rutherford scattering formula breaks down: at any scattering angle this formula predicts a $1/E^2$ energy dependence of the cross section, which is found to be violated above some critical energy. Taking into account quantum mechanical effects, one finds[2] that the heaviest nuclei have a radius not exceeding 10 fm.

Beyond that critical energy the measured elastic scattering angular distributions differ more and more from the predictions of the Rutherford formula[3]: in fact the observed scattering pattern is reminiscent of that of light diffracted by a circular aperture (or equivalently, invoking the Babinet principle, by a circular black disk). This similarity can be understood if one

Springer Series in Nuclear and Particle Physics   **Clustering Phenomena in Atoms and Nuclei**
Editors: M. Brenner · T. Lönnroth · F.B. Malik   © Springer-Verlag Berlin, Heidelberg 1992

postulates that because of the growing importance of the non-elastic scattering processes, incident particle trajectories which hit the target are completely absorbed. This picture can be incorporated within a quantum mechanical frame by describing the scattering as due to a complex interaction potential ("optical potential") with a large imaginary part.

In the beginning of the sixties it became technically feasible to measure elastic scattering angular distributions up to fairly large scattering angles: this had until then been hindered by the smallness of the cross section at large angles and the low intensity of the beams supplied by the early accelerators. In many cases the measurements provided little surprise (the angular distributions remain diffractive and decrease regularly up to large angles); however in some cases an anomalous behaviour consisting in an important backward rise is observed[4]: for low mass targets like ^{12}C or ^{16}O, the measured cross section at extreme back angles can exceed the Rutherford formula prediction by two or three orders of magnitude!

This anomaly, which was coined ALAS (for Anomalous Large Angle Scattering) evaded conventional interpretations for a long time. Several symposia, held in Marburg[5], Cracow[6] and Louvain-la-Neuve[7], were devoted to this subject. It first appeared impossible to reconcile the new data with an optical model description, since the latter had been associated until then with strong absorption (and thus with a diffractive behaviour). Many more or less *ad hoc* models were thus proposed to describe the anomaly. However, a careful investigation of the capabilities of the optical potential description revealed[8] that contrary to earlier expectations it was indeed possible to reproduce the anomalous data in a quantitative way - including their complicated energy behaviour - within the frame of the optical model, provided the absorptive part of the potential is moderate. A dramatic example[9] of the precision achievable is presented in Fig. 1 in the case of $\alpha + {}^{40}$Ca elastic scattering: note how well the intricate energy dependence of the data is reproduced by the model, the more so if one takes into account the fact that there is only one slowly energy-dependent parameter (the imaginary potential well depth).

To elucidate the mechanism of the backward rise the semiclassical approach proposed by Brink and Takigawa[10] proves illuminating. If the real potential well is deep enough, and for not too large angular momenta, the total effective (that is, nuclear + Coulomb + centrifugal) potential displays an internal "pocket" separated from the outside region by a potential barrier. The method of Ref. 10 makes it possible to decompose the elastic scattering amplitude $f(\theta)$ into two contributions corresponding respectively to the part of the incident wave which is reflected at the barrier ($f_B(\theta)$), and the part which tunnels through the barrier and reemerges after penetrating the nuclear interior ($f_I(\theta)$). For "normal" targets strong absorption makes the internal contribution very small and the cross section is dominated on the whole angular range by the diffractive barrier contribution (*cf.* the case of the elastic ^{44}Ca(α,α) angular distribution at 29 MeV incident energy[11], see Fig. 2a). The weaker absorption needed for describing the anomalous cases makes the internal contribution much larger. The case of $\alpha + {}^{40}$Ca scattering at 29 MeV[11] is presented in Fig. 2b: it is seen that f_I exceeds by far the barrier contribution in the backward hemisphere and is responsible for the observed spectacular backward rise;

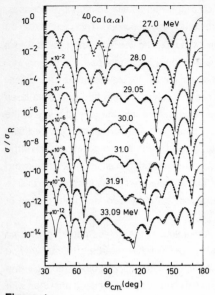

Figure 1

moreover the interference between f_B and f_I generates complicated patterns at intermediate angles with a seemingly hectic behaviour (*cf.* Fig 1).

In the anomalous cases the backangle data thus carry information on the internal part of the potential, which is not available when one analyses "normal" data, or data restricted to the forward hemisphere. It is thus possible in these cases to extract the optical potential with much better accuracy down to rather small distances. This enhanced precision is partly due to the interference pattern generated by f_B and f_I at intermediate angles, which is very sensitive to small changes of the optical potential parameters.

2. Microscopic approaches

The remarkable - and unusual - accuracy with which the potential is determined in the anomalous cases makes it an ideal interface between experiment and the results of microscopic calculations. For systems where small inter-nucleus distances, which correspond to appreciable overlap of the partners, play an important role, one expects that the fermionic nature of the constituents should be taken into account, that is, that the antisymmetrization of the total wave fuction for the interchange of any two nucleons should be cared for properly.

The prototype of such fully antisymmetrized approaches is the resonating group method[12] (RGM), in which the wave function of the system is, in the spirit of a variational approach, written in the single-channel approximation as:

$$\Psi = A \{ \varphi_1(\xi_1) \varphi_2(\xi_2) F(\rho) \}$$

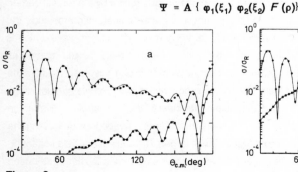 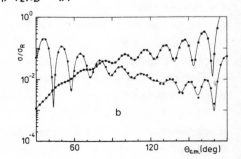

Figure 2

where φ_1 and φ_2 are the internal (ground state) wave functions of the interacting nuclei and $F(\rho)$ is the relative motion wave function; A is the antisymmetrization operator. Besides handling antisymmetrization in an exact way, the RGM has the attractive feature of treating the bound and scattering states of the system on an equal footing. F satisfies a Schrödinger-like equation (RGM equation) with a complicated, strongly non-local, angular momentum and energy dependent interaction. The latter, which can in principle be calculated once an effective nucleon-nucleon interaction has been chosen, bears at first sight very little resemblance to the phenomenological potentials.

This discrepancy has been alleviated by the introduction of the concept of "redundant states" of the RGM. These states, which are exact solutions of the RGM equation at any energy when the interacting nuclei are described by harmonic oscillator wave functions with equal parameters, appear in the numerical solutions as spurious components with undetermined weights; however as they are of short range (they look like bound states) they do not affect the asymptotic behaviour of the scattering wave function and the phase shift is determined in an unambiguous way. The internal part of the wave function can be defined uniquely by orthogonalizing it to the redundant states without affecting its asymptotic behaviour. This procedure underlies the "orthogonality condition model" (OCM) (for a nice didactical account, see Ref.13) and helps to bridge the gap between the phenomenological potentials and the theoretical approaches. Indeed the real parts of the phenomenological potentials are deep and they thus support many bound states, which look at first sight devoid of any physical significance. However if these states are similar to the redundant states of the RGM, the scattering wave function, to which they are automatically orthogonal, will be very similar in the inner region to the orthogonalized microscopic wave function - in particular they will have closely similar nodal structure. If in addition the phenomenological potentials are comparable to the microscopic ones in the tail region - where the latter reduce to the "folding" direct contribution - both approaches will generate comparable phase shifts despite of their very different appearance. Moreover after discarding the bound states of the phenomenological potentials which simulate the redundant states of the RGM, one is left in this interpretation with bound and quasi-bound states which should be susceptible of a physical interpretation.

3. $\alpha + {}^{16}O$ elastic scattering and α-cluster structure in the ${}^{20}Ne$ nucleus

A nice confirmation of these ideas has been provided by the study of α-particle scattering from ${}^{16}O$. Many microscopic calculations have been devoted to this system[14] in connection with the α-particle spectroscopy of ${}^{20}Ne$. Indeed several levels in this nucleus are difficult to describe quantitatively within the frame of the conventional shell model: they appear to correspond better to a picture where an α-cluster orbits an unexcited ${}^{16}O$ core (because of antisymmetrization the distinction between shell-model states and cluster states is actually not so clear-cut as this simple picture seems to imply). Low-energy levels of this type are

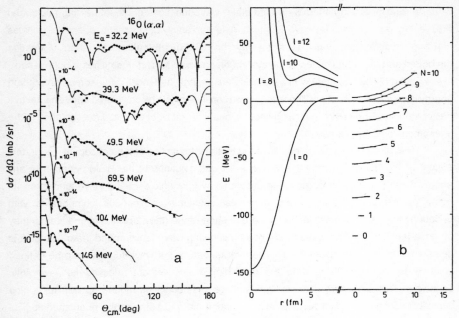

Figure 3

preferentially excited in collisions where the incident channel matches best the cluster configuration under study (in the present case, α + ^{16}O).

The α + ^{16}O system displays a spectacular backward anomaly (Fig. 3a) which disappears progressively with incident energy[15]. A global optical model analysis of the available data made possible the extraction of an unambiguous potential, with only two slowly energy-dependent parameters controlling its real and imaginary parts, respectively; the predictions of this potential[15] also appear in Fig. 3a as full lines.

The real part of this potential is found to bind a number of states whose quantum numbers agree with those of the redundant states of the RGM for that system: these are states with $N = 2n + \ell < 8$, where n and ℓ denote the radial and orbital angular momentum quantum numbers. These states are plotted in Fig. 3b (states with the same value of N are linked with a thin line), together with a few effective potential curves. Bound and quasi-bound states with $N \geq 8$ must be considered as physical states, and indeed their properties are in very good agreement with those of well-known cluster states of the ^{20}Ne nucleus: not only do they group into quasi-rotational bands, with energies agreeing well with their experimental counterparts, but other significant properties like absolute values of intraband quadrupole transition probabilities are reproduced quantitatively. Our phenomenological α + ^{16}O potential has received direct support from microscopic calculations of Wada and Horiuchi[16]: using the WKB approximation, they were able to convert the complicated non-local RGM potential into a local-equivalent version which reproduces very accurately the original microscopic phase shifts; after averaging this slightly ℓ-dependent potential over ℓ, they obtain a local-equivalent potential quantitatively similar to the real part of our optical potential.

4. α-particle elastic scattering from other nuclei and α-cluster structure in the related systems

The predictive power of this simple approach can be used with advantage in cases where the microscopic calculations lack the strong experimental indications which are needed to calibrate the effective nucleon-nucleon interaction.

One of the most striking examples is that of the ^{44}Ti nucleus, which is the fp-shell analog of ^{20}Ne, since the relevant core, that is, ^{40}Ca, is like ^{16}O a doubly closed shell nucleus. All available RGM calculations predict the existence of α-cluster states in ^{44}Ti, but they disagree on the absolute energies of these states; for a recent discussion of the status of the problem the interested reader is referred to Ref. 17. On the experimental side, there has also been considerable debate about the identification of α-particle states in ^{44}Ti.

A calculation[18] of the properties of the bound and quasi-bound states associated with the real part of a global optical potential describing the ^{40}Ca(α,α$_0$) data on broad angular and energy ranges[8] has made possible a considerable clarification of the situation. For this system states with N < 12 have to be discarded in the spirit of an OCM interpretation; states with N ≥ 12 are found to group into several quasi-rotational bands, the first of which, with N = 12, falls below the α + ^{40}Ca threshold and agrees well with the ^{44}Ti ground state band, not only for energies, but also for intraband quadrupole transition probabilities, which are notoriously difficult to reproduce in shell model calculations. This band has however a relatively weak cluster character since the calculated r.m.s. distance between the clusters is found to be less than the sum of their radii, and there is thus significant overlap between their matter densities. Bands with stronger cluster character are predicted at higher excitation energy; the most interesting one to discuss here is the N = 13, negative parity band, starting a few MeV above the α + ^{40}Ca threshold, with distinct α-cluster character. The first members of this band, which at the time of this calculation had no experimental counterparts, have been identified very recently in a ^{40}Ca(^{6}Li,d)^{44}Ti α-transfer experiment carried out at Osaka[19] (the same experiment has at the same time disqualified previous α-cluster candidates which did not fit into our picture). Our picture has also received additional support from recent RGM calculations[20], which have shown that the choice of the nucleon-nucleon interaction is strongly constrained if one insists upon the reproduction of the α + ^{40}Ca data up to high incident energies, and that this interaction then predicts bound and quasi-bound state properties in agreement with our calculation.

We have likewise analyzed ^{36}Ar(α,α) data and calculated the properties of the low-energy bound and quasi-bound states supported by the potential extracted[21]; as ^{36}Ar is not a doubly closed shell nucleus, we are led to identify these states with 4p-4h states of the unified system ^{40}Ca, if we discard, like for the α + ^{40}Ca system, bound states with N < 12. The theoretical N = 12 band is found to agree very well with the experimental 4p-4h band built on the well known $J^\pi = 0^+$, $E_x = 3.35$ MeV deformed state, and intraband quadrupole transition probabilities are in good semi-quantitative agreement with experiment. Here also we predict the existence of a negative parity band of states, with more substantial cluster character, starting near the $\alpha + {}^{36}$Ar threshold. Because these states are predicted to have larger reduced widths than in the $\alpha + {}^{40}$Ca case, and because the density of states in ^{40}Ca is lower than in ^{44}Ti, there is even better prospect to detect these states in a ^{36}Ar(^{6}Li,d) transfer experiment; such an experiment is actually being carried out at Osaka[22].

There are several other systems where this type of approach could reveal fruitful. We have examined in an exploratory search several systems around the sd-shell closure where plausible candidate α-cluster states can be identified, while keeping contact with the few existing α-particle scattering data[21]; this study indicates that investigating the coexistence of spherically symmetric and deformed states in the A $\approx$ 40 mass region in terms of α-particle clustering could prove rewarding.

5. Deep versus shallow potentials

Although microscopic approaches like the RGM substantiate the description of elastic scattering in terms of deep local potentials - provided bound states simulating the redundant states are discarded - one can also justify local equivalent potentials of a quite different nature. In the case of the much studied $\alpha + \alpha$ system, Thompson et al.[23] built, more than 20 years ago, local shallow potentials which are equivalent to the RGM interaction: these potentials possess a hard core for small angular momenta, and are angular momentum and energy dependent. Aoki and Horiuchi[24] built later, using the WKB technique referred to in § 3, a local deep version with redundant states starting from similar ingredients. In the case of $\alpha + \alpha$ scattering, phenomenological potentials belonging to both classes are available: the shallow, repulsive-core, ℓ-dependent potentials of Darriulat et al.[25] and of Ali and Bodmer[26], and the deep, ℓ-independent potentials of Neudatchin et al.[27] and of Buck et al.[28] These seemingly very different interactions are thus substantiated on a common microscopic basis.

To shed more light on the connection between these two descriptions we have implemented a procedure for building explicitly a shallow potential which is phase equivalent to a given deep potential but which lacks the spurious bound states of the latter (for a detailed account of the procedure see Refs. 29 and 30). We will content ourselves of sketching our construction procedure taking the simple $\alpha + \alpha$ system as an example.

The deep $\alpha + \alpha$ potentials of Neudatchin et al.[27] and of Buck et al.[28], which reproduce very well the experimental phase shifts up to about 50 MeV incident energy, bind three states having the quantum numbers of the redundant states of the RGM for that system; indeed these states have $N = 2n + \ell < 4$, and there are thus two $\ell = 0$ and one $\ell = 2$ unphysical bound states. Because of the Levinson theorem, the S-wave and D-wave phase shifts generated by these potentials start from 2π and π, respectively, at zero incident energy, and go to zero at high energy[27]. The microscopic phase shifts turn out to share this property since for non-local potentials like the RGM interaction redundant states must be taken into account along with physical bound states in the generalized version of the Levinson theorem. Because of the ordinary Levinson theorem, it is clearly impossible to build a potential which would be phase-equivalent to the deep version on the whole energy range, but which would lack the unphysical bound states of the former. Both these requirements can however be satisfied if we leave the class of regular potentials, that is, potentials whose first two moments are finite, by introducing a $1/r^2$ repulsive singularity with an appropriate weight near the origin[29,30]. Starting from the deep potential of Buck et al.[28], we succeeded in this way to build a phase-equivalent interaction with no spurious bound states, which turns out to be nearly identical to the shallow α-α interaction of Ali and Bodmer[26], demonstrating in an explicit way the equivalence between the two descriptions[29]. Using an exact alternative procedure based on supersymmetry arguments essentially similar results have been obtained by Baye[31] for the same system.

The same construction procedure was applied[29] to heavier systems like $\alpha + {}^{16}O$ and $\alpha + {}^{40}Ca$. In these cases no shallow phenomenological potential is available, but examination of the shallow interactions built from the deep phenomenological potentials described in §§ 3-4 reveals striking similarities with the angular momentum projected energy curves of the generator coordinate method (GCM), a microscopic approach equivalent to the RGM, and of related methods based on the two-center shell model; in particular the effective potentials reconstructed behave differently for even and odd parity, and the minima of the potential curves shift to smaller r when ℓ increases at low ℓ (antistretching effect)[29].

6. Energy behaviour of the α-nucleus potential near the Coulomb barrier

The very low energy properties of the α-nucleus interaction are worth investigating because several recent theoretical calculations point to a possible rapid energy dependence of the real part of the nucleus-nucleus interaction near the threshold. On one hand RGM calculations like those of Horiuchi et al.[32] indicate that antisymmetrization effects induce a rapid increase of the potential barrier height as energy decreases; this phenomenon, which is dominated by the one-nucleon exchange contribution, is predicted to persist and even intensify at negative energy. On the other hand recent calculations of Nagarajan et al.[33] predict a fast increase of the potential strength above the inelastic threshold, due to a dispersive contribution originating in the rapid opening of non-elastic channels in this region of energy.

We have studied[34] the properties of the $\alpha + {}^{16}O$ interaction at energies close to the Coulomb barrier (3 MeV $< E_\alpha <$ 5 MeV). The choice of this system has several obvious advantages. Indeed it has been thoroughly studied, both from a theoretical point of view (the doubly magic nature of both partners makes microscopic calculations more tractable) and from a phenomenological point of view (a global potential describing the scattering from 20 to 150 MeV incident energy is available[15], cf. § 3); moreover the Coulomb barrier for that system (V_{CB}

≈ 3 MeV) is lower than the energy of the first inelastic threshold and therefore the scattering can be studied without introducing an absorptive potential. The data that we have analyzed are high quality data of Buser[35] which extend on the whole angular range and have been measured at closely spaced energies.

A simple extrapolation towards low energies of the potential of § 3 turns out to predict angular distributions disagreeing with Buser data, even after renormalizing its depth. An excellent agreement with the data, starting from the higher energy global potential, could only be achieved after making the interaction very slightly ℓ-dependent, and after reducing its diffuseness with respect to that needed at higher energy[34]. This last feature results in an increase of the barrier height, in accordance with the predictions of the two models mentioned at the beginning of this paragraph[32,33]; disentangling these two effects will necessitate a careful examination of the model dependences of these two theoretical approaches.

7. Parity dependence of the α-nucleus potential

Another interesting feature has emerged recently in the study of elastic α-particle scattering from light targets. Systematic measurements on the whole angular range have been performed by Abele et al.[36] for ten targets ranging from ^{11}B to ^{24}Mg, at incident energies around 50 MeV. The most striking feature of these data is the observation of nearly identical scattering patterns at small and large angles for most of these nuclei; in contrast ^{12}C, and more especially ^{20}Ne and ^{24}Mg, whose angular distributions do not depart from those of other targets at small angles, show markedkly different behaviours at back angles. ^{20}Ne for example, instead of displaying a mid-angle broad plateau followed by a rainbow-like exponential falloff at larger angles, shows a remarkable enhancement at an energy where ALAS has disappeared in ^{16}O and nearby targets.

A conventional local potential of the type discussed in §§ 3-4 turns out to be unable to reproduce this new anomaly. We have therefore investigated the possibility of describing it in terms of a small parity splitting in the real part of an interaction potential in line with that used for lower mass targets[37]; the parity dependence which was superimposed was chosen to have a gaussian form factor. It was soon found that the introduction of this new degree of freedom induces changes in the calculated cross section which qualitatively match the main trends of the data; adjusting the parameters of the model to the data provided a precise agreement with experiment[37].

The volume integral per nucleon pair of the parity-averaged potential is comparable to that obtained for neighbouring targets. On the other hand, the parity splitting found necessary to fit the data is comparatively small, since it has a peak value of only 600 keV, and its contribution to the volume integral does not exceed 1,5 %. The backward anomaly in the angular distribution is found to be due to a gaussian-like contribution to the S-matrix due to the parity-dependent term, which is centered near the grazing angular momentum; this mechanism is similar to that pinpointed by von Oertzen et al.[38] in their description of elastic transfer between similar mass nuclei. Still elastic transfer is thought to be important only in cases where the mass difference between the interacting nuclei is small; the range of the core-exchange contribution could however be substantially increased when target clustering is important[39]. Before any firm conclusion can be drawn on the mechanism of the backangle anomaly in the present and nearby systems, it would clearly be of great value to repeat the measurements at lower and higher energies on the whole angular range, since the energy evolution of the phenomenon is expected to put a stringent constraint on the model proposed here.

8. Possible effects of core deformation

An extension of the above ideas away from shell closure is likely to imply taking into account the effects of core deformation. Coupled channel effects have indeed been invoked by Staudt et al.[40] to account for the anomaly seen in ^{20}Ne(α,α) scattering that was discussed in the preceding paragraph. The main trends of the data could be reproduced by taking into account the coupling to the first two excited states of the ^{20}Ne ground state rotational band, that is the $J^{\pi} = 2^+$ and 4^+ levels at $E_x = 1.63$ and 4.25 MeV, respectively; however the phasing of the elastic scattering oscillations was in substantial disagreement with experiment at back angles. Here also the lack of additional data at nearby energies precludes any definite conclusion to be drawn, although it is likely that the polarization potential induced by moderate coupling to inelastic channels, which is expected to be located in the surface region, can be incorporated in the optical potential provided use is made of form factors with sufficient flexibility.

Another light system where core deformation is expected to play a significant role in the description of the scattering is $\alpha + {}^{12}$C; the properties of this system in connection with the α-cluster properties of ^{16}O are presently under study[41].

9. Light heavy ion scattering: the ^{16}O + ^{16}O case

The recent measurements by the Hahn-Meitner group[42] of elastic ^{16}O + ^{16}O scattering at $E_{lab} = 350$ MeV display a clear nuclear rainbow pattern at large angles. The importance of nuclear rainbow scattering for eliminating discrete ambiguities in the real potential depth has been pointed out for a long time[43]. It turns out that the very shallow potentials which have nearly exclusively been used to describe low energy resonant phenomena in heavy ion systems - and in the ^{16}O + ^{16}O system in particular - are unable to describe rainbow scattering at this energy. The large angle features can only be reproduced[42,44] using a deep potential with a volume integral per nucleon pair of about 300-400 Mev.fm^3. On the other hand, Kondō et al.[45] have shown that deep potentials of this type can also reproduce successfully the low energy resonant behaviour of the system; a description of $^{16}O + ^{16}O$ cluster states in ^{32}S, using a deep potential compatible with the Pauli principle , admitting unphysical bound states simulating the redundant states of the RGM (which for the present system are given by $N = 2n + \ell < 24$) - and similar to that used for describing α-particle clustering and elastic scattering - thus seems to be at hand.

Acknowledgements

The authors are grateful to Professor R. Ceuleneer for his constant support during the course of the research presented in this report, and to I.I.S.N. and F.N.R.S. for financial support.

References

1) H. Geiger and E. Marsden, Phil. Mag. **25**(1913)604
2) see, e.g., R.M. Eisberg and C.E. Porter, Rev. Mod. Phys. **33**(1961)190
3) see, e.g., B. Fernandez and J.S. Blair, Phys. Rev. **C1**(1970)523
4) J.C. Corelli, E. Bleuler, and D.J. Tendam, Phys. Rev. **116**(1959)1184
5) *Proc. Symposium on Four Nucleon Correlations and Alpha Rotator Structure*, Marburg, Germany, 1972, edited by R. Stock
6) *Proc. First Louvain-Cracow Seminar on the Alpha-Nucleus Interaction,* Cracow, Poland, 1974, edited by A. Budzanowski
7) *Proc. Second Louvain-Cracow Seminar on the Alpha-Nucleus Interaction,* Louvain-la-Neuve, Belgium, 1978, edited by Gh. Grégoire and K. Grotowski
8) F. Michel and R. Vanderpoorten, Phys. Rev. **C16**(1977)142; Th. Delbar *et al.*, Phys. Rev. **C18**(1978)1237
9) F. Michel and G. Reidemeister, Phys. Rev. **C29**(1984)1928
10) D.M. Brink and N. Takigawa, Nucl. Phys. **A279**(1977)159
11) J. Albinski and F. Michel, Phys. Rev. **C25**(1982)213
12) see, e.g., K. Wildermuth and Y.C. Tang, *A Unified Theory of the Nucleus*, Vieweg, Braunschweig, 1977
13) D. Baye, in Ref. 7, p. 330
14) Y. Fujiwara *et al.*, Suppl. Prog. Theor. Phys. **68**(1980)29; and references therein
15) F. Michel *et al.*, Phys. Rev. **C28**(1983)1904
16) T. Wada and H. Horiuchi, Phys. Rev. Lett. **58**(1987)2190
17) F. Michel, in *Proc. Fifth Int. Conf. on Clustering Aspects in Nuclear and Subnuclear Systems*, Kyoto, Japan, 1988, edited by K. Ikeda, K. Katori and Y. Suzuki (J. Phys. Soc. Jpn **58**(1989), suppl., p. 65)
18) F. Michel, G. Reidemeister, and S. Ohkubo, Phys. Rev. Lett. **57**(1986)1215; Phys. Rev. **C37**(1988)292
19) T. Yamaya *et al.*, Phys. Rev. **C41**(1990)2421; Phys. Rev. **C42**(1990)1935
20) T. Wada and H. Horiuchi, Phys. Rev. **C38**(1988)2063
21) G. Reidemeister, S. Ohkubo, and F. Michel, Phys. Rev. **C41**(1990)63
22) T. Yamaya and S. Ohkubo, private communication
23) D.R. Thompson *et al.*, Phys. Rev. **185**(1969)1351
24) K. Aoki and H. Horiuchi, Prog. Theor. Phys. **68**(1982)1658
25) P. Darriulat *et al.*, Phys. Rev. **137**(1965)B315
26) S. Ali and A.R. Bodmer, Nucl. Phys. **80**(1966)99

27) V.G. Neudatchin *et al.*, Phys. Lett. **34B**(1971)581
28) B. Buck, H. Friedrich, and C. Wheatley, Nucl. Phys. **A275**(1977)246
29) F. Michel and G. Reidemeister, J. Phys. **G11**(1985)835
30) F. Michel, in *Proc. Int. Symp. on Developments of Nuclear Cluster Dynamics*, Sapporo, Japan, 1988, edited by Y. Akaishi, K. Katō, H. Noto and S. Okabe
31) D. Baye, Phys. Rev. Lett. **58**(1987)2738
32) K. Aoki and H. Horiuchi, Prog. Theor. Phys. **68**(1982)2028
33) M.A. Nagarajan, C.C. Mahaux, and G.R. Satchler, Phys. Rev. Lett. **54**(1985)1136
34) F. Michel, Y. Kondō, and G. Reidemeister, Phys. Lett. **B220**(1989)479
35) M.W. Buser, Helv. Phys. Acta **54**(1981)439
36) H. Abele *et al.*, Z. Phys. **A326**(1987)373
37) F. Michel and G. Reidemeister, Z. Phys. **A333**(1989)331
38) see, e.g., L.J.B. Goldfarb and W. von Oertzen, in *Heavy Ion Collisions*, Vol. I, North Holland, Amsterdam, 1979, edited by R. Bock, p. 215
39) M. Le Mere and Y.C. Tang, Phys. Rev. **C37**(1988)1369
40) G. Staudt *et al.*, in *Proc. XVII Int. Symp. on Nuclear Physics*, Gaussig, D.D.R., 1987
41) F. Michel, G. Reidemeister, and S. Ohkubo, work in progress
42) E. Stiliaris *et al.*, Phys. Lett. **B223**(1989)291
43) D.A. Goldberg *et al.*, Phys. Rev. Lett. **29**(1972)500
44) Y. Kondō, F. Michel, and G. Reidemeister, Phys. Lett. **B242**(1990)340
45) Y. Kondō, B.A. Robson, and R. Smith, Phys. Lett. **B227**(1989)310

Alpha Clustering in Nuclei and the Structure of ^{44}Ti

P.E. Hodgson

Nuclear Physics Laboratory, University of Oxford, Oxford OX1 3RH, UK

Abstract: The application of the concept of the alpha-particle mean field to unify alpha clustering in nuclei and alpha scattering by nuclei is described. The analysis of the ^{44}Ti $= {}^{40}$Ca $+ \alpha$ system is described in detail.

1. Introduction

The concept of alpha-clustering has found wide application to problems of nuclear structure and nuclear reactions. Basically, the idea of clustering is that it is simpler in some circumstances to treat a group or cluster of nucleons in the nucleus as a single entity rather than as a combination of individual nucleons. This often helps us to think more clearly about the processes taking place. We have no doubt, in principle, that to treat the nucleons individually would be more accurate, and more fundamental, but the mathematical difficulties of doing this may be very great or even insurmountable. We are then willing to sacrifice that unattainable accuracy in favour of practicable calculations that give results of a precision that is still useful. If in addition we are willing to introduce phenomenological parameters, and to adjust them to fit selected data, then the accuracy attainable may be quite high.

The purpose of introducing cluster concepts is thus to assist our physical insight and to make our calculation simpler, without too great a loss of accuracy. Thus for example we can treat ^{6}Li as an alpha-particle bound to a deuteron, and ^{12}C as three alpha-particles. Similarly, alpha-transfer reactions can be analysed with the distorted wave theory by assuming that the alpha-particle remains as an undivided unit throughout the interaction, and alpha decay by evaluating the probability of its transmission through the potential barrier. Such calculations are much easier to perform than those that treat each nucleon individually and often give results of acceptable accuracy.

2. The alpha-particle mean field

The alpha-clustering features of nuclear structure and nuclear reactions can be unified by using the concept of the alpha-particle mean field. As in the more familiar model for the nucleon mean field, we replace all the interactions between the alpha-particle and the remainder of the nucleus by a one-body potential. The parameters of this potential can be adjusted so that its eigenvalues are the bound and unbound alpha-particle states, and it also gives the measured differential cross-section for alpha scattering.

Springer Series in Nuclear and Particle Physics   **Clustering Phenomena in Atoms and Nuclei**
Editors: M. Brenner · T. Lönnroth · F.B. Malik   © Springer-Verlag Berlin, Heidelberg 1992

The main similarities between the nucleon mean field and the alpha-particle mean field are that each are described by a one-body complex, local potential with bound and unbound states whose energies and quantum numbers can be determined by transfer and knockout reactions. The density distributions can be calculated by summing the squares of the wavefunctions, weighted by the occupation probabilities. The single-particle wavefunctions can be used to calculate transition and decay rates. There are however some differences, particularly that the alpha-particle is a transient structure that is formed by the coalescence of four nucleons, orbits for long enough to affect the structure of the nucleus and then breaks up into its constituent nucleons. The alpha particles are confined to surface states, so that the probability of alpha-particle formation peaks quite strongly in the surface region. Alpha decay is more prominent than nucleon decay, and the unbound alpha-particle states have much effect on the scattering cross-section for alpha-particles, particularly for double-closed shell nuclei which have a small value of the imaginary potential.

At positive energies the alpha potential is simply the familiar alpha-particle optical model that describes very well the differential cross-sections for the elastic scattering of alpha-particles by nuclei. This model is able to account for the apparently anomalous back-angle scattering (ALAS) that occurs for scattering from double magic nuclei such as ^{16}O and ^{40}Ca. The observed structure in the cross-sections may be traced to the resonances in particular partial waves that become prominent when the imaginary potential is small. The strength of the real potential at energies near the Coulomb barrier also shows apparently anomalous behaviour that can be understood by applying the dispersion relations.

At negative energies many of the excited states can be obtained as eigenvalues of the alpha-particle potentials, and this model also gives their decay widths and B(E2) transition strengths. These states are of course the ones with predominant alpha-particle structure, and can only be obtained with great difficulty by shell-model calculations as they require the superposition of a large number of single-particle configurations. It is thus the same alpha-particle potential that gives the bound states and the unbound states responsible for the resonance effects in the scattering cross-sections. The parallel with the nucleon single-particle model is exact in almost every respect, except that the alpha-clusters are transient, forming and then breaking up, but living long enough to manifest their presence.

In order to make calculations with the alpha cluster model a suitable form must be chosen for the alpha-nucleus potential. The observable quantities are sensitive to the details of the potential, so it is important to include the relevant features of the structures of the interacting nuclei. This may be done by defining the potential by the double folding integral

$$V(r) = \iint \rho_\alpha(\mathbf{r}_\alpha)\rho_c(\mathbf{r}_c)U(|\mathbf{r} + \mathbf{r}_c - \mathbf{r}_\alpha|)d\mathbf{r}_\alpha d\mathbf{r}_c \qquad (1)$$

where $\rho_\alpha(\mathbf{r}_\alpha)$ and $\rho_c(\mathbf{r}_c)$ are the alpha cluster and core densities and $U(|\mathbf{r}_1 - \mathbf{r}_2|)$ the effective nucleon-nucleon interaction. This potential may be quite accurately represented by the analytical cosh form

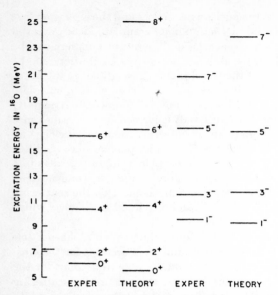

Figure 1. Comparison between the theoretical and experimental energies of the 0^+ and 1^- rotational bands in ^{16}O. The theoretical energies are calculated from an alpha-^{12}C folded potential. The arrow indicates the energy of the $\alpha + {}^{12}$C threshold in ^{16}O. (Buck, Dover and Vary, 1975).

$$V(r) = \frac{V_0(1 + \cosh(R/a))}{\cosh(r/a) + \cosh(R/a)} \tag{2}$$

where V_0 is the depth of the potential, R its radius and a its diffuseness.

The main requirements of the Pauli exclusion principle are satisfied by choosing the quantum numbers of relative motion n (the number of interior nodes in the radial wavefunction) and L (the orbital angular momentum) to obey the Wildermuth condition $2n + L \geq N$. The integer N is chosen large enough to ensure that all the cluster nucleons occupy orbitals above those occupied by the core nucleons.

This model of alpha structure was applied to ^{16}O considered as an alpha-particle moving relative to a ^{12}C core, by Buck, Dover and Vary (1975). As shown in Figure 1 and Table 1 this model accounts very well for the low-lying 0^+ and 1^- rotational bands, and also for the B(E2) transition rates.

Subsequently the model has been extended successfully to other light nuclei from ^{17}O to ^{19}F.

3. Application to ^{44}Ti

It is of interest to see if the model can also be applied to fp shell nuclei, and recently there have been several studies of ^{44}Ti, considered as an alpha particle in the field of a ^{40}Ca core. Pál and Lovas (1980) calculated the positive parity

TABLE 1. Theoretical and experimental alpha widths for ^{16}O

J^π	Γ_α^{exp}(KeV)	Γ_α^{th}(KeV)	$\bar{f}$(fm)
	0$^+$ band		
4$^+$	27	17.5	1.436
6$^+$	125	238	1.432
8$^+$	—	385	1.425
	1$^-$ band		
1$^-$	510	675	1.536
3$^-$	1,200	1,750	1.5575
5$^-$	700	≈2,000	1,539
7$^-$	750	776	1.668

Data from Buck, Dover and Vary (1975).

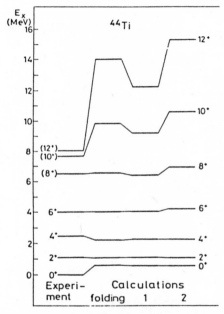

Figure 2. Experimental and calculated states of the $2N + L = 12$ alpha-cluster states in ^{44}Ti. The calculations were made with a folding potential and two cosh potentials. (Pál and Lovas, 1980).

states of the yrast band and were able to fit their energies quite well up to the 8$^+$ state, as shown in Figure 2, and also the B(E2) transition rates in this band (Table 2).

Subsequently, Michel, Reidemeister and Ohkubo (1986ab, 1988) made extensive calculations based on a potential obtained from the alpha-particle — ^{40}Ca elastic scattering cross-sections (Delbar *et al*, 1978). At low energies there

TABLE 2.   B(E2) values for transitions in ^{44}Ti in Weisskopf units (Pál and Lovas, 1980).

Transition	Experiment	Folded	1	2
$2^+ \rightarrow 0^+$	13 ± 4 (1)	9.5	5.5	21.7
$4^+ \rightarrow 2^+$	30 ± 6 (1)	12.6	7.1	29.5
$6^+ \rightarrow 4^+$	17 ± 3 (2)	11.9	6.4	29.6
$8^+ \rightarrow 6^+$	> 1.5 (3)	9.8	4.8	26.5
$10^+ \rightarrow 8^+$	15 ± 3 (3)	7.6	3.0	20.7
$12^+ \rightarrow 10^+$	< 6.3 (4)	3.5	1.4	12.3

References: (1) Dixon *et al*, 1973; (2) Simpson *et al*, (1973); (3) Simpson *et al*, (1975); Kolata *et al*, (1974).

TABLE 3.

Theoretical and experimental B(E2) values for the $J \rightarrow J - 2$ transitions (in W.U.), and intercluster rms radii for the ^{44}Ti $N = 12$ and $N = 13$ states. The local potential depth given for the $N = 12$ states is that reproducing the experimental energies with respect to the $\alpha + {}^{40}$Ca threshold; for the $N = 13$ states U_0 is fixed to the average value $U_0 = 180$ MeV (Michel *et al*, 1986a).

		$N = 12$				$N = 13$	
	U_0	$B(E2)$		$< R^2 >^{1/2}$		$B(E2)$	$< R^2 >^{1/2}$
J^π	(MeV)	Theor.	Expt.	(fm)	J^π	Theor.	(fm)
0^+	184.1	–	–	4.50	1^-	–	5.35
2^+	182.5	11.6	13 ± 4	4.51	3^-	28.7	5.28
4^+	181.2	15.9	30 ± 6	4.47	5^-	30.2	5.15
6^+	180.7	15.2	17 ± 3	4.38	7^-	26.4	4.95
8^+	179.0	12.8	> 1.5	4.27	9^-	20.0	4.71
10^+	181.8	8.1	15 ± 3	4.06	11^-	12.8	4.43
12^+	186.8	3.6	< 6.3	3.83	13^-	5.9	4.14

are well-known ambiguities in this potential, but at high energies only one family gives acceptable fits. Using analyses at intermediate energies, it is possible to trace this unique parameter set down from high energies to low energies and thus obtain a unique optical potential. This potential not only gives an excellent description of the elastic scattering data, but also predicts the existence of several bands of alpha-^{40}Ca cluster states, with the $N = 12$, $J^\pi = 0^+$ band head predicted to lie around 10 MeV below threshold, i.e. in the region of the ^{44}Ti ground state. It has subsequently been renormalised by Michel *et al* to a depth of 180 MeV, so that it generates the $N = 12$ band of cluster states in closer agreement with the yrast band in ^{44}Ti. These predictions are compared with the experimental data in Figure 3 and the transition rates are compared with the experimental values in Table 3.

Further calculations of ^{44}Ti using the alpha-^{40}Ca model have been made by Merchant, Pál and Hodgson (1989). They used the folding model to calculate

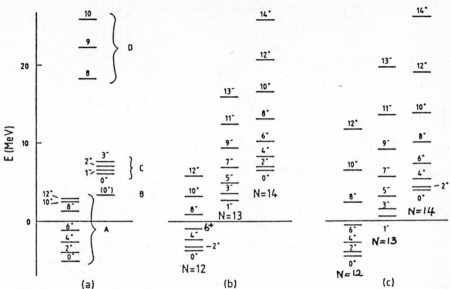

Figure 3.

(a) Experimental data on the possible bound and unbound alpha-cluster states of ^{44}Ti, relative to the alpha-particle — ^{40}Ca threshold at 5.127 MeV. The states are labelled to indicate the experimental methods used to detect them: (A) The ground state band (Simpson *et al*, 1975). (B) A state at 8.54 MeV relative to the ground state strongly populated in the ^{40}Ca(^{6}Li, d) reaction (Strohbusch *et al*, 1974). (C) A group of four mixed-parity states at 11.2, 11.7, 12.2 and 12.8 MeV found from (α, α) scattering (Frekers *et al*, 1976, 1983). (D) The lower members of a band of nine states found from (α, α) scattering at higher energies (Stock *et al*, 1972; Löhner *et al*, 1978).

(b) States of ^{44}Ti calculated from a phenomenological potential (Michel *et al*, 1988).

(c) States of ^{44}Ti calculated from a double-folded potential (Hodgson and Merchant, 1988).

the alpha-^{40}Ca potentials, and found that it is very similar to that used by Michel *et al*. In this calculation, the density distributions were obtained from the charge distributions, assuming that the proton and neutron distributions are the same. For the effective nucleon-nucleon interaction both zero-range and finite-range Gaussian forms were used, and in each case the strength was chosen so as to give the experimental energy of the 2^+ state in ^{44}Ti at 1.083 MeV. The resulting potentials, together with those of the unique potential of Michel *et al* and the 'cosh' potential of Pál and Lovas, are compared in Figure 4. This shows that the two folded potentials are both much closer to the unique optical potential than the 'cosh' potential, and that the finite range folded potential is closest of all. The effective potential deduced by Wada and Horiuchi (1987,

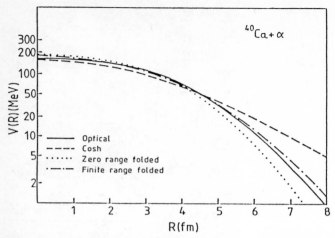

· **Figure 4.** A comparison of the real part of the unique optical potential (Delbar *et al*, 1973) (full curve), zero-range (dotted curve) and finite-range (chain curve) folded potentials and a 'cosh' parametrisation (Pál and Lovas, 1980) (broken curve) for the ^{40}Ca + α system. All of these potentials have been renormalised to place the 2^+ member of the $N = 12$ band at 4.04 MeV below the $\alpha - {}^{40}$Ca breakup threshold in ^{44}Ti (Merchant *et al*, 1989).

1988) from the resonating group model is also very similar to the folded and optical potentials.

The states of ^{44}Ti obtained from the finite-range folded potential of Merchant *et al* are shown in Figure 3, and the B(E2) transition rates in Table 4. These results are very similar to those of Michel *et al*. The ground state yrast band in ^{44}Ti is quite well fitted to the 8^+ state; the compression of the 10^+ and 12^+ states has recently been ascribed by Yamada to the admixture of [31] configurations due to the spin-orbit interaction.

At higher energies there is a state at 8.54 MeV which was strongly excited in the $({}^6\text{Li},d)$ alpha-transfer study of Strohbusch *et al* (1974), and tentatively assigned the spin-parity value of (0^+). However there are other experimental indications that this is incorrect, and that the state actually has a rather high spin (Betts, 1988).

At slightly higher energy is a group of five resonant states, with spins ranging from 0 to 4 and having natural parity. They were first seen in elastic alpha scattering from ^{40}Ca by Frekers *et al* (1976) and have recently been confirmed by Sellschop *et al* (1987). Their energies seem to form a rotational sequence, and there are strong suggestions that at least some of them form the beginning of a band of alpha-cluster states.

Finally, between 20 MeV and 50 MeV, a series of possible natural parity resonances ranging from 8^+ to 16^+ has been proposed by Frekers *et al* (1976) and Löhner *et al* (1978). The basis for this claim is their ability to fit the anomalous sharp backward rise in the angular distributions of the elastic alpha scattering with a single Legendre polynomial at these energies. This a far from conclusive

TABLE 4. Calculated and experimental B(E2↓) reduced transition strengths for the lowest-lying $N = 12$ and $N = 13$ bands of alpha-cluster states in ^{44}Ti (Merchant *et al*, 1989).

	B(E2↓) (Weisskopf units)				
$J_i^\pi \rightarrow J_f^\pi$	Cosh	Squared Saxon-Woods	Zero-range folding	Finite-range folding	Expt
$2^+ \rightarrow 0^+$	21.7	11.9	9.3	12.9	13 ± 4
$4^+ \rightarrow 2^+$	29.5	15.8	12.2	17.3	30 ± 6
$6^+ \rightarrow 4^+$	29.6	14.9	11.5	16.9	17 ± 3
$8^+ \rightarrow 6^+$	26.5	12.1	9.4	14.5	> 1.5
$10^+ \rightarrow 8^+$	20.7	8.3	6.5	10.7	15 ± 3
$12^+ \rightarrow 10^+$	12.3	4.2	3.3	5.8	< 6.5
$3^- \rightarrow 1^-$	48.4	26.8	21.6	28.6	
$5^- \rightarrow 3^-$	55.2	28.3	22.9	31.5	
$7^- \rightarrow 5^-$	55.9	24.9	20.2	30.0	
$9^- \rightarrow 7^-$	54.4	19.0	15.6	25.6	
$11^- \rightarrow 9^-$	49.2	12.2	10.2	18.6	
$13^- \rightarrow 11^-$	27.3	5.7	4.9	9.9	

criterion for identifying a resonance, since a sum of Legendre polynomials corresponding to neighbouring L values, with comparable amplitudes, can easily simulate the behaviour of a single L value. However, it is interesting to note that if one assumes that these states comprise a rotational band whose members have alternating parity, and extrapolates down in energy to find the $L = 0$ band head, then one predicts that it should lie around 11–12 MeV, where the other strongly resonant states were located in elastic alpha scattering.

Examination of Figure 3 shows that the alpha-cluster model predicts a negative parity band at low excitation energies, but there is little indication of its presence in the experimental data shown. This band has now been found experimentally by Yamaya *et al* (1989) using the ^{40}Ca(^{6}Li, d)^{44}Ti reaction; this provides further support for the alpha-cluster model based on the ground state of ^{44}Ti. Their experimental results are compared with several theoretical predictions in Figure 5. The observation of this band excludes the alternative alpha-cluster models of ^{44}Ti based on the 0^+ state at 8.54 MeV (Horiuchi, 1985) and on the 0^+ state at 11.2 MeV (Friedrich and Langanke, 1975). Additional support is provided by the differential elastic scattering of alpha-particles by ^{40}Ca, which cannot be fitted in the rainbow region by these two models. The ground state model fits the spectroscopic properties of the low-lying states and also the elastic scattering from 18 to 100 MeV (Ohkubo, 1989).

Now that the experimental situation concerning the low-lying unbound states is becoming clearer, the details of the comparison between theory and experiment acquire increasing importance. In particular, it becomes interesting to see if the dispersion correction to the real part of the alpha-particle optical

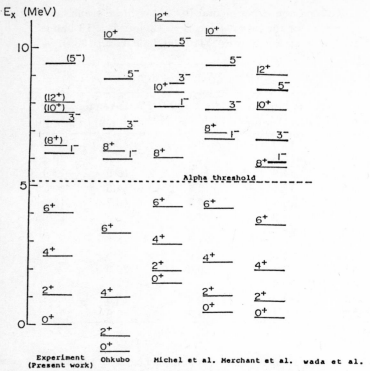

Figure 5. Experimental results of Yamaya *et al* (1989, 1990) on the alpha-cluster states of ^{44}Ti compared with theoretical results (Yamaya, 1989).

potential is comparable with the remaining discrepancies. This is currently being investigated.

References

Betts, R.R., 1988 (Private Communication).

Buck, B., Dover, C.B. and Vary, J.P., 1975 *Phys. Rev.* **C11** 1803.

Delbar, Th., Grégoire, Gh., Paic, G., Ceulcneer, R., Michel, F., Vanderpoorten, R., Budzanowski, A., Dabrowski, H., Freindl, L., Grotowski, K., Micek, S., Planeta R., Strazalkowski, A. and Eberhard, K.A., 1978 *Phys. Rev.* **C18** 1237.

Dixon, W.R., Storey, R.S. and Simpson, J.J., 1973 *Nucl. Phys.* **A202** 579.

Frekers, D., Eickhoff, H., Löhner, H., Poppensicker, K., Santo, R. and Wiezorek, C., 1976 *Z. Phys.* **A276** 317.

Frekers, D., Santo, R. and Langanke, K., 1983 *Nucl. Phys.* **A394** 189.

Friedrich, H. and Langanke, K., 1975 *Nucl. Phys.* **A252** 47.

Hodgson, P.E. and Merchant, A., 1988. Proc. 5th International Conference on Nuclear Reaction Mechanisms. Ed. E. Gadioli, Varenna, June 1988. *Ricerca Scientifica ed Educazione Permanente Supplemento* **66** 375.

Horiuchi, H., 1985 *Prog. Theor. Phys.* **73** 1172.

Kolata, J.J., Olness, J.W. and Warburton, E.K., 1974 *Phys. Rev.* **C10** 1663.

Löhner, H., Eickhoff, H., Frekers, D., Gaul, G., Poppensicker, K., Santo, R., Drentje, A.G. and Put, L.W., 1978 *Z. Phys.* **A286** 99.

Merchant, A.C., Pál, K.F. and Hodgson, P.E., 1989 *J. Phys.* **G15** 601.

Michel, F., Reidemeister, G. and Ohkubo, S., 1986a *Phys. Rev. Lett.* **57** 1215; 1986ab *Phys. Rev.* **C34** 1248; 1988 *Phys. Rev.* **C37** 292.

Ohkubo, S., 1988 *Phys. Rev.* **C38** 2377; 1989 Proc. Int. Nuclear Physics Conference (São Paulo) I (Contributed Papers) 89.

Pál, K.F. and Lovas, R.G., 1980 *Phys. Lett.* **96B** 19.

Sellschop, J.P.F., Zucchiatti, A., Mirman, L., Gering, M.Z.I. and DeSalvo, E., 1987 *J. Phys.* **G13** 1129.

Simpson, J.J., Dixon, W.R. and Storey, R.S., 1973 *Phys. Rev. Lett.* **31** 946.

Simpson, J.J., Dünnweber, W., Wurm, J.P., Green, P.W., Kuchner, J.A., Dixon, W.R. and Storey, R.S., 1975 *Phys. Rev.* **C12** 468.

Stock, R., Gaul, G., Santo, R., Bernas, M., Harvey, B., Hendrie, D., Mahoney, J., Sherman, J., Steyaert, J. and Zisman, M., 1972 *Phys. Rev.* **C6** 1226.

Strohbusch, ,U., Fink, C.L., Zeidman, B., Markham, R.G., Fulbright, H.W. and Horoshko, R.N., 1974 *Phys. Rev.* **C9** 965.

Wada, T. and Horiuchi, H., 1987 *Phys. Rev. Lett.* **58** 2190; 1988 Kyoto Preprint RIFP–747; 1988 *Phys. Rev.* **C38** 2603.

Yamaya, T., Oh-Ami, S., Fujiwara, M., Itahashi, T., Katori, K., Tosaki, M., Kato, S., Hatori, S. and Ohkubo, S., 1989 Osaka University Laboratory of Nuclear Studies Report 112; 1990 *Phys. Rev.* **C42** 1935.

Yamaya, T., 1989. Proceedings of the RIKEN Symposium on Light and Light-Heavy Ion Reactions 141.

General Regularities in Alpha-Particle-Nucleus Interaction and Cluster Structure of Light Nuclei

V.P. Rudakov

I.V. Kurchatov Institute of Atomic Energy, 123182 Moscow, Russia

Today one can say that the cluster or quasimolecular aspect of nuclear structure is as fundamental as the single particle and collective aspects. The most complete experimental information has been obtained about alpha-cluster structure and in particular about alpha-cluster structure of excited states in light nuclei. Now we know many nuclear states with large alpha particle reduced widths usually forming rotational bands.

However, in spite of a rather long history of investigations of alpha-cluster states, one might say that these investigations were in some sense "accidental" ones. This means that in most cases the aim of an investigation was the observation of a new alpha-cluster state, and that there were no attempts to look for some general regularities of their appearance. It seems that this is what we now have to aim at.

The possible existence of some general regularities in the alpha-cluster structure can be seen in the results of two published works. In the work [1] the excitation function for alpha particle elastic scattering on ^{40}Ca has been measured. It was shown that in the ^{44}Ti nucleus there are alpha-cluster states with quantum numbers $0^+, 1^-, 2^+, 3^-$ and 4^+ located in the excitation energy region 11-14 MeV. These states form a rotational band with small parity splitting and they are strongly fragmented into several narrow levels.

Recently the same picture has been obtained in the investigation of alpha-cluster structure of the ^{28}Si nucleus [2]. It was found that in ^{28}Si existed alpha-cluster states $6^+, 7^-, 8^+, 9^-$ and 10^+ with excitation energies from 18 to 30 MeV. These states form a rotational band without parity splitting. Like the states in ^{44}Ti they are strongly fragmented.

Fig. 1 shows the levels of the ^{28}Si nucleus with large alpha particle reduced width. One can say that these levels order themselves into groups of several levels. These groups are clearly separated from each other. The levels of each group have the same spins and parities.

First of all we have to consider the experimental grounds for this picture. A very powerful method to investigate the alpha-cluster states is the measurement of the dα-correlation in the reaction A(^{6}Li, d)B*(α)A. Recently the reaction ^{24}Mg(^{6}Li,d)^{28}Si(α)^{24}Mg has been investigated at the I.V. Kurchatov Atomic Energy Institute [2]. Fig. 2 shows the dα-coincidence spectrum measured for the deuteron detector at an angle of 9° and alpha

Springer Series in Nuclear and Particle Physics   **Clustering Phenomena in Atoms and Nuclei**
Editors: M. Brenner · T. Lönnroth · F.B. Malik   © Springer-Verlag Berlin, Heidelberg 1992

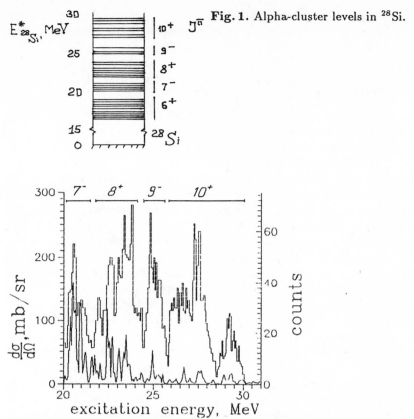

Fig. 1. Alpha-cluster levels in ^{28}Si.

Fig. 2. The $\alpha+^{24}$Mg elastic scattering excitation function (lower) and the ^{24}Mg(^{6}Li,d)^{28}Si(α)^{24}Mg dα-angular correlation function (upper histogram).

particles at an angle of 171°. The peaks in this spectrum reflect the population of excited states of the ^{28}Si nucleus in the (^{6}Li,d) reaction and their α-decay into the ground state of the ^{24}Mg nucleus. The dα-angular correlation functions have been obtained for all parts of this spectrum and angular momenta corresponding to the structures in the spectrum were estimated and indicated in the upper part of figure 2. It is clear that the rather poor energy resolution in the measurement of the dα-coincidence spectrum does not allow to resolve completely the fine structure of the broad peaks in the spectrum.

But there is the second complementary method to investigate the alpha-cluster structure of nuclei, namely the measurement of alpha particle elastic scattering excitation functions. The elastic scattering cross-section at a resonance is proportional to the squared ratio of the alpha particle width to the total width. Alpha-cluster states that have this ratio close to unity are clearly seen in the excitation functions.

361

Figure 2 shows the excitation function of alpha particle elastic scattering by ^{24}Mg measured at CM-angle 179° in the alpha particle energy region from 8 to 23 MeV [3] (lower part of fig. 2). This excitation function has been measured at Saclay quite long ago, but the authors did not analyze these results and only pointed out that some nonstatistical mechanism was responsible for most of the cross-section.

As one can see the excitation function consists of several groups of peaks separated from each other. It is well known that the alpha particle excitation function very closely resembles the dα-coincidence spectrum in the regions where the alpha-cluster states are located [4].

We can compare the results obtained by both methods in figure 2. The comparison shows a close similarity of the structures. Thus, for two nuclei we have a very clear regular picture of the alpha-cluster structure manifestation: the groups of levels with the same spins, the value of the spin increases with increasing excitation energy, and the dependence of the group energy on the spin value corresponds to a rotational band.

First of all one can point out that a similar picture, excitation of the groups of closely located levels with the same spin, has also been observed in other processes. One example is the excitation of the well known isobar-analog resonances [5]. Another example is the excitation functions of the fissioning isomers formation [6].

These pictures are considered as manifestations of the excistence of two interacting subsystems. In the case of the isobar-analog resonance one supposes that a single particle doorway state with isotopic spin $T_>$ lies in the excitation energy region of the compound nucleus where the density of levels with $T_<$ is high. Due to isotopic spin nonconservation the doorway state gets mixed with compound nuclear states. The observed fragmentation reflects the degree of this mixing.

In case for example of ^{240}Pu(n,f) reaction [6] one supposes that as a result of a neutron capture a strongly deformed isomeric doorway state is populated, i.e. a state in the second potential well. Due to finite penetrability of the inner potential barrier this state interacts with the compound nuclear states in the first potential well. Characteristics of the observed fragmentation reflect the penetrability of the inner barrier.

If we follow the spirit of this interpretation we may suppose that in our case, the alpha particle interaction with ^{24}Mg, the doorway states $\alpha+^{24}$Mg are formed . These states lie in the excitation energy region of the ^{28}Si nucleus where the density of the compound nuclear levels is very high. The observed fragmentation indicates that the doorway states interact with the compound nuclear levels. This interaction is rather weak because the alpha particle does not dissipate and the probability to escape back into the entrance (elastic) channel is large enough.

For the $\alpha+^{24}$Mg system one can make quantitative estimations of this interaction and of some parameters of the observed states. We have a possi-

Table 1. Total width, cross-sections at the resonances, spins and parities obtained for states observed in the $^{24}Mg(\alpha,\alpha)$ and $^{24}Mg(^6Li,d)$ reactions.

$E^\star_{^{28}Si}(MeV)^{a)}$	19	21	23	25	28
J^π	6^+	7^-	8^+	9^-	10^+
$kR^{b)}$	6	7	7.5	8.2	9
$\Gamma_\alpha/\Gamma_{tot}$	0.37	0.35	0.25	0.16	0.13
Γ_α (keV)	48	50	32	22	20
$\Gamma^{\uparrow\ c)}$ (keV)	480	250	230	70	130
$\gamma_\alpha^2/\gamma_W^2$ $^{d)}$ (keV)	1.2	0.7	0.8	0.2	0.4
$\Gamma^{\downarrow\ e)}$ (keV)	1700	750	1100	580	2600
N $^{f)}$	10	5	7	3	7
< comp[V] el > keV	17	10	13	10	30

a) Excitation energy indicated corresponds to the middle of the group.
b) Wave number corresponds to the alpha particle energy indicated in the first row. Radius is equal to R = $1.25A^{1/3}$ + 1.6 fm.
c) Alpha particle width of a doorway state.
d) The ratio of the reduced alpha particle width to the Wigner limit. The penetrabilities calculated for the radius indicated above.
e) Spreading width.
f) Number of levels in a group.

bility to do it because there is practically complete experimental information about these states obtained from $^{24}Mg(\alpha,\alpha)$ and $^{24}Mg(^6Li,d)$ reactions: the total width, the cross-sections at the resonances, and the spins and parities. The results of the analysis are shown in Table 1.

The estimates have been obtained for all levels of each group, but in Table 1 the average values for each group are shown. First of all one has to pay attention to the values of the ratio Γ_α/Γ. One can see that this ratio is equal to 0.4 – 0.15. This indicates that the degree of alpha-clusterization of these states is very high. The same conclusion follows from ratios of the reduced alpha particle widths to the Wigner limit. These ratios are close to unity. So the alpha-cluster structure of the considered states that qualitatively manifested itself in selective population of these states in the resonance scattering and direct transfer reactions now obtains a quantitative support.

Now we consider the characteristics of the doorway states. If we compare the values of kR with spins of the states we see that the equality kR = $\sqrt{J(J+1)}$ is well fulfilled if the radius R is equal to the sum of the radii of a nucleus and an alpha particle, i.e. R = $1.25A^{1/3}$ + 1.6 fm. It means that the angular momenta of these states correspond to the grazing partial waves. So we can suppose that the peculiarities we observed in the excitation functions and in the coincidence spectra reflect processes in the surface region of the nuclei. Further, if we plot the excitation energy of these states versus value $J(J+1)$ we see a linear relationship. It suggests that the states belong to a rotational band.

Now one can assume how the alphas particle interacts with a ^{24}Mg nucleus. The alpha particle interaction with ^{24}Mg leads to formation of a doorway state and only one single grazing partial wave contributes. These doorway states order themselves on a rotational band. These states lie in the excitation energy region where the density of levels of the ^{28}Si nucleus is quite high. The coupling of these doorway states to the compound nuclear states leads to a fragmentation. The levels into which the doorway state is fragmented have clear alpha-cluster structure (large reduced alpha particle widths).

What can be said about the generality of this picture? Recently we have measured dα-coincidence spectra and angular correlation functions for ^{28}Si and ^{32}S as target nuclei. The alpha-cluster structure of the ^{32}S and ^{36}Ar nuclei resembles very close that of ^{28}Si and ^{44}Ti. Thus we have now data for the four nuclei ^{28}Si, ^{32}S, ^{36}Ar and ^{44}Ti, and one can suppose that the observed picture is sufficiently general.

The main aim of this talk is to present systematic experimental data concerning of the alpha-cluster structure of the light nuclei excited states. On the basis of this data one can propose a qualitative model of a process of alpha particle interaction with a nucleus. Here it is necessary to point out that the long standing problem of the alpha particle anomalous large angle scattering , that until now has no generally accepted explanation in the light of presented experimental data, cannot be solved in the framework of the optical model. It is clear that the considered data reflects the excistence of the alpha-cluster or quasimolecular structures. The essential role in the observed picture is played, however, by such parameters as the radius of a nucleus and the wavelenght of the alpha particle.

In this work K.P. Artemov, M.S. Golovkov, V.Z. Goldberg, L.S. Daneljan, V.V. Pankratov, I.N. Serikov and V.A. Timofeev from I.V. Kurchatov Atomic Energy Institute and M. Madeja, J. Shmider and R. Wolsky from Institute of Nuclear Physics, Cracow, Poland have participated. The authors are indepted to Prof. D.P. Grechukin for many valuable discussions.

References

1. D. Frekers, R. Santo and K. Langanke, Nucl. Phys. **A394** 189 (1983)
2. K.P. Artemov, M.S. Golovkov, V.Z. Goldberg, V.I. Dukhanov, I.B. Mazurov, V.V. Pankratov, V.V. Paramonov, V.P. Rudakov, I.N. Serikov, V.A. Solovyev and V.A. Timofeev, Yad. Fiz. **51** 1220 (1990)

3.   F. Auger, B.Berthier, P. Charles, B. Fernadez, J. Gasebois and E. Plagnol, Compte rendu d'active du department de physique nucleare 1974-75, Note CEA-N-1861, 38

4.   K.P. Artemov, V.Z. Goldberg, I.P. Petrov, V.P. Rudakov, I.N. Serikov, V.A. Timofeev and P.R. Christensen, Nucl.Phys. **A320** 479 (1979)

5.   J.F. Wimpey, G.E. Mitchell and E.G. Bilpuch, Nucl. Phys. **A269** 46 (1976)

6.   A. Bjornholm and B. Lynn, Rev. Mod. Phys. **52** 700 (1980)

Excitation Functions of Elastic Scattering of Alpha-Particles on Light Nuclei Measured in Reversed Geometry at Zero Degree

V.Z. Goldberg

The I.V. Kurchatov Institute of Atomic Energy, Moscow, Russia

The speculation about α-structure in nuclei has a long history. The ground for the permanent attention to the problem is the well-known stability of the α-particle and high probabilities of the particle emission out of nuclei. Also it is reasonable to assume that the understanding of main features of α-clusterization can help to find an approach to the nuclear clusterization in general.

The current theories predict quasi-rotational bands of states with large reduced α-widths [1]. It is important that the α-cluster states are not real bound but resonance states. So it seems evident that the proper way to study states with the most simple cluster structure (α-cluster + a nucleus) is to use the elastic scattering of α-particles. Really the most complete data on α-cluster states in ^{16}O and ^{20}Ne were obtained by this method.

However only sparse and/or incomplete data are available for a number of nuclei such as ^{17}O, ^{18}O, ^{18}F, ^{19}F, ^{22}Ne, ^{24}Mg...

The reason for this lack of information is straightforward: to reach the nuclei in question it is nessesary to use isotopically pure and/or gaseous targets.

The stripping reactions (^{6}Li,d) and (^{7}Li,t) are of moderate help in the resonance region. The fact is that the suppression of continious spectra of ^{6}Li (or ^{7}Li) break up asks for the measurements of correlation between deutrons (tritons) and α-particles [2]. The measurements are very time-consuming. Besides one has to take into account the defects of current theories of stripping into unbound states and persisting problems of measurements with unconvenient targets.

We have developed a method to measure excitation functions of the elastic scattering with the aim to overcome the above-mentioned difficulties of the experiment.

Springer Series in Nuclear and Particle Physics **Clustering Phenomena in Atoms and Nuclei**
Editors: M. Brenner · T. Lönnroth · F.B. Malik   © Springer-Verlag Berlin, Heidelberg 1992

The Outline of the Method.

The outline of the method is very straithgforward.

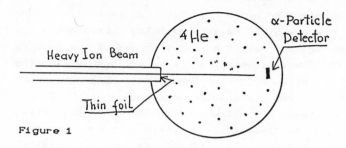

Figure 1

Accelerated heavy ions enter (through a thin foil) into a scattering chamber filled by helium-4 and stop there. The recoil α-particles ejected in the forward direction come out of the interaction region and stop in a detector which is placed in the same chamber. α-particle leave just a small part of its energy due to the specific energy losses of α-particle and heavy ion being very different.

If the energy of heavy ion at some point along its path in the gas corresponds to a resonance in the compound nucleus the probability of the elastic scattering sharply increases. The single resonance cross-section of the α-particle ejection at zero degree, which corresponds to 180° in the usual geometry of the α-particle nucleus elastic scattering is equal to :

$$(d\sigma/d\Omega)_{\circ} = \chi^2 (2J+1)^2 (\Gamma_\alpha/\Gamma)^2 \qquad (1)$$

J being the spin of the resonance (the heavy ion spin is assumed to be zero). Γ_α and Γ are partial and total widths of the resonance.

If $\Gamma \cong \Gamma_\alpha$ the cross section (1) can be extreemly large. For the $\alpha +{}^{12}C$ elastic scattering in the vicinity of the 7^- resonance ($E_{c.m.s.}$ =18.7 MeV) the cross section is about 1 bn/sr [3].

The nonresonance (potential) elastic scattering in the forward direction (the reversed geometry) is about thousand times less. So it seems rather evident that the α-cluster states must manifest itself by strong peaks in the energy spectra of α-particles. Analysis has shown the inelastic resonance scattering could be the most dangerous source of background with the magnitude up to 0.05 of that of the elastic scattering.

Experimental conditions

The experiment was performed at the I.V.Kurchatov Institute isochronous cyclotron by group of physicists [4].

We have used accelerated beams of ^{12}C, ^{15}N [4], ^{13}C, ^{14}N, ^{16}O, ^{19}F and ^{20}Ne; the energy of heavy ions being different, but mainly around 50 MeV and 100 MeV. The energy spread of the beam is of the order of 0.5%. The incident beam entered the scattering chamber through a 3-5 microns thick Havar foil separated the cyclotron vacuum from helium gas chamber. The homogeneity of the Havar foil is better than 10%. The design of the scattering chamber with 45cm internal diameter gave possibility to use only up to 1 atm of the gas pressure. The target gas was chemically pure natural helium. The detector system consisted of three silicon Si(Li) detectors 2.5 mm thick. The first one was fixed at zero degree to the beam in the end of 50cm tube connected with the chamber (Fig1), and detected α-particles when the thickness of the gas in the chamber was not enough to stop heavy ions. The second and the third detectors were mounted on the rotating top and bottom covers of the chamber. For one of these detectors placed at 20° to the beam direction there was a collimator, restricting the observed region of interaction in the vicinity of the chamber center. The other one could detect α-particles from any point of the heavy ion path inside the chamber.

There was the forth detector of the surface-barrier type which was fixed inside the chamber and detected heavy ions elastically scattered from the entrance window to monitor the beam current. The energy resolution of the detectors was better than 50 keV for 8.8 MeV α-particles. The set of the detectors served to measure spectra of α-particles at different angles, to monitor the helium pressure and to get an estimate of the contribution to the α-particle spectra from the reactions in collimator of the beam and in the Havar foil.

Experimental results

Fig 2 and Fig 3 show the experimental spectra of α-particles from interaction of ^{12}C and ^{15}N beams with helium; the angle of registration being zero degree [4]. The spectra were not corrected for different solid angles and specific energy losses along the path of the heavy ions. The excitation functions measured at electrostatic generators [5,6] are given in the inserts to the Figs 2 and 3. The similarity of the spectra and excitation

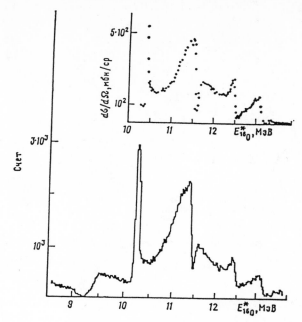

Fig 2. The zero degree α-particle spectrum of the $^{12}C-α$ elastic scattering $Einc$ $^{12}C=28$ MeV. In the insert is given the excitation function of the $α-$ ^{12}C elastic scattering measured at the angle 158.8° c.m. [5].

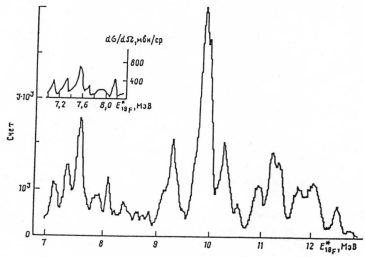

Fig 3. The zero degree α-particle spectrum of the $^{15}N-α$ elastic scattering $Einc$ $^{15}N=45$ MeV. In the insert is given the excitation function of the $α-$ ^{15}N elastic scattering measured at the angle 169.1° c.m. [6].

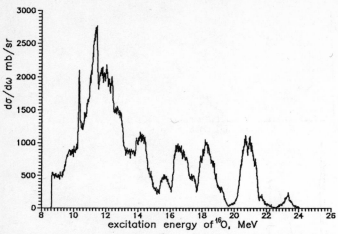

Figure 4: Excitation function of $^{12}C-\alpha$ elastic scattering at $\Theta = 180°$ (Einc ^{12}C=69 MeV)

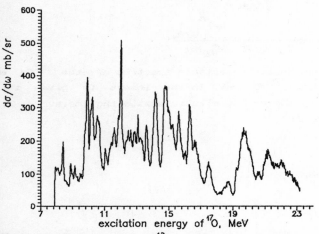

Figure 5: Excitation function of $^{13}C-\alpha$ elastic scattering at $\Theta = 180°$ (Einc ^{13}C=73 MeV)

functions is evident. It is important to note that there are no extra peaks or continious contribution in the low energy parts of the spectra. The energy resolution in the spectra is about 70keV in the c. m. s.

Several factors contribute to this value : the energy spread of incident beam, the spread of energy coused by inhomogeneity of the window, the straggling of the incident beam and α-particles in the gas and the energy resolution of the detector, the second factor being the largest.

Figs 4, 5, 6, 7, 8 and 9 show the excitation functions of the HI-α elastic scattering, which were obtained from transformation

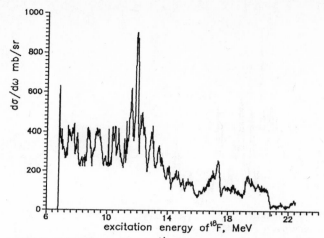

Figure 6 Excitation function of ^{14}N$-\alpha$ elastic scattering at $\Theta = 180°$ (Einc ^{14}N=82.7 MeV)

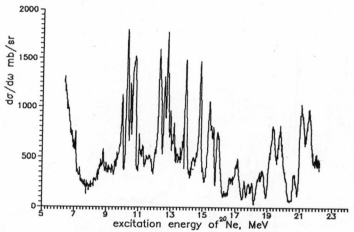

Figure 7: Excitation function of ^{16}O$-\alpha$ elastic scattering at $\Theta = 180°$ (Einc ^{16}O=89 MeV)

of the α-particle spectra. These excitation function were got in the second run and the overall energy resolution was better- about 40keV in the c.m.s. due to change of the window foil for a more thinner one (3μ thick). Most of the intensive peaks in the Figs 3-9 (except for levels in ^{20}Ne) were observed for the first time. The excitation energy calculations of the levels depend upon the specific energy losses of heavy ions and α-particles) in this method. We have used the data of Ziegler for the specific energy losses of the heavy ions. The density of the helium gas was got by means of energy losses measurements of α-particles from the Am +Cu source. Starting from this point we have got a systematic 50keV

371

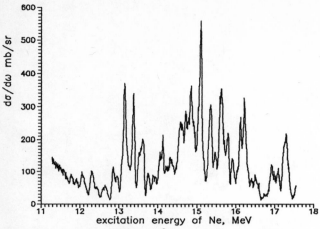

Figure 8: Excitation function of $^{18}O-\alpha$ elastic scattering at $\Theta = 180°$ (Einc ^{18}O=44.8 MeV)

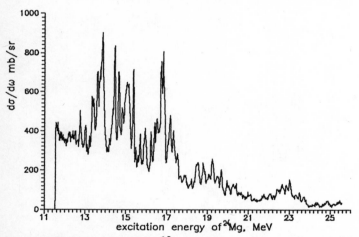

Figure 9: Excitation function of $^{20}Ne-\alpha$ elastic scattering at $\Theta = 180°$ (Einc ^{20}Ne=99 MeV)

error in energy excitation of levels in the low energy part of the spectra. We have related this error with the uncertainly of the data on the specific energy losses and compensated it by 2% increasing the measured values of the helium pressure. Having introduced this correction into the calculation, we get the correspodence between the electrostatic generator [5,6,7] and the present data within the 20 keV margins.

Fig10 presents the angular distributions for several resonances observed in ^{22}Ne nucleus. These raw angular distributions were obtained just by integration of peaks in the spectra. At each angle we took into consideration the changes in

372

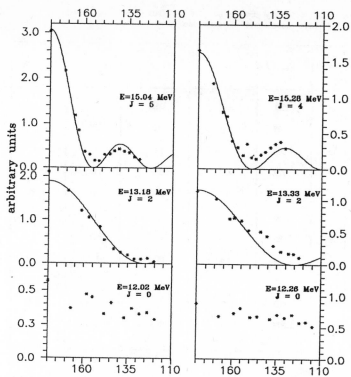

Figure 10. Angular distributions of ^{18}O−α elastic scattering for
several resonances in ^{22}Ne.

target thickness and effective solid angle. The curves in Fig10
are the squared Legendre polinomials normalised to the
experimental data. The entire process was imagined to occur as the
single resonance excitation in fusion of two spinless particles.

Table 1 presents the results of the analysis.

Comments to the Table:

1. The ratios Γ_α/Γ (and then Γ_α and Γ_n) were calculated
from comparison of the observed peak cross sections with the
maximum values given by expression 1.

2. The partial widths were not estimated for the 0^+ levels in
the present work due to large background.

3. The penetrability factors were calculated at the radius
equal to 4.85 F. The same value was used to calculate the Wigner
limit both for α-particle and neutron width.

4. The reduced widths θ_α and θ_n are given as ratios to the
Wigner limit.

Table 1: Parameters of energy levels of ^{22}Ne determined by analysis of the present experiment.

LEVEL	$Eexc$ MeV	J^{π}	Γcm keV	$\Gamma\alpha\ cm$ keV	$\Gamma n\ cm$ keV	θ_{α} %	θ_n %
1 [9]	12.04	0^+	58	29	29	21	1.7
1 [10]	12.04	0^+	65	35	30	25	1.7
1	12.02	0^+	52				
2 [9]	12.26	0^+	73	47	26	16	1.5
2 [10]	12.25	0^+	65.2	57	8.2	20	0.5
2	12.26	0^+	79				
3	13.18	2^+	90	78	12	22.7	0.14
4	13.33	2^+	50	44	6	8.5	0.05
5	15.04	5^-	50	33	17	34.5	0.6
6	15.28	4^+	66	38	28	7.7	0.45
7	17.18	5^-	125	57	68	8.2	1.4

Discussion and conclusion.

The presented excitation functions show rich structure which is evidently related with the previously unknown α-cluster states. The measurements of the excitation functions in reversed geometry must be succeeded by accurate investigations of the peak regions at accelerators with better experimental conditions. Some insight into the future results could be gained from analysis of the ^{18}O-α scattering. Fig11 presents the scheme of the α-cluster levels in ^{20}Ne and ^{22}Ne, the thresholds of the α-decay in the both nuclei being set equal.

The agreement between the same spin values in the both nuclei is perfect. Also, the results for the reduced α-widths of positive parity levels in ^{22}Ne (Table1) are in close agreement with those of the work [8].

Thus, we come to the conclusion that the effective potential of α-particle- nucleus interaction is of "threshold" character and does not depend strongly upon the 4n-structere of the nuclei. However, the experimental cross sections generally are much lower for ^{22}Ne resonances than that for ^{20}Ne. Looking through Table1,

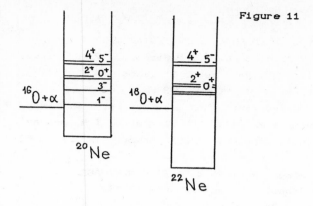

Figure 11

one can find that even small reduced widths of neutron decay
results in large partial widths. That is the evident consequence
of large penetrability factors for a neutron decay from ^{22}Ne. It
seems reasonable to suppose that the extreem cross sections found
in α-^{12}C and α-^{18}O elastic scattering are explained at least
partly by extreemly high thresholds for a nucleon decay. However,
the evident difference between the values of cross sections for
resonance population in ^{19}F (^{15}N+ α) and in ^{22}Ne (^{18}O+ α) nuclei
is probably the reflection of extra shell nucleon influence on the
formation of the α-cluster structure. Also the puzzle is the
doubling of the 0^{+}, 2^{+} (and probably 4^{+}) levels with large reduced
α-width in ^{22}Ne.

We suppose that the thorough investigation of properties of
the α-cluster levels in the neighboring s-d shell nuclei will
bring new possibilities for understanding the correspondence
between the shell and α-cluster structure and about the nature of
α-clusterization in nuclei. The proposed method seems to be very
effective. Just about ten minutes of cyclotron time are needed to
get the excitation functions, the beam current being of some nA.
So it seems the method is very suitable for investigation of the
resonance phenomena with radioactive low intensity beams. The
usage of radioactiv beams will give possibility (apart from other
applications) to study the α-cluster states in the mirror nuclei
and to use the Coulomb shift analysis for understanding the
peculiarities of the α-cluster states wave functions.

References

1. A. Arima, S. Yoshida, Nucl. Phys. A279 (1974) 475

2. K. P. Artemov, V. Z. Goldberg, I. P. Petrov, V. P. Rudakov, I. N. Sericov, V. A. Timofeev, Phys. Rev. Lett. 21 (1968) 39

3. E. B. Carter, Phys. Lett. 27B (1968) 202

4. K. P. Artemov, O. P. Belanin, A. L. Vetoshkin, R. Wolski, M. S. Golovkov, V. Z. Goldberg, M. Maddeja, V. V. Pankratov, I. N. Sericov, V. A. Timofeev, V. N. Shadrin, Y. Szmider, A. E. Pakhomov, Yad. Fiz. 52 (1990) 634

5. T. D. Marvin, P. P. Singh, Nucl. Phys. A180 (1972) 282

6. H. Smotrich, K. W. Jones, Phys. Rev 122 (1961) 232

7. S. R. Riedhauser Phys. Rev C29 (1984) 29

8. V. Z. Goldberg, V. P. Rudakov, V. A. Timofeev, Sov. J. Nucl. Phys 19 (1974) 253

9. D. Powers, J. K. Bair, J. L. C. Ford, Jr, H. B. Willard, Phys. Rev. 134 (1964) B1237

10. S. Gorodetzky, M. Port, J. Craff, J. M. Thirion, G. Chouraque Jour. Phys. 29 (1968) 271

High-Resolution Study of α+^{28}Si Elastic Scattering with a Cyclotron–Fragmented α-Cluster States?

P. Manngård[1], M. Brenner[1], K.-M. Källman[1], T. Lönnroth[1], M. Nygård[1], V.Z. Goldberg[2], M.S. Golovkov[2], V.V. Pankratov[2], and V.P. Rudakov[2]

[1]Department of Physics, Åbo Akademi, Turku, Finland
[2]I.V. Kurchatov Institute of Atomic Energy, 123182 Moscow, Russia

The α + ^{28}Si elastic scattering excitation functions at θ_{lab} = 43, 50, 82, 118, 148, 163 and 173 degrees have been measured at the cyclotron of the Åbo Akademi Accelerator Laboratory in energy steps of 40 keV. The energy resolution of the beam after the analysing magnet was better than 0.1 %. From the excitation function at 173 degrees (fig.1.) it is evident that the elastic cross-section exhibits sharp peaks with widths that seem to be less than or about 70 keV.

Angular distributions measured on and off these peaks are different in magnitude at backward angles as well as in pattern in the whole angular region. This shows that, in this energy region, fitting discrete angular distributions with a pure optical model without knowledge of the excitation functions is not a meaningful procedure. We have compared the on-peak angular distributions at backward angles with single squared Legendre polynomials, $P_\ell^2(cos\theta)$. Spins have been assigned (fig.1.) for peaks where a) the minima and b) the magnitude of successive maxima are correctly reproduced. This

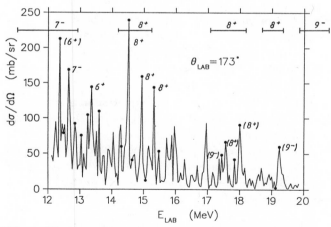

Fig. 1. The α-particle elastic scattering excitation function measured in steps of 40 keV at $\Delta\Theta_{lab}$ = 173° . The dots show the energies at which angular distributions have been measured. Spin assignments from $P_\ell^2(cos\theta)$ (numbers at each peak) and from the corresponding ($^6Li, d\alpha$) correlation experiment [1] (regions in the upper part) are indicated.

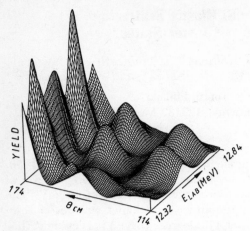

Fig. 2. Angular distributions for $E(\alpha)$ = 12.32 - 12.84 MeV measured with 40 keV steps at backward angles with $\Theta_{lab} = 3°$. The data have been smoothed and interpolated with spline functions for good graphical display.

indicates that the peaks might be due to resonance phenomena, i.e. we assume they are evidence for fragmented α-cluster states in ^{32}S.

If this is indeed the case the angular momenta ℓ should have a correspondance to the spins of the broad states obtained from the d-α correlation in the reaction ^{28}Si(^{6}Li,d)^{32}S*(α)^{28}Si performed at the Kurchatov Institute [1]. Examination of figure 1 shows that this correspondence is indeed rather good especially for the 8$^+$ assignments around 15 MeV. However, some of the peaks (numbers in parenthesis) do not fulfill both conditions a) and b), but are composed of two or more polynomials. This could be due to overlapping states or interference with a background from an underlying optical potential. In order to obtain more detailed information about this, we measured angular distributions (114° < θ_{cm} < 174°) in steps of 40 keV for $E(\alpha)$ = 12.32 - 12.84 MeV as shown in fig.2.

Note in figure 2 that there is structure both as a function of angle *and* as a function of energy. This tells us that the interaction between the α-particle and the target nucleus at these energies is different from the interaction at higher energies, where the optical model gives a good description of the scattering process. This can be seen for example in ref. 2, where the corresponding picture for about 30 MeV $\alpha+^{40}$Ca has been *calculated*. Thus our assumption receives further support.

References

1. K.P. Artemov, M.S. Golovkov, V.Z. Goldberg, V.V. Pankratov, V.P. Rudakov, M. Brenner, P. Manngård, K-M. Källman and T. Lönnroth: submitted to Yadernaya Fizika

2. F. Michel and G. Reidemeister: Phys. Rev. C **29** 1928 (1984)

Semi-microscopic $n\alpha$-Cluster Model for Light Nuclei

K. Fukatsu and K. Katō

Department of Physics, Hokkaido University, Sapporo 060, Japan

A molecular viewpoint of nuclear systems has succeeded in understanding dynamical excitations of light nuclei. The Ikeda diagram[1] schematically shows various kinds of the possible molecule-like structure as appearances of saturating sub-units in α-nuclei. In the diagram, we can see the $n\alpha$-cluster phase as the upper limit in excitation and the phase of the spatially compact shell-model-like structure near the ground state at the lower limit. Between these two phases various combinations of element clusters are expected to form molecule-like structure.

To confirm this picture of the molecule-like structure in nuclei, we investigate $n\alpha$-cluster model for ^{8}Be, ^{12}C and ^{16}O in a unified way by means of a semi-microscopic method of the orthogonality condition model(OCM)[2].

In the OCM of $n\alpha(n \geq 2)$ systems, we first prepare the Pauli-allowed states[3] and then solve the OCM equations in their basis sets of the Pauli-allowed states by using the common α-α potential given by a folding-type potential[2], which well reproduces the observed α-α phase shifts. In Ref.[4], we have discussed that it is difficult to reproduce cosistently the binding energies of ^{12}C and ^{16}O by 3α and 4α models, respectively, without introducing an additional three-alpha potential. Using the three-alpha potential such as $V_3 \exp[-\lambda\{(\boldsymbol{r}_1 - \boldsymbol{r}_2)^2 + (\boldsymbol{r}_2 - \boldsymbol{r}_3)^2 + (\boldsymbol{r}_3 - \boldsymbol{r}_1)^2\}]$, where V_3 and λ are chosen to fit the ^{12}C and ^{16}O binding energies and the ground rotational energies of ^{12}C, we solve 0^+ states of the 4α system and investigate the structure of the obtained wave functions. Their results well reproduce the closed shell structure of the ground state and the ^{12}C(0^+) + α cluster structure of the excited 0^+ state, which are very similar to results of the ^{12}C+α OCM[5].

From those results, we obtain the conclusion that the structure change from the closed-shell structure to the ^{12}C+α structure in ^{16}O is well reproduced within the semi-microscopic 4α model being a natural extension from 2α and 3α models for ^{8}Be and ^{12}C systems.

References

[1] K.Ikeda, N.Takigawa and H.Horiuchi, Prog. Theor. Phys. Suppl. Extra Number (1968), 464.

[2] S.Saito, Prog. Theor. Phys. 40(1968), 893: 41(1969), 705.

[3] K.Katō, K.Fukatsu and H.Tanaka, Prog. Theor. Phys. 80(1988), 663.

[4] K.Fukatsu, K.Katō and H.Tanaka, Prog. Theor. Phys. 81(1989), 738.

[5] Y.Suzuki, Prog. Theor. Phys. 55(1976), 1751: 56(1976), 111.

Springer Series in Nuclear and Particle Physics **Clustering Phenomena in Atoms and Nuclei**
Editors: M. Brenner · T. Lönnroth · F.B. Malik   © Springer-Verlag Berlin, Heidelberg 1992

Large Clusters in Light Nuclei

R. Zurmühle

Department of Physics, University of Pennsylvania,
Philadelphia, PA 19104, USA

The current status of our knowledge about cluster states in ^{48}Cr is reviewed and a summary of several theoretical descriptions of highly deformed states in this mass region is given. The merits of transfer reactions in studies of molecular states are discussed and results from recent investigations of ^{12}C$-^{12}$C molecular states in ^{24}Mg with transfer reactions are presented.

Molecular resonances in the mass 48-56 region

The most compelling evidence for molecular structure in light nuclei comes from the study of systems that consist of identical alpha particle nuclei. The two heaviest symmetric systems that exhibit such structure are ^{24}Mg + ^{24}Mg[1]) and ^{28}Si + ^{28}Si.[2]) Angle averaged cross sections for ^{24}Mg + ^{24}Mg elastic and inelastic scattering as a function of c.m. energy are shown in Fig. 1. Perhaps the most intriguing aspect of these excitation functions is the very high correlation between resonances seen in the

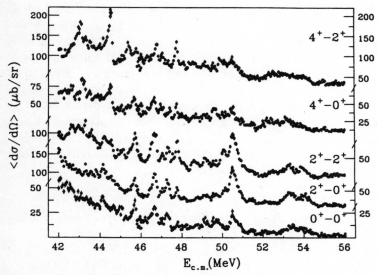

Fig. 1. Excitation-function data for ^{24}Mg + ^{24}Mg scattering averaged over $66° \leq \theta_{c.m.} \leq 93°$.

Springer Series in Nuclear and Particle Physics   **Clustering Phenomena in Atoms and Nuclei**
Editors: M. Brenner · T. Lönnroth · F.B. Malik   © Springer-Verlag Berlin, Heidelberg 1992

three lowest excitations. Some of these resonances also appear at the higher excitations but the higher excitations are on the whole dominated by different structures, structures that are only weakly excited in the elastic channel and the lowest inelastic channels.

For a better understanding of the nature of these resonances it is important to know their detailed properties. Most important is the determination of their spins. The elastic scattering angular distributions can provide some information about the spins. However, the resonances are superimposed on a background that slowly changes with beam energy and that constitutes about 50% of the cross section, even at the peaks of the resonances. Therefore, it is not possible to obtain unambiguous spin assignments from the elastic channel. More definitive information can, however, be gained from particle-gamma correlation measurements for the lowest inelastic excitation.[3] Such measurements have been carried out with the spin spectrometer at Oak Ridge National Laboratory.[4] This device consists of 70 large NaI (Tl) crystals that cover the full sphere and serve as gamma-ray detectors. It can be used to record the complete gamma-ray angular distribution pattern in inelastic transitions such as

$$^{24}\text{Mg} + {}^{24}\text{Mg} \rightarrow {}^{24}\text{Mg}(\text{g.s.}) + {}^{24}\text{Mg}(2^+).$$

The contributions to this quadrupole radiation pattern are, of course, quite different for $|m| = 0,1$, and 2. Consequently one is able to separate the inelastic cross section into the individual contributions from these three spin projections for any chosen quantization direction. Fig. 2 shows the separation for quantization along the beam direction, and one can see that the 45.70 MeV resonance contributes predominantly to the $m = 0$ component of the cross section. This is a consequence of the strong alignment of the spin vector $I = 2$ with the orbital angular momentum l of the relative motion of the fission fragments. This strong alignment suggests that the resonance spin is high and that it can only barely be supported by the system. The energy independent background under the resonance consists mainly of $|m| = 1,2$ components and the correlation measurement is therefore able to separate it from the resonance cross section.

Figure 3 shows angular distributions for the fission decay at the resonance energy for the total cross section and for the $m = 0$ substate. The strong oscillations in the angular distribution for the $m = 0$ substate indicate that the orbital angular momentum is dominated by a single l-value. The curves shown in Fig. 3a are calculated from the expression

$$\sigma_m(\theta_{\text{c.m.}}) = |\sum_{l} A_l(l2-mm \,|\, J0)Y_l^{-m}(\theta_{\text{c.m.}})|^2$$

with $m = 0$. All three possible l-values J - 2, J, and J + 2 are allowed to contribute. The relative amplitudes can be determined from the ratios of the cross sections to the three magnetic substates. The solid and the dashed curves are parameter free calculations for $J = 36$ and 34 respectively. Only the curve for $J = 36$ fits the data and

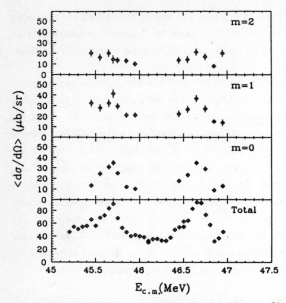

Fig. 2. Magnetic substate excitation functions for $^{24}\text{Mg}(^{24}\text{Mg},^{24}\text{Mg})^{24}\text{Mg}(2^+)$ averaged over $66° \leq \theta_{\text{c.m.}} \leq 93°$.

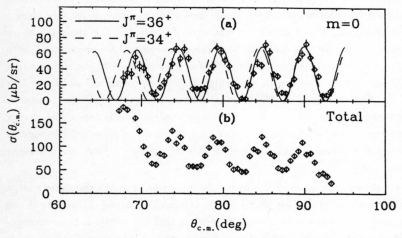

Fig. 3. Angular distributions for $^{24}\text{Mg}(^{24}\text{Mg},^{24}\text{Mg})^{24}\text{Mg}(2^+)$ at $E_{\text{c.m.}} = 45.70$ MeV.

therefore one can assign $J = 36$ to the 45.70 MeV resonance. A similar measurement for the 46.70 MeV resonance leads also to a $J = 36$ assignment. These spin assignments are indicated in Fig. 4a. Also shown in this figure are some additional spin assignments which are tentative, because they are based on a comparison of the angular distributions of elastic scattering between the 45.70 MeV resonance and the other resonances.

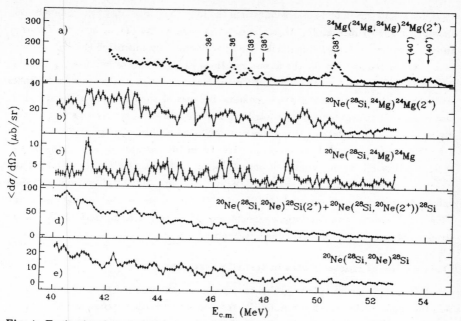

Fig. 4. Excitation functions for several entrance and exit channels in the ^{48}Cr system. The data for the ^{20}Ne + ^{28}Si entrance channel are averaged over $60° \leq \theta_{c.m.} \leq 110°$ and shifted to align all reactions with respect to excitation energy in ^{48}Cr.

From the resonance cross sections for the elastic and inelastic channels one can derive partial decay widths. It turns out that the sum of all observed partial widths is only a small fraction of the measured widths of the resonances. This indicates that the bulk of the decay strength of these resonances is still unaccounted for. Some of the most likely channels to carry a portion of the unobserved strength are those of the ^{20}Ne + ^{28}Si partition. Therefore we have recently undertaken a careful study of the ^{20}Ne + ^{28}Si channel. For this purpose we have built a differentially pumped gas target and investigated the reactions ^{20}Ne(^{28}Si,^{28}Si*)^{20}Ne* and ^{20}Ne(^{28}Si,^{24}Mg*)^{24}Mg*. The results for several low lying excitations are shown in Fig. 4b-e.[5] Much to our surprise we found that the cross sections for the elastic and inelastic channels have very little structure and a rather smooth energy dependence. The transfer channels leading to final states in ^{24}Mg exhibit considerable structure which is, however, not strongly correlated with the resonances in ^{24}Mg(^{24}Mg,^{24}Mg)^{24}Mg* shown in Fig. 4a. The energy region where there is a hint of a correlation is one that was studied by Saini et al.[6] several years ago between 45 and 48 MeV. Saini et al. investigated the inverse reaction ^{24}Mg(^{24}Mg,^{20}Ne*)^{28}Si* and therefore only the ground state transition can be directly compared. These transitions are very weak, and while the two measurements are not in disagreement our much larger sample presents a less convincing case for the presence of correlations than the earlier data. There is an indication of some strength for the

383

45.70 MeV and the 46.65 MeV resonances in the low ^{20}Ne(^{28}Si,^{24}Mg)^{24}Mg channels. An example is shown in Fig. 4b. Other resonances such as the one at 50.5 MeV do not appear at all in the transfer channels. Angular momentum mismatch can be ruled out as a possible explanation for the absence of these decay branches because the Q-value for the transfer is very small. The most plausible explanation is that these resonances strongly favor the symmetric fission channel because of their internal structure. It is also possible that negative parity contributions from odd l values play a part in obscuring some of the correlations that otherwise might be visible. The reactions shown in Fig. 4b, d, and e do not have identical particles in the entrance or exit channels and therefore they allow such contributions. It is, for instance, possible that some of the strength present around $E_{c.m.}$ = 49 MeV in ^{20}Ne(^{28}Si,^{24}Mg)^{24}Mg* is from an odd l value. In summary, the partial widths of ^{24}Mg$-^{24}$Mg molecular resonances in the ^{20}Ne + ^{28}Si elastic and inelastic channels are small and a large part of the decay strength of these resonances remains unaccounted for.

Observation of cluster states in transfer reactions

For many reasons it is desirable to supplement the method of measuring excitation functions in the search for molecular resonances with other techniques. For example, in the ^{24}Mg + ^{24}Mg systems it is quite likely that very narrow resonances with spins J < 36 are present below the 45.70 MeV resonance in ^{48}Cr; because of their narrow widths such resonances are extremely hard to find. The discovery of such states could substantially increase our understanding of highly deformed states in this system. If they are sufficiently narrow one might even contemplate a search for weak gamma decay branches. Such an experiment could provide nearly model independent information on their quadrupole deformation. Another problem with elastic and inelastic scattering measurements lies in the strong contributions from direct processes. At best these contributions interfere with the resonance mechanism and cast doubt on the spin assignments that are derived from angular distribution measurements. The situation is even worse when the structures are wider than those seen in ^{48}Cr, which is the case in some of the lighter systems such as ^{12}C + ^{12}C.[7] Under these circumstances it becomes questionable whether a given structure is the signature of a resonance in the composite system or simply a modulation in the direct cross section. In the ^{12}C + ^{12}C system reaction models have been quite successful in describing the energy dependence and other properties of the elastic and inelastic scattering cross section.[8] The excitation of molecular states in a transfer reaction should be free of such background.

For these reasons frequent attempts have been made to populate molecular resonances in cluster transfer reactions. The subsequent fission of the cluster state then results in a three body final state. Several years ago we undertook such a search for narrow ^{24}Mg$-^{24}$Mg cluster states in ^{48}Cr.[9] We used the reaction ^{24}Mg(^{28}Si,α^{24}Mg)^{24}Mg at a beam energy of 150 MeV. We observed events of the kind ^{24}Mg + ^{24}Mg + α but they were dominated by fission into ^{24}Mg + ^{28}Si* with subse-

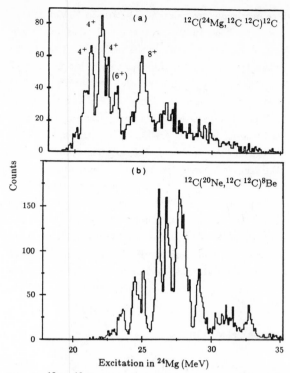

Fig. 5. ^{12}C$-^{12}$C molecular states observed in transfer reactions.

quent alpha decay of the ^{28}Si*. Similar efforts by several groups to identify molecular structure in lighter systems have also failed in the past. A well known example is the reaction ^{12}C(^{16}O,α^{12}C)^{12}C.[10] It appears that this reaction is dominated by the ^{12}C$-\alpha$ final state interaction rather than the ^{12}C$-^{12}$C final state interaction. Recently two new experiments have been reported[11,12] that use different reactions and they have produced remarkable results. The prescription for success in these recent measurements lies in the avoidance of final states which include an alpha particle. Why? Because many excited clusters produced in the fission process are unbound and alpha decay, and this process tends to obscure all others, including the transfer reaction. If one selects a reaction that emits an ^{8}Be or ^{12}C nucleus during the formation of the cluster state, this background poses no problem.

The results of these experiments are summarized in Fig. 5. The ^{12}C(^{24}Mg,^{12}C^{12}C)^{12}C reaction was investigated by Freer et al.[11] with a 170 MeV ^{24}Mg beam at Daresbury Laboratory while the measurements using the ^{12}C(^{20}Ne,^{12}C^{12}C)^{8}Be reaction were carried out by Fulton et al.[12] with a 360 MeV ^{20}Ne beam at Lawrence Berkeley Laboratory. The two experiments used identical detection techniques. Coincidences between pairs of fragments were detected with charged particle detector telescopes that consisted of three surface barrier detectors each. The 10mm × 10mm telescopes covered large solid angles and were position sensitive in two dimensions. Such coincidence measurements completely determine the reaction kinematics. They allow the determination of the total reaction Q value and of the c.m. energy for any two of the three reaction fragments. The spectra in Fig. 5 only include events with all three nuclei in their ground state. They are plotted against the c.m. energy of two ^{12}C nuclei expressed in terms of excitation in ^{24}Mg. The presence of narrow peaks in these spectra clearly proves that the three body decay proceeds predominantly through excited ^{24}Mg configurations and it therefore justifies this choice of abscissa. The spins of these states can be determined from the angular correlations if the detectors cover a sufficient angle range. Spin assignments based on a method developed by Marsh and Rae[13] are given for the dominant groups in Fig. 5a. These spins are near the grazing values, which is consistent with a molecular picture that implies a very shallow potential well. Similar assignments should eventually also be possible for the other reaction study. Results obtained from these new data and from similar future experiments for other systems hold great promise to substantially add to our knowledge and understanding of molecular structure.

Theoretical descriptions of molecular states in ^{48}Cr

Theoretical descriptions based on several different models have recently been made for strongly deformed high spin states in ^{48}Cr and for some of the neighboring alpha particle nuclei. The different calculations complement each other in many respects and together they provide us with a great deal of insight into the structure of the observed resonances. A summary of some of the major successes of these calculations is presented below.

Many-particle-many-hole states in the Hartree Fock method

Many-particle-many-hole calculations using the Hartree Fock approach have recently been carried out by Zheng, Zamick, and Berdichevsky[14] for alpha particle nuclei between ^{40}Ca and ^{56}Ni. Of particular interest to us are their results for ^{48}Cr and ^{56}Ni. In ^{48}Cr they found a 16p-8h band ($\beta = 0.847$) involving the orbits $\Omega nlj\pi$ $\left(\frac{3}{2}0d\frac{3}{2}+,\frac{1}{2}0d\frac{3}{2}+\right)^{-8}\left(\frac{1}{2}0f\frac{7}{2}-,\frac{3}{2}0f\frac{7}{2}-,\frac{1}{2}0g\frac{9}{2}+,\frac{1}{2}1p\frac{3}{2}-\right)^{16}$ that becomes yrast at high spin. Similarly they found a highly deformed 16p-16h band ($\beta = 1.053$) in ^{56}Ni. A comparison between their results and the observed resonance structure in these two nuclei is shown in Table 1.

In ^{56}Ni the agreement between the calculated and the observed energies is excellent. However, in ^{48}Cr the predicted energies are considerably higher than the observed resonances. Moreover the spacing between adjacent spins is overpredicted. This indicates

Table 1: Level energies from Hartree Fock calculations.

J^π	^{24}Mg + ^{24}Mg			^{28}Si + ^{28}Si		
	$E_{c.m.}$ (MeV)	E_x (MeV)	H.F. calc.[14] (MeV)	$E_{c.m.}$ (MeV)	E_x (MeV)	H.F. calc.[14] (MeV)
36^+	45.70	60.66	68.6	53	64	63.5
	46.65	61.50				
	47.25	62.20				
	47.75	62.70				
38^+	50.50	65.45				
40^+	53.3	68.3	80.05	59	70	70.9
	54.0	69.0				

that the observed resonances in ^{48}Cr have a larger deformation than the model states and therefore they should probably not be identified with these configurations. A direct comparison of the energies of the observed structures of a given spin value in ^{48}Cr and ^{56}Ni is also quite instructive. For instance, the lowest 36^+ resonance occurs at $E_{c.m.} = 45.70$ MeV in the ^{24}Mg + ^{24}Mg system, but in the more massive ^{28}Si + ^{28}Si system it is located at a much higher energy, around 53 MeV. The spin values for the ^{56}Ni resonances are of course less certain than those in ^{48}Cr, because they have been deduced from elastic scattering angular distributions. Our experience in ^{48}Cr suggests that they are probably within 2 ℏ of the assumed values. Therefore this trend must be real and we conclude that the observed resonances in ^{48}Cr have even larger deformations than those found for the 16p-16h states in ^{56}Ni, which reproduce those resonances so well. The fact that the Hartree Fock calculations do not generate any $J = 36$ state below the lowest observed $J = 36$ resonance makes it all the more likely that this is an yrast configuration.

Two-center shell model

Calculations based on the Nilsson model and the two-center shell model, both using the Strutinsky prescription, have also been carried out. In view of the considerable fission branch of the high spin resonances, the two-center shell model calculations might be the more suitable approach. Results of such calculations performed by Maass and Scheid[15] are shown in Fig. 6. The contours of the potential energy surface $V(r,\epsilon) + l(l+1)\dfrac{\hbar^2}{2\mu r^2}$ are plotted for the ^{24}Mg + ^{24}Mg system as a function of the separation r between the centers of mass of the two nuclei and of the neck parameter ϵ which controls the barrier height between the two oscillator potentials. The potential energies are normalized to vanish for large separation. The model predicts a shallow potential energy minimum that agrees well with our resonance energies. For $L = 36$ the minimum is at 42 MeV, in good agreement with $E_{c.m.} = 45.75$ MeV observed for our lowest $J = 36^+$ resonance. The calculated shape of the density distribution at the minimum is shown in Fig. 7 for $L = 36$.

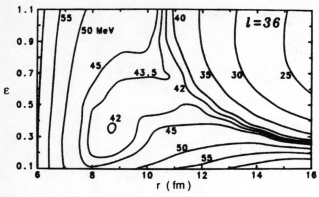

Fig. 6. Potential energy surface for ^{24}Mg + ^{24}Mg from the two-center shell model.

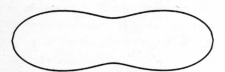

Fig. 7. Shape for ^{24}Mg + ^{24}Mg at the potential energy minimum.

Molecular model calculations

A macroscopic model of two interacting ^{24}Mg nuclei that describes resonances in ^{24}Mg + ^{24}Mg scattering has been introduced by Uegaki and Abe.[16] The coordinates that describe the relative motion between the two nuclei are shown in Fig. 8. The interaction potential is derived from a double folding calculation that also includes a repulsive term which is switched on where the density exceeds twice the normal density. Fig. 9 shows a contour plot of the potential energy for J = 36. The quantity $V(R, \beta, \alpha = 0) + J(J+1)\dfrac{\hbar^2}{2\mu R^2}$ is plotted as a function of the separation distance R and the Euler angles $\beta = \beta_1 = \beta_2$. One important result that emerges from this calculation is the existence of a potential energy minimum for the pole to pole configuration with $\beta_1 = \beta_2 = 0$. This configuration has the largest possible moment of inertia for two touching ^{24}Mg nuclei. The association of the observed resonances with this configuration agrees with the fact that the measured spins are about 4 ħ higher than grazing l-values determined from an optical potential. Excitations of this molecule above the pole to pole configuration can take place in several modes involving one or several of the collective coordinates. Therefore this model can generate a rich spectrum of levels and it can easily explain the observed fragmentation of the J = 36 resonances.

Maass and Scheid[17] have recently published a similar calculation. They take the result from the two-center shell model calculation to fix the energy of the pole to pole

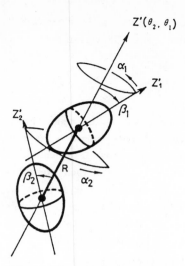

Fig. 8. The coordinates in the rotating molecular frame.

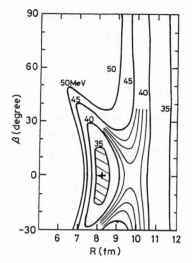

Fig. 9. Potential energy surface for ^{24}Mg + ^{24}Mg with J = 36.

"ground state" energy and use the molecular model only to determine the energy added by the various excitations. They have also calculated partial widths and reduced widths for the fission to low lying states of ^{24}Mg. A comparison between the lowest observed J = 36 resonance and the model "ground state" is shown in Table II.

Table II. Resonance Parameters

Experiment					Theory			
$E_{c.m.}$	Γ_{Total}	Final	l value	Γ	$E_{c.m.}$	assumed	Γ	$\dfrac{\gamma^2}{\gamma^2_{S.P.}}$
(MeV)	(keV)	states		(keV)	(MeV)	l value	(keV)	
45.70	170	g.s.	36	1.8	48.55	36	1.3	.004
		$2^+,0^+$	34	6.1		34	7.3	.014
		$2^+,2^+$	34	7.1		32	11.7	.015

The agreement between data and calculation is very satisfactory. Notice, however, that the calculation for the mutual excitation assumes the dominance of the lowest allowed l-value while the angular correlation measurements suggest the dominance of l = 34. Besides other factors, the calculated reduced widths also take account of the change in deformation of the ^{24}Mg fragments during the fission process. In this respect the ^{24}Mg + ^{24}Mg system is very much favored over other systems because the shape of the mass distribution in the two center shell model differs only slightly from that of two grazing ^{24}Mg nuclei in pole to pole alignment. It would be very interesting to have similar calculations available for other systems such as ^{28}Si + ^{28}Si and ^{20}Ne + ^{20}Ne.

We are particularly interested in the ^{20}Ne + ^{20}Ne system and we plan to investigate it as soon as the Argonne ECR source becomes operational and ATLAS can produce a ^{20}Ne beam.

The alpha cluster model

Finally, I would like to show some recent results based on the alpha cluster model.[18] Some time ago, Marsh and Rae[19] performed extensive calculations based on that model for ^{24}Mg, and they found several stable configurations. In a qualitative comparison, the agreement with known rotational bands in terms of their excitation energy, deformation, and parentage was quite good. Rae has now performed similar calculations for heavier systems including ^{48}Cr. Some very recent results for this nucleus are shown in Fig. 10. The model predicts a highly deformed rotational band with an axis ratio around 3:1 that can be followed up to spin 56 ħ and an excitation energy of nearly 100 MeV. The energies of the rotational states are shown in Fig. 10 along with the cluster configurations. The configurations look very much like a chain of three ^{16}O nuclei and therefore one would expect a considerable ternary fission branch into ^{16}O + ^{16}O + ^{16}O. The alpha cluster configuration does have a finite overlap with ^{24}Mg + ^{24}Mg and that makes it possible to identify the resonances with this configuration. This is a very intriguing thought, particularly in view of the fact that at least 70% of the decay strength of the resonances in Fig. 1 is still unaccounted for.

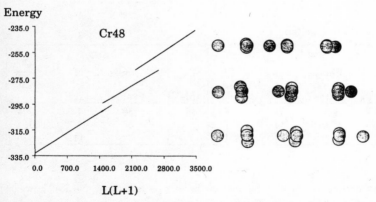

Fig. 10. Angular momentum dependence of the binding energy of highly deformed α cluster states in ^{48}Cr. The configurations are shown to the right.

Acknowledgements

I wish to thank B.R. Fulton and his collaborators for allowing me to present some of their recent unpublished results. I am also grateful to W.D.M. Rae for sharing the results of his recent calculations with us. This work was supported by a grant from the National Science Foundation.

References

1. R.W. Zurmühle, P. Kutt, R.R. Betts, S. Saini, F. Haas, and O. Hansen, Phys. Letters **126B** (1983) 384.

2. R.R. Betts, B.B. Back, and B.G. Glagola, Phys. Rev. Letters **47** (1981) 23.

3. A.H. Wuosmaa, R.W. Zurmühle, P.H. Kutt, S.F. Pate, S. Saini, M.L. Halbert, and D.C. Hensley, Phys. Rev. Letters **58** (1987) 1312, and Phys. Rev. C. **41** (1990) 2666.

4. M. Jääskeläinen, D.G. Sarantites, R. Woodward, F.A. Dilmanian, J.T. Hood, R. Jääskeläinen, D.C. Hensley, M.L. Halbert, and J.H. Barker, Nucl. Instr. and Methods **204** (1983) 385.

5. S. Barrow, D. Benton, Q. Li, Z. Liu, Y. Miao, R. Zurmühle, to be published.

6. S. Saini, R.R. Betts, R.W. Zurmühle, P.H. Kutt, and B.K. Dichter, Phys. Letters **185B** (1987) 316.

7. T.M. Cormier, C.M. Jachinski, G.M. Berkowitz, P. Braun-Munzinger, P.M. Cormier, M. Gai, J.W. Harris, J. Barette. and H.E. Wegner, Phys. Rev. Letters **40** (1978) 924.

8. See e.g. O. Tanimura, R. Wolf, and U. Mosel, Phys. Letters **120B** (1983) 275.

9. A.H. Wuosmaa, S.F. Pate, and R.W. Zurmühle, Phys. Rev. C **40** (1989) 173.

10. See e.g. W.D.M. Rae, R.G. Stokstad, B.G. Harvey, A. Dacal, R. Legrain, J. Mahoney, M.J. Murphy, and T.J.M. Symons, Phys. Rev. Letters **45** (1980) 884.

11. M. Freer, S.J. Bennett, B.R. Fulton, J.Y. Murgatroyd, G.J. Gyapong, N.S. Jarvis, C. Jones, D.L. Watson, J.D. Brown, R. Hunt, W.D.M. Rae, A.E. Smith, and J.S. Lilley, in *Nuclear Structure and Heavy-Ion Reaction Dynamics 1990*, ed. by R.R. Betts and J.J. Kolata (Institute of Physics Conference Series, 1991).

12. B.R. Fulton, S.J. Bennett, J.T. Murgatroyd, N.S. Jarvis, D.L. Watson, W.D.M. Rae, Y. Chan, D. DiGregorio, J. Scarpaci, J. Suro Perez, and R.G. Stokstad, to be published.

13. S. Marsh and W.D.M. Rae, Phys. Letters **153B** (1985) 21.

14. D.C. Zheng, L. Zamick, and D. Berdichevsky, Phys. Rev. C. **42** (1990) 1004.

15. R. Maass and W. Scheid, Phys. Letters **202B** (1988) 26.

16. E. Uegaki and Y. Abe, Phys. Letters **231B** (1989) 28.

17. R. Maass and W. Scheid, J. Phys. G: Nucl. Part. Phys. **16** (1990) 1359.

18. D.M. Brink: Int. School of Physics "Enrico Fermi" **XXXVII** (1966) 185.

19. S. Marsh and W.D.M. Rae, Phys. Letters **180B** (1986) 185.

Molecular Resonances in ^{24}Mg
Revealed in the ^{12}C(^{20}Ne,^{12}C^{12}C)^{8}Be Reaction

B.R. Fulton[1], S.J. Bennett[1], J.T. Murgatroyd[1], N.S. Jarvis[2],
D.L. Watson[2], W.D.M. Rae[3], Y. Chan[4], D. DiGregorio[4], J. Scarpaci[4],
J. Suro Perez[4], and R.G. Stokstad[4]

[1]Physics Department, University of Birmingham, Birmingham B15 2TT, UK
[2]Physics Department, University of York, York YO1 5DD, UK
[3]Nuclear Physics Laboratory, University of Oxford, Oxford OX1 3RH, UK
[4]Building 88, Lawrence Berkeley Laboratory, Berkeley, CA 94720, USA

Since the initial discovery of the ^{12}C+^{12}C scattering resonances by Almquist et al. [1] there has been much work directed towards an understanding of this phenomenon. One recurring theme in these studies has been the extent to which the scattering resonances are linked to states in the ^{24}Mg nucleus and if so what structural properties are associated with the states. This question is difficult to address in the scattering studies since the resonant structures usually appear on top of a large underlying continuum yield from other processes, making analysis and interpretation difficult.

In an effort to circumvent this difficulty there has long been a search for an alternative way of populating the resonances which would avoid the background contribution. Transfer reactions are by their nature much more selective, and indeed this selectivity can often be of use in indicating the structure of the states excited. There was thus considerable excitement when Lazzarini et al. [2] and Nagatani et al. [3] published results which suggested that the ^{12}C+^{12}C resonances could be populated via the ^{12}C(^{16}O,α)^{24}Mg reaction. These experiments resulted in a considerable amount of activity in the following years, as other authors sought to confirm the process and use it to probe the structure of the resonances. However, the results of these experiments were disappointing in that they revealed the peaks in the alpha energy spectrum arose from sequential decay of the ^{16}O projectiles excited in the collision. There was little or no evidence for ^{12}C+^{12}C breakup from the resonant states.

The apparent failure of the (^{16}O,α) reaction to provide an easy route to populate the ^{24}Mg resonances was a disappointment, and the goal of finding a reaction route to populate the ^{24}Mg resonances still exists. To this end we have looked at the ^{12}C(^{20}Ne,^{12}C^{12}C)^{8}Be breakup reaction. In this contribution we present evidence which shows that sequential breakup of ^{24}Mg to ^{12}C+^{12}C occurs following alpha pickup onto the ^{20}Ne projectile, and that this decay occurs from discrete states which appear to overlap the ^{12}C+^{12}C scattering resonances.

The experiment was performed using a ^{20}Ne beam from the 88-Inch cyclotron at the Lawerence Berkeley Laboratory. The setup employed six telescopes mounted three on each side of the beam axis. Each telescope consisted of three silicon surface barrier detectors. The DE and E detectors were position sensitive with their position axes crossed to give both X and Y readout. Hence the in-plane and out-of-plane angle of each fragment could be determined. Data was obtained for 5 angle settings of the telescopes.

Figure 1 shows the E_{tot} spectrum for coincident carbon nuclei in one pair of telescopes at one of the angle settings. E_{tot} is the total energy in the exit channel (the summed energy of the two detected fragments plus that of the unobserved recoiling particle calculated from the missing momentum) and hence is a measure of the Q value of the reaction. The peak labelled Q_{ggg} is at the expected energy for the pickup of an alpha particle onto the ^{20}Ne projectile, followed by the breakup of the excited ^{24}Mg ejectile to two ^{12}C nuclei both in their ground state. Also visible are peaks when one or both of the ^{12}C nuclei are left in the excited 2_1^+ state at 4.44 MeV. Gating on the Q_{ggg} events we can calculate the relative energy (E_{rel}) of the two fragments, which relates directly to the excitation energy of the ^{24}Mg nucleus before it breaks up. This spectrum (figure 2) shows a sequence of narrow states in the ^{24}Mg nucleus. The extent of this structure is striking. It appears that there are narrow states in ^{24}Mg up to an excitation of 33 MeV which have a ^{12}C+^{12}C cluster structure.

Springer Series in Nuclear and Particle Physics **Clustering Phenomena in Atoms and Nuclei**
Editors: M. Brenner · T. Lönnroth · F.B. Malik © Springer-Verlag Berlin, Heidelberg 1992

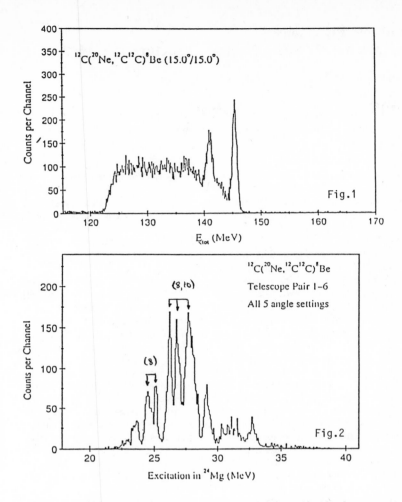

Fig.1

Fig.2

The new data suggests the possibility that the $^{12}C(^{20}Ne,^{12}C^{12}C)^8Be$ reaction proceeds via states in ^{24}Mg which correspond to the $^{12}C+^{12}C$ scattering resonances. If so this would provide a new insight into the nature of these states. For example, although the resonances have been extensively studied in the $^{12}C(^{12}C,\alpha)^{20}Ne$ reaction (the two-body analogue to the breakup reaction), the resonances always appear on top of an enormous continuum which makes interpretation very difficult. The breakup data on the other hand shows peaks with no underlying background and provides unambiguous evidence that the $^{12}C+^{12}C$ yield is from specific states in ^{24}Mg. The large number of angle settings used in the experiment means we have data over a large angular range and it is possible to extract spin assignments from the angular correlation information. The bracketed assignments in figure 2 are a result of a preliminary analysis and should be strengthened when we have wider angular coverage from analysis of the data from the other detector pairs. This information will hopefully lead to new insights into the origin of this large cluster structure and any relationship to the scattering resonances.

(1)   E Almqvist et al.; Phys. Rev. Lett. 4 (1960) 515
(2)   A J Lazzarini et al.; Phys. Rev. Lett. 40 (1978) 1426
(3)   K Nagatani et al.; Phys. Rev. Lett. 43 (1979) 1480

Detection of ^{12}C Fragments in the First Excited 0^+ State–Nuclear Structure in ^{24}Mg?

A.H. Wuosmaa[1], R.R. Betts[1], S. Barrow[2], D. Benton[2], P. Wilt[1], and R.W. Zurmühle[2]

[1]Physics Division, Argonne National Laboratory, Argonne, IL 60439, USA
[2]Department of Physics, University of Pennsylvania, Philadelphia, PA 19104, USA

The nucleus ^{24}Mg has long been a testing ground for many nuclear models. Descriptions in terms of the shell model, the cranked deformed shell model, and the α-cluster model nicely reproduce the low-lying rotational behavior of ^{24}Mg[1-3]. One of the most fascinating predictions of nearly all cluster models, as well as for cranked Nilsson-Strutinsky calculations for ^{24}Mg, is the existence of an exotic structure corresponding to a linear chain of 6 α particles. In ^{24}Mg, this 6α particle chain would be expected to lie well above the threshold for decay into 6 α particles. The identification of this decay mode thus presents an intriguing and difficult experimental challenge.

One decay channel of particular interest is ^{24}Mg$\rightarrow^{12}$C$(0_2^+)+^{12}$C(0_2^+). The first excited 0^+ state of ^{12}C at 7.65 MeV is known to consist primarily of a linear 3-α configuration[4]. A linear 6-α chain structure in ^{24}Mg might then be expected to have a large structural overlap with the exit channel consisting of two ^{12}C nuclei in their first 0^+ excited state. The ^{12}C(0_2^+) level lies, however, 380 keV above the threshold for decay into 3 α particles. In order to study this process in detail, one must therefore contend with the problem of obtaining detailed spectroscopic information about a six-body final state.

We have studied this reaction using an experimental setup consisting of a recoil - coincidence arrangement between a segmented X-Y position-sensitive double-sided silicon strip detector (DSSD), and a gas E-ΔE particle-identification telescope. Beams of ^{12}C from the University of Pennsylvania Tandem Accelerator bombarded ^{12}C targets at two energies, 58 and 70 MeV. The three α particles from one decaying ^{12}C nucleus were detected in the DSSD. To identify the reaction channel, coincident α particles from the other ^{12}C fragment were detected and identified in the gas telescope. A measurement of the angles and energies of the three α particles detected in the DSSD allows for the reconstruction of the excitation energy, scattering angle and kinetic energy of the decaying ^{12}C nucleus. From this information, the two-body scattering Q value can be determined.

Figure 1 shows the Q-value spectrum for the reaction ^{12}C$+^{12}$C$\rightarrow^{12}$C$(3\alpha)+^{12}$C(3α) at a laboratory energy of 58 MeV. The two mutual excitations 0_2^+-0_2^+ and 0_2^+-3^- are clearly observed. The advent of highly segmented, high-resolution detector systems such as the DSSD make it possible for the first time to study these complicated final states, and present the opportunity to examine possible exotic cluster structures such as the chain configuration in ^{24}Mg.

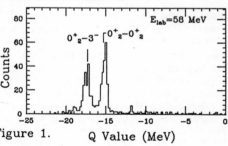

Figure 1.   Q Value (MeV)

Work supported by the U. S. Department of Energy, Nuclear Physics Division, under contract W-31-109-Eng-38 (ANL) and the U. S. National Science Foundation (U. of Pennsylvania)

[1] M. Carchidi and B. H. Wildenthal, Phys. Rev. **C37** (1988) 1681.
[2] R. K. Sheline, et al., J. Phys. **G14** (1988) 1201.
[3] S. Marsh and W. D. M. Rae, Phys. Lett. **180B** (1986) 185.
[4] N. Takigawa and A. Arima, Nucl. Phys. **A168** (1971) 593.

The Properties of ^{28}Si Resonances at the 31.5 MeV and 33.2 MeV Excitation Energies via the ^{24}Mg(α,^{12}C)^{16}O and ^{24}Mg(α, α)^{24}Mg Reactions

R. Wolski, I. Skwirczynska, A. Budzanowski, L. Freindl, J. Jakiel, W. Karcz, and J. Szmider

Institute of Nuclear Physics, ul. Radzikowskiego 152, PL-31342 Krakow, Poland

Our earlier study of the ^{28}Si compound system by means of (α, ^{12}C) reaction in the α-particle energy range from 24.9 MeV to 27.76 MeV[1], revealed the existence of a narrow resonance at 33.2 MeV exc. energy (27.1 MeV in the lab. sys.), not reported earlier, and a structure at 31.5 MeV (25.1 MeV lab). The latter could be identify as known 14.7 MeV CM energy resonance in the ^{12}C $+^{16}$O reaction[2].

In order to obtain reduced widths of these resonances, complementary measurements of (α,α) elastic scattering on nat.Mg target were performed, e.g.the excitation functions at the angle of 178° and the angular distributions at energies close to and off the proposed resonance energies. The 11^{-}, 27.1 MeV resonance in (α,^{12}C) channel corresponds to a maximum in the excitation function of 178° (α,α) scattering, whereas the 25.1 MeV structure (proposed 9^{-} resonance) is seen as a sharp mininum in the (α,α) channel. In futher analysis of elastic scattering data we have treated both resonances on equal footing. It was assumed, that the amplitude of a single isolated resonance interferes with background amplitude smoothly varying with energy[2],[3]. The background phase shifts were generated by optical model fits to the sets of measured angular distributions for each resonance. The phase shift for resonant l=9 or l=11 wave was modified by inclusion the resonant term of the scattering amplitude[3] in order to reproduce the excitation function of elactic scattering at 178° in lab. energy regions 24.8-25.6 MeV and 26.5-28.0 MeV respectively, where l=9 and l=11 resonances were observed.. The resonance parameters : ratio of α-particle partial width to the total width Γ_{α}/Γ , and the relative resonance-background mixing phase β, were obtained from these procedures under constraint that unitarity has to be conserved. We failed to fully reproduce both , the set of the angular distributions and the excitation function for given resonance, however it apears that , α - partial width is weakly dependent on background amplitude parameters. The partial width for $^{12}C+^{16}O$ channel Γ_c was estimated from the excess over the background of the total $(\alpha, ^{12}C)$ reaction cross-section at the resonance energies[1] σ.

(1) $\quad \sigma = 4\Pi / _{k^2} \, (2l+1)\Gamma_\alpha \, \Gamma_c // \Gamma^2 \qquad , \quad k$ is the wave vector

The formula neglects interference effect on reaction cross section, it could be justified in the case 11^- resonance but for the 9^- one it leads to underestimation of Γ_c.

In order to extract the reduced width Θ_i^2 for each channel appropriate penetrability was calculated at the radius R that of the effective potential barrier. Obtained reduced widths exhaust only small fraction of Wigner limit thus indicating unimportance of a simple dinuclear structure.

Table: 25.1 MeV and 27.1 MeV resonances widths

E_α MeV	channel	Γ_i/Γ	Γ_{keV}	Γ_i keV	R_{fm}	Θ_i^2 (%)
25.1	$\alpha+Mg$	0.16	280	44.8	6.7	$9.2 \, 10^{-3}$
9^-	$^{12}C+^{16}O$	0.026	280	7.3	7.5	$5.4 \, 10^{-3}$
27.1	$\alpha+Mg$	0.07	370	26.0	6.5	$8.3 \, 10^{-3}$
11^-	$^{12}C+^{16}O$	0.053	370	20.0	7.3	$1.7 \, 10^{-2}$

References:

1) I.Skwirczynska et al . Nucl.Phys. A452 (1986) 432
2) A.D.Frawley et al . P.R.L. 44 (1982) 1377
3) D.Robson and A.H.Lane, Phys.Rev. 161 (1967) 982

Part V

**Atomic Clusters
and Ions**

Energetics and Dynamics of Solvation and Fission in Clusters

U. Landman, R.N. Barnett, A. Nitzan, and G. Rajagopal*

School of Physics, Georgia Institute of Technology, Atlanta, GA 30332, USA
*Permanent Address: School of Chemistry, Tel-Aviv University,
Tel-Aviv, Israel

I. INTRODUCTION

The nature, properties, and behavior of physical systems depend upon the
identity of the constituents and the nature of interactions between them and
upon the degree of aggregation and the ambient conditions. Investigations of
the energy level structure, elementary excitations, morphology (shape or
crystallographical structure), phase transformations and dynamics of finite
systems, and their dependence upon the degree and form of aggregation, are
common endeavors in the physical sciences spanning a wide spectrum of
interaction forms and strengths, spatial dimensions, and temporal scales.
Thus the evolution of energetic, structural, dynamical and thermodynamic
properties of matter as a function of the degree of aggregation (i.e., size of
the system, or number of particles composing it) is of fundamental interest in
diverse fields cutting across the disciplines of atomic, molecular, nuclear,
intermediate and high-energy particle physics, astrophysics, chemical
dynamics, materials science and condensed matter physics. In this context it
is of interest to note close analogies between some of the properties and
phenomena exhibited by atomic and molecular clusters and those found in atomic
nuclei, despite gross differences in the nature of binding in these systems
and their spatial extent. For example, the electronic shell structure and
magic number stabilities [1], quadrupolar surface deformations [2,3], collective electronic excitations [1a,4-11] (giant dipole, plasma resonances) and
dissociation and fission (i.e., fragmentation of charged clusters, either
spontaneously following ionization or by collisions with rare gas atoms)
studied mostly in alkali-metal clusters [1a,12-27] are properties whose
nuclear analogs have long been studied [28]. Similarly, certain features of
reactions between colliding clusters, or between a cluster and an incident
particle [29], may be analyzed using concepts common to nuclear and heavy-ion
collisions (such as stripping, spectator and door-way state mechanisms).

It is consequently not surprising that theoretical tools which are
developed for and employed in the exploration of such diverse fields and
phenomena share a high degree of common features, as even a brief scan of
papers in these fields indicates. Thus, for example, the use of symmetry

Springer Series in Nuclear and Particle Physics   **Clustering Phenomena in Atoms and Nuclei**
Editors: M. Brenner · T. Lönnroth · F.B. Malik   © Springer-Verlag Berlin, Heidelberg 1992

groups (discrete and continuous), techniques for calculations of energy level spectra (Hartree-Fock based methods, density-functional theories, configuration interaction, correlated wave functions etc.), many-body techniques for calculations of dispersion relations and elementary excitations, statistical mechanics formulations and analysis of energy level structure, and of reactive events and fragmentation phenomena, calculations of phase-diagrams and phase-transformations, methods of discretization on lattices, the use of newtonian dynamics and trajectory analysis, stochastic dynamics, formal scattering methods and time-dependent Hartree-Fock techniques for studies of reactions, and the use of path-integral based techniques for calculations of equilibrium finite temperature properties and time dependent processes are all prevalent, with rather small variations, across disciplinary boundaries.

Finite systems, beyond the very small end of the size spectrum present an immense theoretical challenge since the number of particles in these systems renders the use of molecular science techniques rather cumbersome (or impractical) while their finiteness prohibits the employment of condensed matter methodology based on translational symmetry, and complicates, due to size defects, the adaptation of the analytical framework and techniques of statistical mechanics which are formulated on the premise of the thermodynamic limit. Computer simulations, using either Monte Carlo (MC) or Molecular Dynamics (MD) techniques alleviate certain of the major difficulties which hamper other theoretical approaches, thus opening new avenues for investigations of physical systems [30,31], finite ones in particular [32]. These methods allow simulations, at finite temperatures, of equilibrium as well as non-equilibrium (i.e., time dependent) phenomena. For equilibrium studies both the MC and MD methods [33] sample in an efficient manner, the phase-space of the system. These methods are most useful even for the "modest" objective of the determination of the ground-state configuration of a small cluster at T = 0, since with these methods the search is not restricted to only a few points (representing highly symmetric structures) on the potential energy surface of the system as is the case in most other techniques. Since the number of local minima is expected to increase exponentially with N, the use of these simulation methods offers a decided advantage [34]. The MD techniques allows in addition studies at finite temperatures of dynamical processes in equilibrium and non-equilibrium situations.

Guided by the above considerations we review in this paper Molecular Dynamics Simulations of classical and quantum phenomena in finite systems, illustrating the methods and the richness of microscopic information revealed via these techniques. In the second section we outline the basic elements of classical and quantum molecular dynamics simulations. Selected case studies (chosen particularly in the context of the conference) are presented in Section III.

400

II. METHODOLOGY OF CLASSICAL AND QUANTUM MD SIMULATIONS

A. Classical MD

The classical Molecular-Dynamics (MD) method consists of a numerical generation of the phase-space trajectories for a system of N particles, interacting via a potential function $V(\vec{r}_1,\ldots,\vec{r}_N)$, where $\vec{r}_i$ is the coordinate of particle i.

The starting point of a classical MD simulation is a well-defined microscopic description of the physical system, in terms of a Hamiltonian or a Lagrangian from which the equations of motions are derived. Thus, given a Hamiltonian

$$\mathcal{H} = \frac{1}{2} \sum_i \vec{P}_i \cdot \vec{P}_i / m_i + V(\vec{r}_1,\ldots,\vec{r}_N) \quad,$$

numerical integration of the corresponding equation of motions allows investigations of the dynamics of equilibrium as well as non-equilibrium properties.

In equilibrium studies the properties of the system under investigation are calculated as averages along the dynamical trajectory of the system, instead of the customarily used ensemble averages. The trajectory average $\langle A \rangle$ of the property $A(\vec{r}^N(t), \vec{P}^N(t))$, where $\vec{r}^N(t)$ and $\vec{P}^N(t)$ stand for the collection of position and momentum vectors of the N particles, is defined as

$$\langle A \rangle = \lim_{t \to \infty} \frac{1}{t} \int_0^t A(\vec{r}^N(t'), \vec{P}^N(t'))dt' \quad.$$

The property A can be an equal-time characteristic of the system, such as the average potential energy $\langle V \rangle$ or the average pair (or higher order) distribution function, or a time-correlation function, such as the velocity-velocity correlation function, $\langle \vec{v}(o) \cdot \vec{v}(\tau) \rangle$. When $A \equiv K = \sum_i \vec{P}_i \cdot \vec{P}_i / 2m_i$ and using the equipartition theorem, the temperature of the microcanonical ensemble is obtained from

$$\langle K \rangle = \frac{3}{2} NkT$$

where k is Boltzmann's constant. Non-equilibrium processes are investigated via analysis of the time-dependent phase-space trajectories generated by the simulation.

For detailed descriptions of the MD method and of the numerical algorithms used we refer to recent reviews [30,31,33].

B. Quantum MD

We term as Quantum MD (QMD) simulations situations where all the particles in the system evolve quantum mechanically (such as a droplet of [4]He atoms at low temperature [35]) or where some of the particles in the system evolve quantum mechanically and others (coupled to the first ones) obey classical equations of motion (such situations include for example the solvation of electrons in a classically described medium, or simulations of nuclear dynamics on the Born-Oppenheimer electronic potential energy surface evaluated concurrently with the motion of the nuclei).

We briefly review three QMD simulation methods: (i) Quantum Path Integral Molecular Dynamics (QUPID), (ii) Time-Dependent Self-Consistent Field (TDSCF) and the Adiabatic Simulation Method (ASM), and (iii) Born-Oppenheimer Local Spin Density (BO-LSD) Simulations.

(i) QUPID

The QUPID method rests on the Feynman path-integral formulation of quantum statistical mechanics [36], and provides a convenient method for studies of the equilibrium, finite temperature properties of systems consisting of interacting quantum and classical degrees of freedom [32(a-e)]. In this formulation the expression for the partition function Z for a system consisting of a quantum particle (mass m and coordinate $\vec{r}$) interacting with a set of N classical particles (whose phase-space trajectories are generated by classical equations of motion) via a potential $V(\vec{r}) = \sum\limits_{j=1}^{N} V(\vec{r}, \vec{R}_j)$, is given as

$$Z_p = \left[\frac{mP}{2\pi\hbar^2\beta} \right]^{3P/2} \int d\vec{r}_1 \cdots d\vec{r}_P \, d\vec{R}_1 \cdots d\vec{R}_N \, e^{-\beta V_{eff}} , \tag{1a}$$

where

$$V_{eff} \equiv \sum_{i=1}^{P} \frac{mP}{2\hbar^2\beta^2} (\vec{r}_i - \vec{r}_{i+1})^2 + \frac{1}{P} \sum_{j=1}^{N} \sum_{i=1}^{P} V(\vec{r}_i, \vec{R}_j) + V_c(\vec{R}_1, \ldots \vec{R}_N) , \tag{1b}$$

and

$$Z \equiv \lim_{P\to\infty} Z_p . \tag{1c}$$

V_c is the interaction potential between the classical particles and $\beta = (k_B T)^{-1}$. Generalizations to many quantum particles (obeying Fermi or Bose statistics while rather straightforward, are numerically demanding (see ref. 35 and refs. 18, 19 therein).

Equations (1) establish an isomorphism [36,37] between the quantum problem and a classical one in which the quantum particle is represented by a flexible periodic chain (necklace) of P pseudoparticles (beads) with nearest-neighbor harmonic interactions with a temperature-dependent spring constant, $Pm/\hbar^2\beta^2$. In practice the finite value of P employed in the calculations is chosen to yield convergent results and depends upon the temperature and characteristics of the interaction potential V. The average energy of the system at equilibrium is given by

$$E = \frac{3N}{2\beta} + \langle V_c \rangle + K + \frac{1}{P} \langle \sum_{i=1}^{P} V(\vec{r}_i) \rangle , \qquad (2a)$$

where

$$K = \frac{3}{2\beta} + \frac{1}{2P} \sum_{i=1}^{P} \langle \frac{\partial V(\vec{r}_i)}{\partial \vec{r}_i} \cdot (\vec{r}_i - \vec{r}_P) \rangle , \qquad (2b)$$

and the angular brackets indicate statistical averages over the probability distribution as defined by Eq. (1). The first two terms in Eq. (2a) are the mean kinetic and potential energies of the classical components of the system. The quantum particle kinetic energy estimator [38,39] K (Eq. (2b)) consists of the free particle term ($K_f = 3/2\beta$) and a contribution due to the interaction (K_{int}). Finally, the last term in Eq. (2a) is the mean potential energy of interaction between the quantum particle and the classical field.

The formalism described above is converted into a numerical algorithm by noting the equivalence [39] between the equilibrium statistical averages over the probability distribution given in Eq. (1) and sampling over phase space trajectories, generated by a classical Hamiltonian

$$H = \sum_{i=1}^{P} \frac{m^* \dot{\vec{r}}_i^2}{2} + \sum_{I=1}^{N} \frac{M_I \dot{\vec{R}}_I^2}{2} + \sum_{i=1}^{P} \left[\frac{Pm}{2\hbar^2 \beta^2} (\vec{r}_i - \vec{r}_{i+1})^2 + \frac{V(\vec{r}_i)}{P} \right]$$

$$+ V_c(\vec{R}_1, \ldots, \vec{R}_N) \qquad (3)$$

where m^* is an arbitrary mass, chosen such that the internal frequency of the necklace, $\omega = [mP/m^* \beta^2 \hbar^2]^{1/2}$, will match the other frequencies of the system, and M_I is the mass of a classical particle.

Descriptions of the applications of the method to a wide variety of quantum many-body systems of chemical and physical interest can be found in references 32(a-e) 35,39-50.

(ii) TDSCF and ASM

The quantum time evolution in the TDSCF method [51] is based on a repeated evaluation of the short-time propagation of the wave function (in real or imaginary time) according to [52]

$$\psi(\vec{r}, t+\Delta t) = \exp[-\frac{i}{\hbar}(\hat{K}+\hat{V})\Delta t]\psi(\vec{r},t) = \exp[-\frac{1}{2}\frac{i}{\hbar}\hat{K}\Delta t] \exp[-\frac{i}{\hbar}\hat{V}\Delta t] \times$$

$$\times \exp[-\frac{1}{2}\frac{i}{\hbar}\hat{K}\Delta t] \psi(\vec{r},t) + o((\Delta t)^3) , \qquad (4)$$

where $\hat{K}$ and $\hat{V}$ are the kinetic and potential-energy operators, and an expansion in the plane-wave, free-particle, basis set

$$\psi(\vec{r},t+\Delta t) = \frac{1}{(2\pi)^3} \exp[-\frac{i}{2\hbar}\hat{K}\Delta t]\ \exp[-\frac{i}{\hbar}\hat{V}(\vec{r})\Delta t]\int d^3k\, e^{-i\vec{k}\cdot\vec{r}}\ \exp[-\frac{i\hbar k^2}{4m}\Delta t]$$

$$x\int d^3r'e^{-i\vec{k}\cdot\vec{r}'}\ \psi(\vec{r}',t)\ , \qquad (5)$$

where m is the mass of the quantum particle and V is his interaction potential. The Fast-Fourier Transformation (FFT) algorithm is applied to the discretized version of Eq. (5) on a grid.

Using Eq. (4) for a fixed configuration of the nuclei and transforming to imaginary time, i.e., $t = -i\beta'$, propagation (using the FFT algorithm) from $\beta' = 0$ to a value β large enough so that the expectation value $E = \langle\psi(\beta)H\psi(\beta)\rangle/\langle\psi(\beta)\psi(\beta)\rangle$ converges to a constant value, allows determination of the ground-state energy, E, and corresponding ground-state wave function $\psi(\beta)$. Sequential determination of electronic excited states can be achieved via application of the above imaginary time propagation in conjunction with projection operators which project out of the initial wavefunction the previously determined lower states [32(c),53-56].

In the quantum-classical version of the TDSCF method (for a review see Ref. [52d], for a critical study see ref. [51b], and for recent applications see Refs. [32(c),54-62] the coupled dynamical evolution of the quantum and classical degrees of freedom is described via the coupled equations

$$\frac{\partial\psi(\vec{r};\{\vec{R}\},t)}{\partial t} = -\frac{i}{\hbar}H(\vec{r};\{\vec{R}\})\ \psi(\vec{r};\{\vec{R}\},t)\ , \qquad (6a)$$

$$M_I\ddot{\vec{R}}_I = -\int d\vec{r}\,|\psi(\vec{r};\{\vec{R}\},t)|^2\frac{\partial}{\partial\vec{R}_I}V(\vec{r},\{\vec{R}\}) - \frac{\partial}{\partial\vec{R}_I}V_c(\{\vec{R}\}),\ (I=1,\dots,N) \quad (6b)$$

$$H(\vec{r};\{\vec{R}\}) = H_0(\vec{r}) + V(\vec{r},\{\vec{R}\})\ , \qquad (6c)$$

where $\{\vec{R}\}$ denotes the collection of vector positions of the N atomic (classical) constituents of masses M_I, H_0 is the Hamiltonian of the isolated quantum subsystem, $V(\vec{r},\{\vec{R}\})$ is the interaction potential between the quantum and classical degrees of freedom and $V_c\{\vec{R}\}$ the interatomic potential. In this approximation the classical subsystems evolves in the quantum-averaged interaction potential $\langle V\rangle$ [integral on the right-hand side of Eq. (6b)] and the classical interaction $V_c(\{\vec{R}\})$.

For many situations, where the adiabatic (Born-Oppenheimer) approximation applies, propagation of the classical subsystems can be performed on a single chosen electronic energy surface, by restricting the electronic wave-function to remain in that chosen state throughout the simulation. The Adiabatic Simulation Method - ASM or Ground-State Dynamics (GSD) when propagation on the ground electronic state is studied. Generalization of the method for multi particle quantum systems (Fermi or Bose statistics) is straightforward [63].

(iii) <u>BO-LSD</u>

In principle, complete descriptions of static, dynamic and thermodynamic physical and chemical properties and processes of materials are contained in solutions to the Schrodinger equation. However, for all but the most ideal- ized models, such solutions are not possible without resort to approximate methods, motivated by physical and practical (computational) considerations.

One of the pillars of modern quantum-theoretical treatments of materials is the Born-Oppenheimer (BO) approximation [64], based on the time-scale separation between nuclear and electronic motion. Even within the BO framework, calculations for multi-electron and nuclei system require further approximations. While high-level quantum-chemical computational techniques provide valuable information for small clusters of atoms (for selected nuclear configurations), they are not suitable for investigations of extended systems and of time-dependent phenomena which require repeated evaluation of the electronic energy along the trajectories of the nuclei.

Density functional theory [65,66], and in particular local spin-density [67] (LSD) and the local density (LD) approximations [65], provide accurate (although approximate) and practical methods of solution for quantum many-body problems and have become over the past two decades cornerstones of electronic structure calculations in condensed matter physics. The long conceived goal of combining nuclear molecular dynamics (with the nuclei treated classically) with ground-state electronic structure calculations via the LD method has been implemented successfully by Car and Parrinello [68] (CP) and used to investi- gate several systems [68-70].

The method which we have developed [26,27] combines classical nuclear molecular dynamics with LSD electronic structure calculations, within the framework of the BO approximation. As such, this study is part of continued efforts aimed at the development of effective methods for adiabatic simulations of the dynamics of systems of coupled nuclear and electronic degrees of freedom.

In the LSD theory [67] the total electron density $\rho(\vec{r};\{\vec{R}(t)\})$, for an N electron system, for a given nuclear configuration $\{\vec{R}(t)\}$, is given by

$$\rho(\vec{r};\{\vec{R}(t)\}) = \sum_{j,\sigma} f_{j\sigma} |\psi_{j\sigma}(\vec{r};\{\vec{R}(t)\})|^2 \equiv \rho_\alpha(\vec{r}) + \rho_\beta(\vec{r}) \tag{7}$$

where $\sigma = \alpha$ or β (↑ or ↓) is a spin-label, and $0 \leq f_{j\sigma} \leq 1$ are the orbitals' occupation numbers. The orbitals $\psi_{j\sigma}$ are the single-particle self-consistent solutions (corresponding to eigenvalues $\epsilon_{j\sigma}$) of the Kohn-Sham (KS) LSD equations

$$H_\sigma^{KS}(\vec{r},\rho(\vec{r}),\xi(\vec{r});\{\vec{R}(t)\})\psi_{j\sigma}(\vec{r};\{\vec{R}(t)\}) = \epsilon_{j\sigma}\psi_{j\sigma}(\vec{r};\{\vec{R}(t)\}) \qquad (8a)$$

where

$$H_\sigma^{KS}(\vec{r},\rho(\vec{r}),\xi(\vec{r});\{\vec{R}(t)\}) = -\frac{\hbar^2}{2m}\nabla_{\vec{r}}^2 + V_{eI}(\vec{r},\{\vec{R}(t)\}) +$$
$$V_H(\rho(\vec{r}),\vec{r}) + V_{xc,\sigma}(\rho(\vec{r}),\xi(\vec{r})) \qquad (8b)$$

The first term in Eq. (8b) is the electron kinetic energy operator V_{eI} is the electron-ion pseudopotential, and V_H and $V_{xc,\sigma}$ are the Hartree potential and local, spin-dependent, exchange-correlation potential ($\frac{\partial}{\partial\rho_\sigma}E_{xc}(\rho(\vec{r}),\xi(\vec{r}))$. The spin polarization function $\xi(\vec{r})$ is given by

$$\xi(\vec{r}) = (\rho_\alpha(\vec{r}) - \rho_\beta(\vec{r}))/\rho(\vec{r}) \quad . \qquad (9)$$

In the calculations which we present in Section III.A we have used the norm-conserving electron-sodium ion pseudopotential [71], and for E_{xc} the interpolation formula of Vosko and Wilk [72].

In our studies we have used several methods for solving the KS equations. In earlier studies [26] the solutions were obtained via the fast Fourier Transform (FFT) technique, propagating in imaginary-time with the split-operator method [52]. The manifold of KS orbitals is obtained by successive projections of lower (earlier determined) eigenstates $\psi_{j\sigma}$ ($j < m$), when determining $\psi_{m\sigma}$. The self-consistent solutions are achieved by iterating the process with the input electronic density for the (i+1)th iteration constructed from the densities of previous iterations (i.e., density mixing) [26,27].

More recently [27] the self-consistent solutions were obtained by us via the subspace iteration method [73], or the block Davison method [74].

To overcome instabilities in the iteration scheme, which occur in the event of eigenvalue degeneracy, we use a Fermi-distribution function [21,75] for the occupation numbers $f_{j\sigma}$,

$$f_{j\sigma} = [\exp(\epsilon_{j\sigma} - \mu_\sigma)/k_B T_e + 1]^{-1} , \qquad (10)$$

where the chemical potential μ_σ is obtained by solving the equation $N_\sigma = \Sigma_j f_{j\sigma}$ ($\Sigma_\sigma N_\sigma = N$). We found that choosing $k_B T_e \sim 10^{-3}$ Hartree is adequate.

Given an efficient method for solving the ground-state of the many-electron problem, for a given ionic configuration, the positions and velocities of the ions ($I = 1,...,N$) are propagated for a short time-interval

406

using equations of motion similar to Eq. (6b), with $|\psi|^2$ denoting the electronic density obtained from the KS eigenfunctions. Subsequently, the LSD equations are solved for the new ionic configuration and the process repeated.

III. CASE STUDIES

A. Fission of Small Metal Clusters

Recent studies of metallic clusters (particularly of simple metals) unveiled systematic energetic, stability [1], spectral [11] and fragmentation [1,12-27] trends of intrinsic interest for understanding the properties of matter at the small aggregation limit and the size-dependendent evolution toward bulk behavior. Moreover, several properties of atomic clusters (e.g., electronic shell structure [1], and most recently supershells [1b], portrayed by the occurrence of magic numbers in the abundance spectra and ionization potentials; the influence of shape fluctuations analyzed within the jellium model [1]; giant spectral resonances interpreted as evidence for collective plasma oscillations [1,4-11]; and fragmentation, fission, patterns of ionized clusters [13-16]) bear close analogies to corresponding phenomena exhibited by atomic nuclei. These observations suggest an intriguing universality of the physical behavior of finite-size aggregates, though governed by interactions of differing spatial and energy scales.

These analogies have led to the adaptation of several established concepts used in the context of nuclear phenomena for interpretation of recent studies of atomic clusters [26,76]. In particular, experimental results pertaining to fragmentation patterns (symmetric versus asymmetric fission) and fission barriers have been interpreted [25,76,77] using the framework of the celebrated liquid- droplet model (LDM) of nuclear fission [28], and predictions of fission channels were made using the jellium model [1,21-24].

Using the BO-LSD method described in the previous section we investigated [27] the asymmetric fission of small doubly charged sodium clusters. The KS-LSD equations were solved on spatial cartesian grids, with spacing $\Lambda = 0.8$ a_o, using a plane-wave basis with maximum energy of 210 eV.

In dynamical simulations the classical equations of motion were integrated using a 5th-order predictor-corrector algorithm with a time-step of 1-5 fs to assure energy conservation. The electronic ground-state was redetermined after each classical step. Minimum energy structures were obtained by a steepest-descent-like method, starting from configurations selected from finite temperature simulations. (Note that the existence of a (local) minimum-energy-configuration implies a barrier for fission.) Barrier heights and shapes were obtained by constrained energy minimization, with the center-of-mass distance, R_{cm-cm}, between the fragments specified.

Table I

Potential energies, E_p, for Na_n^{+2} ($4 \leq n \leq 12$) clusters, in the minimum energy configurations.[12] Values marked by * indicate energies for systems which fragment with no barrier, and are calculated for the minimum energy configuration of Na_n^+. Dissociation energies, $\Lambda_m = E(Na_{n-m}^+) + E(Na_m^+) - E(Na_n^{+2})$, for $4 \leq n \leq 12$, ($\Lambda_m < 0$ indicates exothermic fragmentation). IV_k and IA_k (k = 1,2), the vertical and adiabatic ionization energies for $Na_n \rightarrow Na_n^+$ and $Na_n \rightarrow Na_n^{+2}$, respectively; IV^+ and IA^+, the ionization potentials for $Na_n^+ \rightarrow Na_n^{+2}$. Energies in eV.

\n	4	5	6	7	8	9	10	12
E_p	-10.50*	-19.00*	-23.46*	-30.29	-36.95	-43.47	-50.36	-62.83
Λ_1	-2.48	0.10	-1.78	-1.07	-1.08	-0.74	-0.43	-0.26
Λ_2	-2.03	-0.24	-1.70	-1.22	-0.68	-0.82	-0.10	-0.23
Λ_3			-2.50	-1.59	-1.28	-0.87	-0.65	-0.94
Λ_4					-0.85	-0.67	0.10	-0.27
Λ_5							-0.13	-0.44
Λ_6								+0.00
IV_1	4.41	4.33	4.67	4.24	4.62	3.83	4.11	3.93
IA_1	4.38	4.11	4.40	4.07	4.35	3.70	4.04	
IV_2	12.92	12.44	12.94	12.45	12.72	11.40	11.37	10.65
IA_2	10.30*	10.11*	9.80*	11.81	11.61	11.02	10.48	10.28
IV^+	8.40	6.25	7.90	8.01	7.69	7.61	7.09	
IA^+	5.92*	6.01*	5.40*	7.74	7.26	7.32	6.44	

From the energetics of the clusters and of the various fragmentation channels, given in Table I, we observe first that in all cases the energetically favored channel (see Λ) is $Na_n^{+2} \rightarrow Na_{n-3}^+ + Na_3^+$ ($n \leq 12$), i.e., asymmetric fission (except for n = 6), in contrast to results obtained from spherical jellium calculations where fragmentation via ejection of Na^+ is favored. Moreover, for many of these clusters the single-ion fission channel is not even the energetically second-best competing channel. Secondly, the first vertical and adiabatic ionization potentials (IV_1 and IA_1) exhibit an odd-even oscillation [1] in the number of particles in the cluster as well as shell closing effects for systems containing 8 electrons. Similar effects are seen for the other ionization energies though they are complicated by structural changes upon ionization [27]. Finally, for n > 6 fission involves energy barriers. The barriers for n = 8, 10 and 12 have been determined via constrained minimization to be: 0.16 eV, 0.71 eV and 0.29 eV for the energetically favored channel and larger barriers were found for the ejection of Na^+ from these clusters (0.43 eV and 1.03 eV for n = 8 and 10, respectively). The barriers for Na_{10}^{+2} are higher because of the closed-shell structure of this parent cluster, an effect that usually has been discussed only in the context of stability (abundance) and ionization potentials [1].

The potential energies along the reaction coordinates for the energetically favored channel and for Na^+ ejection, in the case of Na_{10}^{+2} are shown in Fig. 1. The most interesting feature seen from the figure is the rather unusual shape of the barrier for the favored fission channel. The origin of the "double-humped" barrier is most clearly revealed from the dynamics of the fission process of Na_{10}^{+2} displayed in Fig. 2. This simulation started from a 600K Na_{10} cluster from which two electrons were removed (requiring 11.23 eV) and additional 0.77 eV was added to the classical ionic kinetic energy. The variation of the center of mass distance with time (Fig. 2a) exhibits a plateau for 750 fs $\leq$ t $\leq$ 2000 fs (see also the behavior of the electronic contribution to the potential energy of the system versus time in Fig. 2c). The contours of the electronic charge density of the system (Fig. 2(d-f)), at selected times, and the corresponding cluster configurations (Fig. 2(g-i)), reveal that the fission process involves a precursor state which undergoes a structural isomerization prior to the eventual separation of the Na_7^+ and Na_3^+ fission products. In this context we remark that examination of the contributions of individual Kohn-Sham orbitals to the total density for

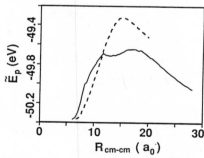

Figure 1: Potential energy ($\tilde{E}_p$, in eV) versus distance (in a_o) between the centers of mass for the fragmentation of Na_{10}^{+2} into Na_7^+ and Na_3^+ (solid) and Na_9^+ and Na^+ (dashed), obtained via constrained minimization of the LSD ground-state energy of the system.

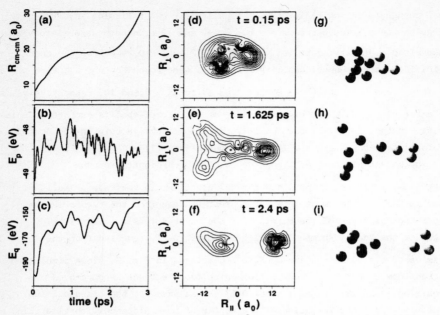

<u>Figure 2</u>: Fragmentation dynamics of Na_{10}^{+2} . (a-c) Center of mass distance
between the eventual fission products (R_{cm-cm}), total potential
energy (E_p) and the electronic contribution (E_q) to E_p, versus
time. (d-f) Contours of the total electronic charge distribution
at selected times calculated in the plane containing the two
centers of masses. The R axis is parallel to $\vec{R}_{cm-cm}$. (g-i)
Cluster configurations for the times given in (d-f). Dark and
light balls represent ions in the large and small fragments,
respectively. Energy, distance, and time in units of eV, Bohr
(a_o), and ps, respectively.

the intermediate stage (Fig. 2e) reveals that the lowest-energy orbital
(s-like) is localized on the Na_7^+ fragment, the next orbital (s-like) is
localized on the Na_3^+ fragment, the third is a p-like bonding orbital
distributed over the two fragments, and the highest orbital is localized on
the larger fragment.

The results for the fragmentation channels, the existence of fission
barriers for Na_n^{+2} with n > 6 (which is in contrast to the conclusion obtained
from an adaptation of the LDM, predicting [77] barrierless fission for n $\leq$
12), and the precursor mechanism for fission illustrate the importance of
non-jellium and dynamical effects in investigations of charged cluster
fragmentation.

B. Electron Localization and Solvation in Clusters

Studies of electron attachment and localization in finite clusters open new avenues for probing size effects of energetics and solvation mechanisms, pertaining to the issues of the modes and dynamics of electron localization, and their dependence on the system size, the minimal cluster sizes which sustain bound states of an excess electron, and the spectroscopic consequences of these phenomena.

Using the QUPID method, and more recently TDSCF and BO-LSD, in conjunction with electron to atom or molecule pseudopotentials and tested interatomic interactions, we have carried extensive investigations of the excess electron attachment localization, dynamics and spectroscopy in polar molecular [32c,46-50,54,55,57-60,78] (water and ammonia) and ionic [32c,42-44, 46,60,79] (alkali-halide) clusters, for a wide range of cluster sizes and temperatures (we limit ourselves here to a brief discussion pertaining to water clusters).

Some of the main results obtained from our studies are:

(1) The localization mode of an excess electron in water clusters depends on the cluster size. For water clusters $(H_2O)_n^-$ in the size range $10 \lesssim n \lesssim 64$ the electron is relatively strongly bound in a __surface state__ while for the larger clusters ($32 \leq n \leq 64$) a gradual transition to __internal solvation__ occurs. For $n \gtrsim 60$ the most stable state of the electron is a solvated state in a cavity of mean radius $\sim 7a_0$, located in the interior of the cluster. Attachment of the excess electron to small clusters, $n < 10$ is in a diffuse weakly bound surface state. The mean radius of gyration of the excess electron distributions varies with cluster size and the mode of localization (e.g., $\sim 11a_0$ and $5.5a_0$ for the surface states in clusters with $n = 8$ and 18, respectively, and $4a_0$ for internal states for $n \gtrsim 60$).

(2) The origin of the mode of localization (surface vs internal states) is the balance between the excess electron binding energy to the cluster and the water reorganization energy associated with the electron attachment. For the smaller clusters the reorganization energy associated with an interior state is larger than the binding energy of this state, resulting in the preferred surface localization which requires less cluster reorganization. The opposite is true for the larger ($n \gtrsim 60$) clusters which make the interior state more stable. In intermediate size clusters ($18 \lesssim n \lesssim 60$) interior states appear as long lived metastable states. Similarly in larger clusters surface states appear as long lived metastables.

The energetics of the systems can be expressed in terms of the electron vertical and adiabatic binding energies. EVBE and EABE, respectively, where EVBE is the electron ionization energy of the negatively charged cluster with no cluster reorganization and EABE = EVBE + E_c, where E_c is the cluster regoranization energy, i.e., the difference between the equilibrium intra- and intermolecular energies of the negatively charged cluster and the corresponding neutral. Energetic stability is inferred from the magnitude and sign of EABE, which at the limit of $n \to \infty$ is the heat of electron solvation in the bulk, and with EABE < 0 corresponding to an energetically stable bound state.

An adaptation [47,49] of a dielectric continuum model [80], where the molecular cluster is described by a dielectric sphere of radius $\bar{R} = r_s n^{-1/3}$ and where r_s is the (mean) radius of the molecular constituent, yields EABE $(\bar{R}) = $ EVBE $(\infty) + An^{-1/3}$ and EVBE $(\bar{R}) = $ EABE $(\infty) + Bn^{-1/3}$, where $A = e^2/2r_s (1 - D_s^{-1})$ and $B = (e^2/2r_s)(1 + D_{op}^{-1} - 2D_s^{-1})$ with D_s and D_{op} being the static and optical dielectric constants of the material, respectively.

The values of the predicted slopes A and B from the dielectric model agree with those obtained from simulation results. Furthermore, an extrapolation of the EABE to $n \to \infty$ yields values in agreement with current experimental estimates (-1.7 eV and -1.1 eV for water and ammonia, respectively) for the bulk heats of solution in these materials. The quantitative analysis of the size dependence of EABE and AVBE for internal states establishes a continuous transition between microscopic solvation effects in finite systems and in the macroscopic polar fluids.

(3) Surface states of an excess electron in water clusters $(H_2O)_n^-$ ($n \geq 18$) support up to three excited states [for larger clusters ($n \geq 64$) the surface localization mode is metastable]. The stable interior ground excess electron states in clusters of size $64 \leq n \leq 256$ correspond to cluster configurations which support at least three-excited (p-like) states [58]. The binding and bound-bound excitation energies [58] associated with surface states are about half the value of these quantities for interior states (see Fig. 3). For the interior states the excitation spectra show very weak or no dependence on the cluster size and are characterized by peak absorption at $\sim$ 2.1 eV and width of $\sim$ 1 eV, compared to experimental results in bulk water, 1.72 eV and 0.92 eV, respectively.

(4) The adiabatic dynamics of the electron solvation process [84] (starting from the electron localized in a pre-existing trap of the neutral water), as well as the relaxation dynamics following excitation of the solvated electron, is not strongly sensitive to cluster size for clusters ($n \geq 60$) that support internally bound states, and is characterized by two time scales: (a) A fast one (20-30 fs), associated with a rotational (librational) motion of the water molecules in the first solvation shell about the electron and (b) a slower relaxation stage ($\sim$ 200 fs) which is the order of the longitudinal dielectric relaxation in water.

412

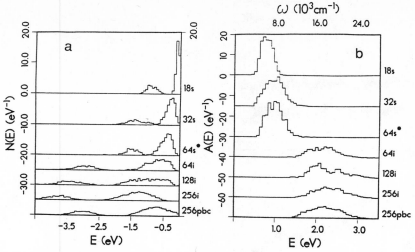

<u>Figure 3</u>: (a) Energy distributions N(E) versus energy for the ground and three lowest excited states of an excess electron in water clusters, $(H_2O)_n^-$, at 300K. For n = 18 and 32 the stable localization mode is in a surface state while for n $\gtrsim$ 64 the electron is localized internally. For n = 64, N(E) for both a metastable surface state (marked 64s*) and the stable internal state (marked 64i) are shown. The result for bulk water (256 water molecules with periodic boundary conditions) is shown at the bottom.
(b) Spectra A(E) for surface and internal states of an excess electron in $(H_2O)_n^-$ clusters, at 300 K as well as for an electron in bulk water These results were obtained by weighting the energy differences from Fig. 3a by the corresponding transition-dipole matrix elements.

Most recently, application of a newly developed method [82] for simulating non-adiabatic quantum processes to calculations of the radiationless transition rate of the hydrated electron from its lowest excited-state (i.e., the "wet" electron state) to the ground solvated state, yielded results in reasonable agreement with recent experimental [83] indications that this process dominates the solvation dynamics of excess electrons in water.

(5) The ground-state dynamics of electron localization, migration and solvation in medium and large water clusters was investigated using the GSD method [55]. The time evolution of energies of the excess electron (ground and excited states and the kinetic and potential contributions), the distance of the center of the excess-electron density from the center of mass of the molecular cluster (r_{e-cm}), and the radius of gyration of the electron distribution (r_g) are shown in Figs. 4a and 4b. The time development of the excitation energies to the first three excited states is shown in Fig. 4c. In

413

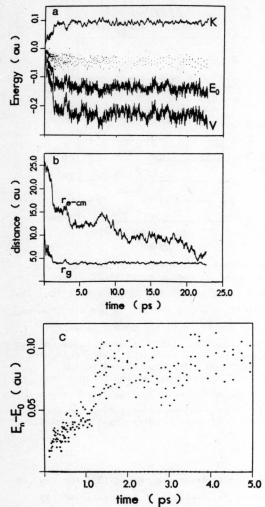

Figure 4:

Time evolution of an excess electron in a $(H_2O)_{256}^-$ cluster at
T = 300K.   (a)   The electron ground state energy E_o and the
potential V and kinetic K energy contributions for $0 \le t \le 23.0$ ps.
The first three excited states are given at selected times by the
dots.   (b)   Width of the electron distribution r_g and distance of
the center-of-mass r_{e-cm} vs time.   Energies and distances are in atomic
units (hartree and bohr radius) and time is in ps.   (c)   The excitation spectrum $E_n - E_o$, n = 1-3, of an excess electron in $(H_2O)_{256}^-$
for $0 \le t \le 5$ ps.   No significant variation occurs beyond the time
span shown.   Note the sharp increase in the excitation spectrum at
$t \simeq 1.2$ ps.

414

this simulation [55] the electron stands at t = 0 in a diffuse surface state, loosly bound to an equilibrium configuration of the cluster (selected from the equilibrium ensemble of a neutral $(H_2O)_{256}$ cluster at T = 300K). This initial state simulates the state of a zero-kinetic-energy incident electron at the instant of attachment to the cluster.

As seen from Figs. 4(a,b) at the very initial stage following attachment $(0 \leq t < 1.2$ ps) a transition to a compact, localized surface state occurs. Furthermore during this stage the excess electron is found to explore various sites on the surface of the cluster accompanied by local molecular reorganization (dipole reorientation) in the vicinity of the electron, culminating in the formation of a well-bound surface state. In the second stage $(1.2 \leq t \leq 1.5$ ps) the first molecular solvation shell is formed and a sudden (in less than 1 ps) increase occurs in the magnitudes of the electronic ground state energy (see E_o in Fig. 4a) and separation between the ground and excited states (see Fig. 4c), achieving values characteristic of the fully solvated electron.

For longer times (t > 1.5 ps) migration of the electron towards the center of the cluster occurs, characterized by gradual buildup of successive solvation shells, and "polaronlike" dynamical evolution where the electron-induced cluster reorganization accompanies the migration of the electron.

In studying the migration of an electron in condensed media which do not possess translational periodicity (i.e., in our case, the penetration of the electron from the external surface to the interior of the cluster), several modes of propagation may be considered. Studies of excess electron migration in molten salts [61a] suggest that the electron transport in these systems is mostly due to short-time jumps between two spatially separated sites. This mechanism is characterized by the occurrence of configurations, where at the intermediate time (between sites) the wave function exhibits splitting (i.e., a bimodal electron density distribution is found), and it appears that a potential barrier separate the initial and final localized states of the electron. In our simulations [55], we did not find evidence for such electron hopping events. In particular, we did not find configurations of the excess electron which are characterized by a bimodal distribution of the electron density distribution. Such events would have been exhibited in large variations in the width of the excess electron distribution r_g, which are absent in our results (see Fig. 4b).

The difference between the electron migration mechanisms in molten alkali halides and in our polar molecular systems may be attributed to the difference in the host reorganization energy in these systems. In the case of molten alkali halides, the solvated electron substitutes for a halide anion and the energy of the region containing the electron is close to that of neighboring regions in the fluid. On the other hand, in a polar molecular system, the

415

energy of a region around the solvated excess electron is much larger in magnitude than that of an equivalent neighboring neutral region. Furthermore, in the latter case, due the sizeable reorganization energy which accompanies the formation of the solvation shells, solvent fluctuations leading to a favorable solvation site in a neutral region are unlikely. Indeed the absence of deep traps in neutral water has been recently demonstrated [84].

Acknowledgement: This work was supporte by the US Department of Energy, Grant No. FG05-86ER45234. Calculations were performed at the Florida State University Computing Center through a DOE grant of computer time.

References

1. (a) See review by W. A. de Heer, W. D. Knight, M. Y. Chou and M. L. Cohen in Solid State Physics 40, 93 (1987); (b) For a recent investigation of the electronic shell structure in large metallic clusters and references to earlier studies see H. Gohlich, T. Lange, T. Bergmann and T. P. Martin, Phys. Rev. Lett. 65, 748 (1990); see also, J. L. Persson, R. L. Whetten, H.-P. Cheng and R. S. Berry (preprint); (c) For recent work on cold sodium clusters and references to earlier studies, see E. C. Honea, M. L. Homer, J. L. Persson, and R. L. Whetten, Chem. Phys. Lett. 171, 147 (1990).

2. K. Clemenger, Phys. Rev. B 32, 1359 (1985).

3. W. Ekardt and Z. Penzar, Phys. Rev. B 38, 4273 (1988).

4. W. A. de Heer, K. Selby, V. Kressin, J. Masuri, M. Volhmer, A. Chatelain, and W. D. Knight, Phys. Rev. Lett. 59, 1805 (1987).

5. C. Brechignac, Ph. Cahuzac, F. Calier, and J. Leygnier, Chem. Phys. Lett. 164, 433 (1989).

6. C. R. C. Wang, S. Pollack, and M. M. Kappes, Chem. Phys. Lett. 166, 26 (1990).

7. M. J. Puska, R. M. Nieminen, and M. Manninen, Phys. Rev. B 31, 3486 (1985).

8. W. Ekardt, Phys. Rev. B 31, 6360 (1985).

9. D. E. Beck, Phys. Rev. B 35, 7325 (1987).

10. C. Yannouleas, R. A. Broglia, M. Brack, and P. F. Bortignon, Phys. Rev. Lett. 63, 255 (1989); C. Yannouleas, J. M. Pacheco, and R. A. Broglia, Phys. Rev. B 41, 6088 (1990).

11. V. Bonacic-Koutecky, P. Fantucci, and J. Koutecky, Chem. Phys. Lett. 166, 32 (1990); V. Bonacic-Koutecky, M. M. Kappes, P. Fantucci, and J. Koutecky, Chem. Phys. Lett. 170, 26 (1990).

12. See a review by O. Echt in Physics and Chemistry of Small Clusters, edited by P. Jena, B. K. Rao, and S. N. Khanna (Plenum, New York, 1987), p. 623 and references therein.

13. C. Brechignac, Ph. Cahuzac, F. Carlier, and J. Leyghier, Phys. Rev. Lett. 63, 1368 (1989); C. Brechignac, Ph. Cahuzac, J. Leygnier, and J. Weiner, J. Chem. Phys. 90, 1492 (1989).

14. C. Brechignac, Ph. Cahuzac, F. Calier, and M. de Frutos, Phys. Rev. Lett. 64, 2893 (1990).

15. For experimental studies of Pb_n^{++} see: P. Pfau, K. Sattler, R. Pflaum, and E. Recknagel, Phys. Lett. 104A, 262 (1984); W. Schulze, B. Winer, and I. Goldenfeld, Phys. Rev. B 38, 12937 (1988).

16. For experimental studies of fission of charged Gold clusters see W. A. Saunders, Phys. Rev. Lett. 64, 3046 (1990).

17. B. K. Rao, P. Jena, M. Manninen, and R. M. Nieminen, Phys. Rev. Lett. 58, 1188 (1987).

18. C. Baladron, J. M. Lopez, M. P. Iniquez, and J. A. Alonzo, Z. Phys. D11, 323 (1989).

19. G. Durand, J. P. Daudley, and J. P. Malrieu, J. Phys. (Paris) 47, 1335 (1986).

20. S. N. Khanna, F. Reuse, and J. Buttet, Phys. Rev. Lett. 61, 535 (1988).

21. S. Saito and M. L. Cohen, Phys. Rev. B 38, 1123 (1988).

22. M. P. Iniguez, J. A. Alonso, A. Rubio, M. J. Lopez, and L. C. Balbas, Phys. Rev. B 41, 5595 (1990).

23. M. P. Iniguez, J. A. Alonso, M. A. Allen, and L. C. Balbas, Phys. Rev. B 34, 2152 (1986).

24. Y. Ishii, S. Ohnishi and S. Sugano, Phys. Rev. B 33, 5271 (1986).

25. S. Sugano, A. Tamura, and Y. Ishii, Z. Phys. D 12, 213 (1989).

26. R. N. Barnett, U. Landman, A. Nitzan and G. Rajagopal, J. Chem. Phys. 94, 608 (1991).

27. G. Rajagopal, R. N. Barnett and U. Landman, Phys. Rev. Lett. (1991).

28. A. Bohr and B. R. Mottelson, Nuclear Structure (Benjamin, London, 1975).

29. H.-P. Kaukonen, U. Landman and C. L. Cleveland, J. Chem. Phys. (1991).

30. See U. Landman, W. D. Luedtke and R. N. Barnett in Many-Atom Interactions in Solids, Eds. R. M. Nieminen, M. J. Puska and M. J. Manninen (Springer, Berlin, 1990), p. 103.

31. F. F. Abraham, Adv. in Phys. 35, 1 (1986); J. Vac. Sci. Technol. B2, 534 (1984).

32. See reviews: (a) U. Landman, R. N. Barnett, C. L. Cleveland, J. Luo, D. Scharf and J. Jortner, in Few-Body Systems and Multiparticle Dynamics (AIP Conf. Proc. 162), edited by D. A. Micha, (AIP, New York, 1987), p. 200; (b) J. Jortner D. Scharf and U. Landman in Elemental and Molecular

Clusters, edited by G. Benedek and M. Pachioni (Springer, Berlin, 1988) p. 148; (c) R. N. Barnett, U. Landman, G. Rajagopal and A. Nitzan, Israel J. Chem. 30, 85 (1990); (d) U. Landman, in Recent Developments in Computer Simulation Studies in Condensed Matter Physics, edited by D. P. Landau, K. K. Mon, and H. B. Schuttler (Springer, Berlin, 1988) p. 144; (e) B. J. Berne and D. Thirumalai, Annu. Rev. Phys. Chem. 37, 401 (1986); (f) R. Car, M. Parrinello and W. Andreoni, in Microclusters, edited by S. Sugano. Y. Nishina, and S. Ohnishi (Springer, Berlin, 1987) p. 134.

33. M. P. Allen and D. J. Tildesly, Computer Simulations of Liquids (Clarendon Press, Oxford, 1987).

34. R. S. Berry, T. L. Beck, H. L. Davis, and J. Jellinek, Adv. Chem. Phys. (1988); see also C. L. Cleveland and U. Landman, J. Chem. Phys. 94, 7376 (1991).

35. C. L. Cleveland, U. Landman and R. N. Barnett, Phys. Rev. B39, 117 (1989).

36. R. P. Feynman and A. R. Hibbs, Quantum Mechanics and Path Integrals (McGraw-Hill, New York, 1965).

37. D. Chandler and P. G. Wolynes, J. Chem. Phys. 79, 4078 (1981); D. Chandler, J. Phys. Chem. 88, 3400 (1984).

38. M. F. Herman, E. J. Bruskin, and B. J. Berne, J. Chem. Phys. 76, 5150 (1982).

39. M. Parrinello and A. Rahman, J. Chem. Phys. 80, 860 (1984).

40. A. Wallqvist, D. Thirumalai and B. J. Berne, J. Chem. Phys. 85, 1583 (1986).

41. M. Sprik and M. Klein, Comp. Phys. Rep. 7, 147 (1988).

42. R. N. Barnett, U. Landman, D. Scharf and J. Jortner, Acct. Chem. Res. 22, 350 (1989).

43. D. Scharf, J. Jortner and U. Landman, J. Chem. Phys. 88, 4273 (1988).

44. U. Landman, D. Scharf and J. Jortner, Phys. Rev. Lett. 54, 1860 (1985).

45. R. N. Barnett, U. Landman, C. L. Cleveland and J. Jortner, Phys. Rev. Lett. 59, 811 (1987).

46. U. Landman, R. N. Barnett, C. L. Cleveland, D. Scharf and J. Jortner, J. Phys. Chem. 91, 4890 (1987).

47. R. N. Barnett, U. Landman, C. L. Cleveland and J. Jortner, J. Chem. Phys. 88, 4421 (1988).

48. R. N. Barnett, U. Landman, C. L. Cleveland and J. Jortner, J. Chem. Phys. 88, 4429 (1988).

49. R. N. Barnett, U. Landman and J. Jortner, Chem. Phys. Lett. 145, 382 (1988).

50. R. N. Barnett, U. Landman, N. R. Kestner and J. Jortner, J. Chem. Phys. 88, 6670 (1988); Chem. Phys. Lett. 148, 249 (1988).

51. (a) P. A. M. Dirac, Proc. Cambridge Philos. Soc. 26, 376 (1930);
    (b) D. Kumamoto and R. Silbey, J. Chem. Phys. 75, 5164 (1981).

52. (a) M. D. Feit, J. A. Feit, Jr. and A. Steiger, J. Comput. Phys. 47,
    412 (1982); (b) M. D. Feit and J. A. Fleck, Jr., J. Chem. Phys. 78,
    301 (1983); 80, 2578 (1984); (c) D. Kosloff and R. Kosloff, J. Comput.
    Phys. 52, 35 (1983); (d) see review by R. Kosloff, J. Phys. Chem. 92,
    2087 (1988).

53. R. Kosloff and H. Talezer, Chem. Phys. Lett. 127, 223 (1986).

54. R. N. Barnett, U. Landman and A. Nitzan, J. Chem. Phys. 89, 2242 (1988).

55. (a) R. N. Barnett, U. Landman and A. Nitzan, Phys. Rev. Lett. 62, 106
    (1989); (b) J. Chem. Phys. 91, 5567 (1989).

56. See review by P. J. Rossky and J. Schnitker, J. Phys. Chem. 92, 4277
    (1988).

57. R. N. Barnett, U. Landman and A. Nitzan, Phys. Rev. A 38, 2178 (1988).

58. R. N. Barnett, U. Landman and A. Nitzan, J. Chem. Phys. 93, 6226 (1990).

59. R. N. Barnett, U. Landman and A. Nitzan, J. Chem. Phys. 93, 8187 (1990).

60. R. N. Barnett et al., Phys. Rev. Lett. 64, 2933 (1990).

61. (a) A. Selloni, P. Carenvali, R. Car and M. Parrinello, Phys. Rev. Lett.
    59, 823 (1987), and Refs. 5-8 therein; (b) see also D. Thirumalai,
    E. J. Bruskin, and B. J. Berne, J. Chem. Phys. 83, 230 (1985).

62. Z. Kotler, A. Nitzan and R. Kosloff, in Tunneling, edited by J. Jortner
    and B. Pullman (Reidel, Boston, 1986), p. 193.

63. Z. Kotler, Ph.D. thesis, Tel Aviv University (1989).

64. M. Born and J. Oppenheimer, Ann. Phys. 84, 457 (1927); M. Born and K.
    Huang, Dynamical Theory of Crystal Lattices (Oxford University, London,
    1954).

65. See articles in Theory of the Inhomogeneous Electron Gas, edited by S.
    Lundqvist and N. M. March (Plenum, New York, 1983).

66. W. Kohn and L. J. Sham, Phys. Rev. 140, A1133 (1965).

67. D. Gunnarson and B. I. Lundqvist, Phys. Rev. B 13, 4274 (1976).

68. R. Car and M. Parrinello, Phys. Rev. Lett. 55, 2471 (1985); for details
    see R. Car and M. Parrinello, in Proceedings of the NATO ARW, NATO ASI
    Series (Plenum, New York, 1989).

69. G. Gali, R. M. Martin, R. Car and M. Parrinello, Phys. Rev. Lett. 62, 555
    (1989); R. Car and M. Parrinello, ibid. 60, 204 (1988).

70. P. Ballone, W. Andreoni, R. Car and M. Parrinello, Europhys. Lett. 8, 73
    (1989).

71. D. R. Hamann, M. Schluter, and C. Chiang, Phys. Rev. B26, 4199 (1982).

72. S. H. Vosko and L. Wilk, J. Phys. C 15, 2139 (1982); S. H. Vosko,
    L. Wilk and M. Nusair, Can. J. Phys. 58, 1200 (1980).

73. A. Jennings, Matrix Computations for Scientists and Engineers (Widy, Chichester, 1977); R. B. Corr and A. Jennings, Int. J. Numer. Methods Eng. 10, 647 (1976).

74. E. R. Davison, J. Comput. Phys. 17, 87 (1975); see review of the method by D. M. Wood and A. Zunger, J. Phys. A 18, 1343 (1985).

75. G. W. Fernando, G.-X. Qian, M. Weinert and J. W. Davenport, Phys. Rev. B 40, 7985 (1989).

76. S. Sugano in Microclusters, edited by S. Sugano et al. (Springer, Berlin, 1987), p. 226.

77. W. A. Saunders, Phys. Rev. Lett. 66, 840 (1991).

78. R. N. Barnett, U. Landman, S. Dhar, N. R. Kestner, J. Jortner and A. Nitzan, J. Chem. Phys. 91, 7797 (1989).

79. G. Rajagopal, R. N. Barnett and U. Landman, Phys. Rev. Lett. (1991).

80. J. Jortner, J. Chem. Phys. 30, 839 (1959).

81. R. N. Barnett, U. Landman and A. Nitzan, J. Chem. Phys. 90, 4413 (1989).

82. E. Neria, A. Nitzan, R. N. Barnett and U. Landman, Phys. Rev. Lett. (1991).

83. A. Migus, Y. Gaudel, J. L. Martin and A. Antonetti, Phys. Rev. Lett. 58, 1529 (1987); F. H. Long, H. Lu and K. B. Eisenthal, Phys. Rev. Lett. 64, 1469 (1990).

84. J. Schnitker, P. J. Rossky and G. A. Kenney-Wallace, J. Chem. Phys. 85, 2986 (1986).

Role of Valence Electrons in the Structure and Properties of Microclusters

S.N. Khanna, C. Yannouleas, and P. Jena

Physics Department, Virginia Commonwealth University,
Richmond, VA 23284, USA

One of the most exciting developments in clusters has been the discovery of magic numbers in the mass spectra of metallic clusters in beam experiments.[1-3]   For Na_n clusters,[1] these experiments show that the clusters containing 2, 8, 18, 20, 40... atoms are more dominant than other sizes.   A simple model[1-3] based on replacing the actual cluster ions by a Jellium of appropriate charge density and filling the electronic levels by the valence electrons shows that the magic numbers correspond to sizes where the valence electrons completely fill the jellium electron levels. The existence of magic species is therefore explained to be a purely electronic effect.   It is also suggested that the magic species are geometrically spherical and the Clusters containing electrons which do not completely fill the jellium electronic levels undergo topological distortions to prolate or oblate shapes[4] similar to the case of nuclei in the nuclear shell model.[5]   It is therefore clear that the number of valence electrons govern the shape, stability and the properties of the simple metal clusters. While the number of valence electrons can be changed by changing the number of atoms, one can also change the number of electrons[6] by alloying a given cluster with a foreign atom having different valency and/or by adding or removing electrons from a given cluster.[7] The purpose of this paper is to focus on the changes on the structure and stability of cluster caused by valence electrons. We divide the paper into three portions.   (1) In the first section we show the relation between geometry and electronic spectrum of homo-nuclear clusters. In particular, we show how the geometrical distortions change the electronic levels. (2) We then consider the behavior of clusters as one changes the number of valence electrons by alloying with heteroatomic atoms. We discuss this by considering the case of K_n clusters containing Mg. (3) We finally consider the behavior of clusters as one changes the number of valence electrons by adding or removing electrons. In particular we discuss the behavior of doubly ionized clusters. Here, the coulomb repulsion between positive holes left by ionization can fragment the cluster. This phenomenon called the "coulomb explosion" has been extensively investigated and one of the basic questions has been the minimum cluster size, N_c, at which the cluster would be stable against double ionization. Recent experiments[8] have shown that doubly ionized dimers or trimers such as Mo_2^{++}, Au_2^{++}, Pb_3^{++} etc. are observable. We would discuss how these small clusters survive the coulomb repulsion and become observable.

All our studies are based on a linear combination of atomic-orbitals-molecular orbital (LCAO-MO) approach.[9] The cluster wave function is expanded as a sum of non-contracted Gaussian basis sets centered at the atomic sites. The core effects are incorporated via norm-conserving non-local pseudopotentials[10] and the exchange correlation contributions are approximated within the density functional approach.[11] The form of the exchange correlation potential we have employed is that proposed by Ceperley and Alder. For each cluster geometry, the Kohn-Sham equations[12] are solved self-consistently and the total energy is determined via the usual density functional formula. The details of the calculations are published elsewhere and the reader is referred to our earlier papers for details.[7,13]

I. Electronic Shell Structure and Geometry

One of the early evidence that the electrons determine the geometry of a cluster came from the analysis of the magic numbers of Na_n clusters in terms of a jellium model.[1] In this model a cluster was assumed to have a spherical shape. Consequently, the angular momentum of the electrons is a good quantum number and the electronic energy levels can be characterized in terms of the orbital angular momentum. A complete closing of each of these shells gave rise to an unusually stable cluster which were identified by pronounced peaks in the mass spectra.

Clusters that do not contain enough electrons to fill a complete shell, however are not spherical and undergo topological distortions to minimize energy. This indeed was shown to be the case by Clemenger[4] who calculated the energetics of Jellium clusters distorted into ellipsoidal forms. He was able to account for peaks in mass spectra that occur in between the magic species. Similar distortions have been known to exist in nuclei with unfilled shells. It is this distortion which is also responsible for the low spin of the clusters as well as the nuclei. The topological distortion splits the energy levels of otherwise degenerate electron orbitals. If the splitting is large compared to the gain in energy due to Hund's rule[14] coupling among parallel spins the Pauli principle would dictate a lower spin structure for the cluster.

While the above analysis gives a qualitative description of the role of electronic structure on geometrical shape, accurate picture of the geometry can only be derived from atomistic calculations. For example, consider a cluster consisting of 3 monovalent atoms. Two of these electrons will go into a s-type bonding orbital which will require the cluster to be an equilateral triangle. However, the third electron that occupies the antibonding state would distort the cluster from the equilateral

423

triangular shape.[15] The trimer can thus assume either acute or obtuse form depending upon whether the antibonding electron is distributed on the apex bonds or on the basal bond. In contrast, if one has a singly ionized trimer, the geometry indeed will be equilateral. This description of geometry based upon quantum mechanics of electrons is indeed borne out by self consistent calculations. Similar analysis can be done for larger clusters although keeping track of bonding and antibonding states becomes rather cumbersome.

While the geometry of the clusters is determined by the electronic structure, the stability of the cluster, at least for small sizes may not be strongly influenced by the geometry i.e. a cluster of eight monovalent atoms need not be spherical as demanded by Jellium model to exhibit unusual stability. What is important for the stability of small clusters is that the subshells are completely filled. This can be shown succinctly by taking a cluster of 4 mono-valent atoms in tetrahedral, square, rhombus and linear forms. The tetrahedral cluster is close to being spherical in shape while the linear chain is farthest from a sphere. As shown in Fig. 1 the change from the tetrahedral to the linear structure results in a degenerate p-orbital to split into px, py, and pz states. However, the splitting between these levels are small compared to the energy gap between the successive angular momentum states. In addition splitting does not result in any rearrangement of the levels. Thus the tetramer whether spherical or not will have its sub shell filled and as such will be more stable than a trimer irrespective of its topological form. Thus the validity of the Jellium model arises not because clusters are spherical but because their distorted geometry is unable to alter the shell structure otherwise derived from the spherical Jellium model.[16] This situation however gets critical as one approaches large clusters consisting of thousands of atoms. In this size

424

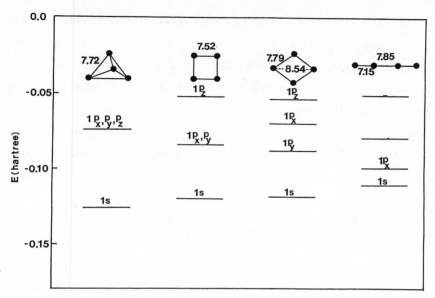

Fig. 1. Molecular orbital levels in K_4 confined to tetrahedral, square, rhombus and linear geometries. For each case, the optimized bond lengths are given in atomic units.

range, the electronic shells at the highest occupied levels are very close to each other and a small distortion of the geometry can easily affect the ordering of the energy shells and consequently the magic numbers. Thus contrary to one's expectations, the magic numbers in very large clusters can not always be predicted from the Jellium model. One has to take into account other competing mechanisms such as the packing of the atoms in the cluster. This argument is now supported by experiment.[17]

II. **Hetero-Atomic Clusters**

In the above section we saw that the stability of the clusters is determined by the shell structure. Clusters with enough valence electrons to completely fill the electronic subshells are more stable than clusters with partially filled shells. One also expects clusters with filled shells to be spherical while clusters with unfilled shells to undergo geometric distortions. These considerations suggested that the number of valence electrons determines the behavior of clusters. One can vary the number of electrons either by increasing the cluster size but also by mixing elements with different valency. Keeping this in mind, experiments[6] were performed to see if mixtures of simple metals like K with divalent elements like Mg would produce peaks in the mass spectra at sizes where K_nMg has the same number of valence electrons as a pure K_n cluster. Surprisingly, it was found that the mass spectra of K_nMg had a peak at K_8Mg and not at K_6Mg even though one would have expected K_6Mg to be magic since it has the same number of valence electrons as K_8 which is a magic size. These experiments indicated that the behavior of clusters is determined not only by the number but also by the chemical nature of electrons.

To further investigate the above behavior we carried out ab-initio calculations on K_6Mg and K_8Mg clusters. Since Mg interacts strongly with K in these clusters, the structures of K_6Mg and K_8Mg are characterized by a central Mg surrounded by K atoms. In K_6Mg, K atoms form an octahedric cage around Mg while in K_8Mg, the eight K atoms sit at the corners of a cube with Mg at the body center. Simple Jellium picture would predict a spherical shape for K_6Mg since there are eight valence electrons, and a distorted structure for K_8Mg because of ten valence electrons. However both K_6Mg and K_8Mg show compact symmetric structures. A study of the electronic spectrum indicates that the Mg 2s valence levels lie about 2 eV below the K valence manifold. The 2 valence electrons contributed by Mg therefore lie in a deep impurity like state and do not contribute to the valence manifold. It is for this reason that the magic numbers for K_n clusters are unchanged upon the addition of a Mg atom and K_8Mg appears as a magic species instead of K_6Mg.

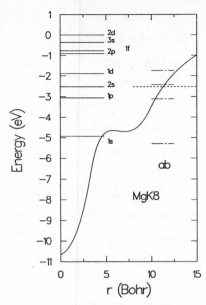

Fig. 2. One electron levels in a Jellium containing a central Mg surrounded by a shell of Na atoms along with the corresponding levels based on ab-initio calculations.

One can also look at the above situation from a Jellium viewpoint. The Mg atom at the center of the K_nMg cluster provides a strong attractive potential and the resulting potential has a double minima as shown in Fig. 2. This potential rearranges the ordering of levels in spherical jellium and the electronic shells now appear in order 1s, 1p, 2s, 1d... The number of electrons required to fill 1s, 1p, and 2s levels is 10 and consequently K_8Mg appears as the stable species. The reordering of the jellium electron levels due to the introduction of Mg can be compared to the nuclear levels where the introduction of the spin orbit coupling mixes energy levels belonging to different angular momentum.[18]

III. **Doubly Ionized Clusters**

We have so far considered the behavior of clusters as the number of electrons is changed by changing the cluster size or by adding impurity atoms. One can also change the number of electrons for a given cluster by simply adding or removing electrons. Of particular interest is the behavior of clusters under multiple ionization. Consider for simplicity a doubly ionized cluster X_n of n atoms. The separation between the positive holes left behind by the ionization is limited by the cluster size. Consequently the cluster experiences a coulomb repulsion that is inversely proportional to its size and can fragment if the size is small. This phenomenon called the "coulomb explosion" has been the subject of considerable theoretical and experimental interest.[19,20] One of the initial interests[19] was to determine the minimum cluster size N_c at which the cluster would be stable against the double ionization. Initial experiments had indicated $N_c \sim 30$ which could be theoretically explained as the size at which the coulomb repulsion between positive holes is comparable to the cohesive energy per atom. Later experiments,[8] however, reported observing small doubly ionized clusters such as Mo_2^{++}, Au_2^{++}, Pb_5^{++} etc. in suitably carried out experiments. It is surprising how these clusters survive the enormous coulomb repulsion and become observable. A second issue relates to the most probable fragmentation channel and if it can be predicted on the basis of energetics of the parent and fragmented cluster.

Over the past few years we have been engaged in a detailed study of the behavior of doubly charged clusters.[7,20,21] We refer the reader to our original papers for details and focus here on two issues. The first one is concerned with the stability and the observability of these clusters. Here we show that many of the doubly ionized clusters are metastable and can be observed in carefully carried out experiments. Secondly we show that the

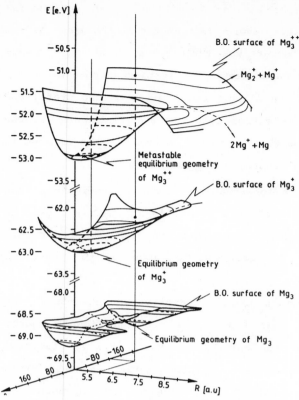

E [e.V]

- 50.5
- 51.0

B.O. surface of Mg_3^{++}

$Mg_2^+ + Mg^+$

- 51.5
- 52.0
- 52.5
- 53.0

2$Mg^+ + Mg$

Metastable
equilibrium geometry
of Mg_3^{++}

- 53.5

B.O. surface of Mg_3^+

- 62.0
- 62.5
- 63.0

Equilibrium geometry
of Mg_3^+

- 63.5
- 68.0

B.O. surface of Mg_3

- 68.5
- 69.0

Equilibrium geometry of Mg_3

- 69.5

-80   -160

160   80   0   5.5   6.5   7.5   8.5   R [a.u]

Fig. 3.   Born-Oppenheimer surfaces of neutral, singly,
and doubly ionized Mg trimers corresponding to various
isosceles triangular configurations. E correspond to total
energies in the pseudopotential scheme.   Distances are in
atomic units and energies in eV.

energetics alone is not enough to determine the fragmentation
channels and that the barriers play an important role.

We start by considering the observability of doubly charged
clusters.   In Fig. 3 we show the Born oppenheimer energy
hypersurfaces[7] of Mg_3, Mg_3^+, and Mg_3^{++}. Mg_3 has a ground state which
is equilateral triangle. Mg_3^+ is a linear chain. Mg_3^{++} is
energetically unstable but has a deep minimum in the energy
hypersurface corresponding to a linear configuration. Note that a

429

single step double ionization of Mg_3 leads to an equilateral triangle Mg_3^{2+} cluster, whose energy is higher than the dissociation threshold and the cluster would fragment into Mg_2^+ + Mg^+ spontaneously. However, a sequential ionization where one removes a single electron, let the cluster cool to the ground state, and then remove the second electron leads to a Mg_3^{++} trapped in the metastable minimum. The barrier height protecting the minimum is about 1eV and hence the cluster would have long

lifetime and would be observable. This shows that the observability of doubly charged clusters is intimately linked to the details of the ionization process and that smaller doubly ionized clusters can be observed in experiments involving sequential ionization.

We finally consider the role of barriers in determining the fragmentation patterns. In Fig. 4, we show the lowest energy configurations of doubly ionized clusters Mg_n^{++} (n ≤ 7) clusters, their atomisation energies, and the dissociation channels along with the dissociation energies and the height of the potential barriers. It is interesting to note that Mg_n^{++} clusters for n ≤ 6 are all linear chains. Mg_7 is the smallest doubly ionized cluster with a three dimensional germ. All doubly ionized clusters (n ≤ 7) are energetically unstable. It is therefore interesting to study the fragmentation channels of these clusters. Realistic studies of these require calculations of the reaction paths and the barriers involving the parent cluster breaking into the fragments. These studies are complex and various authors[22] have used the energies of the parent and the fragment clusters to predict the dissociation channels. It is therefore important to inquire if the energetics alone can be used to analyze fragmentation patterns. From Fig 4 we note that if one considered the energetics alone, the lowest fragmentation channel for breaking Mg_4^{++} would be Mg_2^+

Cluster size	Atomization energies (eV)	Equilibrium geometry	Dissoc. channel	Dissoc. energy (eV)	Barrier (eV)
2	-1.27	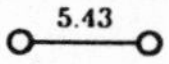 o—5.43—o	$Mg^+ + Mg^+$	-2.54	0.48
			$Mg^{++} + Mg$	4.88	
3	0.07	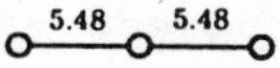 o—5.48—o—5.48—o	$Mg_2^+ + Mg^+$	-1.30	0.77
			$Mg^+ + Mg^+ + Mg$	0.22	
4	0.53	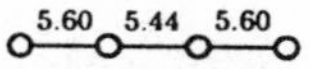 o—5.60—o—5.44—o—5.60—o	$Mg_2^+ + Mg_2^+$	-0.92	0.90
			$Mg_3^+ + Mg^+$	-0.41	0.82
			$Mg_2^+ + Mg^+ + Mg$	0.61	
5	0.72	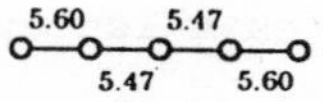 (5.60, 5.47, 5.47, 5.60)	$Mg_3^+ + Mg_2^+$	-0.48	0.98
			$Mg_4^+ + Mg^+$	0.25	
6	0.80	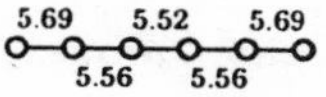 (5.69, 5.52, 5.69, 5.56, 5.56)	$Mg_3^+ + Mg_3^+$	-0.29	1.03
			$Mg_4^+ + Mg_2^+$	-0.05	1.03
			$Mg_5^+ + Mg^+$	0.15	
7	0.84	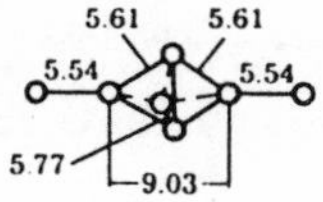 (5.61, 5.61, 5.54, 5.54, 5.77, 9.03)	$Mg_5^+ + Mg_2^+$	-0.27	
			$Mg_4^+ + Mg_3^+$	0.02	

Fig. 4. Equilibrium geometry, atomization energy, and dissociation channels of doubly ionized clusters. The atomization energy is defined as $(n - 2) E + 2E^+ - E_n^{2+}$. The dissociation energy measures the difference in energy between the fragments and the parent cluster, a minus sign indicates that Mg_n^{2+} is metastable with repect to the corresponding fragments. The barrier height has been calculated in assuming that the interatomic distances of the separated fragments correspond to the equilibrium geometry of isolated fragments. Only linear configurations have been considered. Distances are given in atomic units, and energies in eV.

+ Mg_2^+. On the other hand, a detailed study of the reaction path shows that the channel $Mg_3^+ + Mg^+$ has a lower barrier than breaking in to $Mg_2^+ + Mg_2^+$. For Mg_6^{++}, energetics alone would prefer breaking into $Mg_3^+ + Mg_3^+$ while a study of barrier heights shows that this channel is as probable as $Mg_4^+ + Mg_2^+$. These examples indicate that energetics alone may not be sufficient and the reaction barriers play an important role in determining fragmentation patterns.

To summarize we have studied the role of valence electrons in determining the properties of clusters. We have shown that the electron shells in small neutral clusters do not rearrange under geometric distortions. The magic species which arise due to the filling of the electron subshells are therefore insensitive to geometrical distortions. This accounts for the validity of the Jellium model despite the fact that the geometries for small clusters are significantly different from spherical shape assumed in the Jellium model. We then considered the effect of heteroatomic atoms on the shell fillings, in particular the K_n clusters containing Mg atoms. These clusters show K_8 Mg (containing 10 e) as the magic species as opposed to a simple Jellium which would produce shell fillings at 2, 8, 18, 20...... electrons. This feature arises due to the fact that the introduction of Mg at the cluster center generates a strong attractive potential[23] which rearranges the Jellium levels to 1s, 1p, 2s, 1d, ... thereby producing a shell filling at 10 (1s + 1p + 2s). Finally we discussed the doubly ionized clusters. Here we showed that small doubly ionized clusters are energetically unstable. However, their Born oppenheimer surface is characterized by deep minima protected by barriers. If the cluster could be made to lie in the metastable potential minima, it will have long life and hence would be observable. We have also shown that the potential barriers play an important role in the fragmentation patterns of the ionized clusters and that the energetics of the ground states of the parent and fragments is not enough to determine the fragmentation channels.

This work was supported by a grant from the Department of Energy (DE - FG05 - 87ER45316).

References

1. W. D. Knight, K. Clemenger, W. A. de Heer, W. A. Saunders, M. Y. Chou, and M. L. Cohen. Phys. Rev. Lett. 52, 2141 (1984).

2. H. Gohlich, T. Lange, T. Bergmann, and T. P. Martin. Phys. Rev. Lett. 65, 748 (1990).

3. S. Bjornholm, J. Borggreen, O. Echt, K. Hansen, J. Pederson, and H. D. Rasmussen. Phys. Rev. Lett. 65, 1627 (1990).

4. K. Clemenger. Phys. Rev. B32, 1359 (1985).

5. S. G. Nilsson, Mat. - Fys. Medd. K. Dan. Vidensk. Selsk. 29, No. 16 (1955).

6. M. M. Kappes, P. Radi, M. Schar, and E. Schumacher. Chem. Phys. Lett. 119, 11 (1985).

7. F. Reuse, S. N. Khanna, V. de Coulon, and J. Buttet. Phys. Rev. B41, 11743 (1990).

8. W. Schulze, B. Winter, and I. Goldenfeld. Phys. Rev. Lett. 62, 1037 (1989).

9. W. J. Hehre, L. Radom, P.v.R. Schleyer, and J. A. Pople. "Ab-initio Molecular Orbital Theory", John Wiley, New York, 1986.

10. G. B. Bachelet, D. R. Hamann, and M. Schluter. Phys. Rev. B26, 4199 (1982).

11. D. M. Ceperley and B. J. Alder. Phys. Rev. Lett. 45, 566 (1980).

12. W. Kohn and L. J. Sham. Phys. Rev. 140, A 1133 (1965).

13. J. L. Martins, J. Buttet, and R. Car. Phys. Rev. B31, 1804 (1985).

14. B. K. Rao, S. N. Khanna, and P. Jena. Chem. Phys. Lett. 121, 202 (1985).

15. B. K. Rao, S. N. Khanna, and P. Jena. Sol. St. Commun. 56, 731 (1985).

16. S. N. Khanna and P. Jena. Chem. Phys. Lett. (In press).

17. T. P. Martin, T. Bergmann, H. Gohlich, and T. Lange. Chem. Phys. Lett. $\underline{172}$, 209 (1990).

18. M. Goeppert Mayer. Phys. Rev. $\underline{78}$, 16 (1950).

19. K. Sattler, J. Muhlbach, O. Echt, P. Pfau, and E. Recknagel. Phys. Rev. Lett. $\underline{47}$, 160 (1981).

20. S. N. Khanna, F. Reuse, J. Buttet. Phys. Rev. Lett. $\underline{61}$, 535 (1988).

21. L. Yi, S. N. Khanna, and P. Jena. Phys. Rev. Lett. $\underline{64}$, 1188 (1990).

22. B. K. Rao, P. Jena, M. Manninen, and R. M. Nieminen. Phys. Rev. Lett. $\underline{58}$, 1188 (1987).

23. W. D. Knight, W. A. de Heer, W. A. Saunders, K. Clemenger, M. Y. Chou, and M. L. Cohen. Chem. Phys. Lett. $\underline{134}$, 1 (1987).

Atomic Clusters and Inelastic Sputtering of Condensed Matter by Ions

I. Baranov[1], B. Kozlov[2], A. Novikov[1], V. Obnorskii[1], I. Pilyugin[2], and S. Tsepelevich[1]

[1]Khlopin Radium Institute, Röntgen Str. 1, 197022 St. Petersburg, Russia
[2]Ioffe Physics-Technical Institute, Polytechnitsheskaya Str. 26, 194021 St. Petersburg, Russia

Abstract: The interest in microclusters stems from the fact that the understanding of their properties may provide insights to the fundamental problems of solid state physics and chemistry. Their wide practical use is expected [1-3]. Therefore, it is necessary to find ways of obtaining such clusters in free state. We suppose that free-state microclusters involving the number of atoms $\sim 10^3$ (or less) to $\sim 10^5$ can be obtained by irradiation of some metals with high Z or of semiconductive actinide oxides prepared as islet-like or ultradispersed layers with fission fragments. Such a conclusion can be made on the basis of our recent results of inelastic sputtering of Au and actinide (U, Pu, Am, Cm and Cf) oxides by fission fragments [4]. These results are discussud here. Experimental check of the assumption is being performed now with the specially developed original version of a dynamic mass spectrograph using an electrostatic mirror. Tentative results on mass measurements of Au clusters up to $\sim 10^7$ amu, as well as the mass spectrograph itself, are also discribed in this work.

1. Some Characteristics of the Inelastic Sputtering of Islet-Like Films of Au and Actinide Oxides by Heavy Multi-Charged Ions

Sputtering of matter occurs due to elastic and inelastic collisions between ions and surface atoms. The amount of sputtered matter depends on the nuclear and electronic stopping power. That is why maximum values of elastic sputtering yields are observed at ion velocities of $\sim 10^8$ cm/s and those of inelastic sputtering yields at $\sim 10^9$ cm/s. Elastic sputtering has been studied for more than 130 years ans its yields can be estimated with an accuracy of a factor of 2, except for special cases [4-8]. Inelastic sputtering of solids by ions has been discovered not long ago in ITEP (Moscow) and in the Khlopin Radium Institute (Leningrad) [9,10].

This work concerns only the results of the inelastic sputtering of Au and actinide oxides by heavy multicharged ions of $\sim$1MeV/nucleon which allow us to make conclusions on the masses of knocked-out particles. Fission fragments of ^{252}Cf nuclei are used as heavy multicharged ions.

A. Inelastic sputtering of Au ions occurs only when Au matter represents small islet-clusters less than 150-200 Å in size on a substrate [11-13].

Analogous results are obtained for actinide oxides. In these cases sputtering yields reach very high values, about 10^4 atoms per ion or more. If the sputtered metal or semiconductor has a coarse-grain structure with grain size more than 200 Å, inelastic sputtering will not be observed [13,14] but only elastic sputtering will take place. In this case the sputtering yield is less than 10 atoms per ion. Thus, inelastic sputtering of metals and actinide oxides is connected with a size effect. This is, however, not the case for insulators [4,7].

B. Inelastic sputtering of, for example, UO_2 or Au has a threshold character. If the ion energy deposited in the cluster is less than a critical value, inelastic sputtering does not occur [15,16]. The critical value of deposited energy depends on the cluster size, its atomic binding energy and on the adhesion of the cluster to a substrate.

C. It is important to note that clusters are sputtered due to inelastic processes only completely but not partially [13,15,17].

These results do not allow us to clonclude the nature of the sputtered particles. A cluster can be evaporated as separate atoms [12,18-20], but it can also fly as a whole. It is difficult to measure the masses of, say, Au clusters of $\sim 10^5 - 10^7$ amu since the velocities of such heavy particles are too small in usual time-of-flight mass spectrometers ($10^4 - 10^5$ cm/s) to induce the secondary electron emission necessary for their detection [21].

Some attempts have been made to estimate particle masses by studying thin collectors of sputtered matter on a transmission electron microscope or by radiography. In some cases sputtered Au of UO_2 are observed on the collector films as aggregates of atoms, the sizes of which amount to tens and hundreds of Å [10,22-25]. Sometimes such "clusters" are even larger than the initial grains on the substrate [22]. It should be noted that these methods are not direct ones for the mass and size measurement of particles flying out from the substrate.

D. The angular distributions of Au and actinide oxides inelastically sputtered by fission fragments are peaked along the normal of the target surface. The higher the sputtering yield, i.e. the larger the size of the clusters on the substrate (but not more than 200 Å), the stronger is the spread in the angular distribution [4,17,26]. In that way large clusters appear to fly away from the substrate wholly, whereas only small clusters are evaporated [17]. Actually, if all the clusters had been evaporated as separate atoms, the angular distributions would have been described by a cosine law in contrary to the experimental data.

However, narrow angular distributions can be explained in terms of properties of the Knudsen layer which can arise as a result of the evaporation of clusters, if the evaporation occurs in a similar fashion as the evaporation of condensed matter by laser beam irradiation [27].

E. When islet-like actinide oxide layers are sputtered inelastically, a considerable part (tens of per cent) of sputtered matter is produced in a charged state [28]. In the case of Au 50-80% of sputtered matter flies in a negatively charged state [26,29], with sputtering yields reaching very high values, $5 \cdot 10^3 - 5 \cdot 10^4$ per ion. If the Au clusters were evaporated as separate atoms it would mean that 50-80% of the Au atoms fly with negative charge. But each Au cluster on the substrate involves $10^3 - 10^5$ atoms. This would imply that during the evaporation $10^3 - 10^5$ electrons must be removed from the substrate in the cluster region, and this seems unreasonable.

For the measurement of the charge state of sputtered particles, points A-D above have to be taken into account, and the results obtained apparently indicate that the cluster leaves the substrate as an entity. In the case of Au it captures one or several electrons [17]. The sign of the cluster charge must be determined as a difference between the energies of breaking away of an electron from the cluster and from the substrate, while the value of charge must be determined by the cluster size. The latter can be considered as a method of studying the properties of microclusters themselves.

It follows from ref. [30] that the work function for Au clusters of 50-150 Å in size can be both greater and smaller than that of Au in bulk. Therefore the cluster charge depends on its size.

2. Mass Spectrograph and Tentative Results on Measuring Masses of Au Clusters

Masses of Au particles knocked out from the substrate by fission fragments due to inelastic processes were measured directly with a specially developed original dynamic mass spectrograph. The instrument has a space mass scanning, which is approximately scaled in accordance with the relation $X_M = X_0 + k \cdot \sqrt[3]{M}$, where X_M is the current coordinate of the mass spectrograph scale, X_0 is the zero point of the scale, M is the ion cluster mass and k is a coefficient of proportionality. The choice of such a scale allows to register ion masses within several orders of magnitude during the same experiment. Although the mass scale is essentially non-linear, and there are difficulties for light mass separation ($10^2 - 10^3$ amu), we can receive information on the order of magnitude of cluster masses in this experiment.

Figure 1 shows the mass spectrograph schematically. It consists of a thin layer of fissioning ^{252}Cf (4), a thin Ni film on which Au clusters are deposited (5), an acceleration gap (2) and an electrostatic mirror (1). The voltage applied to the acceleration gap is a combination of a low-voltage constant (-50 V) and a linearly changing sawtooth component (up to -500 V). The negatively charged sputtered Au particles, both atoms and clusters, are collected on an aluminum film of high purity (6), whereas the neutral Au particles are collected on another film (7). The setup is placed in a chamber with an ultra-high ($\sim 5 \cdot 10^{-8}$ torr) oil-free vacuum.

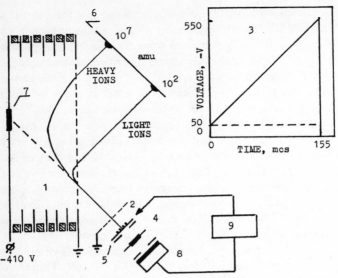

Fig. 1. Schematic view of the mass spectrograp. The numbering denotes: 1. Electrostatic mirror. 2. Accelaration gap. 3. Accelaration voltage. 4. ^{252}Cf layer. 5. Au layer on a thin film. 6. Collector of cluster ions. 7. Collector of neutral species. 8. Semiconductor detector. 9. Generator of sawtooth voltage.

The mass spectrograph works in the following way: From the spontaneous fission of ^{252}Cf in the thin film (4) either of the two fission fragments hits the semiconductor detector (8). The pulse from this detector starts the voltage generator (9). The other fission fragment sputters an Au cluster either by breaking it away from the substrate (5) as an entity or by evaporation as separate atoms. (This latter ambiguity is presently being the subject of research.)

The light Au clusters or atomic ions having negative charge are affected by the low constant potential applied to the Au layer (5) and pass rapidly through the acceleration gap and acquire only a small part of the energy at the initial region of the sawtooth scanning. The heavier the sputtered particle, the greater is the acquired energy, and thus the chosen form of the acceleration voltage allows us to establish a correspondance between the ion mass and its energy. This correspondance is single-valued for equally charged particles. The voltage rise time of 155 ms is chosen equal to the time of passage through the acceleration gap of the heaviest Au cluster - 10^7 amu, and only this mass then acquires the total possible energy.

Further, both light and heavy ions enter the electrostatic mirror, and the depth of their entering depends on their energy. Then the ions are reflected from the mirror, and following different trajectories move to the collector (6) and finally are deposited on different positions. We thus achieve a spatial separation of sputtered particles according to their masses. The location of

438

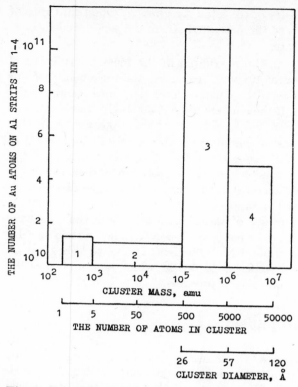

Fig. 2. Mass distribution of Au clusters on an Al collector.

Au clusters on the collector along its length depends on their masses and is determined by calculation, which presumes that each Au particle has one negative charge.

After three weeks of exposition the Al collector with deposited Au was activated by thermal neutrons from a reactor. The strip was sectioned and the amount of Au on each Al strip was determined with a γ—ray spectrometer.

Figure 2 shows a mass distribution of Au atoms on four sectioned Al strips after subtraction of the background. The horisontal axis gives the mass intervals of the Au clusters. The appropriate number of atoms in the clusters as well as the cluster diameters are plotted along the same axis. If one supposes that all the Au clusters on strip 3 have the same diameter of 57 Å, the same mass of 10^6 amu, which means that the number of atoms is $5 \cdot 10^3$, the number of such clusters in strip 3 is $2 \cdot 10^7$. Similarly, assuming a diameter of 120 Å and a mass of 10^7 amu for each Au cluster on strip 4, i.e. the number of atoms is $5 \cdot 10^4$, one has $4 \cdot 10^6$ clusters on strip 4. It is necessary here to point out the existance of a large background which is connected to the small quantity of collected Au clusters. The count rate of the 412 keV gamma quanta of the activated Au nuclei amounted to only 15-35 % of the total count rate.

As can be seen from fig. 2, the maximum in the mass distribution of Au clusters lies in the region of 10^6 amu. However, the uncertainty of our knowledge of the charge states of the clusters gives some uncertainty in the measured mass distribution. If the clusters are not all singly charged as was assumed, the mass distribution will be shifted towards higher masses. One could make a better estimate of the charge states involved. For this it is necessary first to measure the size distribution of the Au clusters on the target substrate with a transmission electron microscope, and then to measure the mass distribtuion which must be compared to the mass distribution measured with the mass spectrograph.

Despite the qualitative nature of these data it is clear that we have demonstrated that a single nuclear particle (fission fragment), such as a Kr or Xe ion with an energy of about 1 MeV/nucleon, is able to knock out an Au cluster with a number of atoms up to $\sim 10^5$, which amounts to $\sim 10^7$ amu. Further research of this kind is in progress.

3. Discussion and Conclusions

The mechanism of breaking away whole metal clusters involving $10^3 - 10^5$ atoms from a subtrate affected by a single nuclear particle has been considered by us previously [4,17]. Here we want only to note that the energy deposited by an ion in the electronic subsystem of a cluster will be transferred after thermalization of the hot electrons (within a time of $\sim 10^{-15}$ s) to the Au atoms via electron-phonon interaction initially at the cluster surface. It decreases the time taken to transfer energy from the electronic subsystem to the lattice in comparison with bulk metal. This is one manifestation of a size effect.

Another manifestation is connected to the initial energy deposited in the cluster, namely, the ion energy deposited in a cluster is proportional to its diameter, whereas the number of atoms in the cluster is proportional to the cube of the diameter. Consequently, a fission fragment deposits on the average about 10^5 eV on a Au cluster with a diameter of 100 Å, and the number of atoms in such a cluster is $\sim 3 \cdot 10^4$. Thus there is insuffient energy to evaporate the cluster to its constituent atoms taking into account the possible partial dissipation of the deposited energy. Therefore, Au clusters with diameters of 70-200 Å will only warm up, and as a result of pulse expasion will jump away from the substrate normally to the surface. This effect was estimated in [23]. Contrary to this, Au clusters with diameters less than 50 Å evaporate as separate atoms after the fission fragment passes through them. However, such small clusters can apparently be raised from the substarte intact, if one uses ions the stopping power of which is less than that of the fission fragments. This is probably realized when fission fragments pass through a small path in small clusters, but the argument has to be verified experimentally.

The proposed method can be used in addition to other methods of obtaining free-state clusters. For example, there is a method using two lasers and helium jets [3], which is very effective for obtaining clusters involving up to 100 atoms.

We would like to note that some properties of microclusters can be studied with storage ring ion beams. For example, it is possible to measure the atomic binding energy in the clusters and its dependence on the cluster size as well as to measure the adhesion of clusters to a substrate and its dependence on different factors. This possibility is based on the high precision with which the storage ring ion energy can be changed. Also, the ion beam density in such a ring is vry high, $\sim 10^{16} \text{cm}^{-2} \text{s}^{-1}$.

Finally, let us point out one possible practical use of microclusters, which may not be the most important, but, nevetheless, of interest. There is a possibility to develop a source of cluster ions, and to use it for various purposes. For example, one may accelerate heavy cluster ions and study their interactions with matter, or deposit them on substrates or use microclusters as targets which can be an alternative to gas targets.

References

1. S. Nepiiko: "Physical properties of small metal particles", Kiev, Naukova Dumka (1985)
2. J. Phillips: Chemical Rev. **86** 619 (1986)
3. M. Duncan, D. Rouvray: Sci. Am. **261** 6 (1989)
4. I. Baranov, Yu. Martyrenko, S. Tsepelevich, Yu. Yavlinskii: Usp. Fiz. Nauk **156** 417 (1988) (English translation: Sov. Phys. Usp. **31** 11 (1988))
5. R. Behrisch: "Sputtering by particles bombardment", vol. 1, Springer (1981)
6. R. Behrisch: "Sputtering by particles bombardment", vol. 2, Springer (1983)
7. K. Wien: Rad. Eff. and Defects in Solids **109** 137 (1989)
8. P. Sigmund: Phys. Rev. **184** 383 (1969)
9. F. Lapteva, B. Ershler: Atomnaya Energiya **4** 63 (1956)
10. B. Alexandrov, I. Baranov, A. Krivokhatskii, G. Tutin: Atomnaya Energiya **33** 821 (1972)
11. B. Alexandrov, I. Baranov, N. Babadzhanyants, A. Krivokhatskii, V. Obnorskii: Atomnaya Energiya **41** 417 (1976)
12. H. Andersen, H. Knudsen, P. Petersen: J. Appl. Phys. **49** 5638 (1978)
13. I. Baranov, V. Obnorskii: Atomnaya Energiya **54** 184 (1983)
14. I. Baranov, V. Obnorskii, S. Tsepelevich: Atomnaya Energiya **62** 137 (1988)
15. I. Baranov, V. Obnorskii: in Vosprosy Atomnoi Nauki i Techniki, Ser. FR-PiRM (Kharkov) **5(28)** 50 (1983)
16. I. Baranov, V. Obnorskii: Rad. Eff. **79** 1 (1983)
17. I. Baranov, V. Obnorskii, S. Tsepelevich: Nucl. Instr. Meth. B **35** 140 (1988)

18.   A. Goland, A. Paskin: J. Appl. Phys. **35** 2188 (1964)
19.   K. Izui: J. Phys. Soc. Japan **20** 915 (1965)
20.   M. Rogers: J. Nucl. Mater. **16** 298 (1965)
21.   R. Beuhler: J. Appl. Phys. **54(7)** 4118 (1983)
22.   M. Rogers: J. Nucl. Mater. **15** 65 (1965)
23.   I. Vorob'eva, Ya. Gegusin, V. Monastyrenko: Poverkhnost **4** 141 (1986)
24.   J. Biersack, D. Fink, P. Mertens: J. Nucl. Mater. **53** 193 (1974)
25.   I. Vorob'eva, Ya. Gegusin, V. Monastyrenko: Izv. Acad. Nauk. SSSR, Ser. Fiz. **50** 1597 (1986)
26.   I. Baranov, S. Tsepelevich: Atomnaya Energiya **61** 265 (1986)
27.   R. Kelly, R. Dreyfus: Nucl. Instr. Meth. B **32** 341 (1988)
28.   I. Baranov, S. Tsepelevich: Poverkhnost **7** 73 (1989)
29.   I. Baranov, S. Tsepelevich: Atomnaya Energiya **60** 82 (1986)
30.   A. Garron: Ann. Phys. **10** 595 (1965)

Precision Measurements of Atomic Masses Using Highly Charged Ions and Atomic Clusters[1]

I. Bergström[1], G. Bollen[2], C. Carlberg[1], R. Jertz[1,2], H.-J. Kluge[2], and R. Schuch[1]

[1]Manne Siegbahn Institute for Physics, Stockholm, Sweden
[2]Physics Department, Johannes Gutenberg University,
   W-6500 Mainz, Fed. Rep. of Germany

Abstract: A high precision Penning trap will be connected to the beam of highly charged ions from the electron beam ion source CRYSIS at the Manne Siegbahn Institute for Physics (MSI) in Stockholm. The first series of experiments aim at accurate mass measurements by exploiting the increase of the cyclotron frequency with the charge state of the trapped ion. Using charged states of about 50 it should be possible to achieve relative mass accuracies for mass doublets better than 10^{-9}. For this high accuracy a Penning trap with low imperfections is needed, as well as a sophisticated beam handling and retardation system for controlled injection of the ions into the trap. In order to minimize the effect of residual trap imperfections the ion motion has to be cooled. In addition the use of mass doublets will contribute to the highest accuracies. We also intend to investigate singly charged cluster ions which offer a new method of achieving mass doublets. So for example a $^{28}Si_3$ -cluster has the same mass number 84 as a $^{12}C_7$ cluster. Using such singly charged clusters, produced in an external source, we hope to achieve a relative mass accuracy of about 10^{-8} for $^{28}Si/$ ^{12}C. This measurement is part of an attempt to search for a new atomic kilogram standard. Using fully stripped ions the accuracy should be improved by at least another factor of ten.

I. Introduction

The principle used in a Penning trap mass spectrometer is based on the determination of the cyclotron frequency v_c of the trapped ion. As well known this frequency is given by:

$$\omega_c = 2\pi\, v_c = q\, B\, /\, m$$

One step on the way to highest accuracy is the use of highest possible resolving power (R).

$$R = v_c\, /\, \Delta\, v_c$$

Δv_c is the full width at half maximum. The line width Δv_c is inversely proportional to the observation time of the ion motion. So for example for a singly charged ion with mass number 76 in a magnetic field of 5T the cyclotron frequency is 1 MHz. Such an ion observed during one second gives $R = 10^6$ (1).

The resolving power can be further increased by either increasing the observation time or by increasing the charged state of the stored ion. Increasing the observation time considerably requires highest possible vacuum (10^{-12} mbar) in order to minimize the number of collisions of the stored ions with the residual gas. In addition the total measuring time increases.

The use of highly charged ions does not only increase the resolving power but has additional advantages (2). The high charge state leads to a reduced effect of the electric field imperfections, reduces the cooling time if resistive cooling is applied and increases the sensitivity if mirror currents are used for detection of the ion motion. The use of different charged states allows furthermore a new way of finding systematic errors.

For achieving highest accuracy it is of course necessary to minimize electrical and magnetic imperfections. Thus a carefull design of a compensated trap is needed. Furthermore the ions have to be injected into the trap in a well controlled, reproducable way and to be cooled. Residual systematic errors in mass comparison, which are due to trap imperfections, can be minimized by using mass doublets. Such mass doublets can be found in many applications where highly charged ions are used. Clusters offer a new way and more flexible way of obtaining mass doublets.

II. The Stockholm-Mainz ion trap facility

Beams of slow, highly charged ions can nowadays be produced by recoil-ion sources, Electron Cyclotron Resonance ion sources (ECR) or Electron Beam Ion Sources (EBIS). The Stockholm EBIS (3) named CRYSIS is thus designed to produce pulses of highly charged ions containing up to about 10^9 charges per pulse with a repetition frequency of a few Hertz . In an EBIS the DC electron beam is confined in the axial field of a strong magnet ($B_{max} = 5$ T), in the case of CRYSIS a 1.6 m long superconducting coil, where the external field is closed by an iron yoke. Axially in this field 25 electrodes are placed, each one about 9 cm long and with an internal diameter of about 9 mm (figure 1). These electrodes can be set at suitable trapping potentials during a predetermined confinement time which can be varied from tens of microseconds to minutes.

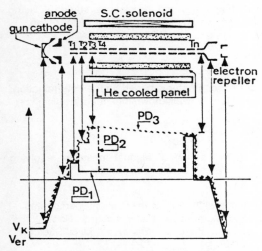

Figure 1: An electron beam ion source is in principle (ref 3; Arianer et al) a large ion trap in which highly charged ions are produced by bombardment of a very high density (about 1000 A/cm²) electron beam. In addition to "end caps" there are some 25 identical cylindric electrodes each one 9 cm long with an inner diameter of about 0.9 cm. The electrode potentials can be set at arbitrary voltages for any wanted time between tens of microseconds to minutes. PD₁, PD₂ and PD₃ are three different potential settings during confinement until ejection. The maximum electron energy is about 50 keV sufficient to produce beams of fully stripped xenon ions during confinement (trapping) times 1 s - 1 min.

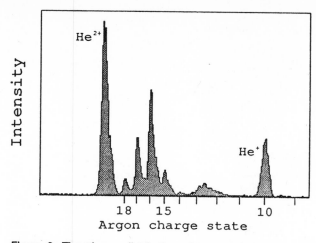

Figure 2: The charge distribution of argon ions (ref 3; Liljeby et al) produced by an electron beam of an energy about 12 keV and a current of 200 mA. The confinement time was about 1 s. The number of Ar^{18+}-ions per pulse was about 10^6.

Figure 2 shows the charge distribution of argon ions produced by a 12 keV electron beam of about 200 mA after a confinement time of about 1 s (4).

CRYSIS is designed for an electron energy of 50 keV sufficient for producing naked xenon nuclei. Already at 12 keV xenon ions of charge state q = 44 have been produced (5) by introducing neutral xenon gas into the magnet bore region. Tests will soon be done to introduce singly charged monoisotopic ions from an external isotope separator.

CRYSIS is going to be used for five low energy experiments (figure 3) including the Stockholm-Mainz Penning ion trap facility with beam energies of q x (3-5) keV as well as being the injector for a storage ring.. The Penning trap, mass spectrometer including pumps, slits for differential pumping, beam optics and time-of-flight equipment etc. is under construction at Mainz, ready to be tested in September 1991. This part of the trap facility is very similar to the Mainz trap (6a) which has been used for several years at ISOLDE (6) with great succes. The beam transportation system from CRYSIS to the trap, including a 90 degree double focusing magnet, is designed and built in Stockholm and will be ready for tests in the beginning of 1992. The trap now at Mainz will be installed at MSI in the first part of 1992 after satisfactory measurements of the relative masses of ^{28}Si and ^{12}C using suitable clusters as mentioned above. For space and stability reasons the ion trap has a vertical position.

Figure 5 shows a schematic drawing of the entire Stockholm-Mainz trap facility when connected to the CRYSIS beam at MSI. For details we refer to this figure.

For beam handling reasons the ions when leaving CRYSIS will go through a potential drop of at least 3-5 kV giving the ions a kinetic energy of q x (3-5) keV. Therefore the trap itself has to be set a a retardation voltage of 3-5 kV. The distance between the end caps of the trap is about 20 mm. When the ion pulse enters the trap the front cap is electrically open and the ions reflected at the other end cap. At the time the reflected ions again reach the entrance end cap it receives its operation voltage and the trap is

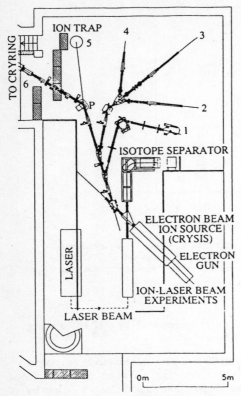

Figure 3: The low energy beams of highly charged ions from CRYSIS serves 6 experimental stations. One of the beams (6) can be accelerated by a radiofrequency quadrupole and directed to a storage ring named CRYRING. 5 refers to the position of the vertical part of the trap facility (compare figure 1).

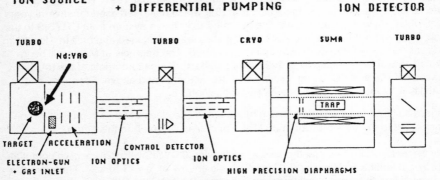

Figure 4: A scetch of the Penning trap test set up during the construction at Mainz where a precision measurement of the mass ratio $^{28}Si/^{12}C$ will be made using singly charged cluster ions produced by laser light in order to achive identical mass numbers (84).

446

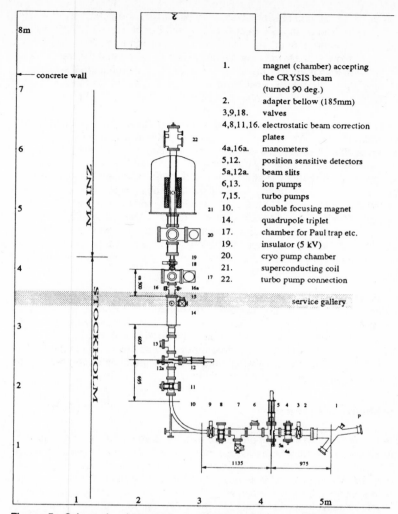

8m

concrete wall

7

6

5

4

3

2

1

MAINZ

STOCKHOLM

service gallery

1135   975

1   2   3   4   5m

1.   magnet (chamber) accepting the CRYSIS beam (turned 90 deg.)
2.   adapter bellow (185mm)
3,9,18.   valves
4,8,11,16.   electrostatic beam correction plates
4a,16a.   manometers
5,12.   position sensitive detectors
5a,12a.   beam slits
6,13.   ion pumps
7,15.   turbo pumps
10.   double focusing magnet
14.   quadrupole triplet
17.   chamber for Paul trap etc.
19.   insulator (5 kV)
20.   cryo pump chamber
21.   superconducting coil
22.   turbo pump connection

Figure 5: Schematic view of the Stockholm-Mainz ion trap facility with the connection to the electron beam ion source CRYSIS at the Manne Siegbahn Institute for Physics.

An electrostatic quadrupole triplet is positioned to the right of the 10° bending magnet (1) and will focus the beam from CRYSIS (position P is indicated both in figures 1 and 2) at the entrance slit of 90° analyzing double focusing magnet. At the entrance and exit slits of the magnet position sensitive detectors are placed (5,12). Another electrostatic triplet (14) focuses the beam at the entrance aperture (upper part of 17) of the Penning trap system. This slit consists of a 10 cm long tube with a diameter of 1 cm which thus is part of the differential pumping. At several positions (4, 8, 11, 16) pairs of electrostatic beam correction plates are placed. The entire trap facility consists of three parts separated by three valves (3, 9, 18). The part 3-9 can be exchanged by an ion source for singly or modestly charged ions or a cluster source. The trap system can also be provided with a 5-10 keV electron gun for current less than 10 mA producing ions of average charges in the very trap. An extra chamber (17) offers a future possibility to install this electron gun as well as a radiofrequency trap of the Paul type. Here a beam of laser light can also be introduced and directed into the trap.

447

closed. This means that the trap only receives a fraction 2/L of the pulse train (L is the ion pulse length in cm). The distance L depends on a number of operational conditions of CRYSIS which have to be determined experimentally. The ion trap acceptance efficiency is probably below 1 %. However, it has to be remembered that the trap will normally be operated with less than 100 ions and in measurements requiring the highest precision with only 1 ion. Therefore considerable losses can be tolerated in the beam line from CRYSIS as well as in the trapping procedure.

The cyclotron resonance frequency will be determined with the time-of-flight method introduced by Gräff et al (8). When the hyperbolic ring of the trap split in two equal parts only dipole excitation of the ions is in principle possible and the ions pick up angular momentum according to their q/m and the magnetic field. In such a case the so called modified cyclotron frequency is strongly excited. After excitation the ions are extracted from the trap as a slow bunch. In the inhomogenous magnetic field at the end of the coil the radial energy related to the orbital magnetic moment is transferred into axial kinetic energy. The ions are detected after a flight distance of about 50 cm by a channel plate detector. For each excitation frequency the time-of-flight is measured. The difference of the off- and in-resonance flight iimes gives a typical resonance curve with a pronounced minimum exactly at resonance where the flight-time has a maximum.

One observes usually the so called modified cyclotron frequency which is shifted to lower frequences compared to the true cyclotron frequency (1). However, in Gräffs pioneering experiment the true cyclotron frequency was also weakly observed, probably due to mechanical imperfections of the trap. Thus it is indeed possible to exploit this fact for a determination of the true cyclotron frequency. It was shown (7) that by splitting the hyperbolic ring in four parts allowing quadrupole excitation, a strong resonance peak can be observed at the true cyclotron frequency.

III. Cooling of highly charged ions

In an EBIS the highly charged ions are produced by succesive coillisions of atoms with electrons of relatively high energies, a process during which the ions are heated. It has been found (10) that the ions after almost complete stripping have an energy spread as large as $10 \times q$ eV. The cyclotron frequency is independent of the ion velocity as seen by the formula in section 1. However since $w = v / r$ the radius r of the cyclotron orbit varies with the ion velocity. In an assemble of heated ions introduced in the trap by a beam from CRYSIS the ions move in different orbits. If they are too large the individual ions are exposed to different magnetic and electric fields. This means that an unwanted frequency line broadening may appear. Thus there is a need for a preselection of ions in an as narrow energy region as possible and even for further cooling of the ions in the trap itself. The first energy selection can take place in CRYSIS itself, by properly adjusting the trapping voltage of the last electrode. Furthermore it is possible to precool the ions trapped in CRYSIS by introducing a proper gas or a cloud of protons. In the transport from the CRYSIS a squeezing of the energy spread can further be done by time-of-flight discrimination and momentum filtering in the 90 degree bending magnet. In the trap itself the most energetic ions can be driven to the end caps and the remaining ones treated by so called resistive cooling (10). The oscillating ions are via the ring and the two end caps part of an oscillatory circuit. The energy of the ions can then be dumped in an impedance cooled by liquid helium. In the case of protons this kind of cooling leads to cooling times of the order of hours . However, it proves that the cooling time is inversely proportional to q^2. Therefore this method is likely to be usable for ions of very high charge states. The method has the drawback that the circuit has to be tuned to one frequency only and is therefore applicable to one value of q / m.

In a CERN-experiment (11) aiming at a very accurate determination of the ratio of the masses of the proton and the antiproton electrons have been very sucessfully used for cooling the antiprotons. In a similar way positrons could be used for cooling positively highly charged ions (11). A sequence of experimental checks of the ions temperature when they leaving CRYSIS and after entering the trap have to be performed before we make a conclusion of the type of further cooling we are going to apply.

IV. On the use of clusters for mass precision determinations

Already Maxwell (13) pointed out the need of defining time, distance and weight in units based on atomic properties. Thus the second is nowadays defined by the cesium clock to an accuracy of 10^{-13} (1s = 9192631770 oscillations). As well known the meter is related to a very narrow atomic yellow-green spectral line in krypton (1m = 1650763.63 wave lenghts). However, the kilogram is still defined by a number of platinum-iridium cylindrical prototypes at the International Bureau of Weights and Measures near Paris. These prototypes are dependent on cleaning procedures and are not suitable standards if a precision better than 10^{-6} is wanted. Physicists at PTB in Germany have suggested (14) that a macro-crystal of pure ^{28}Si may be a better prototype for the kilogram. This suggestion is based on the assumption that it will be possible to determine with a high accuracy (less than10^{-7}) the number of ^{28}Si atoms in a given crystal. Our contribution will be a better measurement of the mass of ^{28}Si relative to ^{12}C. In case singly charged ions are used q / m are different for the two ions and thus they do not move in the same region in the trap which introduces systematic errors in the determination of the cyclotron frequency which are hard to correct for. A doublet measurement can, however, be achieved by using suitable clusters of silicon and carbon. Thus the clusters $^{28}Si_3$ respectively $^{12}C_7$ give mass number 84 in both cases. Figure 6 shows the results of a preliminary test of laser light production of ^{28}Si and ^{12}C clusters performed with the Mainz cluster Penning trap (15). Thus there will be no problem of producing the wanted clusters in sufficient amounts. The mass measurements will be performed by cycling the cyclotron frequency determinations of the $^{28}Si_3$ and $^{12}C_7$ ions with a time interval not longer than an hour, when the drift of the magnetic field is less than10^{-8}. We aim at an accuracy better than 10^{-8}. In this case q / A = 1 / 84. Using fully stripped carbon and silicon ions q / A = 1 / 2, giving 42 times higher cyclotron frequency and thus a corresponding possibility to increase the accuracy. The clusters are produced in an external source. A silicon respectively carbon surface is heated with an intense laser light pulse. The produced clusters are roughly mass separated by their time-of-flight to the trap. Remaining contaminations, ions with improper masses, can be removed from the trap by driving their axial or their modified cyclotron motioin until they hit an electrode.

Clusters will thus provide mass-spectroscopists with a new flexible method of achieving mass doublets in a very wide mass region. In the following we will as an illustration mention a few cases related to some mass measurements which we have in mind. Mass doublets are frequently of interest in fundamental physics. Their relative masses can be measured with an accuracy better than 10^{-9} using fully stripped or highly charged ions:

1. There are some confusing reports about the possibility of the existence of a 17 keV neutrino from studies of the beta-decay of ^3H, ^{14}C and ^{32}S embedded in a surface bar-

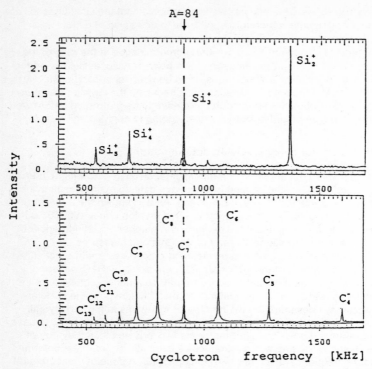

Figure 6: Preliminary results of producing clusters of silicon and carbon by laser bombardment of silicon and carbon wafers in the Mainz cluster Penning trap. The use of silicon and carbon clusters of the same mass number (84) means that the cluster ions move the same field regions in the trap, which is necessary for a high precision in the relative mass determination.

rier detector. An accurate knowledge of the relative masses of the initial and final nuclei ^{3}He, ^{14}N and ^{35}Cl is part of the problem .

2. Double beta-decay without the emission of neutrinos is a lepton number forbidden transition which has been searched for in ^{76}Ge. If existing the process would manifest itself as an electron peak at an energy given by the mass difference between ^{76}Ge and ^{76}As. The search for extremely weak peaks is very much simplified if the peak position can be exactly fixed. It is therefore justified to measure the mass difference of these two isotopes accurately. It should be noticed that a flourine cluster with cluster number 4 has the same mass number 76 which may offer a link to lighter masses.

3. An accurate knowledge of the masses of several stable atoms is important for certain problems in the lead region. ^{205}Tl has been suggested as a detector material for low energy neutrinos. Therefore it is important to know accurately the mass difference between ^{205}Tl and ^{205}Pb; a typical mass doublet measurement. These masses can be related to lower masses by comparing them with a potassium cluster with cluster number 5.

4. High spin states in the lead region offer a possibility to obtain information about the possible influence of effective three-body forces in nuclear matter . It is then very im-

portant to know the relative masses of the stable thallium and lead isotopes as well as the masses of some longlived nuclei of thallium, lead, bismut, polonium and astatine atoms with a high accuracy. Several mass doublets are possible for example ^{207}Pb-^{207}Tl and ^{206}Pb-^{206}Bi. There is a cluster link between mass number 207 and lower masses by for example 9 atoms of ^{23}Na. Furthermore, a carbon cluster of 17 atoms has mass number 204 i.e. the same as that one of the lightest stable lead isotope.

The flexibility of using carbon clusters for mass doublet calibrations would increase if also ^{13}C atoms would be present in carbon clusters. So for example carbon clusters of cluster number 17 could be used for doublet measurements of ^{206}Pb , ^{207}Pb and ^{208}Pb if respectively 2, 3 and 4 of the carbon atoms were ^{13}C and 15, 14 and 13 atoms were ^{12}C.

In this context the correction due to the cluster binding energies has to be considered. Let us consider the cluster C_{17} , which has a binding energy less than 50 eV. Suppose we are able to measure the mass of ^{204}Pb with an accuracy of 10^{-9} . This corresponds to about 200 eV. We thus see that one certainly has to worry about the cluster binding energies if the accuracy is improved by another order of magnitude. It would of course be most interesting if the mass determination of cluster doublets could reach the accuracy when it would be possible to measure the mass difference for example due to geometry isomeric states, but this would need an accuracy at the level of 10^{-11}, which some trap groups (16) dream about, but which has not been convincingly realized so far.

V. Acknowledgements

This project has been generously supported by several Swedish and German Foundations. We are very greatful to Knut och Alice Wallembergs stiftelse för vetenskaplig forskning, CarL Tryggers stiftelse för vetenskaplig forskning, Riksbankens Jubileumsfond and die Deutsche Bundesministerium fur Forshcung und Technologie. Furthermore we have recieved, together with Prof. Nico Nibbering from the University of Amsterdam, a four-year grant from the Common Market SCIENCE program for the exploration of the possibilities to study physics and chemistry problems by using highly charged ions in ion traps. We appreciate very much the competent work performed by the crew of the workshop of the Physics Department of the Johannes Gutenberg University at Mainz and the technical work of preparing the beam handling equipment by Adam Soltan at MSI.

References

1. Kluge, H.-J.: Physica Scripta ,T22 85, 1988

2. Bergström, I.: Proposal for a facility Being a combination of an Electron Beam Ion Source and a Fourier Transform Ion Cyclotron Resonance Spectrometer and a Penning Trap: MSI 89-05, ISSN-1100-214X

3. Arianer, A., Cabrespine, A. and Goldstein, C.: Nucl. Instr. and Meth.193, 401, 1982.

4. Liljeby, A. and Engström, Å: International Symposium On Electron Beam Ion Sources and their Applications; Conf. Proc.No 188, Particle and Fields Series 38, AIP

5. Cederquist et al; Proceedings of the Seventh International Conference on the Physics of Electronic and Atomic Collisions, Brisbane 1991

6. Stolzenberg, H., Becker, S. Bollen, G., Kern, F., Kluge, H.-J., Otto, T., Savard, G., Schweikhard, L., Audi, G., and Moore, R. B. : CERN-PPE/90-110; Phys. Rev. Lett. 65, 25, 3104, 1989;

6a. Becker et al., Int. J. Mass Spec. Ion Proc. 99, 53,3104, 1990

7. Gräff, H., Kalinowsky, H and Traut, Z: Z. Phys. A 297, 35 , 1980

8. Bollen et al; J. Appl. Phys. 6819, 4355, 1990

9. Cederquist, H. (MSI): Private Communication; Levin, M. A., Marrs, R. E., Henderson, J. R., Knapp, D. A. and Schneider, M. B.: Physica Scripta, 22 157, 1988

10. Holtzscheiter, M. H.: Physica Scripta, T22, 73, 1988

11. Gabrielse, G., Fei, X., Orozco, L.A., Tjoelker, R. L., Haas, J., Kalinowsky, H., Trainer, T. A. and Kells ,W.: CERN-PPE/90-98, 4 July 1990; Phys. Rev. Lett. 63, 13, 1360, 1989

12. Bollen, G., Kluge, H.-J. and Werth, G.: Application to GSI (Hochschulprogram 1992-1994)

13. Maxwell, J.C.: Report Brit.Association Adv. Science XI. Math. Phys. Sec. 215, 1870

14. Seyfrid, P.: PTB-Mitteilungen 99 5/89

15. Lindinger et al ; Proc. 5th Int. Symp. on Small Particles and Inorganic Cluster; ISSPIC, Konstanz, Sept 1990, Z. Phys. D, 1990

16. Cornell, E. A., Weisskopf, M., Boyce, K. R,, Flanagan, Jr., R. W., Lafyatis G. P. and Pritchard, D. E.: Phys. Rev. Letters 63, 16, 1674, 1989

Carbon Clusters from a Cs-Sputter Ion Source

H.-A. Synal[1], H. Rühl[1], W. Wölfli[1], M. Döbeli[2], and M. Suter[2]

[1]Institute of Intermediate Energy Physics, ETH-Zürich,
CH-8093 Zürich, Switzerland
[2]Paul Scherrer Institut, c/o Institute of Intermediate Energy Physics,
ETH-Zürich, CH-8093 Zürich, Switzerland

Negatively charged carbon clusters have been extracted from a cesium sputter ion source using graphite target material. The energy of the sputtering Cs-beam has been 40 keV and secondary particles are extracted from the ion source at the same energy. The mass distribution of the sputtered particles has been analysed using a high resolution mass spectrometer [2]. Its accessible mass range enables the identification of negative ions with masses up to 330 [amu]. A background current of a few 10^{-13}A and typically 10μA ^{12}C$^-$ current limit the seven decade dynamic range of the system.

Carbon clusters with sizes up to $n = 20$ atoms have been identified, similar to results of former experiments [1]. Characteristic odd/even alterations in formation probabilities of the different cluster sizes were found. For cluster sizes $n < 10$ they are in agreement with alterations of electron affinities of linear chain molecules [3]. The isotopic sequence ^{12}C$_{n-x}$ ^{13}C$_x$ with $0 \leq x \leq 2$ has been resolved for clusters which are not interfered by other negative ions. Overall the formation probability shows an exponential decrease toward higher cluster sizes. For $n > 13$ the most intense current peaks are shifted by 1 amu toward higher masses. This can be explained if one ^{133}Cs atom is part of these cluster molecules.

In order to get more information on the structure of these clusters the feasibility of "coulomb explosion" experiments is investigated. Prior condition for such experiments are carbon clusters at MeV energies. Therefore the possibility of accelerating cluster molecules in a tandem accelerator have been studied. In a first step C$_5^-$ clusters have been injected and accelerated with the EN-tandem accelerator. A mass spectrometer at the high energy side of the accelerator has been used to separate the C$_5$ ion beam (0.16-0.24 MeV/amu). The main question was, whether cluster molecules could survive the charge exchange process at low gas preasure ($\approx 10^{-3}$ Torr) inside the terminal stripper or not. The high energetic C$_5$ clusters were identified with a gas ionisation chamber by measuring the energy loss characteristic of the clusters fragments. In first measurements it could be demonstrated that a fraction of C$_5^+$ and C$_5^{2+}$ clusters survive the stripping process. Further experiments will be focused on acceleration and detection of carbon clusters of other sizes.

References

[1] R. Middleton
     Nucl. Instr. Meth. 144 (1977) 373-399

[2] H.-A. Synal, G. Bonani, R. C. Finkel, M. Suter and W. Wölfli
     Nucl. Instr. Meth. B56/57 (1991) 864-867

[3] S. Yang, K.J. Taylor, M.J. Craycraft, J. Conceicao, C.L. Pettiette, O. Cheshnovsky and R.E. Smalley
     Chem. Phys. Lett. Vol. 144 no. 5,6 (1988) 431-436

Atomic Structure and Magic Numbers for Cu Clusters at Elevated Temperatures

O.B. Christensen

Laboratory of Applied Physics, Technical University of Denmark,
DK-2800 Lyngby, Denmark

Thermodynamical and ground state properties of Cu clusters have been studied
with the effective medium theory including a tight-binding description of the one-
electron spectrum. Ground state energy and structure for sizes 3–29 have been
determined with simulated annealing. Finite temperature ensembles have been
generated for a range of temperatures. The sizes 8, 18, and 20 are magic, still at
temperatures close to 1000K, where the clusters are liquid. The Jahn–Teller effect
stabilizes even–sized clusters relative to odd–sized. At elevated temperature the
thermodynamical behavior of clusters depend very much on the size.

In theoretical models of metallic clusters describing the magic numbers, the atomic
structure is often neglected or included through a few parameters. Calculations including atomic structure have normally included ground state calculations of rather
small clusters or with very simple models. It is, however, important to consider
the atomic degrees of freedom quantitatively if one aims at describing ground state
structure, phase transitions, chemical properties etc.

In this paper we present a systematic study of the stability, structure, and finite
temperature properties of all neutral Cu clusters with up to 29 atoms. In spite of a
strong coupling between structure and stability of the cluster we find that the shell
structure is largely independent of the temperature (in the phonon system) of the
cluster.

The results of this investigation are based on extensive Monte Carlo simulations
applying an approximate total energy method, the effective medium theory[1]. It is
based on density functional theory and has proven very reliable in simulations and
structure determinations for bulk metals and for metal surfaces[2]. The interaction
between an atom and its metallic surroundings is expressed as a function only of
the average electron density contributed by surrounding atoms in a neutral sphere
around the atom. The cohesive function is calculated once and for all within the local
density approximation by embedding one Cu atom in a homogeneous electron gas.
A correction describes electrostatic interaction between neutral spheres that overlap.
The calculation of these two energy terms is described elsewhere[1].

Another correction term describes one-electron energy differences between the real
system and the embedded atom. Here we have included the discreteness of the one-
electron spectrum with a tight-binding Hamiltonian with distance-dependent matrix
elements. This dependence is derived from EMT, and the magnitude has been determined by fitting the s–band width to LMTO calculations.

Springer Series in Nuclear and Particle Physics   **Clustering Phenomena in Atoms and Nuclei**
Editors: M. Brenner · T. Lönnroth · F.B. Malik   © Springer-Verlag Berlin, Heidelberg 1992

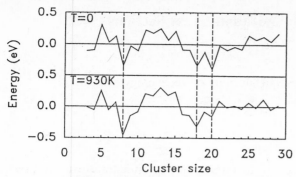

FIG. 1. Deviation of the calculated total energy from a soft fit. Upper curve: ground state. Lower curve: T=930K. The dashed lines indicate the magic numbers 8, 18, and 20.

In Fig. 1 we plot the deviation of the calculated equilibrium energies from a 3-parameter soft fit. It is clear from the figure that the clusters of the magic sizes 8, 18 and 20 are anomalously stable in our calculation, in agreement with experiment. Also the calculated equilibrium structures are in good agreement with previously obtained results for alkali metals, at the sizes available[3]. The pronounced even–odd fluctuation is attributed to efficient Jahn–Teller distortions of the even-sized clusters.

It is also seen from Fig. 1 that the magic numbers remain magic at a temperature where the clusters are observed being liquid.

When looking at the ground state shapes it is observed that 8, 13, and 20 are almost spherical. The s–electron band energy induces spherical ground state structures at the magic numbers, but non–spherical geometries away from the magic numbers. This effect competes with the tendency to close packing of atoms. It seems that none of these effects completely dominate the other.

We have calculated heat capacities for several sizes and temperatures. The close-packed 13–cluster has a pronounced peak at around 400K, which is *not* observed for non–close-packed clusters. Thus it seems that close-packed clusters may not be typical when investigating phase transitions for clusters.

---

[1]K. W. Jacobsen, J. K. Nørskov and M. J. Puska, Phys. Rev. B **35**, 7423 (1987).

[2]K. W. Jacobsen and J. K. Nørskov, in *The Structure of Surfaces II*, eds J. F. van der Veen and M. A. Van Hove (Springer Series in Surface Sciences 11, 1988). P. Stoltze, J. K. Nørskov, and U. Landman, Phys. Rev. Lett. **61**, 440 (1988).

[3]J. L. Martins, J. Buttet, and R. Car, Phys. Rev. B **31**, 1804 (1985). P. Ballone, W. Andreoni, R. Car and M. Parrinello, Europhys. Lett. 8, 73 (1989). M. Manninen, Phys. Rev. B **34**, 6886 (1986).

Calculation of the Photoionization Cross Sections for Small Metal Clusters

B. Wästberg and A. Rosen

Department of Physics, Chalmers University of Technology and
University of Göteborg, S-412 96 Gothenburg, Sweden

Alkali metal clusters have in a number of experiments during the last year been shown to provide good testing systems for analyzing effects of valence electron delocalization as a function of cluster size [1, 2]. A fundamental question that arises in these investigations is how the metallic character will develop and how it will depend on the geometry of the cluster, i.e. the delocalization of the valence electrons with respect to the ion cores of the cluster. Electron structure calculations treating the clusters within the framework of a Fermi gas of electrons in a spherical or deformed potential well, i.e. the jellium model, have been very successful in explaining observed magic numbers and optical response as plasma oscillations [3, 4]. Calculations within the jellium model suffer, however, from the drawback of not including the geometrical structure of the clusters as in traditional quantum chemistry methods, i.e. techniques based on the MO-LCAO approach, as the Hartree-Fock (HF), Configuration Interaction (CI) and Local Spin Density (LSD) methods. Reviews of different calculations have been given by Fantucci, Bonačić-Koutecký and Koutecký [5], and by Halicioglu and Bauschlicher [6].

Recently we applied the Continuum Multiple-Scattering method (CMS) for evaluating energy dependent photoionization cross sections of Na_{2-8} and K_{2-8} [7], Al_2, Al_4, Cu_2, Cu_4 [8] and X-ray Absorption Near-Edge Spectra (XANES) of Mn_2, Co_2, Ni_2 [9] and Fe_2, Ni_2 and Ni_3 [10]. The ground state properties of these clusters were first obtained by solving the Schrödinger equation self-consistently with the LSD method and by assuming some appropriate geometry. The molecular potential was then radially averaged to obtain a muffin-tin potential for evaluating the wave functions for the initial state and the outgoing electron with the CMS method. The photoionization cross sections were then evaluated using these wave functions.

Unfortunately, experimental photoionization cross sections are only known today for a few clusters and in a very limited energy range above the ionization threshold [11, 12]. However, quite recently Bréchignac and coworkers [13] have observed broad resonances when they extended their measurements on ionized potassium clusters to some few eV above the ionization threshold. Our calculations show that the magnitude of the cross section in general depends on the geometrical structure of the cluster and is rapidly decreasing above the ionization threshold. For clusters of potassium

Springer Series in Nuclear and Particle Physics   **Clustering Phenomena in Atoms and Nuclei**
Editors: M. Brenner · T. Lönnroth · F.B. Malik   © Springer-Verlag Berlin, Heidelberg 1992

we also find strong shape resonances about 2-3 eV above the threshold, which are in qualitative agreement with the observations of Bréchignac.

Shape resonances are a well-known phenomenon in molecular photoionization [14] and can be thought of as a temporary trapping of the photoelectron in a centrifugal barrier at the atomic center. Such barriers are present in clusters of potessium but not for the other investigated systems. This means that measured energy-dependent photoionization cross sections together with theoretical calculations may be used for determination of the geometrical structure of clusters. Our calculations also show that knowledge of the absolute magnitude of photoionization cross sections is crucial, since otherwise peaks interpreted as excitations from valence levels may in fact be ascribed to shape resonances.

References

1. W. Knight, K. Clemenger, W. de Heer, W. Saunders, M. Chou, and M. Cohen, Phys. Rev. Lett. **52**, 2141 (1984)
2. M. L. Cohen and W. D. Knight, Physics Today, December, p. 42 (1990)
3. W. A. de Heer, W. D. Knight, M. Y. Chou and M. L. Cohen, Solid State Physics **40**, 93 (1987)
4. W. Eckardt, Phys. Rev. B **31**, 6360 (1985)
5. P. Fantucci, V. Bonačić-Koutecký and J. Koutecký, Z. Phys. D **12**, 307 (1989)
6. T. Halicioglu, C. W. Bauschlicher, Rep. Prog. Phys. **51**, 883 (1988)
7. B. Wästberg and A. Rosen, Z. Phys. D **18**, 267 (1991)
8. B. Wästberg and A. Rosen, Proc. ISSPIC-5, Z. Phys. D
9. B. Wästberg, A. Rosen and D. E. Ellis, Z. Phys. D **12**, 377 (1989)
10. B. Wästberg, A. Rosen and D. E. Ellis, Z. Phys. D **13**, 153 (1989)
11. W. A. Saunders, K. Clemenger, W. A. de Heer, W. D. Knight, Phys Rev. B **32**, 1366 (1985)
12. M. M. Kappes, M. Schär, U. Röthlisberger, C. Yeretzian and E. Schumacher, Chem. Phys. Lett. **143**, 251 (1988)
13. C. Bréchignac, Proc. ISSPIC - 5, eds. Recknagel, Z. Phys. (1991)
14. J. L. Dehmer and D. Dill, Phys. Rev. Lett. **35**, 213 (1975)

Part VI

Reactions and Clusters

Alpha Cluster Emission in Heavy Ion Fragmentation

A. Budzanowski[1], H. Homeyer[2], H. Fuchs[2], L. Jarczyk[3], A. Magiera[3],
K. Möhring[2], C. Schwarz[2], R. Siudak[3], T. Srokowski[1], A. Szczurek[1],
and W. Terlau[2]

[1]Institute of Nuclear Physics, ul. Radzikowskiego 152,
PL-31342 Krakow, Poland
[2]Hahn-Meitner Institute, W-1000 Berlin 39, Fed. Rep. of Germany
[3]Institute of Physics, Jagellonian University,
ul. Reymonta 4, PL-30059 Krakow, Poland

1. Introduction

Heavy ion reactions at the energies higher than 10 MeV/n
are usually dominated by large cross sections for alpha
particle emission. These cross sections can be partly
explained by the evaporation of alpha particles from the
excited fusion like nuclei (FLN), target like fragments (TLF)
and projectile like fragments (PLF). A large fraction of
alpha particles emitted at forward angles with velocities
comparable to that of the beam cannot be explained by
sequential evaporation. Several preequilibrium models like
exciton model [1], moving source model [2] have been evoked in
order to explain the inclusive and exclusive spectra of alpha
particles. These models describe the energy and angular
distributions of the inclusive cross sections but usually fail
to reproduce the observed α–α or α–fragment correlations. The
drawback of these models is that they contain some adjustable
parameters like the initial exciton number and coalescence
radius [1] or source velocity and its temperature [2]. In
case of the moving source model it is also difficult to
understand how the source can be in thermal equilibrium when
its cooling time due to high heat conductivity in nuclear
matter is comparable to the formation time [3].

Recently, new classical α–cluster molecular dynamical model
has been proposed by Möhring et al. [4] in order to describe
in a microscopic way the alpha emission for various projectile
target combinations.

Two versions of this model have been elaborated. In its
original version [4] the incoming projectile is treated like a

bunch of alpha particles interacting with themselves and a
structureless target. A second version of the model in which
two alpha cluster nuclei are colliding was also developed [5].

2. Outline of the model

The classical alpha particle dynamical model of Möhring et
al. [4] can be summarized as follows. The projectile is
assumed to consist of alpha clusters interacting via two-body
forces.   A potential extracted from the adiabatic
time-dependent Hartree-Fock (ATDHF) calculation [6] is used.
At $\alpha-\alpha$ distances below d_w = 3.5 fm a stochastic force
corresponding to the isotropic elastic scattering of alpha
particles is introduced.   The interaction of the alpha
particle with the target is represented by the real part of
the optical model potential supplemented by a friction force.
Classical equations of motion are then solved.   The initial
internal configuration is obtained by random sampling of
positions and momenta of n α particles which are constituents
of the projectile. Extraction of the cross section is done by
Monte Carlo simulation according to the following procedure:
1. sampling impact parameter b; 2. sampling initial $\hat{r}$ and $\hat{p}$
for the n α particles in the projectile (for fixed binding
energy E_{bind} and total momentum $\hat{P}$); 3. classical trajectory
calculations for all n α particles interacting via $\alpha-\alpha$ and
target α-potentials; 4. when an alpha is trapped in the
α-target potential further calculation is continued with fused
nucleus.; 5. identification of outgoing fragments, E, Θ and
correlation distributions.

3. Comparison of the model predictions with the experimental
   data

So far the interaction of 5α system i.e. [20]Ne and 8α system
i.e. [32]S have been studied at 20 MeV/n and 26.2 MeV/n,
respectively. Experiments were performed at the VICKSI Heavy
Ion Cyclotron of the Hahn Meitner Institute. In the case of
[20]Ne + [197]Au it was shown [7], [8], [9] that the dynamical
cluster model reproduces very nicely inclusive cross sections,

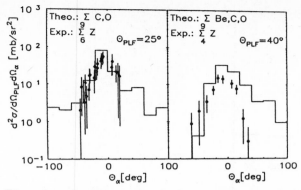

Fig. 1. Comparison of the α–PLF direct break–up cross section for the ^{22}Ne + ^{197}Au at 20 MeV/n reaction with the predictions of the classical α–cluster dynamical model according to ref. [9].

Wilczyński's plot, α–PLF correlation in the deep inelastic part and α–α angular correlation. An example of the fits to the α–deep inelastic PLF double differential cross section is shown on fig. 1. In fig. 2 the decomposition of the α–α angular correlation into various parts corresponding to the sequential decay from the compound nucleus, excited PLF and TLF and direct two 2α emission predicted by the dynamical model is presented. As can be seen from this last figure there is no component left which may be ascribed to the moving source emission.

Recently, an effort has been made to extend exclusive PLF – α measurements for the ^{32}S + ^{197}Au system at 26.2 MeV [10]. Measurements were performed at the VICKSI cyclotron using the multidetector phoswich system Argus. Charged PLF coincidences were measured. It was shown that α – quasi elastic PLF angular correlation after substraction of the sequential decay part can be well reproduced the dynamical model (see fig. 3). However, experimental alpha particle coincident energy spectra show up somewhat broader distribution extending towards higher energies than predicted by the model. This may indicate that the Fermi motion is not properly accounted for by the model and extension towards quantum molecular dynamical theory is required.

Let us finally present results on the application of the classical α–cluster dynamical model to the analysis of the

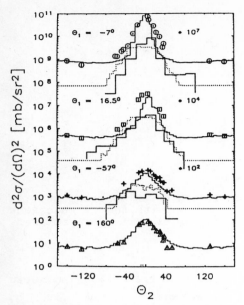

Fig. 2. Energy integrated cross sections for simultaneous emission of the α-particles into angles Θ_1, Θ_2 plotted for for fixed angles Θ_1. Points: experiment; thin solid histograms: contribution of fusion like processes and sequential projectile decay; dashed histogram: fitted moving source contribution; thick solid histogram; dynamical model prediction (according to ref. [8]).

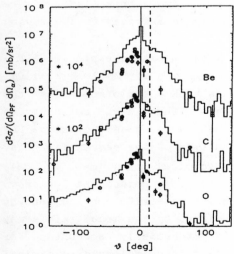

Fig. 3. Comparison of the α-PLF direct break-up cross sections for the $^{32}S + ^{197}Au$ at 26.2 MeV/n reaction with the prediction of the classical α-cluster dynamical model according to ref. [10]. The dashed line indicate the angular position of the PLF detector.

α-spectra from the $^{12}C - ^{12}C$ reaction [5]. Measurements were performed at the Jülich isochronous cyclotron using the beam of ^{12}C at the energy of 28.7 MeV/n. The total inclusive cross section for alpha particles in the measured energy and angular range is 1800 mb. It is much larger than the total reaction cross section calculated from the optical potential (about 1200 mb), what indicates the registration of events with more than one alpha particle emitted.

To analyse the collision of two alpha cluster nuclei we used the classical dynamical molecular model assuming only α-α interaction by conservative forces. The following set of equations has been solved:

$$m_\alpha d\hat{\upsilon}_{i/CM}/dt = \sum_j \hat{F}_{ij}, \quad d\hat{r}_{i/CM}/dt = \hat{\upsilon}_{i/CM}, \tag{1}$$

where $\hat{F}_{ij}$ is the force acting on the i-th alpha particle originating from the j-th one. The $\hat{r}_{i/CM}$ and $\hat{\upsilon}_{i/CM}$ are the distances and velocities, respectively, of i-th particle in the center of mass, m_α is the alpha-particle mass.

Equations (1) are integrated by the Runge-Kutta method. The numerical integration is continued long enough to be sure all interactions between particles are negligible.

The inclusive cross section for a given initial energy E_{CM} is calculated by counting alpha particles for which the direction of flight lies inside the given angular range, $\Delta\Omega$(CM or LAB) and energy is contained in the interval ΔE. It is given by

$$\frac{d\sigma}{d\Omega dE} = \frac{\pi\hbar^2 L^2_{max}}{2\mu E_{CM}} \frac{N_\alpha}{N_{event}\Delta\Omega\Delta E} \tag{2}$$

where N_α is the number of alpha particles recorded in the solid angle $\Delta\Omega$ and in the energy interval ΔE while N_{event} is the total number of events. The angular acceptance of detectors $\Delta\Omega$ has been chosen similar to the experimental one. The maximal value of the angular momentum L_{max} was calculated from $L_{max} = 0.219 R_c \sqrt{\mu(E_{CM} - V_c)} + L_0$ where $R_c = 1.44$ $(A_P^{1/3} + A_T^{1/3}$ fm and $V_c = 1.44 Z_P Z_T/R_c$ and $L_0 = 3$. In the case under study $L_{max} = 48$. The number of generated events was $2 * 10^4$ events. The energy spectra obtained from the model are compared with the experimental data in Fig. 4. Contributions

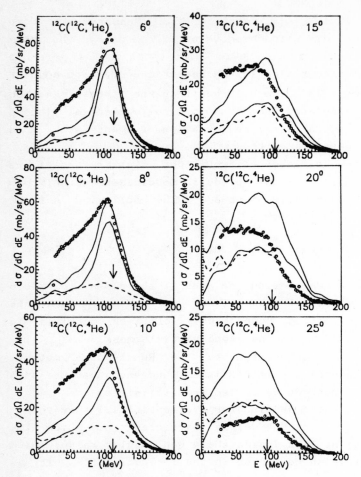

Fig. 4. Calculated energy spectra of alpha particles (solid line) from the $^{12}C + ^{12}C$ reaction at 28.7 MeV. compared with the experimental data (circles). Contributions from the projectile (solid line) and target (dashed line) are plotted additionally. Arrows indicate the position of the alpha particle energy corresponding to beam velocity. From ref. [5].

from the projectile and the target as well as the sum of both contributions are shown. As can be seen the spectra at forward angles are well described by the model. This suggests that the alphas emitted at forward angles are mostly due to direct projectile break up. It should be emphasised that the model does not contain any adjustable parameters. Calculation with the evaporation code Lilita have shown [5] that at larger angles contribution from the compound nucleus are more important.

4. Conclusion

It is shown in terms of the classical α-cluster dynamical model that spectra of alpha particles emitted in heavy ion collisions at the energies above 10 MeV/n are dominated at forward angles by direct fragmentation processes. The model reproduces many observables like alpha particle inclusive cross sections and α-PLF correlations as well as single and coincident Wilczyński's plot. Further improvements of the model e.g. extension toward quantum molecular models are highly desirable.

References

1. Machner H.: Phys. Rev. C29, 109 (1984)
2. Awes T.C., Saini S., Poggi G., Gelbke C.K., Cha D., Legrain R., Westfall G.D.: Phys. Rev. C25, 2361 (1982)
3. Tomonaga S.: Z. Phys. 110, 573 (1938)
4. Möhring K., Srokowski T., Gross D.H.E., Homeyer H.: Phys. Lett. B203, 210 (1988)
5. Szczurek A., Budzanowski A., Jarczyk L., Magiera A., Möhring K., Siudak R., Srokowski T.: Z. Phys. A — Hadrons and Nuclei 338, 187 (1991)
6. Provoost D., Grümmer F., Goeke K., Reinhard P.G.: Nucl. Phys. A431, 139 (1984)
7. Terlau W., Bürgel M., Budzanowski A., Fuchs H., Homeyer H., Röschert G., Uckert J., Vogel R: Z. Phys. A — Atomic Nuclei 330, 303 (1988)
8. Uckert J., Möhring K., Budzanowski A., Bürgel M., Fuchs H., Homeyer H., Terlau W.: Proceedings of the Third International Conference On Nucleus–Nucleus Collisions, Contributed Papers, Saint Malo, June 6–11, 150 (1988)
9. Budzanowski A., Fuchs H., Homeyer H., Möhring K., Terlau W., Uckert J.: Proceedings of the Symposium on Nuclear Dynamics and Nuclear Disassembly, Dallas, Texas April 1989, World Scientific, p. 58
10. Schwarz C.: PhD Thesis, Freie Universität Berlin, 1991

Peculiarities in Partial Cross Sections
of Light Heavy-Ion Reactions –
Evidence for Nuclear Molecules

B. Cujec

Département de Physique, Université Laval, Québec, G1K 7P4, Canada

In this paper, an attempt is being made to present evidence for nuclear molecules by investigating their disintegration products. We first precise which disintegration products are expected, considering the dynamics of molecular states, and then discuss the partial cross sections measured for the collisions of the p-shell nuclei at energies around and below the Coulomb barrier.

1. REACTION DYNAMICS

In the low-energy heavy-ion reactions three processes can be distinguished.

(1) When the colliding nuclei are still appart, they can interact via electromagnetic and nuclear interactions, the result being elastic or inelastic scattering or a direct transfer reaction. The cross sections for these processes are easily computed using the complex interaction potentials (of Woods-Saxon form), which determine all the dynamics of the reaction (e.g. the energy and angular dependence of the cross section), while the nuclear structure effects are factored out. For a direct transfer reaction $A + B \rightarrow C + D$, in which a particle c is transfered from $A = (C+c)$ to B, the cross section is given as $S_1 S_2 \sigma_{DWBA}$, where σ_{DWBA} is computed in the DWBA approximation and S_1 and S_2 are the socalled spectroscopic factors, which measure the overlaps $< A|C+c>$ and $<D|B+c>$, respectively.

(2) The colliding nuclei, with the kinetic energy ($E_{c.m.}$) typically around the Coulomb barrier, are trapped together, forming a nuclear molecule in which the two nuclei are vibrating and rotating around each other. After a while, the nuclear molecule either decays in the original components (elastic or inelastic scattering) or evolves to a more complex state by exchange of energy. Some clusters may be formed and transfered between the two components. In the next step, the nuclear molecule may decay in the two newly formed fragments (nondirect transfer

Springer Series in Nuclear and Particle Physics **Clustering Phenomena in Atoms and Nuclei**
Editors: M. Brenner · T. Lönnroth · F.B. Malik © Springer-Verlag Berlin, Heidelberg 1992

reaction) or evolve further into a more complex state, e.g. a largely deformed nucleus. As in this stage most of the energy is stored in deformation, only some low energy particles (nucleons or alpha) can sequentially be emitted. An other possibility for the nuclear molecule is to continue evolution into more and more complex states until it finally reaches a completely amalgamated and equilibrated state, the socalled compound nucleus.

(3) The compound nucleus, in which all nucleons are in a complete thermal equilibrium, deexcites by statistical evaporation of particles. For this process, the partial cross sections for the different emission channels are computed under assumption that each final state is populated with equal probability, taking into account the penetration of the Coulomb barrier and the conservation of energy and angular momentum in the ingoing and outgoing channels. In as much as the emission of the first particle leaves the final nucleus in a particle-unstable state, the statistical evaporation of the latter is again assumed. The critical quantities in the computation are the level densities, in particular of the nuclei reached after the emission of the first particle.

The intermediate stage in the heavy-ion reaction dynamics, the formation and the evolution of the nuclear molecule, needs more investigation. Actually, the existence of nuclear molecules has been clearly established for the $^{12}C + ^{12}C$ system and suggested for a few other light heavy-ion systems on the basis of observation of about 100 keV wide resonances. It is, however, very natural to expect that, at the incident energies around the Coulomb barrier, the nuclear molecules, as a kind of doorway state [1], are a general phenomenon, while the resonances may be seen only in special circumstances. Other well known doorway states are the one-particle-one-hole states formed in the nucleon (and alpha) inelastic scattering at higher energies, and the giant dipole resonances observed in photonuclear reactions. The resonances, "giant" (a few MeV wide) and those of finer structure (about 100 keV wide), are seen when the reaction selectively populates states of a given J^π value, the best example being the photonuclear reactions which populate just the $J^\pi = 1^-$ states.

In the present paper, an attempt is being made to discern the molecular states by their disintegration effects. In as much as the experimental cross sections can not be attributed either to the direct reaction or to the statistical evaporation of the compound nucleus, which can both be computed, they should be attributed to the disintegration of the molecular state. Moreover, the molecular state is expected to contribute just to specific reaction channels. If ,e.g., one of the colliding nuclei is composed of alpha clusters, one expects that the molecular state favours the transfer and the emission of an alpha cluster.The transfer in the molecular state could be distinguishable from the direct transfer process by the magnitude or energy dependence of the cross section. An other signature of the highly deformed molecular state would be the sequential emission of the low energy particles in as much as it is not in the statistical equilibrium with the single particle emission.

Especially instructive should be the comparison of partial cross sections between two reactions which form, with different colliding nuclei, the same compound system with similar excitation energies and J^π values. In this case, the statistical evaporation of the compound nucleus is the same,but the molecular states are different and their contribution to the different channel cross sections is expected to be different. The advantage of such comparison is that it is based solely on the experimental data and does not depend on the parameters entering in the theoretical evaluation of the cross section. It is mainly this method which we are using in the present discussion.

2. EXPERIMENTAL DATA

The partial cross sections have been measured, at the incident energies around and below the Coulomb barrier, for many light heavy-ion reactions. Table 1 presents the cases where the same compound system, in the same excitation energy range, is reached by two different heavy-ion reactions. The compound systems extend from ^{18}O to ^{32}S and are typically formed with the excitation energy between 20 and 30 MeV. As the two heavy-ion reactions have similar mass excess (Table 1), the compound system is formed, at a given excitation energy E_x, with similar incident energies and thus also with similar angular momenta.

The colliding nuclei extend from ^7Li to ^{20}Ne and differ widely in the disintegration and particle-separation energies (Table 2). Some consist of loosely bound clusters, as ^9Be, in which the n-alpha-alpha clusters are bound by only 1.57 MeV, and ^7Li,

470

Table 1 - Reactions considered in the present work

C.N.	Reaction	Δ(MeV) [a)]	Detection	Ref.
^{18}O	$^9Be + {}^9Be$	23.48	γ-rays	[2]
	$^7Li + {}^{11}B$	24.36	γ-rays	[2]
	$\alpha + {}^{14}C$	6.23	γ-rays	[2]
	$\gamma + {}^{18}O$	0	neutrons	[3]
^{20}F	$^9Be + {}^{11}B$	20.03	γ-rays	[4]
	$^7Li + {}^{13}C$	18.05	γ-rays	[5]
	$n + {}^{19}F$	6.60	γ-rays	[6]
^{22}Ne	$^{11}B + {}^{11}B$	25.36	γ-rays	[4]
	$^9Be + {}^{13}C$	22.50	γ-rays	[4]
^{25}Mg	$^{12}C + {}^{13}C$	16.32	γ-rays	[7]
	$^9Be + {}^{16}O$	19.80	γ-rays	[8]
^{26}Mg	$^{13}C + {}^{13}C$	22.46	γ-rays	[9]
	$^{12}C + {}^{14}C$	19.23	γ-rays	[10]
^{32}S	$^{16}O + {}^{16}O$	16.54	γ-rays	[11]
	$^{12}C + {}^{20}Ne$	18.97	γ-rays	[11]
^{24}Mg	$^{12}C + {}^{12}C$	13.93	p, α	[12]
			$^8Be \rightarrow 2\alpha$	[13]
	$^{24}Mg(\alpha,\alpha')$	-	p, α	[14]

a) Mass excess

in which the t-alpha clusters are bound by 2.47 MeV, while others (^{11}B, ^{14}C) are tightly bound. For the ^{18}O and ^{20}F compound systems, also the measurements with simple projectiles ($\alpha + {}^{14}C$, $\gamma + {}^{18}O$ and $n + {}^{19}F$) are available for comparison.

The measurements were mainly performed by the detection, in germanium detectors, of the γ-rays emitted by the residual nuclei, left in the excited states. The directly measured γ-ray cross sections, associated with the different emission channels, are being here compared between the two reactions. When the barrier penetration effects for the two incident chanels are very different (what is often the case at subbarier energies), the comparison of the relative γ-ray cross sections is preferred.

Table 2 - Characteristics of the colliding nuclei

Nucleus	Separation energy (MeV)			Cluster config.	Q(MeV)
	S_n	S_p	S_α		
^{7}Li	7.25	9.99	2.47	t-α	2.47
^{9}Be	1.67	16.89	2.47	n-α-α	1.57
^{11}B	11.46	11.23	8.67		
^{12}C	18.72	15.96	7.37	α-α-α	7.27
^{13}C	4.95	17.53	10.65		
^{14}C	8.18	20.83	12.01		
^{16}O	15.67	12.13	7.16		
^{20}Ne	16.88	12.84	4.73		

Finally, the disintegration of ^{24}Mg, formed in the ^{12}C+^{12}C reaction, is compared with the disintegration of ^{24}Mg excited by inelastic scattering of 120 MeV alpha particles. The partial cross sections for the proton and alpha-particle emissions (to the different excited states of ^{23}Na and ^{20}Ne) have been measured by particle detection. The ^{24}Mg excited states are, in both cases, mainly $J^\pi = 2^+$ states. With ^{12}C+^{12}C, they are produced at subbarrier energies where the L=2 wave absorption predominates. With ^{24}Mg(α,α'), the disintegration products are measured in coincidence with the scattered α'-particles, which are detected at an angle at which the differential cross section for the L=2 angular momentum transfer has maximum.

We now discuss the important features resulting from the comparison of partial cross sections of the reactions listed in Table 1.

(1) α-transfer reactions from ^{9}Be.

For the neutron-transfer reactions from ^{9}Be, there is ample evidence that they are direct reactions. A direct α-particle transfer has also been suggested to explain the enhancement observed in the nα channel of the ^{9}Be+^{16}O reaction [15]. There

Table 3 - The relative γ-ray cross sections for ^{9}Be + ^{13}C and ^{11}B + ^{11}B at E_x (^{22}Ne) ≈ 28 MeV

Emission channel	^{9}Be+^{13}C $E_{c.m.}$ = 4.82 MeV	^{11}B+^{11}B $E_{c.m.}$ = 2.80 MeV$^{a)}$
n^{21}Ne 0.35 → 0 p^{21}F(β$^-$)^{21}Ne 0.35 → 0	= 1	= 1
np^{20}F(β$^-$)^{20}Ne 1.63 → 0 nn^{20}Ne 1.63 → 0	3.7	3.6
nn^{20}Ne 4.25 → 1.63	0.9	0.6
nα^{17}O 0.87 → 0 3.06 → 0.87 3.84 → 0	3.2 0.6 0.8	0.8 0.6 0.9

a) Average value of the measurements at $E_{c.m.}$ = 2.67 and 2.92 MeV.

exists, however, convincing evidence that the α-transfer reactions from ^{9}Be, are not attributable only to a direct transfer process.

In Table 3, which compares the relative γ-ray cross sections for the ^{9}Be+^{13}C and ^{11}B+^{11}B reactions, forming the ^{22}Ne compound system, we see that the relative cross sections for the different emission channels are similar in the two reactions, except for the nα ^{17}O (0.87 MeV) emission channel. The latter is substantially larger in the ^{9}Be+^{13}C reaction than in the ^{11}B+^{11}B reaction, what is, of course,due to an α-transfer process. Table 4 compares the cross sections for the ^{13}C(^{9}Be,nα)^{17}O reaction, leading to the different excited states of ^{17}O, with those for the ^{13}C(^{7}Li,t)^{17}O and ^{13}C(^{6}Li,d)^{17}O reactions, which have been measured [16] at incident energies well above the Coulomb barrier, where the process is direct. The comparison of these reactions is particularly suitable because of the similar α-binding energies and reaction Q-values. The α-particle binding energies in ^{9}Be and ^{7}Li are the same (E_α = 2.46 MeV), the reaction Q-values for (^{9}Be,^{5}He) and (^{7}Li,t) are, therefore, also the same, while the ^{6}Li E_α = 1.47 MeV. In the case of a direct transfer reaction, where the cross section is expressed as a product of spectroscopic factors, describing the nuclear structure effects, and σ_{DWBA}, describing the dynamical effects, the relative population of the ^{17}O states is

Table 4 - Cross sections (in mb) for the α-transfer reactions from the ^{9}Be and ^{6,7}Li nuclei

Reaction:	^{13}C(^{9}Be,^{5}He)^{17}O Q_o = 3.89 MeV $E_{c.m.}$=4.0 MeV		^{13}C(^{7}Li,t)^{17}O Q_o = 3.89 MeV $E_{c.m.}$=11.1 MeV	^{13}C(^{6}Li,d)^{17}O Q_o = 4.88 MeV $E_{c.m.}$=12.3 MeV
Final state	a)	b)	c)	c)
^{17}O 0.87 1/2$^+$	17.9	11.1	Small	Small
3.06 1/2$^-$	4.8	3.4	0.49	0.43
3.84 5/2$^-$	5.0	3.1	1.01	1.15

Reaction:	^{9}Be(^{9}Be,^{5}He)^{13}C Q_o = 8.18 MeV $E_{c.m.}$ = 3 MeV		^{9}Be(^{7}Li,t)^{13}C Q_o = 8.18 MeV E = 6 MeV
Final state	a)	b)	d)
^{13}C 3.09 1/2$^+$	42	34	1.3
3.68 3/2$^-$	35	22	~1.4
3.85 5/2$^+$	44	30	~4.8

a)   Cross section for the nα emission channel. Ref. [2] and [4].
b)   Cross section a) with the statistical evaporation ($\sigma_{cascade}$) subtracted.
c)   Cross section from Ref. [16].
d)   Cross section from Ref. [17].

expected to be similar in the (^{9}Be,^{5}He), (^{7}Li,t) and (^{6}Li,d) reactions. This, however, is not what is observed (Table 4). While the (^{7}Li,t) and (^{6}Li,d) reactions populate the ^{17}O states with similar cross sections, and that for the ^{17}O (0.87 MeV) state is very small, the (^{9}Be,nα) reaction populates the ^{16}O (0.87 MeV) state with the largest cross section, and this in spite of the fact that the reaction Q-value favours the population of the ^{17}O (3.84 MeV) state (Q ≈ 0).

A similar situation exists for the ^{9}Be(^{9}Be,nα)^{13}C and ^{9}Be(^{7}Li,t)^{13}C reactions, also shown in Table 4: the different excited states of ^{13}C are differently populated in the two reactions.

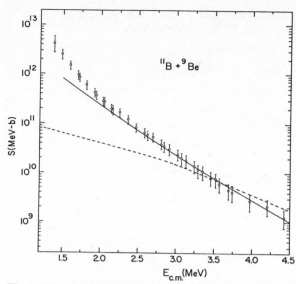

Fig.1. The total alpha transfer cross section for the ^{11}B(^{9}Be,^{5}He)^{15}N reaction, displayed in the form of S-factor, $S = E\sigma e^{2\pi n}$. The solid curve shows the DWBA calculation, normalized to the experimental points at the higher energy end. The dashed curve shows the total fusion cross section, calculated with the "standard" optical model potential. Ref [4].

An other evidence for the nondirect character of the (^{9}Be,^{5}He) α-transfer reaction is provided by the energy dependence of the total α-transfer cross section for the ^{11}B(^{9}Be,^{5}He)^{15}N reaction, shown in Fig.1. At low incident energies, the measured cross section largely exceeds the DWBA prediction.

(2) Enhancement of two nucleon emission with respect to a single nucleon emission.

The γ-ray cross sections measured for the ^{9}Be+^{9}Be, ^{7}Li+^{11}B and α+^{14}C reactions, forming the ^{18}O compound state (at 28 MeV excitation), are compared in Table 5. The incident energies (c.m.) for the three reactions are 5.0, 4.1 and 22.2 MeV, respectively, and their impact parameter values are similar. The γ-ray cross sections have the "plateau" values, except for the $nn\alpha$ ^{12}C emission channel which is still increasing with incident energy. We note that the $n\alpha$ and $\alpha\alpha$ channels have substantially larger cross sections in ^{9}Be+^{9}Be than in ^{7}Li+^{11}B, what is, of course, explicable by the α- and n-transfers from ^{9}Be. The important new result, however,

Table 5 - The γ-ray cross sections (in mb) for the ^{9}Be + ^{9}Be, ^{7}Li + ^{11}B and α + ^{14}C reactions at $E_x(^{18}O) = 28.5$ MeV

Emission channel	^{9}Be + ^{9}Be $E_{c.m.} = 5.0$ MeV	^{7}Li + ^{11}B $E_{c.m.} = 4.1$ MeV	α + ^{14}C $E_{c.m.} = 22.2$ MeV
p ^{17}N 1.37 $\to$ 0	3.	3.	9.
np ^{16}N 0.297 $\to$ 0	17.	15.	-
0.397 $\to$ 0.120	3.6	4.5	2.5
nn ^{16}O 6.13 $\to$ 0	45	30	-
6.92 $\to$ 0	24	15	4
7.12 $\to$ 0	11	8	4
nα ^{13}C 3.09 $\to$ 0	120	32	34
3.68 $\to$ 0	105	62	75
3.85 $\to$ 0	55	34	33
$\alpha\alpha$ ^{10}Be 3.37 $\to$ 0	125	36	76
nnα ^{12}C 4.43 $\to$ 0	28	14	-

is the large enhancement observed in the two nucleon emission channels, np^{16}N and nn^{16}O, in both ^{9}Be+^{9}Be and ^{7}Li+^{11}B reactions with respect to the α+^{14}C reaction. While in the α+^{14}C reaction, the cross section for a single proton emission (p^{17}N channel) is larger than are those for the two nucleon emissions, in the ^{9}Be+^{9}Be and ^{7}Li+^{11}B reactions the situation is inverse : the cross sections for the nn and np emission channels are 5 to 10 times larger than for the proton emission channel.

A similar result provides the comparison with the γ+^{18}O reaction : while the p^{17}N channel contains 25 % of the total cross section in the γ+^{18}O reaction, it contains only 2 % in the ^{9}Be+^{9}Be reaction.

The comparison of the ^{9}Be+^{11}B, ^{7}Li+^{13}C and n+^{19}F reactions, forming the ^{20}F compound system (at 22.5 MeV excitation), gives similar results (Table 6). Also here, the cross section for the single proton emission (p^{19}O channel) is very small in the ^{9}Be+^{11}B and ^{7}Li+^{13}C heavy-ion reactions, while it is large in the n+^{19}F reaction.

Table 6 - The relative γ-ray cross sections for the ^{9}Be + ^{11}B, ^{7}Li + ^{13}C and n + ^{19}F reactions at E_x (^{20}F) = 22.5 MeV

Emission channel	^{9}Be + ^{11}B $E_{c.m.}$ = 2.46MeV	^{7}Li + ^{13}C $E_{c.m.}$ = 4.5MeV	n + ^{19}F E_n = 17-20MeV
p^{19}O(β)^{19}F 1.55 $\rightarrow$ 0.197	-	0.16	1.2
np^{18}O 1.98 $\rightarrow$ 0	2.9	2.3	2.0
nn^{18}F 0.94 $\rightarrow$ 0	= 1	= 1	= 1
nα^{15}N 5.27, 5.30 $\rightarrow$ 0	7.0	4.8	1.7
$\alpha\alpha^{12}$B 0.95 $\rightarrow$ 0	7.6	0.07	-

Table 7 - The relative γ-ray cross sections for the ^{13}C + ^{13}C and ^{12}C + ^{14}C reactions at E_x(^{26}Mg) = 27.2 MeV

Emission channel	^{13}C + ^{13}C $E_{c.m.}$ = 4.77 MeV	^{12}C + ^{14}C $E_{c.m.}$ = 8.0 MeV
n ^{25}Mg 0.59 $\rightarrow$ 0	0.27	0.14
p ^{25}Na 0.98 $\rightarrow$ 0	0.09	-
α ^{22}Ne 1.28 $\rightarrow$ 0	= 1	= 1
nn ^{24}Mg 1.37 $\rightarrow$ 0	4.1	3.3
np ^{24}Na 0.47 $\rightarrow$ 0	0.66	0.05
nα ^{21}Ne 0.35 $\rightarrow$ 0	0.89	0.81

The comparison of the ^{13}C+^{13}C and ^{12}C+^{14}C reactions, forming the ^{26}Mg compound system, also provides an interesting result (Table 7). While in the ^{13}C+^{13}C reaction all neutron emission channels are slightly favoured (with respect to the α^{22}Ne channel), the relative cross section for the np^{24}Na emission channel is ~ 10 times larger than in the ^{12}C+^{14}C reaction. A similar disproportion is observed for the

$^{12}C+^{20}Ne$ and $^{16}O+^{16}O$ reactions, forming the ^{32}S compound state: the ratio $\sigma_{\alpha p}/\sigma_{\alpha}$ is ~ 10 times larger in $^{12}C+^{20}Ne$ than in $^{16}O+^{16}O$. From this observation and the fact that the colliding nuclei are more easily deformable in $^{13}C+^{13}C$ and $^{12}C+^{20}Ne$ than in $^{12}C+^{14}C$ and $^{16}O+^{16}O$, the following conclusion can be made: The sequential emission of low-energy particles is particularly favoured when the two colliding nuclei are easily deformable.

(3) Reactions with carbon nuclei

We are not going to discuss here the well known resonance structure of the $^{12}C + ^{12}C$ system, but the peculiarities observed in the ^{8}Be and α-particle emission channels at the incident energies well below the Coulomb barrier.

(i) ^{8}Be (--> 2α) emission.

the $^{12}C+^{12}C$ -->$^{8}Be+^{16}O$ reaction (Q=0.21 MeV) has been investigated, by the detection of the two α-particles resulting from the ^{8}Be-->2α disintegration (Q=0.09

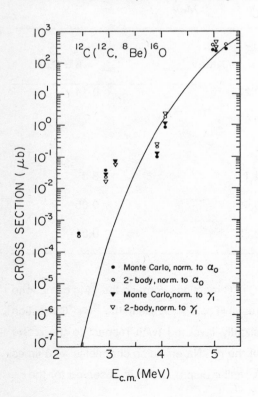

Fig.2. The $^{12}C(^{12}C,^{8}Be)^{16}O$ cross section. The curve shows the DWBA calculation, normalized to the experimental points at the higher energy end. Ref. [13].

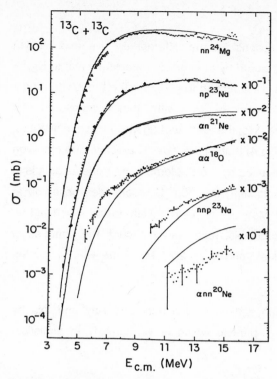

Fig.3. Cross sections for the two (and three) particle emission channels in the ^{13}C+^{13}C reaction. The curves show the statistical-model (CASCADE) calculations. Notice the enhancement in the $\alpha\alpha$ ^{18}O emission channel at low energy. Ref.[9].

MeV), down to $E_{c.m.}$ = 2.4 MeV. The cross sections, which show an L=2 angular distribution, are relatively large, particularly at the lowest measured energies, what is in contradiction with both the statistical evaporation of the compound nucleus and the direct α-transfer reaction (Fig 2.).

An enhancement in the $\alpha\alpha$ emission channel at the low incident energies has also been observed in the ^{12}C+^{13}C and ^{13}C+^{13}C reactions (Fig.3). The ^{13}C+^{13}C --> ^{8}Be+^{18}O reaction has a negative Q-value (Q=-2.09 MeV); to be a transfer reaction, it would require the transfer of a ^{5}He particle from ^{13}C which is, however, very unlikly because of the large binding energy (13.21 MeV).

Therefore, we suggest that the enhancement observed in the $\alpha\alpha$-channel of the ^{12}C+^{12}C, ^{12}C+^{13}C and ^{13}C+^{13}C reactions arises from the contribution of a molecular state, in which a ^{8}Be cluster is formed and emitted.

479

(ii) α-particle emission.

For $^{12}C+^{12}C$, the partial cross sections for the different $\alpha_i+^{20}Ne$ and $p_i+^{23}Na$ emission channels have been measured by the particle detection down to $E_{c.m.} = 2.4$ MeV ($E_x(^{24}Mg) = 16.3$ MeV). It is interesting to compare the branching ratios, $\sigma_\alpha/\sigma_{tot}$, measured for $^{12}C+^{12}C \longrightarrow \alpha_i+^{20}Ne$ with those measured for $^{24}Mg(\alpha,\alpha') \longrightarrow \alpha_i+^{20}Ne$. The comparison for the $\alpha_2+^{20}Ne(4.25$ MeV$)$ and $\alpha_3+^{20}Ne(4.97$ MeV$)$ emission channels shows (Fig.4) that, in the $^{12}C+^{12}C$ reaction, the emission of low-energy α-particles is not inhibited by the Coulomb barrier effects as it is in the $^{24}Mg(\alpha,\alpha')^{24}Mg$ reaction. In particular, the $^{12}C+^{12}C$ resonances observed at $E_{c.m.} = 2.55$ MeV and $E_{c.m.} = 3.12$ MeV decay preferentially to the $\alpha_2+^{20}Ne(4.25$ MeV$)$ and $\alpha_3+^{20}Ne$ (4.97 MeV) channels, respectively, though the energies of emitted α-particles ($E_{\alpha 2} = 2.92$ MeV and $E_{\alpha 3} = 2.77$ MeV) are well below the $\alpha+^{20}Ne$ Coulomb barrier.

The fact that a particular resonance decays in a particular way and this by emission of such a low-energy α-particle which is very unlikly in statistical

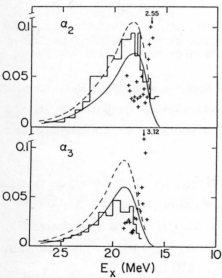

Fig.4. The branching ratios for the α_2 and α_3 emissions from the ^{24}Mg excited states, formed via the $^{24}Mg(\alpha,\alpha')^{24}Mg$ (histograms) and $^{12}C+^{12}C$ reactions (crosses), in dependence on the ^{24}Mg excitation energy. The curves show the statistical-model evaluation for a completely T-mixed decaying state (the full curve) and for a pure T=0 state (the dashed curve). Ref.[14] and data from ref. [12].

evaporation, indicates that the emission of the α-cluster occurs, in these cases, from a molecular state.

3. CONCLUSIONS

The evidence for the following processes, related to nuclear molecules, has been presented:

(i)   nondirect α-cluster transfer,

(ii)   nonstatistical sequential emission of nucleons,

(iii)   nonstatistical emission of α- and ^{8}Be clusters.

REFERENCES

1.   H. Feshbach, A.K. Kerman and R.H. Lemmer, Ann. Phys. (N.Y.) 41 (1967) 230

2.   B. Dasmahapatra, B. Cujec, G. Kajrys and J.A. Cameron, Nucl.Phys. to be published

3.   J.G. Woodworth, K.G. McNeill, J.W. Jury, R.A. Alvarez, B.L. Berman, D.D. Faul and P. Meyer, Phys. Rev C19 (1979) 1667

4.   B. Cujec, B. Dasmahapatra, Q. Haider, F. Lahlou and R.A. Dayras, Nucl. Phys. A453 (1986) 505

5.   B. Dasmahapatra, B. Cujec, F. Lahlou, I.M. Szöghy, S.C. Gujrathi, G. Kajrys, and J.A. Cameron, Nucl. Phys. A509 (1990) 393

6.   J.K. Dickens, T.A. Love and G.L. Morgan, Oak Ridge Nat. Lab. Report ORNL-TM-4538 (1974)

7.   R.A. Dayras, R.G. Stokstad, Z.E. Switkowski and R.M. Wieland, Nucl. Phys. A265 (1976) 153

8.   Z.E. Switkowski, S.-C. Wu and C.A. Barnes, Nucl. Phys. A289 (1977) 236

9.   J.L. Charvet, R.A. Dayras, J.M. Fieni, S. Joly and J.L. Uzureau, Nucl. Phys. A376 (1982) 292

10. B. Dasmahapatra and B. Cujec, unpublished

11. G. Hulke, C. Rolfs and H.P. Trautvetter, Z. Phys. A297 (1980) 161

12. M.G. Mazarakis and W.E. Stephens, Phys. Rev C7 (1973) 1280

13. B. Cujec, I. Hunyadi and I.M. Szöghy, Phys. Rev. C39 (1989) 1326

14. F. Zwarts, A.G. Drentje, M.N. Harakeh and A. Van der Woude, Nucl. Phys. A439 (1985) 117

15. B. Cujec, S.-C. Wu and C.A. Barnes, Phys. Lett. 89B (1979) 151

16. K. Bethge, D.J. Pullen and R. Middleton, Phys. Rev. C2 (1970) 395

17. F.D. Snyder and M.A. Waggoner, Phys. Rev. 186 (1969) 999

Multiply Charged Atomic Clusters

O. Echt

Department of Physics, University of New Hampshire,
Durham, NH 03824, USA

Introduction

The stability of charged droplets is of great importance to
a variety of phenomena, such as thunderheads, electrospray
techniques, and liquid-metal ion sources. A thorough theoretical analysis was published over a century ago.[1] The subject received renewed interest after the discovery of nuclear fission[2-4] and, a decade ago, after the discovery of lower size
limits in mass spectra of multiply charged atomic and molecular
clusters.[5] Early studies focussed on the relation of these size
limits ("appearance sizes") to other properties of the material
in question, and on their dependence on the charge state z. A
coherent picture has emerged for weakly bound systems, even
though the static picture suggested initially is inadequate.
For metallic clusters, appearance sizes have been found to
strongly depend on experimental conditions, varying by as much
as a factor of ten for the "historical" case of Pb_n^{2+}. Ab-initio calculations on metals suggest that several doubly charged
clusters comprising as few as 2 or 3 atoms are, indeed,
metastable, but their atomic configurations are not accessible
by vertical ionization of neutral precursors.

Recent experimental studies have focussed on the dynamics
of multiply charged clusters: The size dependence of the fission rate of thermally excited clusters, fission following collisional excitation, competition of fission with evaporation of
atoms, and the size distribution of fission fragments have been
investigated. Some exciting features have emerged but, by and
large, these studies are still at their infancy. A major obstacle is the inability to prepare cold multiply charged clusters,
or to measure and vary their internal energy in a controlled
manner.

Springer Series in Nuclear and Particle Physics   **Clustering Phenomena in Atoms and Nuclei**
Editors: M. Brenner · T. Lönnroth · F.B. Malik   © Springer-Verlag Berlin, Heidelberg 1992

In this contribution I shall highlight some general trends, some discrepancies, and some findings pertaining to fission of atomic clusters. For a much more detailed discussion, the reader is referred to a recent review.[6] A brief comparison with nuclear clusters will conclude this chapter.

Terminology and Rules of Thumb

In studies of multiply charged clusters, terms such as "critical size" or "stability" are frequently used, but their meaning varies from author to author. Fig. 1 defines and labels some quantities that are pertinent to the present discussion. Let us consider clusters in their rovibrational ground state. For technical reasons, multiply charged atomic clusters P_n^{z+} are usually studied in a given, relatively low charge state (mostly z = 2 or 3). Fission of these species into two charged fragments may be impeded by a *fission barrier* E_{bn}^{z+} (also written ten E_b, if n and z are specified) if n is small. E_b will be zero at the *critical size* $n_c(z)$, which marks the transition from *unstable* to *metastable* (or *kinetically stable*) clusters. (This definition implies non-integral n-values. In practice, the smallest integral n-value for which E_{bn}^{z+} is positive is usually defined to be $n_c(z)$). At and slightly above this size, fission would still be exothermic, i.e. the *heat of fission*

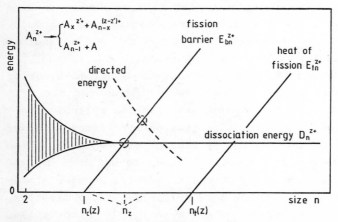

Figure 1: Schematic size dependence of some relevant properties of multiply charged clusters. n = size, z = charge state, $n_c(z)$ = critical size, n_z = appearance size.

E_{tn}^{z+} (or, in short, E_t) will be negative.[7] The transition to *thermodynamically stable* (often called *intrinsically stable* or *truly stable*) clusters occurs at $n_t(z)$, where $E_t = 0$. *Stable* clusters are either metastable or thermodynamically stable. Note that the momentary excitation energy in the cluster does not enter our definitions.

For a charged drop of an incompressible fluid, with a uniform charge confined to the surface, the well-known relation

$$n_t(z) = n_c(z)/0.351 \qquad (1)$$

is obtained. Likewise, the barrier height E_b can be calculated from the fissility parameter n_c/n.[2,8] However, the applicability of these results to metal clusters is highly questionable, because the surface charge density will become non-uniform once the spherical droplet starts to deform.[9] Several theoretical studies have addressed the thermodynamic stability of metal clusters, and the value of $n_t(z)$. For example, $n_t(2) \approx 25$ has been reported for sodium and cesium,[10,11] while Mg_n^{2+} and Be_n^{2+} are predicted to be thermodynamically stable beyond $n = 6$.[12,13]

However, E_t and $n_t(z)$ are of no relevance to the *observability* of multiply charged clusters. Let us call the smallest species being observed the *appearance size* n_z, because this quantity may depend on the experimental conditions. Obviously, $n_z \geq n_c(z)$. If prepared without excess energy, all clusters $n \geq n_c(z)$ will have a lifetime much longer than any reasonable experimental time scale (the latter typically being 10^{-5} s for conventional mass spectrometers, and a few orders of magnitude longer for ion traps). Tunneling through fission barriers can be safely ignored for these species.[11,14]

Appearance sizes of doubly charged clusters range from 2 to 30 for metal clusters, and from 3 to 100 for van der Waals and hydrogen bound clusters. For most of the latter systems, the appearance size translates into a radius of (10 ± 1) Å (e.g. for systems as different as Ar, Kr, Xe, N_2, N_2O, SF_6, C_2H_6, C_3H_8, C_4H_{10}, C_6H_6 [15]), if the bulk density and a spherical shape is assumed. The kinetic energy released upon fission may then be estimated from the heat of fusion of two non-polarizable spherical fragments to be somewhat less than 1 eV.[16] In some systems, the appearance size is remarkably reproducible even though the multiply charged clusters are initially hot. To facilitate a discussion of these findings, Fig. 1 also displays the size-dependence of the *directed energy* and of the *dissociation energy* D_n^{z+}. The former is the potential energy in the compression mode of P_n^{z+}, at the instant of its formation from a neutral precursor.[18,19] The dissociation energy is the energy required for removal of the most weakly bound neutral atom or monomeric unit from P_n^{z+}.[20] It will approach the heat of vaporization H_{vap} for macroscopic clusters (independent of z). However, for n approaching 2, D_n^{z+} may be strongly size dependent (e.g., D_n^{1+} decreases with decreasing n for several metals, while it strongly increases for van der Waals clusters).

D_n^{z+} is intimately connected to the effective temperature of the cluster, because on the time scale of 10^{-5} s the cluster will cool to

$$k_B T \approx D_n^{z+}/25 \tag{2}$$

by successive evaporations of atoms,[21] provided the initial excess energy was large, and no other decay channels (fission!) compete. As $D_n^{z+} \approx H_{vap}$ for large aggregates, Trouton's rule together with the boiling point of the material in question already provides a crude estimate:

$$T \approx 0.4\, T_{boil} \tag{3}$$

which has been confirmed by electron diffraction studies of several neutral van der Waals clusters.[22]

Lower Size Limits of Multiply Charged Clusters

Well-defined lower size limits in mass spectra of multiply charged clusters were first reported by Sattler et al. in 1981.[5] They observed Xe_n^{2+} down to n = 53, $(Na_n \cdot I_{n-2})^{2+}$ down to n = 21, and Pb_n^{2+} down to n = 31. Fig. 2, e.g., displays a mass spectrum of argon cluster ions. The steep rise of the Ar_n^{2+} signal at n = 93 is clearly visible,[44] while Ar_n^{3+} can be traced down to n = 226. To what extent are these appearance sizes dependent on the characteristics of the cluster source and of the mass spectrometer? Exploiting a double focussing magnetic spectrometer with greatly enhanced resolution and signal-to-noise ratio, Märk and coworkers detected Xe_n^{2+} as small as n = 47, and Ar_n^{2+} down to 91.[23,24] Similarly, the appearance sizes of other multiply charged van der Waals clusters have passed the test of time relatively well.[15]

In contrast, lower size limits of doubly charged *metal* clusters suffer from poor reproducibility. In the above-men-

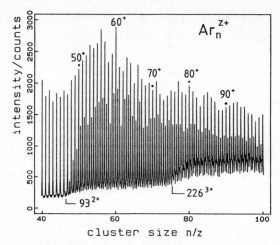

Figure 2: Time-of-flight mass spectrum of argon clusters, ionized by electron impact ionization at 400 eV. Peaks arising from Ar_n^{z+} are labelled n^{z+}. The appearance size of doubly and triply charged clusters is marked below the spectrum (adapted from ref. 52).

tioned experiment on lead clusters, a gas-aggregation source (*smoke source*) was employed, combined with electron-impact ionization of neutral clusters emerging from the source. The initially reported size limit $n_2 = 31$ was reproduced in several experiments,[25] but eventually clusters down to Pb_7^{2+} were detected, even though the stepwise increase of ion abundance beyond Pb_{31}^{2+} was still discernible.[26] These results were assigned to the occurrence of two different structural isomers in the cluster beam: linear and 3-dimensional, compact clusters. The former would give rise to a much smaller appearance size than the latter ones.[27] However, later investigations, using similar experimental techniques, have placed the appearance size at $n_2 = 3$.[28,29]

Is lead special? Appearance sizes below 9 are rather exceptional under conditions where neutral, pre-formed clusters in a beam are ionized by electron impact or by (multi-)photon absorption, although Hg_5^{2+} and Bi_5^{2+} have been observed[28,30] (cf. Fig. 3, to be discussed later). However, if clusters are desorbed from the surface under the influence of a very strong electric field (e.g. laser assisted field evaporation or liquid-metal ion sources), doubly charged metal dimers (Au, Mo, Mg, Sb) and trimers (Sn, Ni, W, Bi) have been reported.[15] For comparison, investigations involving free neutral precursors have failed so far to produce clusters smaller than Au_9^{2+}, Mo_{19}^{2+}, or W_7^{2+}.[15]

Recent ab-initio calculations have shed some light on this behaviour. Several doubly charged metal dimers (Be, Mg, Hg, V, Cr, Fe, Mo, W) are predicted to be metastable.[12-14,31,32] Likewise, larger clusters are stable, but (for Be and Mg, at least) their ground state configuration significantly differs from that of their neutral analogs, and vertical double ionization will generate highly excited species which undergo prompt fission into singly charged fragments.[12,13] Ultimately, however, for large n, prompt fission may be quenched by intra-cluster collisions. Neutrals rather than charged fragments will escape,[33] and the cluster will cool down as predicted by the model of evaporating cooling (eqs. 2,3). Starting with a wide size distribution, we will always, i.e. under all experimental conditions and at all times, have available multiply charged clusters which just boiled down to a size where the fission barrier E_{bn}^{z+} equals[34] the heat of evaporation D_n^{z+} (cf. Fig. 1). Hence, on all experimentally accessible time scales, clusters P_n^{z+} of size n_z will exist, but they are unstable with respect to thermally activated (spontaneous) fission.

Spontaneous fission has, indeed, been detected in case of doubly charged sodium and potassium clusters,[35] and triply charged carbon dioxide, ethylene, and ammonia clusters.[36-39] The effective fission rate is very high if n is barely larger than n_2, it decreases rapidly with increasing n, providing strong evidence for the correctness of the above model. This model also allows to estimate the ratio $n_z/n_c(z)$ provided the size dependence of the barrier E_{bn}^{z+} on the fissility parameter $n_c(z)/n$ can be calculated, and provided D_n^{z+} can be approximated by H_{vap} or by some other known quantities. For doubly charged sodium, the appearance size is $n_2 = 27$,[35] which may be compared with $n_c(2) \approx 12$ or 17, obtained from the liquid drop model (cf. below) for two different values of the surface tension.[6,40] Furthermore, the predicted equality of the fission barrier and the heat of vaporization at the appearance size is reproduced by the liquid drop model. However, a recent ab-initio study of Na_n^{2+} calculates $n_c(2) = 8$.[41]

A ratio $n_z/n_c(z) \approx 1.3$, i.e. much closer to 1.0 than for sodium, is obtained from theoretical studies of $(CO_2)_n^{z+}$ (z =

2,3) and Xe_n^{2+}.[19,37] This assumes that the basic idea of the model is correct, although there is evidence that this may not be so for doubly charged van der Waals clusters: Attempts to identify fragments from spontaneous fission of $(CO_2)_n^{2+}$, $(NH_3)_n^{2+}$, and Ar_n^{2+}, have failed.[37-39,42] Instead, these species evaporate neutrals for all n, down to n_2, indicating that the fission barrier is significantly *larger* than the dissociation energy D_n^{z+}. Why, then, are clusters just below n_2 non-observable, i.e. why do they undergo fission before they can be detected in the mass spectrometer? Introducing two localized charges (electronic holes) into a neutral, equilibrated van der Waals cluster will result in an ion which features a well-defined potential energy with respect to its non-spherical equilibrium shape. This "directed" energy may drive the cluster over the fission barrier within less than a period of the collective mode, i.e. it acts much more efficiently than an equivalent amount of thermal energy.[18,19,37] Theoretical studies suggest that the directed energy exceeds the dissociation energy (cf. Fig. 1), hence a cluster which is just large enough to survive the first few periods of the strongly damped collective oscillation will evaporate atoms. Whether or not it undergoes fission at much later times, after it has evaporated several atoms, will depend on the *total* excess energy initially available, and on the distance between the two relevant crossing points (Fig. 1).

We have discussed two models which postulate a correlation between the appearance size and the critical size. Fig. 3 demonstrates the existence of such a correlation for van der Waals systems. The critical radius is calculated from the Rayleigh criterion

$$r_c = (z \cdot e_o)^2 / (4\pi\epsilon_o \cdot 16\pi\sigma) \tag{4}$$

which establishes the limit where a charged metallic sphere becomes unstable with respect to quadrupolar deformation.[1,2] Hence, we obtain $n_c(2)$ as a function of $\sigma \cdot v$ (solid line in Fig. 3). The appearance sizes of van der Waals clusters (full dots) closely follow this line, although they apparently violate the trivial relation $n_2 \geq n_c(2)$. This suggests a numerical deficiency of eq. 4 (note that eq. 4 assumes a *uniform surface charge!*), but may also be traced to our taking σ values from the boiling point of the material in question. The effective

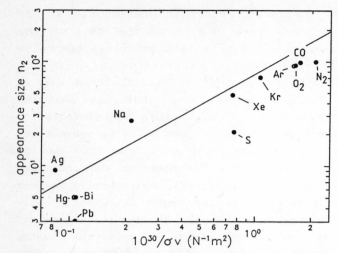

Figure 3: Appearance sizes of doubly charged clusters[15] plotted versus $(\sigma \cdot v)^{-1}$ (σ = surface tension, v = molecular volume). The solid line indicates the critical size calculated from the Rayleigh criterion.[1]

temperature is likely to be much less (eq. 3), hence we underestimate σ and overestimate $n_c(2)$. It is also interesting to note that the appearance sizes of van der Waals clusters in higher charge states exhibit the z-dependence predicted by eq. 4. The average ratio of the values reported so far is $n_c(2)$: $n_c(3)$: $n_c(4)$ = 1 : 2.22 : 4.08.[15]

On the other hand, the appearance sizes of *metallic* clusters do not scale as $(\sigma \cdot v)^{-1}$. Of course, the liquid drop model is bound to fail if the critical cluster comprises just a few atoms. Sticking to the language of continuum models, there will be no universal correlation between the effective surface tension of the charged drop and that of the neutral, because the nature of bonding will depend on z (recall[32] the metastability of He_2^{2+}!). Also note that the large scatter of n_2 for metal clusters (Fig. 3) may be viewed as a consequence of the poor reproducibility of experimental values as mentioned earlier. Neither of our models account for this phenomenon.

Space does not permit to dicuss other models; let us merely mention that fragmentation-less ionization of Pb_n by electron impact has been invoked as a possible explanation.[28,29] This implies that the smallest observable species Pb_n^{z+} is simply given by the smallest precursor being generated in the source.

It has some experimental support,[28,29] but fails to account for other observations.[15] Also recall that ab-initio calculations predict that small doubly charged Be or Mg clusters, formed by vertical ionization from neutral precursors, will exhibit strong fragmentation, because of grossly different equilibrium geometries. On the other hand, these studies suggest that the local ground state of metastable, doubly charged species may be reached by vertical ionization of singly charged precursors. This would explain the frequent occurrence of doubly charged dimers or trimers under conditions where post-ionization is a likely mechanism (field evaporation, liquid-metal ion sources). It would also explain the recent observation of Nb_2^{2+} in an in-beam experiment, because neutral as well as singly charged clusters were present under those conditions.[43]

Size Distribution of Fission Fragments

Under typical experimental conditions, multiply charged cluster ions traversing a mass spectrometer are relatively hot, hence they may undergo (low energy) fission. Is the size distribution of fission fragments symmetric or asymmetric? Does fission produce fragments of a specific size? A first answer was obtained in experiments on triply charged carbon dioxide clusters, which feature an appearance size $n_3 = 108$.[36,37] The probability for the reaction

$$(CO_2)_n^{3+} \quad \longrightarrow \quad (CO_2)_m^{2+} + (CO_2)_x^+ + y \cdot (CO_2) \tag{5}$$

where $m+x+y = n$, is displayed in Fig. 4. The signal has been integrated over the range $108 \leq n \leq 120$, and the abscissa shows the relative fragment size, m/n. The size distribution is strikingly narrow and asymmetric, it peaks at 92%. However, only the (large) doubly charged fragments, but not the small singly charged fragments, could be identified due to technical difficulties (background from fragmenting singly charged clusters and increased divergence of the ion beam due to the recoil energy), hence the number of prompt neutrals lost upon fission remains unknown. Experiments on triply charged ammonia clusters, with definite mass identification of fragment ions, lead to similar results,[38] while the size distribution of doubly charged fragments from $(C_2H_4)_n^{3+}$ is considerably less asymmetric.[37] Simple model calculations agree reasonably well with the

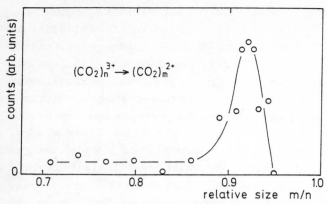

Figure 4: Size distribution of $(CO_2)_m^{2+}$ originating from thermally excited $(CO_2)_n^{3+}$ (integrated over $108 \leq n \leq 120$) by spontaneous fission (adapted from refs. 36,37)

appearance size and size distribution.[37] Generally, the asymmetry is predicted to become even more pronounced for larger sizes (i.e. for systems featuring larger appearance sizes n_3), while low-energy fission of doubly charged clusters is expected to be more symmetric. The reasoning behind this is as follows: The surface tension generally favors emission of atoms or other small species, but the solvation energy of ions in van der Waals systems is large compared to the surface tension, hence escaping ions want to be solvated by one or two layers.

Spontaneous (low energy) fission of doubly charged alkali clusters has been the subject of several recent studies; some of them are discussed in other contributions to this conference.[11,35,41] One interesting issue has been the preference for magic fragments. Calculations within the spherical jellium model indicate that magic fragments (e.g., Na_3^+, Na_9^+, Na_{21}^+,...) would be strongly favored on energetic grounds.[45] Mapping of the potential energy surface of deformed Na_n^{2+} also suggests the crucial role of shell corrections, although those studies were restricted to deformations that are symmetric under inversion.[46] A local-spin-density molecular dynamics of Na_n^{2+} (n = 8, 10, 12) clearly indicates preference for fission into Na_3^+, and it reveals a particularly high fission barrier for the closed-shell cluster Na_{10}^{2+}.[41]

Experimental evidence for preferential fission of doubly charged clusters of monovalent metals into magic fragments exists, indeed. Katakuse and coworkers have analyzed the unimolecular decay of Ag_n^{2+} ($n = 12 - 22$), generated by ion bombardment of silver under vacuum.[47] For technical reasons, only heavy fragments Ag_m^+, $m > n/2$, could be detected (lighter fragments are buried under the background of decaying singly charged cluster ions, $Ag_{n/2}^+$). Up to $n = 16$, and for $n = 18$, the fragment of size $m = n-3$ is dominant; for some parent ions, Ag_9^+ is another preferred fragment. In other words, fission into the closed-shell clusters Ag_3^+ and Ag_9^+ is strongly preferred. For heavier clusters ($n = 17$, $19 - 22$), the size of the most intense heavy fragments is but slightly larger than $n/2$. In some cases, either Ag_9^+ or Ag_{n-9}^+ is a dominant fragment.

Another intriguing series of experimental studies involve collision-induced dissociation. Not surprisingly, the distribution of fragments from high-energy fission of $(CO_2)_n^{2+}$ (center-of-mass collision energy $\approx 10^2$ eV) is very broad and featureless,[48] while a trend towards fissioning of Au_n^{2+} into magic fragments is apparent for collision energies in the eV-range.[49] Even so, the preferred fission channel may change if the collision energy is tuned from 0 to 10 eV.[50] Another interesting outcome of this study is the size dependence of the total fission rate versus the rate of monomer evaporation, which is found to be in good agreement with an adapted version of the nuclear liquid drop model.[49]

Outlook

Nuclear fission has been under investigation for half a century, the essential features of spontaneous fission, low energy fission and superasymmetric fission are believed to be well understood.[3,4,51] Studies of multiply charged clusters, mostly performed over the past 5 years, have not yet reached a similar degree of maturity. On the theoretical side, the energy surface of fissioning clusters has been mapped in just a few cases, limited to rather small systems.[12,13,41] Analysis of the dynamics of larger systems, relevant to, e.g., van der Waals clusters, still involve a number of grossly simplifying assumptions.[18,19,37] One obstacle in the calculation of fission barriers within continuum models is that the range of attractive and repulsive forces is not grossly disparate, as it is in case of nuclei. There is no simple approximation for the cluster-cluster interaction once they are "separated", because there is no clear-cut point of scission.

On the experimental side, one has not yet succeeded in detecting all fission fragments in a coincidence experiment, hence we do not know the number of "prompt neutrons" (neutrals). The kinetic energy of charged fragments has been determined in a few cases, involving high-energy collisions.[43,48] However, the Q-value (energy balance) of the reaction remains elusive as long as only one fragment per event is detected (unless the number of emitted neutrals is zero). Also note that the internal excitation energy of the precursors is poorly known, although rules of thumb allow us to estimate the excitation energy in case of *spontaneous* fission. For not too small precursors, this energy grossly exceeds the evaporation energy, in stark contrast to the situation encountered in radioactive nuclei, where photon (gamma) emission provides a very efficient cooling mechanism on the time scale of the experiment. Furthermore, it is difficult to vary the excitation of atomic clusters in a controlled way, unless the clusters are completely thermalized in a buffer gas, and the distribution of excitation energies will be broad. Hence, analysis of experimentally obtained fission rates has not yet provided quantitative information on fission barriers. It is possible that recent advances in the technology of cluster cooling, guided ion beams, and ion traps, need to be applied before our understanding of multiply charged clusters can be significantly advanced.

References

1. Lord Rayleigh, Phil. Mag. 14, 185 (1882)
2. Bohr, N., and Wheeler, J.A., Phys. Rev. 56, 426 (1939)
3. Bjornholm, S., and Lynn, J.E., Rev. Mod. Phys. 52, 725 (1980)
4. Brosa, U., Grossmann, S., and Müller, A., Phys. Rep. 197, 167 (1990)
5. Sattler, K., Mühlbach, J., Echt, O., Pfau, P., and Recknagel, E., Phys. Rev. Lett. 47, 160 (1981)
6. Echt, O., and Märk, T.D., Chapter 5.8 in *Clusters of Atoms and Molecules* (H. Haberland, ed.), to appear in *Graduate Texts in Contemporary Physics*, Springer Verlag
7. The heat of fission also depends on the reaction channel, i.e. the size and charge state of one of the fragments (cf. insert in Fig. 1). E_{tn}^{z+} simply denotes the minimum of $E_{tn}^{z+}(x,z')$, i.e. it refers to the thermodynamically most favorable channel.
8. Lipparini, E., and Vitturi, A., Z. Phys. D17, 57 (1990)
9. Schmidt, R., private communication
10. Baladrón, C., López, J.M., Iñiguez, M.P., and Alonso, J.A., Z. Phys. D11, 323 (1989)
11. M. Barranco, contribution to this conference; Garcias, F., Alonso, J.A., López, J.M., and Barranco, M., Phys. Rev. B43, 9459 (1991)

12. Durand, G., Daudey, J.-P., and Malrieu, J.-P., J. de Phys. 47, 1335 (1986)

13. S.N. Khanna, contribution to this conference; Khanna, S.N., Reuse, F., and Buttet, J., Phys. Rev. Lett. 61, 535 (1988)

14. Liu, F., Press, M.R., Khanna, S.N., and Jena, P., Phys. Rev. Lett. 59, 2562 (1987)

15. For an exhaustive compilation of references, consult ref. 6

16. Hence the term *Coulomb explosion* was used to characterize fission of multiply charged clusters.[5] However, this term is also used in conjunction with the much more violent *Coulomb explosion imaging technique*, which was developed at nuclear research facilities for determining geometric structures of small molecules and clusters [17]

17. Vager, Z., Feldman, H., Kella, D., Malkin, E., Miklazky, E., Zajfman, J., and Naaman, R., Z. Phys. D 19, 413 (1991)

18. Gay, J.G., and Berne, B.J., Phys. Rev. Lett. 49, 194 (1982)

19. Casero, R., Saenz, J.J., and Soler, J.M., Phys. Rev. A37, 1401 (1988)

20. Removal of a larger neutral fragment would usually cost more energy, with the possible exception of C, Si and Ge clusters.

21. Klots, C.E., J. Phys. Chem. 92, 5864 (1988)

22. Farges, J., de Feraudy, M.F., Raoult, B., and Torchet, G., Surf. Sci. 106, 95 (1981)

23. Scheier, P., Walder, G., Stamatovic, A., and Märk, T.D., J. Chem. Phys. 90, 4091 (1989)

24. Scheier, P., and Märk, T.D., J. Chem. Phys. 86, 3056 (1987)

25. Pfau, P., Ph.D. Thesis, University of Konstanz, 1984 (unpublished), also see: Sattler, K., Surface Science 156, 292 (1985)

26. Pfau, P., Sattler, K., Pflaum, R., and Recknagel, E., Phys. Lett. 104A, 262 (1984)

27. Mukherjee, S., Tomanek, D., and Bennemann, K.H., Chem. Phys. Lett. 119, 241 (1985)

28. Schulze, W., Winter, B., and Goldenfeld, I., Phys. Rev. B 38, 12937 (1988)

29. Hoareau, A., Melinon, P., Cabaud, B., Rayane, D., Tribollet, B., and Broyer, M., Chem. Phys. Lett. 143, 602 (1988)

30. Bréchignac, C., Broyer, M., Cahuzac, Ph., Delacretaz, G., Labastie, P., and Wöste, L., Chem. Phys. Lett. 133, 45 (1987); ibid, 118, 174 (1985)

31. Strömberg, D., and Wahlgren, U., Chem. Phys. Lett. 169, 109 (1990)

32. Actually, the kinetic stability of He_2^{2+} was predicted long ago (L. Pauling, J. Chem. Phys. 1, 56 (1933)). However, somewhat *larger* doubly charged He clusters will be *unstable*

33. These events, if occurring within $\ll 10^{-5}$ s, will go undetected, because a wide size range of neutral precursors is usually present in the beam of neutral precursors.

34. The relevant point may be slightly shifted from the crossing point, due to differences in pre-exponential factors ("attempt frequencies") and in geometric degeneracies

35. Ph. Cahuzac, contribution to this conference; Bréchignac, C., Cahuzac, Ph., Carlier, F., and de Frutos, M., Phys. Rev. Lett. 64, 2893 (1990)

36. Kreisle, D., Echt, O., Knapp, M., Recknagel, E., Leiter, K., Märk, T.D., Saenz, J.J., and Soler, J.M., Phys. Rev. Lett. 54, 1551 (1986)

37. Echt, O., Kreisle, D., Recknagel, E., Saenz, J.J., Casero, R., and Soler, J.M., Phys. Rev. A38, 3236 (1988)

38. Kreisle, D., Leiter, K., Echt, O., and Märk, T.D., Z. Phys. D3, 319 (1986)

39. Leiter, K., Kreisle, D., Echt, O., and Märk, T.D., J. Phys. Chem. 91, 2583 (1987)

40. Saunders, W.A., Phys. Rev. Lett. 66, 840 (1991)

41. U. Landman, contribution to this conference; Barnett, R.N., Landman, U., and Rajagopal, G., Phys. Rev. Lett., in print

42. Stace, A.J., private communication

43. Radi, P.P., von Helden, G., Hsu, M.T., Kemper, P.R., and Bowers, M.T., Chem. Phys. Lett. 179, 531 (1991)

44. Mass spectrometers analyze the mass-to-charge ratio, hence the size-to-charge ratio n/z. The contribution from multiply charged clusters with *integral* n/z values is buried under the signal of singly charged clusters.

45. Iñiguez, M.P., Bálbas, L.C., and Alonso, J.A., Physica 147B, 243 (1987); Iñiguez, M.P., Alonso, J.A., Aller, M.A., and Bálbas, L.C., Phys. Rev. B34, 2152 (1986)

46. Nakamura, M., Ishii, Y., Tamura, A., and Sugano, S., Phys. Rev. A42, 2267 (1990)

47. Katakuse, I., Ito, H., and Ichihara, T., Int. J. Mass Spectrom. Ion Proc. 97, 47 (1990)

48. Gotts, N.G., and Stace, A.J., Phys. Rev. Lett. 66, 21 (1991)

49. Saunders, W.A., Phys. Rev. Lett. 64, 3046 (1990)

50. Saunders, W.A., and Fedrigo, S., Chem. Phys. Lett. 156, 14 (1989)

51. Sandulescu, A., J. Phys. G15, 529 (1989)

52. Miehle, W., Diploma Thesis, University of Konstanz, 1987

Analytical Determination
of the Microscopic N-N Interaction Potential–Applications to Cluster Transfer and Fusion Reactions

R. K. Gupta

Institut für Theoretische Physik der J.W. Goethe-Universität,
W-6000 Frankfurt, Fed. Rep. of Germany
Permanent Address: Physics Department, Panjab
University, Chandigarh-160014, India

Abstract: The energy density model based on the energy density functional of the Skyrme effective interaction is reviewed briefly. The method is extended to include the spin density contribution in the heavy-ion interaction potentials, especially for unclosed shell nuclei. Analytical expressions are obtained for the complete interaction potential V(R) as well as for the interaction barrier height and position, in terms of simply the masses, charges and shell model configurations of the colliding nuclei. Applications of the model are made to cluster transfer resonance data and the fusion cross-sections for colliding nuclei from sd and $f_{7/2}$-shells.

I. Introduction

Some effective nucleon-nucleon (n-n) interactions, such as the short range Skyrme interaction[1], the finite-range density-dependent Brink-Boeker interaction[2] and the modified δ-interaction of Moszkowski[3], allow us to construct the simple energy density functionals (EDF). These functionals are useful not only in nuclear structure calculations, like the binding energies, charge radii and other related nuclear properties, but also have been used for calculating the heavy-ion nucleus-nucleus (N-N) interaction potentials. For other forces, one has to look for a simple and approximate parametrization of the density functionals and reproduce the Hartree-Fock results with such parametrized energy densities[4]. Then, there is an EDF due to Brueckner et al[5], written phenomenologically as the total energy of a fermion system in terms of the one-body density. The aim of this work is to present a simple analytical expression for the N-N interaction potential obtained from the Skyrme interaction energy density functioinal. This energy density functional is derived by Vautherin and Brink[6] for a system whose ground state is described by a Slater determinant.

Two different approaches have been used to calculate the real part of the N-N interaction potential. The simplest one, called the double folding model, is to fold in the effective n-n interaction with the nucleon density distributions of the two interacting nuclei[7]. Also, there have been attempts to use the single folding proceedure, where the phenomenological n-N optical potential is folded with the density distribution of one of the two interacting nuclei[8]. The other, most extensively used, approach is the energy density model (EDM), where the potential is defined as the difference between the total energy of the composite system and that of the individual nuclei. Such potentials are calculated in both the adiabatic[9] and sudden[5,10-21] approximations for the nucleon densities. Using the sudden approximation[5,10,11], Ngô et al[12-14] calculated the N-N interaction potential in EDM for the Brueckner phenomenological EDF and Brink and Stancu[15-17] and the present author and collaborators[18-21] for the Skyrme interaction EDF. The effects of saturation, antisymmetrization and relative motion between the two nuclei are also studied reasonably well.

In this work, we confine ourselves to the energy denity model based on the Skyrme interaction energy density functional. This is described very briefly in Sec.II. The Skyrme interaction EDF is an algebraic function of nucleon density ρ, kinetic energy density τ and the spin density $\vec{J}$. The zero spin density interaction potential is determined in the proximity theorem treatment of Chattopadhyay and Gupta[18] and the spin density contribution is derived for both the closed and unclosed shell nuclei[19]. Section III gives our calculated interaction potentials and their complete analytical determination in terms of either the nuclear and Coulomb potentials or simply the interaction barrier heights and their positions. Applications of the calculated N-N interaction potentials are made in Sec. IV to the cluster transfer resonances and the fusion cross-sections. A summary of our results is presented in Sec. V.

II. The Skyrme interaction Energy Density Model (SEDM)

Using the Skyrme interaction, Vautherin and Brink[6] have derived the energy density functional $H(\vec{r})$ for a system whose ground state is represented by a Slater determinant and the subspace of occupied single-particle states is invariant under time reversal. For an even-even spherical nucleus, this has the form (the subscripts n and p refer to neutrons and protons, respectively)

$$H(\rho, \tau, \vec{J}) = \frac{\hbar^2}{2m}\tau + \frac{1}{2}t_0[(1 + \frac{1}{2}x_0)\rho^2 - (x_0 + \frac{1}{2})(\rho_n^2 + \rho_p^2)] + \frac{1}{4}(t_1 + t_2)\rho\tau + \frac{1}{8}(t_2 - t_1)$$

$$(\rho_n\tau_n + \rho_p\tau_p) + \frac{1}{16}(t_2 - 3t_1)\rho\nabla^2\rho + \frac{1}{32}(3t_1 + t_2)(\rho_n\nabla^2\rho_n + \rho_p\nabla^2\rho_p)$$

$$+ \frac{1}{4}t_3\rho_n\rho_p\rho + \frac{1}{16}(t_1 - t_2)(\vec{J}_n^2 + \vec{J}_p^2) - \frac{1}{2}W_0(\rho\vec{\nabla}\cdot\vec{J} + \rho_n\vec{\nabla}\cdot\vec{J}_n + \rho_p\vec{\nabla}\cdot\vec{J}_p). \quad (1)$$

Here, t_0, x_0, t_1, t_2, t_3 and W_0 are the Skyrme interaction parameters, obtained by different authors[1,6,22−27] to fit the various ground state properties of the nuclei and $\rho = \rho_n + \rho_p$, $\tau = \tau_n + \tau_p$ and $\vec{J} = \vec{J}_n + \vec{J}_p$.

Knowing $H(\vec{r})$, the energy density model defines the N-N interaction potential $V(R)$ as the difference between the energy expectation value $E = \int H(\vec{r}) \, d\vec{r}$ of the colliding nuclei at a finite separation R and at infinity,

$$V_N(R) = E(R) - E(\infty) = \int \{H(\rho, \tau, \vec{J}) - [H_1(\rho_1, \tau_1, \vec{J}_1) + H_2(\rho_2, \tau_2, \vec{J}_2)]\} d\vec{r}. \quad (2)$$

This means that the two nuclei form a composite system at R and are completely separated at infinity. Also, $\rho = \rho_1 + \rho_2$, $\tau = \tau_1 + \tau_2$, $\vec{J} = \vec{J}_1 + \vec{J}_2$ refer to sudden approximation, which means neglecting exchange effects due to antisymmetrization. Brink and Stancu[16,17] have shown that such antisymmetrization effects can be assimilated reasonably well by using for τ, the Thomas Fermi (TF) kinetic energy density τ_{TF} corrected for additional surface effects. We take

$$\tau = \tau_{TF} + \lambda \frac{(\nabla \rho)^2}{\rho}, \quad (3)$$

where, $\tau_{TF} = \frac{3}{5}(\frac{3}{2}\pi^2)^{\frac{2}{3}}\rho^{\frac{5}{3}}$ and the surface correction is due to von Weizsäcker[28], with λ having values between $\frac{1}{4}$ and $\frac{1}{36}$. Eq.(3) makes the energy density (1) a functional of ρ and $\vec{J}$ only, which allows us to calculate (2) simply as:

$$V_N(R) = V_P(R) + V_J(R). \quad (4)$$

Here, $V_P(R)$ is the $\vec{J}=0$, ρ-dependent part of the interaction potential and $V_J(R)$, the spin density dependent part. Notice that $V_J(R)$ also depends on ρ. We have solved[18,19] for $V_P(R)$ directly in the proximity force theorem as

$$V_P(R) = \int \{H(\rho) - [H_1(\rho_1) + H_2(\rho_2)]\} d\vec{r} = 2\pi \frac{C_1 C_2}{C_1 + C_2} \Phi(s), \quad (5)$$

where C_i are the Süssman central radii and the universal function[29]

$$\Phi(s) = \int_{s_0}^{\infty} e(s) ds = \int \{H(\rho) - [H_1(\rho_1) + H_2(\rho_2)]\} dZ. \quad (6)$$

Here, s=R-C_1-C_2, with its minimum value s_0, and e(s) is the interaction energy per unit area between two flat slabs of semi-infinite nuclear matter with surfaces parallel to the XY-plane, moving in Z-direction. For the nucleon density ρ, we have used the TF density or the two parameters Fermi density distributions. In the following, however, we use only the Fermi density distribution, (i=1,2),

$$\rho_i(Z_i) = \rho_{0i}\left[1 + exp\left(\frac{Z_i - C_i}{a_i}\right)\right]^{-1}, \quad -\infty \le Z \le \infty \qquad (7)$$

with $Z_2 = R - Z_1$ for motion in Z-direction in a plane. This density gives results that are identical with the microscopic shell model density and is more realistic for heavy-ion collisions because it does not drop sharply to zero like the TF density distribution. We obtain[19]

$$\Phi(s) = \frac{3}{5}\frac{\hbar^2}{2m}\left(\frac{3}{2}\pi^2\right)^{\frac{2}{3}}I_{[\frac{5}{3}]} + \frac{3}{8}t_0 I_{[2]} + \frac{1}{16}t_3 I_{[3]} + \frac{1}{16}(3t_1 + 5t_2)\frac{3}{5}\left(\frac{3}{2}\pi^2\right)^{\frac{2}{3}}I_{[\frac{8}{3}]} + \Phi_{s\rho} + \Phi_{s\tau_\lambda} \,(8)$$

with

$$I_{[n]} = \int\{\rho^n - [\rho_1^n + \rho_2^n]\}dZ_1; \qquad \Phi_0 = \int\{\rho'^2 - [\rho_1'^2 + \rho_2'^2]\}dZ_1; \qquad \rho' = |\frac{\partial\rho}{\partial Z_1}|, \text{etc.}$$

$$\Phi_{s\rho} = \frac{1}{64}(9t_1 - 5t_2)\Phi_0; \qquad \Phi_{s\tau_\lambda} = \frac{1}{16}(3t_1 + 5t_2)\lambda\Phi_0 + \frac{\hbar^2}{2m}\lambda\left(\int\{\frac{\rho'^2}{\rho} - [\frac{\rho_1'^2}{\rho_1} + \frac{\rho_2'^2}{\rho_2}]\}dZ_1\right).$$

This equation is solved numerically for the Fermi density (7), using $\rho_n = \rho_p$.

For spin density dependent part of the interaction potential, using $\rho_n = \rho_p = (\frac{1}{2}\rho)$, we get from (1) and (2):

$$V_J(R) = \int\{H(\vec{J}) - [H_1(\vec{J_1}) + H_2(\vec{J_2})]\}d\vec{r} = -\frac{3}{4}W_0\int(\rho_2\vec{\nabla}\cdot\vec{J_1} + \rho_1\vec{\nabla}\cdot\vec{J_2})d\vec{r}. \qquad (9)$$

Notice that here we have not included the term $\frac{1}{16}(t_1 - t_2)(\vec{J_n}^2 + \vec{J_p}^2)$, since its contribution is found to be small[30]. Actually, until recently, the contribution of spin density in heavy-ion potentials was completely neglected by either studying only the spin-saturated nuclei (i.e., the nuclei with major shell closed for both protons and neutrons, having $\vec{J}=0$) or taking its contribution to be small. Vautherin and Brink[6] derived an expression for $\vec{J}$ only for closed j-shell nuclei (i.e., nuclei with either $j = l + \frac{1}{2}$ or $l - \frac{1}{2}$ shell filled). Recently, we have extended this formalism and obtained[19] a general expression of $\vec{J}$ for nuclei with one or more pairs of valence particles (or holes) outside the closed shells. This is important, in particular, for processes like the transfer reactions.

In terms of the single-particle orbitals ϕ_i, that define a Slater determinant, the spin density $\vec{J}$ is defined as

$$\vec{J}_q(\vec{r}) = (-i)\sum_{i,s,s'}\phi_i^*(\vec{r}, s, q)[\vec{\nabla}\phi_i(\vec{r}, s', q)\times < s|\vec{\sigma}|s' >]. \qquad (10)$$

Here, s and q (=n or p) represent the spin and isospin indices, respectively, and the summation i runs over all the occupied single-particle orbitals. For ϕ_i, we use the ansatz[6,19]

$$\phi_i(\vec{r}, s, q) = \frac{R_\alpha(r)}{r} \sum_{m_l m_s} < l\frac{1}{2} m_l m_s | jm > Y_l^{m_l}(\hat{r}) \chi_{m_s}(s) \chi_q(t), \tag{11}$$

where $\alpha \equiv q, n, l$ and $R_\alpha(r) = C_{nl} r^{l+1} e^{-\nu r^2} v_{nl}(2\nu r^2)$ is the shell model, normalized radial wave function, with $\nu = \frac{m\omega}{2\hbar}$ (fm^{-2}). The constant C_{0l} varies smoothly with mass number A of the nucleus, only within a shell, which means that the analytical formula for $V_J(R)$ would be different for different shells.

For an even-even nucleus with valence particles (or holes) outside the closed shells, we divide[19] $\vec{J}_q(\vec{r})$ into two parts (for q=n or p): one due to the core consisting of closed shells and another due to the valence n_v particles (+sign) or holes (-sign),

$$\vec{J}(\vec{r}) = \vec{J}_C(\vec{r}) \pm \vec{J}_{n_v}(\vec{r}). \tag{12}$$

Considering that valence nucleons couple to zero angular momemtum,we get

$$\vec{J}_{n_v}(\vec{r}) = \frac{n_v \vec{r}}{4\pi r^4}[j(j+1) - l(l+1) - \frac{3}{4}]R_l^2(r). \tag{13}$$

Here, l and j refer to the shell containing the n_v nucleons. Since a core consists of α-closed shells and for a closed shell n_v=2j+1, we get from (13)

$$\vec{J}_C(\vec{r}) = \frac{\vec{r}}{4\pi r^4} \sum_\alpha (2j_\alpha + 1)[j_\alpha(j_\alpha + 1) - l_\alpha(l_\alpha + 1) - \frac{3}{4}]R_\alpha^2(r). \tag{14}$$

Notice that this equation gives $\vec{J}$=0 for a nucleus with completely filled major shells i.e. both $j = l \pm \frac{1}{2}$ shells filled.

Similarly, it is possible to write[20] (q=n or p)

$$\vec{\nabla} \cdot \vec{J}_q = (\vec{\nabla} \cdot \vec{J})_C \pm (\vec{\nabla} \cdot \vec{J})_{n_v}. \tag{15}$$

Also, within the nuclear shell model, we have written[20] an expression, similar to that of (12) or (15), for the nucleon density $\rho(\vec{r})$. In the following, however, we use the Fermi density (7) also for $V_J(R)$ in (9) since, as stated before, the shell model density match very well the Fermi density, at least in the physically interesting tail region. The calculated $V_J(R)$ are also identical[19] for the two density distributions.

III. The N-N Interaction Potential– Analytical Formulae

We have seen that for the SEDM, the N-N interaction potential splits into the proximity potential V_P and the spin density dependent potential V_J. In this section, we attempt first to parametrize the calculated V_P and V_J potentials. In other words, we obtain[20] analytical expressions for V_P and V_J, which allow us to calculate the N-N interaction potential $V_N(= V_P+V_J)$ simply from the knowledge of masses and shell model configurations of the colliding nuclei. Next, we add Coulomb potential to our calculated nuclear potential V_N, which gives us the position R_B and height V_B of the interaction barrier. Since the knowledge of this single point (R_B, V_B) on $V(R)$ is useful for calculating quantities like the fusion cross-sections, we also obtain[21] analytical expressions for both R_B and V_B in terms of masses and charges of the colliding nuclei.

Figure 1 shows a plot of $V_P(R)$ for a typical reaction $^{40}Ca+^{56}Ni$, using different Skyrme interactions with $\lambda = 0$, and the function $\Phi(s)$ for the force SII($\lambda = 0$) for several target-projectile combinations. We notice in Fig.1(a) that some forces show a repulsive core, characteristic of sudden collisions, whereas the others remain attractive througout which might be representative of surfaces with larger curvatures compared to diffuseness of surface region (like in generalized proximity potential[31]).

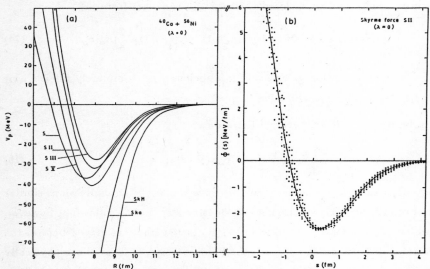

Fig. 1. (a) Proximity potential $V_P(R)$, using Fermi density and Skyrme forces. (b) Universal function $\Phi(s)$, using Skyrme force SII. Solid line here represents the mean curve (16).

Figure 1(b) shows that $\Phi(s)$ is indeed a universal function and its mean behaviour can be parametrized as

$$\Phi(s) = \begin{cases} -2.64 exp[-0.3250(s-0.2)^2] & \text{for } s \geq s_0 \\ -2.64 + 2.15(s-0.2)^2 & \text{for } s \leq s_0. \end{cases} \tag{16}$$

Brink and Stancu[17] have also carried out a similar type of parametrization for SEDM and Ngô and co-workers[13,14] for Brueckner EDM. The difference, however, lies in that we use a priori the proximity theorem in our SEDM whereas these authors prove the validity of the proximity theorem in their, respective, methods. We obtain a formula, similar to (16), for $\lambda = \frac{1}{36}$ and find that only the position of the minimum of $\Phi(s)$ remains the same .

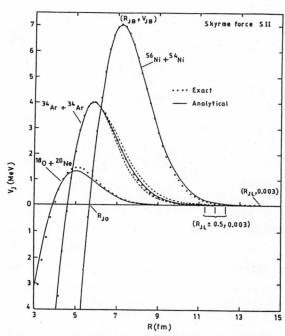

Fig.2. Spin density dependent potential $V_J(R)$, using Fermi density and Skyrme force SII.

For spin density potential V_J, as stated before, the shell structure of colliding nuclei is also important. Therefore, we consider here only the reactions between two nuclei belonging to the same shell; namely the (1s0d)+ (1s0d) and $(0f_{7/2})+(0f_{7/2})$ shell nuclei. Figure 2, dots, marked *"exact"*, illustrates the calculated $V_J(R)$ for two (sd)+(sd) sytems ($^{18}O+^{20}Ne$ and $^{34}Ar+^{34}Ar$) and one $(f_{7/2})+(f_{7/2})$ system ($^{56}Ni+$ ^{54}Ni), using the Skyrme force SII. We find that in terms of four points R_{J0}, R_{JB}, V_{JB} and R_{JL}, marked on one of the curve, $V_J(R)$ can be expressed analytically as

$$V_J(R) = \begin{cases} V_{JB} exp[ln(\frac{V_{JL}}{V_{JB}})(\frac{R-R_{JB}}{R_{JL}-R_{JB}})^{\frac{5}{3}}] & \text{for } R \geq R_{JB} \\ V_{JB} - V_{JB}(\frac{R-R_{JB}}{R_{J0}-R_{JB}})^2 & \text{for } R \leq R_{JB}. \end{cases} \tag{17}$$

Here, R_{JL} is the limiting distance where $V_{JL} = V_J(R = R_{JL}) \to 0$. For practical purposes, however, we take $V_{JL} = 0.003 \mp 0.001 MeV$, instead of zero. The variations of the four constants of Eq.(17) with, respective, relevant quantities are shown in Fig.3 for over 100 reactions each of (sd)+(sd) and $(f_{7/2})+(f_{7/2})$ nuclei. These can be represented by the following simple expressions ($P_0=0.95493$):

$$V_{JB} = 1.3375\frac{P_S}{P_0}; \qquad R_{J0} = 3.58 + (8.70 \times 10^{-4})A_1A_2 \equiv (1.15 \pm 0.3)A^{\frac{1}{3}}$$

$$R_{JB} = 4.58 + (1.11 \times 10^{-3})A_1A_2; \qquad R_{JL} = (12.77 \pm 0.50) - 1129(A_1A_2)^{-1} \qquad (18)$$

and,

$$V_{JB} = 0.94\frac{P_S}{P_0}; \qquad R_{J0} = 4.52 + (3.776 \times 10^{-4})A_1A_2 \equiv (1.15 \pm 0.03)A^{\frac{1}{3}}$$

$$R_{JB} = 5.77 + (4.75 \times 10^{-4})A_1A_2; \qquad R_{JL} = 16.67 - 8621(A_1A_2)^{-1}, \qquad (19)$$

respectively, for (sd)+(sd) and $(f_{7/2})+(f_{7/2})$ shell nuclei. Here, we have introduced a new variable, called the "particle strength" P_S,

$$P_S = \sum_{\alpha} \frac{2j_\alpha + 1}{4\pi}[j_\alpha(j_\alpha + 1) - l_\alpha(l_\alpha + 1) - \frac{3}{4}] \pm \frac{n_v}{4\pi}[j(j+1) - l(l+1) - \frac{3}{4}], \qquad (20)$$

which is related to shell model configurations of total A nucleons of composite nucleus. The scaling factor P_0 represents the strength of six particles filling the $0d_{\frac{5}{2}}$ shell.

The result of calculation using Eqs. (17) and (18) for (sd)+(sd) and Eqs. (17) and (19) for $(f_{7/2})+(f_{7/2})$ collisions is shown as solid lines in Fig.2, which compares nicely with the exact calculation. The dashed lines in the case of ^{34}Ar+^{34}Ar show the effect of changing R_{JL} by $\pm 0.5 fm$, which is apparently a small effect for potential.

Finally, we have added Coulomb potential $E_C(= \frac{Z_1Z_2e^2}{R})$ to the above calculated nuclear potential V_N for the Skyrme force SIII ($\lambda = 0$). This gives us the height V_B and position R_B of the interaction barrier for each reaction, whose variations with the relevant quantities in terms of masses and charges of colliding nuclei are shown in Fig.4 and parametrized as follows:

$$V_B = 0.845\frac{Z_1Z_2}{A_1^{\frac{1}{3}} + A_2^{\frac{1}{3}}} + 1.3 \times 10^{-3}\left(\frac{Z_1Z_2}{A_1^{\frac{1}{3}} + A_2^{\frac{1}{3}}}\right)^2$$

$$R_B = 7.359 + 3.076 \times 10^{-3}(A_1A_2) - 1.182 \times 10^{-6}(A_1A_2)^2 + 1.567 \times 10^{-10}(A_1A_2)^3. \quad (21)$$

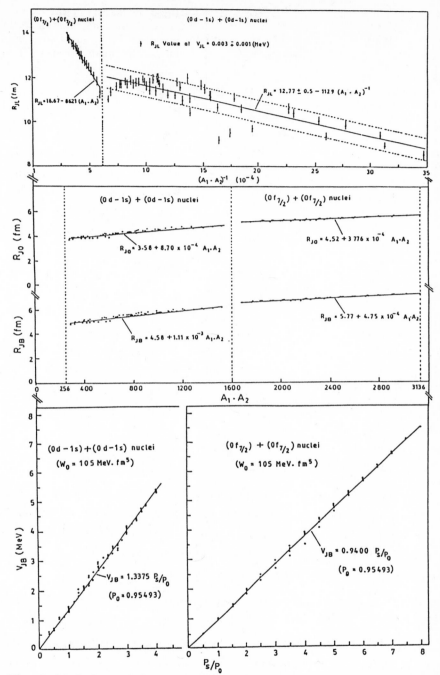

Fig. 3. Variations of the constants V_{JB}, R_{JB}, R_{J0} and R_{JL} for many (sd)+(sd) and $(f_{7/2})+(f_{7/2})$ reactions. Solid lines represent the mean curves (18) and (19).

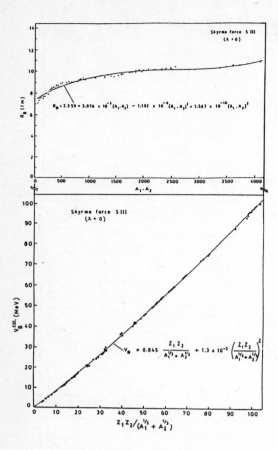

Fig.4 Variations of the calculated barrier heights V_B and positions R_B. Solid lines represent the mean curves (21).

These formulas are good for light nuclei, upto $0f_{7/2}$-shell. Ngô et al[12] and later Stancu and Brink[16] have also suggested a similar but graphical method to calculate V_B and R_B. However, these authors used a parameter $r_B = \dfrac{R_B}{A_1^{\frac{1}{3}}+A_2^{\frac{1}{3}}}$ instead of R_B and V_N at $R=R_B$ instead of V_B, which give a larger scatter of the calculated data around the suggested mean curves. Thus, Eqs.(21) give almost for the first time an analytical method to calculate the interaction barrier.

IV. Applications

We have seen that analytical determination of N-N interaction potential $V(=V_N + E_C)$ is possible not only for the interaction barrier (R_B, V_B) but also for the complete potential V as a function of R. In the following, however, we use the actually calculated V(R) for estimating the yields of cluster transfer resonances as well as the fusion cross-sections.

A. Cluster transfer resonances: We study this phenomenon for the reactions ^{24}Mg$+$ $^{24,26,28}Mg \rightarrow ^{48,50,52} Cr^*$, ^{16}O$+$ $^{40,42,44}Ca \rightarrow ^{56,58,60} Ni^*$ and ^{40}Ca$+^{40}Ca \rightarrow$ $^{80}Zr^*$ at incident energies 1.5 to 2 times the Coulomb barrier where the transfer process is very weak with respect to the inelastic scattering. Nevertheless, some data[32,33] is observed which clearly show an explicit preference of α-cluster transfer in the reaction ^{16}O$+^{40}$Ca at 75 (and 80.6) MeV and its gradual suppression on adding 2 and 4 neutrons to the target nucleus. Similarly, the resonance-like structure observed in the excitation functions for elastic and inelastic scattering of $^{28}Si+^{28}Si$ gets suppressed on adding neutrons to either or both of these α-particle nuclei.

Figure 5(a) illustrates our calculated fragmentation potentials for the $^{56,58,60}Ni^*$ systems at a fixed length $R = R_1 + R_2$, using Skyrme force SIII($\lambda = 0$):

$$V(\eta) = -\sum_{i=1}^{2} B_i(A_i, Z_i) + V(R). \tag{22}$$

Here, $\eta = \frac{A_1-A_2}{A_1+A_2}$ is the mass-asymmetry coordinate. Using this potential and the classical mass parameters[34] $B_{\eta\eta}$, the fragmentation (or transfer) yields Y are calculated[21,35,36] by solving the stationary Schrödinger equation in η:

$$[-\frac{\hbar^2}{2\sqrt{B_{\eta\eta}}} \frac{\partial}{\partial\eta} \frac{1}{\sqrt{B_{\eta\eta}}} \frac{\partial}{\partial\eta} + V(\eta)\,]\,\Psi_R^{(\nu)}(\eta) = E_R^{(\nu)}\,\Psi_R^{(\nu)}(\eta), \tag{23}$$

with

$$Y(A_1) = |\,\Psi_R(\eta)\,|^2\,\sqrt{B_{\eta\eta}(\eta)}\,\frac{4}{A}; \quad |\Psi_R|^2 = \sum_{\nu=0}^{\infty} |\Psi_R^{(\nu)}|^2 exp(-E_R^{(\nu)}/\theta). \tag{24}$$

This is shown in Fig. 5(b) for $^{56}Ni^*$ and compared with experimental data. We

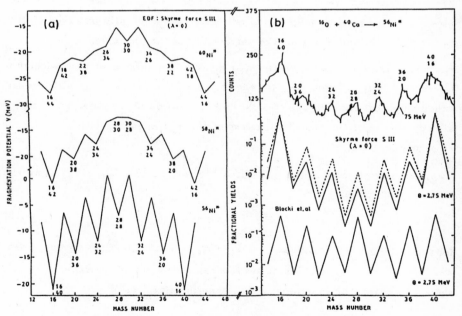

Fig. 5. (a) Fragmentation potentials at $R=R_1+R_2$, using Skyrme force SIII. (b) Calculated yields compared with experimental data for $^{16}O+^{40}Ca \rightarrow ^{56}Ni$.

notice in Fig. 5(a) that potential energy surfaces already give the results of experiment. The minima at α-particle nuclei in ^{56}Ni are already reduced in the case of ^{58}Ni and this is more so in ^{60}Ni. Fig. 5(b) shows that the calculated yields for ^{56}Ni correspond reasonably well with the experimental data. In this figure, we have also shown yields calculated for use of V_N from the nuclear proximity potential of Blocki et al[29]. We notice that the SEDM gives a better representation of the data as compared to the pocket formula of Blocki et al[29]. A closer analysis shows that much of the resonance-like structure is due to the binding energies of the nuclei but the spin density dependent potential $V_J(\eta)$ also contain the α-clustering effects[21]. These are mainly in terms of the closing of shells for protons and neutrons during the transfer process. The α-clustering effects exhibit themselves in $V_J(\eta)$ when due to an α-particle transfer, at least, in one of the transfer products the same j-shell gets closed for both protons and neutrons. This happens only for transfer between the colliding N=Z, A=4n nuclei.

An extension of the above calculation shows[37] that for composite systems with $N \gg Z$, the α-clustering picture breaks down completely and new clusters, like $^{10}Be, ^{14}C, ^{18}O, ^{24}Ne, ^{28}Mg$ etc., are formed preferentially during the transfer process. These clusters might have relevance to the recently observed cluster radioactivity[38], and hence for the universality of the phenomenon.

B. Fusion cross-sections: Figure 6(a) gives a comparison of our calculated interaction barrier heights V_B for the Skyrme force SIII($\lambda = 0$), with the so-called empirical data[39]. The straight line description of the data in this figure already speaks of the expected success of the SEDM in predicting the fusion cross-section data.

The fusion cross-sections are calculated[21,40] for energies $E_{cm} > V_B$, using the sharp cut-off model,

$$\sigma_f = \pi R_B^2 (1 - \frac{V_B}{E_{cm}}), \tag{25}$$

for the reactions $^{12}C + ^{18}O, ^{16}O + ^{16}O, ^{16}O + ^{48}Ca, ^{26}Mg + ^{34}S, ^{40}Ca + ^{40}Ca, ^{40}Ca + ^{48}Ca$ and $^{40}Ca + ^{60}Ni$. Fig. 6(b) illustrates this calculation for the reaction $^{12}C + ^{18}O$, showing a reasonable comparison with experimental data[41].

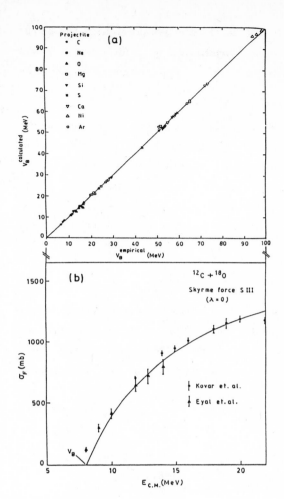

Fig.6 (a) Calculated barrier heights V_B using Skyrme force SIII, compared with empirical data. (b) Calculated fusion cross-sections compared with experimental data for the reaction $^{12}C+^{18}O$ at $E_{cm} > V_B$.

We have also analyzed the role of the spin density term V_J on the fusion cross-sections. As already stated in Sec.II, this term is added here for the first time. We find that this term diminishes the fusion cross-sections by as much as ~50mb, which could be important for comparisons with the experimental data.

V. Summary

The aim of this work was to present a simple analytical formula of the N-N interaction potential obtained in the EDM, using the EDF of the Skyrme effective n-n interaction. Within the sudden approximation, such a nuclear potential splits into the proximity potential $V_P(R)$ and the spin density dependent potential $V_J(R)$. The universal function of the proximity potential, independent of the geometry of colliding nuclei, is shown to be indeed universal, represented analytically by a Gaussian function plus a parabola. A similar parametrization of $V_J(R)$ is found to depend strongly not only on the product of masses A_1 and A_2 of the two colliding nuclei but also on their shell model configurations through a new variable, called the "particle strength".

509

Addition of Coulomb potential to N-N interaction potential gives the height V_B and position R_B of the interaction barrier. Analytical formulas of V_B and R_B are also presented simply in terms of masses and charges of the colliding nuclei.

Applications of the SEDM are made to cluster transfer resonance phenomenon and fusion cross-sections. In addition to giving a better representation of observed data, the spin density effects in the potential are found to exhibit the α-clustering effects during the transfer of nucleons in colliding N=Z, A=4n, α-nuclei and reduce the fusion cross-sections by as much as $\sim$50mb. The spin density effects are included here for the first time, especially for unclosed shell nuclei.

Acknowledgement

Most of the work presented here is contained in the Ph.D thesis of Dr. Rajeev Kumar Puri, carried out at Panjab University, Chandigarh, India, under my supervision. I am also thankful to Professors Walter Greiner and Werner Scheid for useful discussions and their kind hospitality at present at their respective institutes.

References

1) T.H.R. Skyrme, Nucl. Phys. 9 (1959) 615
2) D.M. Brink and E. Boeker, Nucl. Phys. A91 (1967) 1;
   B. Behra and R.K. Satpathy, J. Phys. G:Nucl. Phys. 5 (1979) 85
3) S.A. Moszkowski, Phys. Rev. C2 (1970) 402
4) J. Treiner and H. Kirvine, J. Phys. G 2 (1976) 285
5) K.A. Brueckner, J.R. Buchler and M.M. Kelly, Phys. Rev. 173 (1968) 944
6) D. Vautherin and D.M. Brink, Phys. Rev. C5 (1972) 626
7) G.R. Satchler, Phys. Lett. 59B (1975) 121; Proc. Int. Conf. on Reactions between Complex Nuclei, Nashville, Tenn., USA, 1974
8) D.M. Brink and N. Rowley, Nucl. Phys. A219 (1974) 79;
   D.H.E. Gross and H. Kalinowski, Phys. Lett. 48B (1974) 302
9) J.D. Perez, Nucl. Phys. A191 (1972) 19
10) L. Wilets, A. Goldberg and F.M. Lewis, Phys. Rev. 2 (1970) 1576
11) W. Scheid and W. Greiner, Z. Phys. 226 (1969) 364
12) C. Ngô, B. Tamain, J. Galin, M. Beiner and R.J. Lombard, Nucl. Phys. A240 (1975) 353
13) C. Ngô, B. Tamain, M. Beiner, R.J. Lombard, D. Mas and H.M. Deubler, Nucl. Phys. A252 (1975) 237
14) H. Ngô and C. Ngô, Nucl. Phys. A348 (1980) 140
15) D.M. Brink and Fl. Stancu, Nucl. Phys. A243 (1975) 175

16) Fl. Stancu and D.M. Brink, Nucl. Phys. **A270** (1976) 236
17) D.M. Brink and Fl. Stancu, Nucl. Phys. **A299** (1978) 321
18) P. Chattopadhyay and R.K. Gupta, Phys. Rev. **C30** (1984) 1191
19) R.K. Puri, P. Chattopadhyay and R.K. Gupta, Phys. Rev. **C43** (1991) 315
20) R.K. Puri and R.K. Gupta, Chandigarh preprint 1991; to be published
21) R.K. Puri, Ph.D. thesis, Panjab University, Chandigarh, India, 1990; unpublished
22) M. Beiner, H. Flocard, N. van Giai and P. Quentin, Nucl. Phys. **A238** (1975) 29
23) S. Köhler, Nucl. Phys. **A162** (1971) 385; **A170** (1971) 88
24) H. Kirvine, J. Treiner and H. Bohigas, Nucl. Phys. **A336** (1980) 155
25) M. Brack, C. Guet and H. Håkansson, Phys. Rep. **123** (1985) 275
26) M. Rayet, M. Arnould, F. Tondeur and G. Pauless, Astron. Astro. Phys. **116** (1982) 183
27) C.M. Ko, H.C. Pauli, M. Brack and G.E. Brown, Nucl. Phys. **A236** (1974) 269
28) C.F. von Weizsäcker, Z. Phys. **96** (1935) 431
29) J. Blocki, J. Randrup, W.J. Swiatecki and C.F. Tsang, Ann. Phys. (NY) **105** (1977) 427
30) K.C. Panda and T. Patra, J. Phys. G:Nucl. Part. Phys. **16** (1990) 593
31) J. Blocki and W.J. Swiatecki, Ann. Phys. (NY) **132** (1981) 53
32) R.R. Betts, Proc. Conf. on Resonances in Heavy Ion Reactions, Bad-Honnef, West Germany ed. K.A. Eberhardt (Lecture Notes in Physics, Vol.156, Springer: Berlin, 1981), p.185
33) R.R. Betts, Fundamental Problems in Heavy Ion Collisions-Proc. 5th Adriatic Int. Conf. on Nucl. Phys., Hvar, Croatia, Yugoslavia, ed. N. Cindro et al , World Sc.: Singapore, 1984, p.33
34) H. Kröger and W. Scheid, J. Phys. G:Nucl. Phys. **6** (1980) L85
35) D.R. Saroha, N. Malhotra and R.K. Gupta, J. Phys. G:Nucl. Phys. **11** (1985) L27
36) S.S. Malik and R.K. Gupta, J. Phys. G: Nucl. Phys.**12** (1986) L161
37) S.S. Malik, S. Singh, R.K. Puri, S. Kumar and R.K. Gupta, Pramāna-J. Phys. **32** (1989) 419
38) P.B. Price, Ann. Rev. Nucl. Part. Sci. **39** (1989) 19
39) L.C. Vaz, J.M. Alexander and G.R. Satchler, Phys. Rep. **69** (1981) 373
40) R.K. Puri and R.K. Gupta, Chandigarh preprint 1991;to be published
41) D.G. Kovar, D.F. Geesaman, T.H. Braid, Y. Eisen, W. Henning, T.R. Ophel, M. Paul, K.E.Rehm, S.J. Sanders, P. Sperr, J.P. Schiffer, S.L. Tabor, S. Vigdor, B. Zeidman and F.W. Prosser, Jr., Phys. Rev. **C20** (1979) 1305;
Y. Eyal, M. Beckman, R. Chechik, Z. Frankel and H. Stocker, Phys. Rev. **C13** (1976) 1527.

Formation and Dissolution of Clusters Studied with Antisymmetrized Molecular Dynamics

H. Horiuchi[1], *T. Maruyama*[1], *A. Ohnishi*[1], *and S. Yamaguchi*[2]

[1]Department of Physics, Kyoto University, Kyoto 606, Japan
[2]Yukawa Institute for Theoretical Physics, Kyoto University, Kyoto 606, Japan

Formulation of the antisymmetrized molecular dynamics (AMD) is briefly explained and the frictional cooling method is shown to be very efficient to construct the ground state wave function in the AMD framework. Since the frictional cooling does not depend on any model assumption, we can discuss the formation and dissolution of clusters in the ground states. In the case of self-conjugate 4N nuclei, we have found that Brink's cluster model can be justified. In the application to Be isotopes, we have found very interesting tendency as a function of neutron number about the formation and dissolution of the $\alpha - \alpha$ core structure.

1.Introduction

In studying clustering correlation in nuclei, it is usual to use models in which the existence of clusters is assumed[1]. However, in order to investigate the formation and dissolution of clusters, it is necessary to develop and use models which do not assume the existence of clusters. Formation and dissolution of clusters are an important problem both in nuclear structure study and in the study of nucleus-nucleus collisions.

The purpose of this paper is to show that the antisymmetrized molecular dynamics (AMD)[2] is very useful for the study of formation and dissolution of clusters. The AMD is originally aimed to be a microscopic simulation framework for nucleus-nucleus collisions. In the AMD, the nuclear system is described by a Slater determinant of nucleon wave packets. The time-development of position and momentum coordinates of each nucleon wave packet is determined by applying the time-dependent variational principle to the AMD Slater determinant.

The AMD can be regarded as being an extended version of the time-dependent cluster model [3] in which one adopts the Slater determinant of Brink's cluster model [4] and the position and momentum coordinates of each cluster center-of-mass are determined by applying the time-dependent variational principle.

Springer Series in Nuclear and Particle Physics **Clustering Phenomena in Atoms and Nuclei**
Editors: M. Brenner · T. Lönnroth · F.B. Malik   © Springer-Verlag Berlin, Heidelberg 1992

Recently Feldmeier has proposed a similar but more general framework which he calls fermionic molecular dynamics (FMD)[5]. The FMD is more general than our AMD since the width parameters of nucleon wave packets are also time-dependent and furthermore the time-evolution of spin-isospin wave functions is explicitly treated. However, in [5] what is discussed in detail is limited to two-nucleon problem and many-nucleon problem is only briefly sketched. Corianò et al. have discussed Feldmeier's FMD from the view point of the canonical formulation, but they also have concentrated to two-nucleon problem[6].

Both in structure study and in collision study with the AMD, one starts with the construction of the ground state of the nucleus. For this purpose we have developed and used the frictional cooling method[2]. By this method we can find the set of optimum values of position and momentum coordinates of all the nucleon wave packets which give the minimum binding energy. The frictional cooling method gives us a method to construct the ground state wave function without any model assumption within the functional space spanned by Slater determinants of nucleon wave packets. Therefore we can study whether clusters are formed or not in each nuclear ground state.

Both the time-dependent variation and the frictional cooling method can be applied not only to a single Slater determinant but also to a linear combination of Slater determinants. Therefore, in constructing ground state wave functions we have used parity projected wave functions. Since the obtained ground state wave functions are not good eigenstates of angular momentum in general, we have projected good angular momentum states in order to compare with experiments.

In this paper we report the results of our studies of some self-conjugate 4N light nuclei and of neutron-rich Be isotopes.

2. Antisymmetrized Molecular Dynamics (AMD)

The Slater determinant of the AMD is given as

$$\Phi \equiv \frac{1}{\sqrt{A!}} \det \left[\prod_{\alpha=1}^{4} \prod_{i \epsilon A_\alpha} A(\vec{r_i}, \vec{Z_i}) \chi_\alpha(i) \right] \ ,$$

$$A_\alpha \equiv \left(\sum_{\beta<\alpha} a_\beta + 1, \ \ldots \ , \ \sum_{\beta \leq \alpha} a_\beta \right) \ (\alpha = 1 \sim 4) \ , \ \ \sum_{\alpha=1}^{4} a_\alpha = A \ , \tag{2.1}$$

where χ_α stands for the spin-isospin wave function; $\chi_1 = \chi_{p\uparrow}, \chi_2 = \chi_{p\downarrow}, \chi_3 = \chi_{n\uparrow}$, and $\chi_4 = \chi_{n\downarrow}$. The single particle orbital wave function $A(\vec{r}, \vec{Z})$ with complex vector parameter $\vec{Z}$ is the Guassian wave packet or the coherent state defined as

$$A(\vec{r}, \vec{Z})$$

$$\equiv \left(\frac{2\nu}{\pi}\right)^{\frac{3}{4}} \exp\left[-\nu\left(\vec{r} - \frac{\vec{Z}}{\sqrt{\nu}}\right)^2 + \frac{1}{2}\vec{Z}^2\right] ,$$

$$= \left(\frac{2\nu}{\pi}\right)^{\frac{3}{4}} \exp\left[-\nu(\vec{r} - \vec{D})^2 + i\frac{\vec{K}}{\hbar}(\vec{r} - \vec{D})\right] \cdot \exp\left[\frac{1}{2}\left(\vec{Z}^* \cdot \vec{Z} + i\frac{\vec{K}}{\hbar}\vec{D}\right)\right] ,$$

$$\vec{Z} \equiv \sqrt{\nu}\left(\vec{D} + \frac{i}{2\hbar\nu}\vec{K}\right) , \tag{2.2}$$

Eq.(2.1) means that we treat the system with the total mass number A and the α-th spin-isospin state is occupied by a_α nucleons. The time developments of the nucleon coordinate parameters $\vec{Z_i}$ namely $\vec{D}_i$ and $\vec{K}_i$, are determined by the time-dependent variational principle;

$$\delta \int_{t_1}^{t_2} dt\, \widehat{L}(\vec{Z}^*, \vec{Z}, \dot{\vec{Z}}^*, \dot{\vec{Z}}) = 0 ,$$

$$\widehat{L}(\vec{Z}^*, \vec{Z}, \dot{\vec{Z}}^*, \dot{\vec{Z}}) \equiv \frac{i}{2}\hbar \frac{< \Phi \mid \frac{\partial}{\partial t}\Phi > - < \frac{\partial}{\partial t}\Phi \mid \Phi >}{< \Phi \mid \Phi >} - \frac{< \Phi \mid H \mid \Phi >}{< \Phi \mid \Phi >} ,$$

$$\dot{\vec{Z_i}} \equiv \frac{d}{dt}\vec{Z_i} . \tag{2.3}$$

Variation of the action yields the equations of motion for $\vec{Z}_i$,

$$i\hbar \sum_{j \epsilon A_\alpha, \tau} C^\alpha_{i\sigma,\, j\tau} \dot{Z}_{j\tau} = \frac{\partial \widehat{H}}{\partial Z^*_{i\sigma}} \quad \text{and} \quad c.c. ,$$

$$C^\alpha_{i\sigma,\, j\tau} \equiv \left(\frac{\partial}{\partial Z^*_{i\sigma}}\right)\left(\frac{\partial}{\partial Z_{j\tau}}\right) \log \widehat{N}_\alpha \quad (i, j\epsilon A_\alpha) ,$$

$$\widehat{N} \equiv < \Phi \mid \Phi > , \quad \widehat{H} \equiv < \Phi \mid H \mid \Phi > / \widehat{N} ,$$

$$\widehat{N} = \prod_{\alpha=1}^{4} \widehat{N}_\alpha , \quad \widehat{N}_\alpha = det\,(B_\alpha) ,$$

$$(B_\alpha)_{ij} \equiv \exp(\vec{Z_i}^* \cdot \vec{Z_j}) \quad (i, j\epsilon A_\alpha) , \tag{2.4}$$

where σ and τ specify three components (x,y,z) of the vectors $\vec{Z}_i^*$ and $\vec{Z}_j$, respectively.

514

3.Frictional Cooling Method

In order to construct the ground state of a given system, we first choose randomly the positions $\{\vec{D}_i\}$ and momenta $\{\vec{K}_i\}$ of all A nucleons. The initial AMD wave function Φ^{init} with this initial choice of $\{\vec{Z}_i\}$ represents in general a highly excited configuration. The construction of the ground state is made by cooling down this Φ^{init}. Our cooling method of [2] is analogous to those of [7] and [8] and can be called frictional cooling method.

We introduce the following equations of motion

$$\dot{Z}_{j\tau} = (\lambda + i\mu)\frac{1}{i\hbar} \sum_{k\epsilon A_\alpha,\sigma} D^\alpha_{j\tau,k\sigma}\frac{\partial\widehat{H}}{\partial Z^*_{k\sigma}} \quad \text{and} \quad c.c. \; ,$$

$$D^\alpha_{j\tau,k\sigma} \equiv (D^\alpha)_{j\tau,k\sigma} \; , \quad D^\alpha = (C^\alpha)^{-1} \; ,$$

$$(C^\alpha)_{k\sigma,j\tau} \equiv C^\alpha_{k\sigma,j\tau} \; , \tag{3.1}$$

where λ and μ are real numbers. Since we are not interested in the total center-of-mass motion, we here use the internal Hamiltonian $\widehat{H}_I$ instead of $\widehat{H}$. An important property of the Hermitian matrix D^α is that it is positive definite. A proof of this fact is given in [2]. From Eq.(3.1) there follows

$$\frac{d}{dt}\widehat{H}_I = \sum_{j=1}^A \left(\frac{\partial\widehat{H}_I}{\partial\vec{Z}_j}\dot{\vec{Z}}_j + \frac{\partial\widehat{H}_I}{\partial\vec{Z}^*_j}\dot{\vec{Z}}^*_j\right)$$

$$= \frac{2\mu}{\hbar}Y$$

$$Y \equiv \sum_{\alpha=1}^4 \sum_{k,j\epsilon A_\alpha,\sigma,\tau} \left(\frac{\partial\widehat{H}_I}{\partial Z_{j\tau}}\right) D^\alpha_{j\tau,k\sigma} \left(\frac{\partial\widehat{H}_I}{\partial Z^*_{k\sigma}}\right) \; . \tag{3.2}$$

Since D^α is positive definite, the quantity Y is positive ($Y > 0$) and we have

$$\frac{d\widehat{H}_I}{dt} > 0 \quad \text{for} \quad \mu > 0 \; ,$$

$$\frac{d\widehat{H}_I}{dt} = 0 \quad \text{for} \quad \mu = 0 \; ,$$

$$\frac{d\widehat{H}_I}{dt} < 0 \quad \text{for} \quad \mu < 0 \; . \tag{3.3}$$

Thus by solving Eq.(3.1) with negative μ value we can cool down the initial state Φ^{init}.

As far as the cooling process is concerned, we can replace the matrix D^α by any other positive definite Hermitian matrix in Eq.(3.1). The simplest choice is to replace

D^α by unit matrix, which results in the following frictional cooling equation:

$$\dot{Z}_{jr} = (\lambda + i\mu)\frac{1}{i\hbar}\frac{\partial \widehat{H}_I}{\partial Z_{jr}^*} \quad \text{and c.c.} \quad . \tag{3.4}$$

In the applications discussed in section 5, we have constructed energy-minimum intrinsic configurations with definite parity. In this case we have used Eq.(3.4) by replacing $\widehat{H}_I$ by $\widehat{H}_I^\pm$ defined as

$$\widehat{H}_I^\pm = \frac{<\Phi\,|\,H_I\,|\,\Phi> \,\pm\, <\Phi\,|\,H_I P\,|\,\Phi>}{<\Phi\,|\,\Phi> \,\pm\, <\Phi\,|\,P\,|\,\Phi>} \quad , \tag{3.5}$$

where P is the parity operator and H_I is the internal Hamiltonian operator.

4.Application to Self-Conjugate 4N Nuclei

We report here results of the application of our frictional cooling method to several self-conjugate 4N nuclei. We have adopted the Volkov No.2 force[9] for ^{20}Ne and the Volkov No.1 force[9] for other nuclei as effective two-nucleon force and we have not included the Coulomb force for the sake of simplicity.

Fig.1 displays how the binding energy of ^{8}Be converges to its minimum value when our cooling method is applied. Here we have used $\lambda = 1.0$ and $\mu = -0.7$ in Eq.(3.1). The Majorana exchange mixture m of the Volkov No.1 force has been chosen to be m=0.56 and the size parameter ν of the Gaussian nucleon wave packet has been given the value $\nu = 0.2675$ fm^{-2}. In this figure we have also displayed the time variation of the number of the total oscillator quanta $N_{tot} = \sum_{\alpha=1}^4 < N_\alpha^{op} >$. The operator N_α^{op} is defined as

$$N_\alpha^{op} \equiv \sum_{i=1}^A \overrightarrow{a_i}^+ \cdot \overrightarrow{a_i}\,\delta(s_{zi} - \sigma_\alpha)\delta(t_{zi} - \tau_\alpha) \,,$$

$$\overrightarrow{a_i} \equiv \sqrt{\nu}(\overrightarrow{r_i} + \frac{1}{2\nu}\frac{\partial}{\partial \overrightarrow{r_i}}) \,, \tag{4.1}$$

where s_{zi} and t_{zi} are z-components of the spin and isospin operators of the i-th nucleon, respectively. The values of $(\sigma_\alpha, \tau_\alpha)$ are $(1/2, 1/2)$, $(-1/2, 1/2)$, $(1/2, -1/2)$ and $(-1/2, -1/2)$ for $\alpha = 1, 2, 3$ and 4, respectively. In the final stage of the convergence process of the binding energy, say for $t > 30$ fm/c, there holds $< N_\alpha^{op} >= N_{tot}/4$ for any $\alpha(\alpha = 1 \sim 4)$ with error smaller than 0.01.

In the system of ^{8}Be the possible lowest number of N_{tot} is 4 and $N_{tot} = 4$ means that the configuration is just the p-shell shell model configuration $(0s)^4(0p)^4$. In Fig.1 we see that, although N_{tot} approaches closely to the value 4 in the intermediate

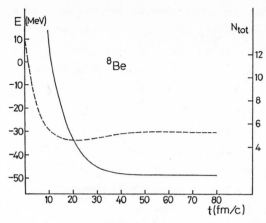

FIG.1   Time variation of the binding energy $(-E)$ of ^{8}Be (solid curve) under
the operation of the frictional cooling. Here is also displayed time variation of the
number of total oscillator quanta N_{tot} of ^{8}Be (dashed curve) under the operation
of the frictional cooling.

time step of the energy convergence process, the final value of N_{tot} is sizably larger
than 4, which means that the ground state of ^{8}Be does not have a simple p-shell shell
model configuration.

We have studied in detail the converged values of the position and momentum
parameters, $\vec{D}_i$ and $\vec{K}_i$ of all the nucleon wave packets and have found firstly that the
values of $\vec{K}_i$ are all negligibly small and secondly that four nucleons with different
spin-isospin variables $\alpha(\alpha = 1 \sim 4)$ gather to the same spatial point while the
remaining four nucleons also with different spin-isospin variables gather to another
common spatial point. This result means that the ground state wave function is just
the Brink-type two-alpha intrinsic wave function[4]. The inter-alpha distance has
proved to be 2.72 fm for the present choice of the Majorana parameter m=0.56 and
it of course increases if we choose larger value for m.

More interesting result has been obtained for the ground state of ^{12}C. Like in the
case of ^{8}Be, all the values of $\vec{K}_i$ have been found to be negligibly small. As for the
values of $\vec{D}_i$, twelve nucleons have been divided into three groups in each of which
there are contained four nucleons with different spin isospin variables $\alpha(\alpha = 1 \sim 4)$
but with the common value for $\vec{D}_i$. Therefore this result again means that the
ground state wave function is just the alpha-cluster model intrinsic wave function
of Brink type which is now composed of three alpha-clusters. Furthermore we have

found that the three inter-alpha distances are the same one another, namely that the three alphas form a regular triangle geometrically. When the size parameter ν of the nucleon wave packet and the Majorana parameter m of the Volkov No.1 force are given values of $\nu = 0.16$ fm^{-2} and $m = 0.65$, respectively, the inter-alpha distance of the regular triangle takes the value of 1.43 fm and the number of the total oscillator quanta N_{tot} takes the value of 8.26 which is slightly larger than the value of 8, the possible lowest number of N_{tot} corresponding to the p-shell shell model configuration. If we choose the value m=0.60, the resulting inter-alpha distance of the 3α regular triangle is only 0.027 fm and the N_{tot} value is 8.0 within the accuracy of 10^{-3}, which means the ground state wave function is almost the same as the intrinsic state wave function of the Elliott SU(3) model[10], $(0s)^4(0p)^8, (\lambda, \mu) = (0, 4)$. The calculated binding energy is 60.1 MeV for m=0.65 and 70.0 MeV for m=0.60.

We have also studied ^{16}O and ^{20}Ne. In ^{16}O, the size parameter ν of the nucleon wave packet is given the value of 0.16 fm^{-2}, and the Majorana parameter m of the Volkov No.1 force has been varied between 0.60 and 0.65. We have found that the N_{tot} value is 12.0 within the accuracy of 10^{-3} for any m value in the above-mentioned range. Therefore the ground state wave function is almost the same as the doubly-closed-shell wave function of $(0s)^4(0p)^{12}$. The calculated binding energy is 127.6 MeV for m=0.60 and 107.4 MeV for m=0.65.

In ^{20}Ne system we have found the same result as in ^{8}Be and ^{12}C that the ground state wave function obtained with the frictional cooling method is just the alpha-cluster model intrinsic wave function of Brink type. What is very interesting in ^{20}Ne is the geometrical arrangement of the five alpha-clusters of the ground state configuration. Fig.2 shows the result obtained with our frictional cooling method. In this figure, three alpha-clusters in the center form a regular triangle. Each of two remaining alpha-clusters is located in such a way that its distances from the three alpha-clusters of the regular triangle in the center are the same one another; namely remaining two alpha-clusters are located on the straight line which passes through the center-of-mass of and is perpendiclular to the regular triangle in the center. We here discuss in detail the results obtained under the choice of $\nu = 0.19$ fm^{-2} and the Volkov No.2 force with m=0.60 which is in accordance with [11]. In [11] the ^{20}Ne structure is studied by adopting the Brink-type wave function with five alpha-clusters arranged just like in Fig.2. In the present calculation the inter-alpha distance of the regular triangle in the center which is denoted as d_3 in Fig.2 has proved to be 0.007 fm while in [11] this distance is fixed to be 0.3 fm. The distances d_1 and d_2 have

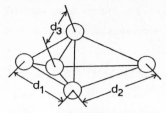

FIG.2 Spatial arrangement of five alpha-clusters of ^{20}Ne ground state obtained by the frictional cooling method. Three alpha-clusters in the center form a regular triangle with inter-cluster distance d_3. Remaining two alpha-clusters are located on the straight line which passes through the center-of-mass of and is perpendicular to this regular triangle.

proved to be 1.46 fm and 1.35 fm, respectively, and the binding energy is obtained to be 183.0 MeV. An interesting point of the ^{20}Ne results reported above is the fact that d_1 and d_2 are not the same each other.

5.Application to Neutron-Rich Be Isotopes

We have also applied the frictional cooling method to the structure study of Be isotopes. A main purpose of this study is to investigate how the $\alpha - \alpha$ cluster structure of ^{8}Be changes when neutrons are added in going from ^{8}Be to neutron-rich side of Be isotoes. This problem was already partly investigated in [12] where the molecular orbital model around $\alpha - \alpha$ core was adopted. In this approach of [12], the degree of α-clustering is measured in terms of the distance $R_{\alpha\alpha}$ between two α clusters which is determined by the energy variation. In our AMD approach we can judge whether the existence of the $\alpha - \alpha$ core assumed in [12] can be justified or not.

The frictional cooling has been applied to parity-projected states by the use of Eq.(3.4). The two-nucleon central nuclear force has been chosen to be Volkov No.1 force with the Majorana parameter m=0.56, while the two-nucleon spin-orbit force v^{LS} has been chosen to be G3RS force [13]

$$v^{LS} = u^{LS} \left(\bar{e}^{\alpha r^2} - \bar{e}^{\beta r^2} \right) P(^3O) \overrightarrow{L} \cdot \overrightarrow{S} \, ,$$
$$u^{LS} = 900 \text{ MeV} \, , \quad \alpha = 5.0 \text{ fm}^{-2} \, , \quad \beta = 2.778 \text{ fm}^{-2} \, , \tag{5.1}$$

where $P(^3O)$ is the projection operator to triplet-odd states. We have not included the Coulomb force. The oscillator parameter ν has been determined by energy variation for each parity state and for each isotope. After the energy-minimum intrinsic state with good parity is constructed with the frictional cooling method, we have projected good angular-momentum states in order to compare the calculation with experiments.

The calculated binding energies of ^{A}Be with $A = 6 \sim 14$ are 24.3, 34.6, 54.3, 54.1, 59.8, 61.2, 68.8, 64.2 and 66.6, respectively, while the corresponding experimental values are 26.9, 37.6, 56.5, 58.2, 65.0, 65.5, 68.8, 66.1 and 69.9, respectively. Fig.3 shows the comparison of the calculated spectra with experimental ones for the cases of ^{7}Be, ^{9}Be, ^{10}Be and ^{11}Be. Except the fact that the calculated energy gaps between positive and negative parity states are larger than the observed ones, the agreement between the theory and experiments is seen to be good. The calculated energy gaps between different parity states can be made smaller and comparable with experiments when we use density-dependent effective nuclear force. This fact is seen again in Fig.3 where we have also shown the spectra of the non-normal parity states which are obtained by the use of the case 3 of MV1 force of [14]. This force is constructed by reducing the strength of the short-range repulsive part of the Volkov No.1 force and instead by adding the repulsive zero-range three-body force $t_3 \delta(\vec{r}_1 - \vec{r}_2)\delta(\vec{r}_2 - \vec{r}_3)$ with $t_3 = 4000$ MeV $\cdot fm^6$. The Majorana parameter m has been chosen also to be 0.56. Here the spectra has been calculated by evaluating the expectation values of the MV1-case 3 force with the AMD wave function obtained with the Volkov No.1 force. We have found that the binding energies of the normal parity states with the MV1-case 3 force are close to those with the Volkov No.1 force with the difference less than 1 MeV. We will see below that the non-normal parity states have lower density than the normal parity states, which causes the reduction of the energy gaps between different parity states.

As for the converged values of $\{\vec{D}_i\}$ and $\{\vec{K}_i\}$, we have found that all the $|\vec{K}_i|$ values are very small although they are larger than those in the cases of self-conjugate 4N nuclei. The position vectors $\vec{D}_i$ of four protons have been found always to cluster into two groups. In each group, two protons with spin up and down locate at (almost) the same spatial point.

Neutrons which form two α-clusters in ^{8}Be by locating in the vicinity of protons tend to separate from protons as we go from ^{8}Be to ^{12}Be; namely α-clusters tend to dissolve in going from ^{8}Be to ^{12}Be. Although the formation of α-clusters is not complete, we can define the inter-α distance $R_{\alpha\alpha}$ as the distance between two groups of protons. In Fig.4 we show the variation of $R_{\alpha\alpha}$ and $P_{\alpha\alpha}$ as a function of the mass number A. The quantity $P_{\alpha\alpha}$ is a measure of the survival probability of the $\alpha - \alpha$

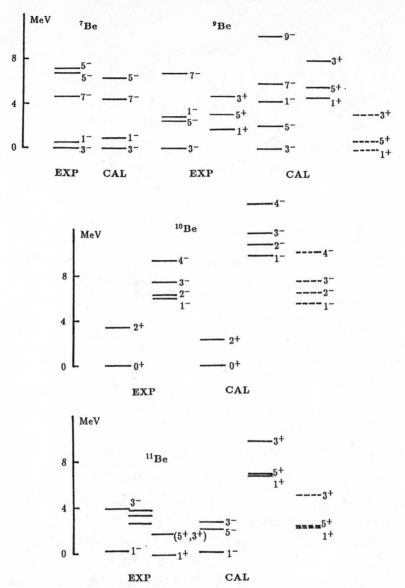

FIG.3 Comparison of calculated energy spectra with experiments. Calculated spectra given with dotted lines are obtained by the use of the MV3 force. For odd-mass isotopes doubled spin values are shown.

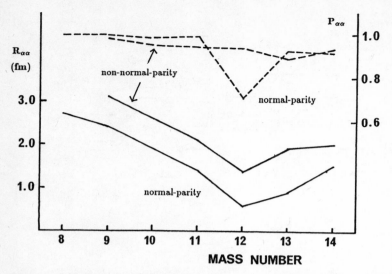

FIG.4   Mass-number dependence of the $\alpha - \alpha$ distance $R_{\alpha\alpha}$ (solid lines) and the $\alpha - \alpha$ -core survival probability $P_{\alpha\alpha}$ (dotted lines) of the normal-parity and non-normal-parity intrinsic states of Be isotopes.

core structure and is defined as

$$P_{\alpha\alpha} = \|y(\overrightarrow{r}_9, \cdots \overrightarrow{r}_A)\|^2 \ ,$$

$$y(\overrightarrow{r}_9, \cdots \overrightarrow{r}_A) = \sqrt{\binom{A}{8}} \langle \psi^B_{\alpha\alpha}(\overrightarrow{r}_1, \cdots \overrightarrow{r}_8) \mid \widehat{\Phi}(\overrightarrow{r}_1, \cdots \overrightarrow{r}_A)\rangle \ ,$$

$$\Psi^B_{\alpha\alpha}(\overrightarrow{r}_1, \cdots \overrightarrow{r}_8) = \text{normalized Brink's } \alpha-\alpha \text{ wave function} \qquad (5.2)$$

with $\alpha-\alpha$ distance equal to $R_{\alpha\alpha}$ of $\widehat{\Phi}$.

Here $\widehat{\Phi}$ stands for the normalized intrinsic AMD wave function. It can be easily proved that when the single-particle wave functions of 8 nuleons contained in $\widehat{\Phi}$ are identical with those in $\psi^B_{\alpha\alpha}$, the quantity $P_{\alpha\alpha}$ is exactly equal to unity. We see in Fig.4 that both $P_{\alpha\alpha}$ and $R_{\alpha\alpha}$ become smaller in going from ^{8}Be to ^{12}Be but tend to become larger in going from ^{12}Be to ^{14}Be. This behavior of $R_{\alpha\alpha}$ is the same as that reported in [12].

When we compare the $R_{\alpha\alpha}$ value of the normal parity intrinsic state with that of the non-normal parity intrinsic state, we find that the latter value is larger than the former value in every Be isotope. This means that the density of the non-normal-parity intrinsic state is lower than that of the normal-parity intrinsic state in every Be isotope.

522

6.Summary and Some Future Prospects

We explained briefly the formulations firstly of the antisymmetrized molecular dynamics (AMD) and secondly of the frictional cooling method which is very efficient to construct ground states of nuclei. Since the construction of ground states with the frictional cooling method is free from any specific model assumption, we can discuss the formation and dissolution of clusters with the AMD framework.

The application to the self-conjugate 4N nuclei has shown that Brink's alpha cluster model can be justified by the present AMD approach. In the application to Be isotopes, we have found very interesting feature of the formation and dissolution of α-α core structure as a function of the neutron number. This feature shows an importance of the cluster model viewpoint in discussing neutron-rich Be isotopes.

The frictional cooling method can be extended so as to be used under constraints [2]. For example we can construct the minimum-energy state under the constraint that the root-mean-square radius has a given artificial value. In [2] an AMD study of monopole vibration is reported which makes use of this constraint. Another example is the construction of the minimum-energy state under the cranking condition. Formation and dissolution of cluster structure along the yrast line in light nuclei are now being studied by the use of this cranking constraint.

Formation and dissolution of clusters during the collision process are now also under study in the AMD framework. In the case of the collision problem we incorporate the two-nucleon-collision process in the AMD framework. This approach can thus be regarded as being an antisymmetrized version of the quantum molecular dynamics [15] or the quasi particle dynamics [16] which has proved to be successful even in rather low incident energy region [17] in addition to successes in medium and high energy region [15,16].

References

[1]   K. Wildermuth and Y.C. Tang, A. Unified Theory of the Nucleus (Vieweg, Braunschweig, 1977);
        H. Horiuchi and K. Ikeda, Cluster Model of the Nucleus, International Review of Nuclear Physics Vol.4 (World Scientific, 1986)

[2]   H. Horiuchi, Nucl. Phys. **A522**(1991), 257c; H. Horiuchi, T. Maruyama, A. Ohnishi, and S. Yamaguchi, preprint KUNS 1028(Kyoto Univ.)

[3]   S. Drożdż, J. Okolowcz, and M. Ploszajczak, Phys. Lett. **109B**, 145(1982);
E. Caurier, B. Grammaticos and T. Sami, Phys. Lett. **109B**, 150(1982);
W. Bauhoff, E. Caurier, B. Grammaticos and M. Ploszajczak, Phys. Rev. **C32**, 1915(1985).

[4]   D.M. Brink, Proc. Int. School of Phys. "Enrico Fermi" **course** 36(1965), ed. C. Bloch, p.247.

[5]   H. Feldmeier, Nucl. Phys. **A515**, 147 (1990).

[6]   C. Corianò, R. Parwani and H. Yamagishi, **A522**, 591 (1991).

[7]   L. Wilet, E.M. Henley, M. Kraft and A.D. MacKellar, Nucl. Phys. **A282**, 341(1977).

[8]   D.H. Boal and J.N. Glosli, Phys. Rev. **C38**, 1870(1988).

[9]   A.B. Volkov, Nucl. Phys. **74**, 33(1965).

[10]   J.P. Elliott, Proc. Roy. Soc.(London)**A245**, 128, 562(1958).

[11]   F. Nemoto, Y. Yamamoto, H. Horiuchi, Y. Suzuki and K. Ikeda, Prog. Theor. Phys. **54**, 104(1975).

[12]   M. Seya, M. Kohno and S. Nagata, Prog. Theor. Phys. **65**, 204 (1981).

[13]   H. Furutani et al., Prog. Theor. Phys. Supple. No.68, Chapt. III (1980).

[14]   T. Ando, K. Ikeda and A. Tohsaki, Prog. Theor. Phys. **64**, 1608 (1980).

[15]   J. Aichelin and H. Stöcker, Phys. Lett. **176B**, 14(1986);
J. Aichelin, G. Peilert, A. Bohnet, A. Rosenhauer, H. Stöcker and W. Greiner, Phys. Rev. **C37**, 2451 (1988);
G. Peilert, H. Stöcker, W. Greiner, A. Rosenhauer, A. Bohnet and J. Aichelin, Phys. Rev. **C39**, 1402 (1989);
J. Aihcelin, C. Hartnack, A. Bohnet, L. Zhuxia, G. Perlert, H. Stöcker and W. Greiner, Phys. Lett. **224B**, 34 (1989);
A. Bohnet, N. Ohtsuka, J. Aichelin, R. Linden and A. Faessler, Nucl. Phys. **A494**, 349 (1989).

[16]   G. E. Beauvais, D. H. Boal and J. C. K. Wong, Phys. Rev. **C35**, 545 (1987);
D. H. Boal and J. N. Gloshi, Phys. Rev. **C37**, 91 (1988).

[17]   T. Maruyama, A. Ohnishi and H. Horiuchi, Phys. Rev. **C42**, 386, (1990).

Cluster Degree of Freedom and Nuclear Reaction and Decay

S.G. Kadmensky and V.G. Kadmensky

Voronezh State University, Voronezh, Russia

Abstract: The application of the shell model with residual interactions allows us successfully to describe not only the main statical and dynamical properties of nuclei, but also the cluster degree of freedom, which emerge in nuclear reactions. On the basis of the given conception the following problems are investigated: 1) The effective numbers W_x of clusters(x) with $2 < A_x < 40$ and their spatial, momentum and energy distributions in nuclei are calculated. It is found that the number of clusters considerably exceeds the number of nucleons ($W_d = 430$, $W_t = 507$, $W_{16O} = 2800$ for ^{208}Pb). 2) The theory of α decay of ground and excited states of nuclei (including neutron resonances) and the theory of heavy-cluster radioactivity with production of ^{14}C, ^{24}Ne and other fragments are constructed, which enables us to calculate the absolute probabilities of α decays, the relative yields of heavy-ions and the hindrance factors. The classification of these decays are made by taking into account of the supper fluid correlations. 3) The formalism to describe the multi-step direct nuclear reactions with the participation of compound particles is developed. The spectra and absolute cross sections of reactions of quasielastic knock-out of d, t, α by fast protons are successfully explained. An important role of six quark structures in these reactions is observed. 4) We have developed the cascade model of nuclear reactions with the participation of d, t, α and other clusters in the internal region of nuclei for the purpose of conceiving the multi-fragmentation processes. The calculations of effective nucleon-cluster cross sections in nuclei are made by taking into account the antisymmetrization effects for both nucleons and clusters.

1. Introduction

With a high degree of precision the atomic nucleus can be considered as a system of interacting nucleons. Thus, one can explain the main statical and dynamical properties of nuclei [1-3] by using the one-nucleon degree of freedom, which is described through quasiparticles for the account of their interactions. Here a certain question arises, namely: *Is it possible to describe the formation of compound particles (clusters(x)) in nuclei, for example, such as deuterons, tritons, α particles and multi-charge ions with $A_x > 4$, which emerge in nuclear reactions and decays?* In other words, is it necessary to get out of the frames of the shell model for residual interactions in order to describe the cluster degree of freedom of nuclei?

Springer Series in Nuclear and Particle Physics   **Clustering Phenomena in Atoms and Nuclei**
Editors: M. Brenner · T. Lönnroth · F.B. Malik   © Springer-Verlag Berlin, Heidelberg 1992

From the theoretical point of view, one must give the negative answer to the above formulated question. Indeed, deuterons, tritons and other compound particles, as well as the heavier nuclei, can be treated as a system of interacting nucleons. Thus, the complete shell basis for the account of the continuous spectra and residual interactions is sufficient, in principle, for describing all the cluster modes in nuclei. Nevertheless, in realistic calculations such a program encounters essential difficulties due to the convergence of the method in the region of the light nuclei ($A < 20$), where a certain number of characteristics, for example the excitation spectrum, indicate the existence of spatially and energetically isolated cluster formation, for instance, in α cluster nuclei. As far as other nuclei are concerned, their main properties can be explained without using the conception of the distinguished clusters and their cluster degrees of freedom can be successfully described [4,5] on the basis of the shell model with the limited basis. The purpose of the present paper is the demonstration of the possibilities of this approach for analyzing nuclear reactions and decays with the participation of compound particles.

2. The Spectroscopic Factors and Effective Numbers of Clusters in Nuclei

Let us consider the process of decomposition of the parent nucleus A with the spin J_i, its projection M_i and other quantum numbers σ_i, described by the wave function $\psi_{\sigma_i}^{J_i M_i}$, into the channel c, corresponding to the formation of the daughter nucleus $(A - x)$ with the wave function $\psi_{\sigma_f}^{J_f M_f}$ and the cluster x with the spin J_x, its projection M_x and internal wave function $\chi_{J_x M_x}$. This process can be real, if the energy Q_c of the relative motion of the cluster x and the daughter nucleus $(A - x)$ is positive and virtual if $Q_c < 0$ and the channel of decay is closed. For characterization of the indicated process, we must introduce the conception [4-5] of the formfactor $\psi_{xc}(R)$ of the compound particles x in the channel c:

$$\psi_{xc}(R) = \left\langle \hat{A}\left[\frac{U_c\,\delta(R - R')}{R'}\right] \middle| \psi_{\sigma_i}^{J_i M_i} \right\rangle \tag{1}$$

where $\hat{A}$ is the antisymmetrization operator and U_c is the channel function:

$$U_c = \left\{\left\{\psi_{\sigma_f}^{J_f}\chi_{J_x M_x}\right\} Y_{L M_L}(\Omega_R)\right\}_{J_i M_i} \tag{2}$$

Here the term in the bracket in (2) denotes the vector addition of momenta; $Y_{L M_L}(\Omega_R)$ is the spherical function describing the angular part of the relative motion of the cluster x and the daughter nucleus $(A - x)$ with the orbital momentum L. In the formula (1), R is the radial variable corresponding to the distance between the center of masses of the decay products x and $(A - x)$.

We must remark that the form factor $\psi_{xc}(R)$ takes into account the combinatoric multipliers connected with the different ways of constructing the decay fragments x and $(A - x)$ from the identical nucleons of the parent nucleus. Due to this fact the form factor $\psi_{xc}(R)$ must not be interpreted only as a radial wave function of the relative motion of the fragments.

Now we determine the spectroscopic factor W_{xc} of the compound particle x in the channel c as:

$$W_{xc} = \int_0^{R_i} \psi_{xc}^2 \, dR, \tag{3}$$

where the definition of the radius R_i depends on the concrete type of decay. Form factors and spectroscopical factors of compound particles are the basic concepts used for describing nuclear reactions and decays with the participation of compound particles. In particular, the amplitude of the reduced width, used in the R-matrix approach, is merely proportional to the formfactor (1).

In order to obtain the formfactor $\psi_{xc}(R)$ concretely, it is convenient to divide the whole variational region of the variable R in the nucleus A into three subregions [5]. In the region, $R \leq R_{sh}$ (the shell region) the parent nucleus can be well described by the multi-particle wave function of the shell model, using the limited and discrete basis as a rule. In the region $R \geq R_{cl}$ (the cluster region) one can consider that the decay fragments are really formed. In the intermediate region, $(R_{sh} \leq R \leq R_{cl})$ both the concepts of the shell model with the limited basis and that of the cluster region are proven to be invalid. This region is the most difficult for the theoretical analysis and in fact it has not been seriously investigated up to the present time. Thus, the spectroscopic factor W_{xc} can be presented as a sum of three spectroscopic factors, which are determined by the formfactors of the particles x in the shell, intermediate and cluster regions.

For the validity of the cluster region concepts the following conditions must be satisfied: 1) The internal wave functions of the fragments must be inconsiderably distorted due to the antisymmetrization effects. 2) In principles the polarization influence of nuclear and Coulomb interactions of the fragments on their internal functions, must be comparatively small and should be taken into account through the connection of the limited number of channels. 3) The renormalization of interaction between the nucleons inside one of the fragments, for instance, due to the influence of the nucleons of the other fragments, must be small. From the physical point of view, it is clear that these conditions are satisfied only in the case when the fragments overlap inconsiderably. In the region where the nucleon density $\rho(R)$ of a heavy fragment is essentially smaller than its value $\rho(0)$ in the center of the nucleus, one may hope on the applicability of the gas approximation over the density $\rho(R)$, in the frames of which all the conditions mentioned above are satisfied. At present it is difficult to determine exactly the value R_{cl}, the border radius of the cluster region $R_{cl} \leq R$. Due to this fact it is natural to estimate this magnitude phenomenologically, basing upon the conditions of the optical model for the particles (x), in which the multi- particle problem of interaction between the fragment x and the nucleus $(A-x)$ can be reduced to the one-particle problem of the motion of the center of mass of the particle x in the complex optical potential $V_x^{opt} = V_x(R) + iW_x(R)$. The optical model for compound particles x is proven to be internally coordinated, since in this model all the calculated values are determined by the surface region of the nucleus only, where the conditions mentioned above are satisfied. This fact is fully connected with the imaginary part of the optical potential, which is responsible for the exponential decrease of the modulus of the optical wave function in the internal region of the nucleus.

In the cluster region the wave function of the parent nucleus can be presented in the form[5]:

$$\psi_{\sigma_i}^{J_i M_i} = \sum_{cx} \hat{A} \left(\left[\frac{\psi_{xc}^{cl}(R)}{R} \right] u_c \right) \tag{4}$$

where $\psi_{xc}^{cl}(R)$ is the cluster form factor which in the absence of channel connection satisfies the radial Schrödinger equation:

$$\left[-\frac{\hbar^2}{2m_x} \frac{d^2}{dR^2} + \frac{\hbar^2}{2m_x} \frac{L(L+1)}{R^2} + \frac{2Ze^2}{R} + V_x(R) - Q_c \right] \psi_{xc}^{cl}(R) = 0 \tag{5}$$

where m_x is the reduced mass. Using the boundary condition corresponding to the decay like (for $Q_c > 0$) or bound (for $Q_c < 0$) sate of the parent nucleus in the channel c and integrating the Schrödinger equation (5) outside, towards the direction of the point $R = R_{cl}$, one can restore the magnitude of the cluster formfactor $\psi_{xc}^{cl}(R)$ in the whole cluster region up to a constant. The value of this constant can be calculated by using experimental data of the widths of the decay of the nucleus A into the channel c ($Q_c > 0$, as well as the cross sections of the direct reactions with the participation of the particle x in the channel c ($Q_c > 0$ and $Q_c < 0$).

After calculating the cluster form factor $\psi_{xc}^{sh}(R)$ and spectroscopic factor W_{xc}^{cl}, one can perform a comparison of their values with the values of analogous quantities $\psi_{xc}^{sh}(R)$ and $W_{xc}^{sh}(R)$ in the shell region of the nucleus. Then, in the general cases one can distinguish between the limits of the two situations. The first situation corresponds to the case when the shell form factors and spectroscopic factors reasonably reflect the scale of cluster ones. In other words, in this case the decay channel of the parent nucleus is described sufficiently well by the shell form factors, the determination of which corresponds physically to the *freezed* situation. It is widely used in the scheme of direct nuclear reactions, when the parent nucleus is not noticeably reconstructed during the transition to the final channel. This case is apparently realized for α decay and direct nuclear reactions with the participation of enough light compound particles. In particular, in the case of α decay from the ground and low-lying states, with detailed investigation, the magnitude of $W_{\alpha c}^{cl}$ for all the open and closed channels with $L < 6$ turns out to be commensurable with the magnitudes of $W_{\alpha c}^{sh}$ and smaller than $10^{-2} - 10^{-3}$ for nuclei with $A \geq 60$. The small values of $W_{\alpha c}^{cl}$, as well as the smoothness of the A dependence of the α particle strength function of neutron resonances, allows us to make a conclusion [5] on the absence of α cluster levels corresponding to quasi-molecular states like $\alpha + (A - \alpha)$ in all sufficiently heavy nuclei. With the growth of the orbital momentum L of an α particles the magnitude $W_{\alpha c}^{cl}$ considerably decreases and for $L \geq 10$ becomes essentially smaller than the magnitude $W_{\alpha c}^{sh}$. An analogous situation arises for $W_{\alpha c}^{cl}$ with the increase of the modulus of the separation energy Q_c, of the α particles, for the closed channels of α decay ($Q_c < 0$). At $Q_c < -20$ MeV the contribution of the α particles cluster spectroscopic factor $W_{\alpha c}^{cl}$ into the complete spectroscopic factor $W_{\alpha c}$ can be neglected.

In the case of the other light particles x, such as d, t, ^{3}He, the situation with $W_{\alpha c}^{cl}$ is qualitatively close to that with the α particle [5,6]. The difference is exclusively of quantitative character. The situation in the second limit corresponds to the case when the shell form factors and spectroscopic factors are essentially less than the cluster ones. This means that the decay nucleus is strongly reconstructed during the transition from the shell region to the cluster one. This situation is most distinctively realized in the process of spontaneous and constrained fission of heavy nuclei, when the decay fragments x and $(A - x)$ are close to $A/2$ [1].

It is clear that between these two extreme cases there is a certain transitional region, the boundary of which with respect to x is desired to be ascertained. Below we shall discuss the situation of the first type. It is important to stress that, for describing the cluster degree of freedom in α decay and exclusive direct nuclear reactions with light compound particles ($A_x < 4$), an important role is played by the cluster region of the nucleus, the properties of which cannot be described in principle within the frames of the shell model with the limited basis. The amplitude of a cluster form factor can be restored comparatively easily by using the semiphenomenological way of analytical continuation of this form factor into the shell region. This process ensures the possibility of a sufficiently good description of absolute widths of α decay and the cross sections of direct nuclear reactions with the participation of light compound particles [5].

As it is shown above, with the increase of the excitation energy of the daughter nucleus and simultaneously with the decrease of energy Q_c the role of the cluster region for compound particles with $A_x < 4$ decreases and the shell region of the nucleus becomes more significant. Thus, for inclusive direct nuclear reactions with the participation of the indicated compound

Table 1

The effective numbers W_x of deutrons, tritons, ^{3}He
α particles in atomic nuclei

W_x	Nucleus						
	^{8}Be	^{12}C	^{16}O	^{40}Ca	^{56}Ni	^{118}Sn	^{208}Pb
W_d	6.0	11.5	18.6	60.6	89.6	222	433
W_t	4.9	11.2	19.1	71.6	100	326	670
$W_{^3\text{He}}$	4.9	11.6	20.3	76.6	109	261	524
W_α	2.0	4.3	9.6	91	130	290	507

particles, where a wide excitation spectrum of the daughter nuclei is observed, the cluster degree of freedom of the parent nucleus, based on the one-nucleon shell motion, begins to play the main role. In connection with this, it is natural to begin the global analysis of shell and cluster spectroscopic factors for different clusters x in a wide region of atomic nuclei. It is convenient to conduct such an investigation with the help of the concept [4] of the effective number of compound particles W_x in the nucleus A, which is define as a sum the shell spectroscopic factor W_{xc}^{sh} over all the open and closed decay channels c of the nucleus A:

$$W_x = \sum_c W_{xc}^{sh} \tag{6}$$

Using the completeness condition for the daughter nucleus states and the second quantized representation, a method [5,6] was developed for the calculation of effective numbers and their distribution for d, t, ^{3}He and α particles.

Table 1 illustrates the large values of the magnitudes of W_x, which, as a rule exceed the nucleon numbers A in sufficiently heavy nuclei. One should remark that the values W_x are determined by the global properties of the self-consistent nuclear potential and they weakly depend on such details as normal and super-fluid nucleon-nucleon correlations, the transition from the Wood-Saxon shell basis to the oscillator one, spin-orbit splitting, small excitation energies and small deformations of the nucleus A. Figs. 1 and 2 illustrate, using deuterons as an example, the typical energy and momentum distribution of effective numbers in the ^{208}Pb nucleus.

In the paper [7], the consideration of effective number of clusters with $A_x > 4$, from ^{8}Be to ^{16}O, is carried out. The results of the calculations confirms all the basic regularities discovered earlier for lighter clusters. In particular, the effective number for the ^{16}O cluster in the oscillator double magic nucleus ^{228}A turned out to be equal to 1200. Meantime, an interesting tendency of the decrease of the number of clusters x having small separation

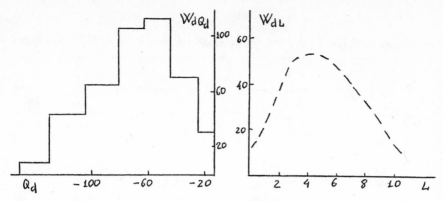

Fig. 1. The distribution of effective numbers of deutrons W_{dQ_d},
depending on the separation
energy Q_d, in the ^{208}Pb nucleus.

Fig. 2. The distribution of deutron
effective numbers W_{dL} over momenta
in the ^{208}Pb nucleus.

energies Q with the increase of the cluster mass A_x was found. For example,
in the case of the ^{208}Pb nucleus one tenth of the total effective number for
d, t, ^{3}He , α, ^{16}O is accumulated at the separation energies Q_x greater
than -30, -40, -45, -60, -180 MeV, respectively. This result allows us
to understand the threshold character of the cross sections of knock-out of
sufficiently heavy clusters from the target nucleus by fast particles, which
brightly displays in the multi-fragmentation reaction cross sections.

Now we can make the next step for more qualitative comprehension of
the situation with the effective numbers of compound particles. Let us make
use of the fact that the oscillator shell basis reproduces the cluster effective
number fairly well and expand the shell form-factor $\psi_{xc}(R)$ into a series over
the radial oscillator functions $\phi_{NL}(R)$ describing the motion of the cluster
center of mass in the oscillator field:

$$\psi_{xc}^{sh}(R) = \sum_N \phi_{NL}(R)\langle\phi_{NL}(R)|\psi_{xc}^{sh}(R)\rangle \qquad (7)$$

Using the orthonormalization of the function $\phi_{NL}(R)$ and the completeness of the daughter nucleus states, we obtain:

$$W_x = (2J_x + 1) \sum_{NLM} W_x(N, L, M) \qquad (8)$$

where the quantity $W_x(N, L, M)$ is defined as a sum over all the occupied
shell states of the nucleons forming the cluster x. The physical sense of
the formula (8) is obvious. From the view point of effective numbers of

Table 2

The values of $W_x(N, L, M)$ for the case of deutron in the ^{208}Pb nucleus.

N	L										
	0	1	2	3	4	5	6	7	8	9	10
0	0.7										
1		0.8									
2	0.9		0.8								
3		0.9		0.8							
4	1.0		0.9		0.8						
5		0.9		0.9		0.8					
6	0.9		0.8		0.7		0.8				
7		0.7		0.7		0.7		0.7			
8	0.5		0.5		0.5		0.5		0.6		
9		0.3		0.3		0.3		0.3		0.4	
10	0.02		0.02		0.02		0.05		0.07		0.2

compound particles x, the nucleus is a potential well with the oscillator levels characterized by the quantum numbers N, L, M, each of which can include $(2J_x+1)W_x(N, L, M)$ particles x. The energy gap between the levels, the basic quantum numbers N of which differ by the magnitude ΔN, are determined as $\hbar\omega\Delta N$, where $\hbar\omega = 41A^{-1/3}$ Mev is the oscillator energy. For each compound particle the well depth changes in order to comprise all the oscillator levels with $N_x \leq N_x^{max}$, the number of these levels increasing with the growth of the number of neutrons and protons forming the particles x. In the frames of such a representation (we may call it the cluster model of the nucleus), no necessity arises in fact to treat the nucleon degree of freedom of the nucleus.

Let us remark that, although the compound particles x can have an integer spin J_x, for instance, the α particle or d, the limit of the value of $W_x(N, L, M)$ cannot exceed unity. This result is wholly connected with the Fermi statistics of nucleons forming the compound particles.

In Table 2 the calculated quantities $W_x(N, L, M)$ for deutrons in the ^{208}Pb nucleus are presented. As it is seen from the table, for $L = 0$ with the growth of N the magnitude $W_x(N, 0, 0)$ initially increases, reaching the value of 0.98 for $N = 4$ and then it monotonicaly decreases to 0.017 for $N = 10$. On the other hand, for $N = 10$ the magnitude of $W_x(10, L, M)$ monotonously increases as a function of L, reaching the value of 0.21 for the greatest possible $L = N + 10$ due to the *alignment* effect [4].

In the frames of the cluster model of the nucleus it is possible to calculate the density of cluster levels g_x in the vicinity of the Fermi surface. For deutrons, tritons, ^{3}He and α particles the densities are equal to 3g, 9g/2, 9g/2 and 4g respectively, where g is the full density of one-nucleon states near the Fermi surface.

Let us notice that at present the models with preliminary formed clusters (see, for instance ref. [10]) are widely used for describing inclusive nuclear reactions with the production of compound particles. It is assumed in these models that the clusters x (for example, the α particle ones) exist in nuclei in a ready form. In the framework of the pre-equilibrium model, these clusters, as well as the nucleon quasiparticles play the role of elementary excitations or excitons, their level density being determined by the parameter $\tilde{g}_x$ and the disclosure probability at each levels by the parameter Φ_x. In paper [8], for the analyzing the (p, α) reactions on the basis of the α particle knock out mechanism, the magnitude $\tilde{g}_\alpha$ is taken to be equal $g/4$ and the magnitude Φ_α was treated as a phenomenological parameter and is found by means of fitting the theoretical spectra of the investigated reaction to the experimental one. The magnitudes Φ_α is proven to take large values reaching unity. The cluster model developed above enables us to give a quantitative basis to the magnitudes $\tilde{g}_x$ and Φ_x, since in this model they coincide with the magnitudes g_x and $W_x(N, L, M)$ determined above.

3. Classification of Direct Nuclear Reactions and Decays with the Participation of Compound Particles

The shell effective numbers of compound particles can be used as a general measure for classifying the decays and direct nuclear reactions with the production of compound particles in a wide range of energy. We must note that in the case of exclusive direct nuclear reactions, determined by the cluster region, the shell spectroscopic factors are close to the corresponding cluster ones and as a consequence possess the classifying ability.

The investigation of α decay of ground and excited states of a wide class of atomic nuclei including neutron resonances, on the basis of the conception of the closeness of the shell and cluster spectroscopic factors makes it possible, with the account of super-fluid correlations to reproduce both the absolute probabilities of α decay and the hindrance factors, as well as to perform the classification of α transitions according to the degree of facilitation satisfactorily[9].

The open channels of α decay from the ground states of atomic nuclei are connected with the occupation of the ground and low-lying excited states of daughter nuclei due to the small values of α particles separation energy ($Q_\alpha < 10$ MeV) and the high Coulomb barrier. Besides, the appearance of a noticeable centrifugal barrier with the increase of the momenta L of α particles suppresses the channels with large values of L, which can be essentially intensified due to the effect of *alignment* of nucleon momenta.

All these circumstances give rise to the situation when only an insignificant part of the total sum rule for the spectroscopic factors of α particles contributes into the α decay of nuclei from ground states. Thus, as shown by the calculation [5], the values of spectroscopic factors of α particles emerging in α decay do not exceed the value of 10^{-2} for all nuclei with $A > 100$, despite the presence of facilitated α transitions suffering the maximal super fluid intensification.

The situation may change qualitatively in the case of α decay from high-lying excited nuclear states. In recent years considerable yields of α particles from the giant electric quadruple resonance E2 in the nuclei ^{16}O, ^{40}Ca $^{58-62}$Ni were experimentally observed [10]. The explanation of this phenomenon [5] may be connected with the fact that during the α decay of these resonances, lying at the excitation energies of the order of $60A^{-1/3}$ MeV, a large number of the levels of the daughter nucleus can be occupied among which some high spin *aligned* states emerge. The calculations of spectroscopic factors of α particles presented in paper [11] for α transitions from E2 states of the ^{40}Ca nucleus to the main band of the daughter nucleus ^{36}Ar, demonstrate distinctively the *alignment* effect. The spectroscopic factors for α transitions to the levels of ^{36}Ar with $J^\pi = 6^+$ ($L_\alpha = 8$) and $J^\pi = 8^+$ ($L_\alpha = 10$) give more that 60 percent of the total sum of the calculated spectroscopic factors.

The situation is still more favorable in the reactions of pick up of compound particles, when due to the large energy of an incident particles a possibility emerges to extract d, t, ^{3}He, α from the deeply lying shell states and as a result to display a noticeable part of the sum rule for the spectroscopic factors of compound particles in these reactions. In the paper [5] a formalism was developed for describing one-step and multi-step direct inclusive reactions with the participation of compound particles, based on the technique of second quantization and the introduction of the *entrance* multi-particle states. The calculations done for the reaction ^{12}C(d,^{6}Li)^{8}Be at the incident deutron energy of 50 Mev has made it possible to describe [12] the angular and energy distribution of ^{6}Li in the wide range of angles and energies, which is conditioned by the contribution of a noticeable number of α states.

At last, one must argue on the reactions of knock-out of compound particles, in which all the states of daughter nuclei included into the effective numbers of particles W_x, d, t, ^{3}He, α can be occupied at sufficiently high energies of the incident particles. In paper [13], on the basis of Glauber formalism and the inclusive approach, the reactions of quasi-elastic knockout at the incident particles energies of the order of 300 MeV – 5 GeV, such as $(p, p'd)$, $(n, n'n)$ $(p, p't)$, $(p, p'^3\mathrm{He})$, $(p, p'\alpha)$, $(\alpha, 2\alpha)$, are analyzed. It is found that, the technique of effective numbers makes it possible to describe the angular, momentum and energy spectra of the indicated reactions in a wide range of target nuclei, namely $4 < A < 208$. The formalism of effective numbers of compound particles has been generalized for the case of six quark *bag* and a possibility has been demonstrated [14] for the quantitative description of quasi-elastic reactions (p, pd) in the wide range of target nuclei and proton energies, using the concept of the leading role of six-quark structures in these reactions.

4. The Heavy-Cluster Decay of Atomic Nuclei

The technique of spectroscopic factors of compound particles developed above is applied[15] to the description of the recently discovered phenomenon, called the heavy-cluster radioactivity of nuclei and is connected with the spontaneous emission of particles like ^{14}C, ^{24}Ne, ^{28}Mg, ^{32}Si and others by

Table 3

Half lives for the spontaneous heavy cluster decay

Type of decay	$lgT_{1/2}(c)$	$lgT_{1/2}(c)$ $(expt)$
$^{221}Fr(^{14}C)$	16.2	>15.8
$^{221}Ra(^{14}C)$	15.0	>14.4
$^{222}Ra(^{14}C)$	11.8	11.0 ± 0.6
$^{223}Ra(^{14}C)$	16.0	15.2 ± 0.05
$^{224}Ra(^{14}C)$	16.6	15.9 ± 0.12
$^{225}Ac(^{14}C)$	19.7	>18.3
$^{226}Ra(^{14}C)$	21.6	21.2 ± 0.2
$^{230}Th(^{24}Ne)$	25.6	24.6 ± 0.07
$^{231}Pa(^{24}Ne)$	24.4	23.4 ± 0.08
$^{232}Th(^{26}Ne)$	30.4	>27.9
$^{232}U(^{24}Ne)$	21.1	21.0 ± 0.01
$^{233}U(^{24}Ne)$	25.6	24.9 ± 0.15
$^{234}U(^{24}Ne)$	26.0	25.3 ± 0.05
$^{234}U(^{28}Mg)$	25.4	25.7 ± 0.06
$^{236}Pu(^{28}Mg)$	21.1	21.7
$^{238}Pu(^{28}Mg)$	26.0	25.7 ± 0.25
$^{238}Pu(^{32}Si)$	25.9	25.7 ± 0.16
$^{241}Am(^{34}Si)$	28.0	>25.3

nuclei. In these cases, an analogy of the new type of radioactivities with the α decay, where the latter is not accompanied by any noticeable reconstruction of the parent nucleus and the decay process is of a non-adiabatic character. The alternative theoretical approaches [16] refers to the analogy of heavy-cluster radioactivity with the spontaneous fission of heavy nuclei accompanied by the strong and adiabatic (with respect to the one-nucleon motion) reconstruction of the parent nuclei.

In papers [15, 16] complicated objects like the spectroscopic factors for sufficiently complicated compound particles, for examples ^{14}C, ^{24}Ne and others, have been calculated in detail for the first time. An important role of super-fluid correlations, lead to the intensification of the shell spectroscopic factors up to 10^{18} times for ^{24}Ne was observed. On the basis of the performed calculations the half-lives for the spontaneous heavy-cluster decay are obtained, and they are shown to be in a reasonable accordance with the experimental ones (see Table 3). The developed theoretical approach has made it possible to carry out a classification of all types of heavy-cluster decays according to their facilitation degree and to make a number of important predictions in the direction of further searches for new decay of nuclei and new heavy clusters ejection.

In a recent paper [17] the discovery of several spectrum lines of ^{14}C during the decay of ^{223}Ra is reported. Let us note that the prediction of the fine structure of heavy cluster spectra has been made for the first time in a paper [15] and consequently the discovery of such a structure may serve as a decisive confirmation of the analogy between the α decay mechanisms and heavy-cluster radioactivity.

5. The Intranuclear Cascade Mocel with the Participation of Clusters

The model of an intranuclear cascade, suggested more that 40 years ago [18] and mathematically substantiated in paper [19], meets with essential success [20] in describing the differential and integral cross sections, correlation functions, multiplicities of secondary particles and other characteristics of the collision of a sufficiently fast nucleon with the nucleus. In this model the drawing of quasi-free collisions of the cascade nucleons with the nucleons of the target nucleus along the quasi-classical trajectories takes place with the use of the Monte-Carlo method.

Unfortunately the indicated model does not take into account the collisions of nucleons with intranuclear clusters, and thus the yield of compound particles in the reactions of fast nucleon collisions with nuclei is described in a rather artificial way on the basis of the idea of nucleon coalescence [21].

The results obtained above allow us to generalize the model of intranuclear cascade for the case of taking into account nucleon collisions with compound particles. Indeed, the knowledge of energetic spatial and momentum distributions of effective numbers of clusters in nuclei makes it possible to introduce nucleon-cluster and cluster-cluster collisions into the Monte-Carlo scheme. The contribution of these processes into the total cross section of

537

the reaction of fast nucleon collision with nuclei is not to be small, since the cross sections of the nucleon interaction with clusters are commensurable with the nucleon-nucleon interaction cross sections and the effective number of clusters in nuclei exceeds the number of nucleons. Let us note here that the experimentally found small values of the cross sections for the knock-out of clusters by fast nucleons from nuclei, as compared with the analogous cross sections of nucleon knock-out does not testify against the insignificance of nucleon-cluster impacts in the nucleus, because it can be explained through the essentially stronger absorption of clusters in the nucleus as compared with nucleons.

In order to calculate the cross sections of interaction of clusters with nucleons and clusters with clusters inside the nucleus, one can renormalize the corresponding cross sections of free interactions through the effective account of antisymetrization by reducing the phase volume of the final states of the reaction, connected with the impossibility of exceeding unity for the occupation numbers of clusters in the nucleus $W_x(N, L, M)$, as it is mentioned above.

The cluster distribution of $W_x(N, L, M)$ may be transformed into the momentum distribution of clusters corresponding to the smoothed Fermi step in the quasi-classical approximation. In the frames of this approach no additional parameters arise and one can take into consideration the nucleon-nucleon impacts as well as the nucleon-cluster and cluster-cluster ones, in the unified scheme.

6. Conclusion

The review of the results of investigation of cluster degrees of freedom of nuclei made in the present work allows us to make a conclusion about the great possibilities of the shell model with residual interactions for describing a wide range of nuclear reactions and decays connected with different compound particles. We believe that the variant of the intranuclear cascade model proposed above gives rise to the hope for conceiving the main features of the multi-fragmentation phenomenon on the basis of the dynamical characteristics of the nucleus.

References

1. Bohr O., Mottelson B.: *The Structure of the Atomic Nucleus*. Mir, Moscow, 1971.
2. Migdal A.B.: *The Theory of Finite Fermi-systems and the Properties of Atomic Nuclei*. Nauka, Moscow, 1965.
3. Solovyov V.G.: *The Theory of Complex Nuclei*. Nauka, Moscow, 1971.
4. Neudachin V.G., Smirnov Yu.F.: *Nucleon Association in Light Nuclei*. Nauka, Moscow, 1969.
5. Kadmensky S.G., Furman V.I.: *Alpha-decay and Congenial Nuclear Reactions*. Energoatomizdat, Moscow, 1985.
6. Kadmensky S.G., Kadmensky V.G.: *Materials of* XX LINP *Winter School*, p 104–132, Leningrad (1985).
7. Kadmensky S.G., Tchuvilsky Yu.M.: Sov. J. Nucl. Phys. **38**, 6, 1483 (1983).
8. Milazzo-Colli L. et al.: Nucl. Phys. A **218**, 274 (1971).
9. Vahtel V.M. et al.: Particles and Nuclei **18**, 4, 778 (1987).
10. Volkov Yu.M. et al.: Preprint LINP, 536 (1979).
11. Kharitonov Yu. I., Smirnov Ju.I., Sliv L.A.: *Materials of XXX conference of Nuclear Spectroscopy*, Nauka, Leningrad (1980).
12. Goryunov O.U. et al.: Sov. J. Nucl. Phys. **49**, 2, 421 (1989).
13. Kadmensky S.G., Ratis Yu. L.: Izv. AN USSR, Ser. Phys. **47**, 11, 2254 (1983); Sov. J. Nucl. Phys. **38**, 5, 1325 (1983).
14. Kurgalin S.D., Tchuvilsky Yu.M.: Sov. J. Nucl. Phys. **49**, 1, 126 (1989).
15. Kadmensky S.G.,Furlan V.I., Tchuvilsky Yu.M.: Preprint JINR, P4-85-386 (1985).
16. Zamyatnin Yu.S. et al.: Particles and Nuclei **21** 2 537 (1990).
17. Kadmensky S.G. et al.: Sov. J. Nucl. Phys. **51**, 1, 50 (1990).
18. Goldberger M.: Phys. Rev. **74**, 1269 (1948).
19. Bunakov V.E., Matveyev F.V.: Z. Phys. A **31**, 332 (1985).
20. Barashenkov V.S. et al.: International School, JINR, Dubna, E2-5813 (1971).
21. Macher H.: Phys. Rev. C **21**, 2695 (1980).

Competition Between Light Cluster and Constituent Multinucleon Emission in Heavy-Ion Nuclear Reactions

A.C. Xenoulis

Institute of Nuclear Physics, National Center for Scientific Research "Demokritos", GR-153 10 Aghia Paraskevi Attiki, Greece

Isoproduct competition, i.e. competition between cluster and multinucleon emission leading to the same residual nucleus, emerges as a general, interesting and useful characteristic of nuclear reactions. Common properties and factors underlying the competition between pn and d as well as between p2n, dn and t evaporation are recognized and discussed. The application of the isoproduct-competition method in the delineation of the mechanisms involved in ^{7}Li-induced reactions suggests that an additional mechanism, breakup-fusion, is involved even at very low energies. Finally, the competition associated with alpha emission in the $^{12}C + ^{16}O$ reaction demonstrates a strong contribution from composite ^{4}He emission which cannot be accounted for by either the systematics or standard statistical calculations.

1. Introduction

Perhaps the most interesting characteristic of heavy-ion induced nuclear reactions is the large number of exit channels which open up as a result of the large values of energy and angular momentum introduced by the heavy projectile into the entrance channel.

Although, however, general is the consensus that the *number* of emitted particles is large, the issue as to what *kind* of particles, and under what conditions, can be emitted with significant intensity remains outstanding. In that respect it should be noted that an aspect of nuclear reactions, the implications of which have not been fully realized, is that all reaction modes, other that single n and p emission, involve at least potentially more than one exit channel competing for the production of the same residual nucleus. For instance, a typical heavy-ion nuclear reaction can be symbolized as follows

$$
\begin{aligned}
^{12}C + ^{16}O \quad &\rightarrow \quad ^{27}Si + n \\
&\rightarrow \quad ^{27}Al + p \\
&\rightarrow \quad ^{26}Al + (pn/d) \\
&\rightarrow \quad ^{25}Al + (p2n/dn/t) \\
&\rightarrow \quad ^{24}Mg + (2p2n/dpn/dd/.../\alpha) \\
&\rightarrow \quad \text{etc.}
\end{aligned}
$$

The notation used means that all exit channels included in a parenthesis can potentially contribute to the production of the commensurate residual nucleus. Thus, in any nuclear reaction almost all residual nuclei can be produced by cluster and / or constituent - multiparticle emission.

Let us, for the sake of brevity, introduce two abreviations related to the above discussion. For self evident reasons, we may call the exit channels leading to the same residual nucleus *isoproduct exit channels*. Similarly, the relative probabilities with which

Springer Series in Nuclear and Particle Physics   Clustering Phenomena in Atoms and Nuclei
Editors: M. Brenner · T. Lönnroth · F.B. Malik   © Springer-Verlag Berlin, Heidelberg 1992

these exit channels participate in the production of the same residual nucleus may be called *isoproduct competition*.

According to the traditional wisdom isoproduct competition does not exist or at best it is insignificant. In evaporation reactions, this is because conventional assumption holds that a compound nucleus can emit only p, n and alpha particles. On the other hand, in direct reactions it is usually assumed that only composite particle emission is relevant.

The above tidy picture, however, has started to change. At low energies, for instance, it has been shown that compound nuclei can substantially emit deuterons [1], tritons [2] or even complex fragments [3], while on the other hand non-statistical multinucleon emission has been identified in direct, ^{4}He-transfer, reactions [2].

Apparently isoproduct competition constitutes a rather general property of nuclear reactions. The pertinent problem is to recognize the dynamical conditions which in each case determine the relative probability for either the cluster or the constituent-multiparticle emissions leading to the same residual nucleus. It is hoped that the delineation of the factors underlying isoproduct competition associated with various light-cluster emissions will offer not only a better understanding of nuclear reactions but also a better perspective to consider the emission of different clusters within a unified framework.

Below, the already existing data on the competition between pn and d as well as p2n, dn and t emission will be critically reviewed with the purpose to summarize the essential understanding obtained therein. Furthermore, involvement of isoproduct competition in direct alpha-transfer reactions will be evaluated and finally some preliminary results associated with isoproduct competition in the alpha-particle exit channel will be discussed.

2. The experimental method

In cases of isoproduct competition, although the entrance channel and the residual nucleus are the same, the competing exit channels are distinctly different reaction modes, demanding appropriate treatment for either their experimental identification or their theoretical description. The presence and effect, however, of competing isoproduct channels cannot be distinguished by the usual methods of cross-section measurements, such as detection of single γ rays or recoiling ions. These, instead, provide the sum of the cross sections of all the channels participating in the production of a specific residual nucleus. On the other hand, in direct reactions only partial cross sections, usually associated with emitted composite particles, are measured.

An experimental method suitable to quantify isoproduct competition, based on light charged particle-gamma coincidence techniques has been proposed [4] and extensively tested [1,5]. The discrete coincident γ rays are used on the one hand to identify the heavy residual nucleus of interest, and on the other hand to provide the intensities with the help of which the *relative cross sections* of the competing isoproduct exit channels are extracted. For particle detection and identification a DE-E counter telescope of silicon detectors of various thicknessess is used. That technique restricts the observation of the competition to the production of specific *individual* excited states, taking nevertheless into account both side- and cascade-feeding to these states. Clearly, however, information concerning the production of the ground state cannot be obtained.

3. Isoproduct competition in compound-nucleus reactions

It is expected that the mechanism of a nuclear reaction will strongly influence the manner as well as the relative cross sections with which clusters and constituent nucleons are emitted.

The heavy ion reactions at the energies employed in the relevant competition studies [1,2,5,6] are generally considered to proceed via evaporation from a compound nucleus. A few of those, however, such as ^{6}Li- and ^{7}Li- induced reactions, are as well expected to proceed via a direct interaction. The mechanism of the employed reactions was conclusively ascertained via a comparison with statistical Hauser-Feshbach calculations [2,6] which implicitly assume that a compound nucleus has been formed in the reaction as an intermediate unstable species. In such cases, the energy and angular momentum introduced by the projectile are shared among many nucleons of the intermediate compound nucleus. The latter subsequently emits those nucleons or clusters which as a result of random fluctuations concentrate enough energy to "evaporate" through the surface barrier.

In what follows the competition between pn and d as well as between p2n, dn and t evaporation in compound nuclear reactions will be discussed.

3.1 Competition between pn and d evaporation

With the development of the particle-gamma coincidence technique [4], the pn over d emission competion was very soon mapped in about 20 nuclear reactions, each at several bombarding energies, for target-projectile combinations resulting in compound nuclei $12 \leq A \leq 71$ [1,5,6].

The relevant data demonstrated that for a given nuclear reaction the relative probability for the pn contribution increases with increasing bombarding energy. Otherwise, however, an interconnection among the magnitude of the competition associated with different reactions was not discernible. On the contrary, at similar bombarding energies the ratio σ_{pn}/σ_d assumes widely different values for different target-projectile combinations.

However, with an association of the σ_{pn}/σ_d ratio with the maximum excitation energy, $E_{CM}+Q_{pn}$, available to the commensurate residual nucleus, shown in Figure 1, a systematic trend emerges, illustrating a nearly linear dependence for almost all of the reactions investigated. Specifically, 18 reactions were found to comply with the systematics, while the competition in the reactions ^{6}Li (^{6}Li, pn/d) and ^{28}Si (^{12}C, pn/d) was found to deviate significantly. In all 20 reactions, nevertheless, the relative probability for multinucleon emission consistently increased with increasing bombarding energy.

It has been found instructive to visualize the competition in terms of the phase space of residual states participating in the multinucleon and cluster evaporations. The number of states available to pn and d emission were calculated for several reactions as a function of bombarding energy using Newton's formula [7]. Some typical results are plotted in Figure 2 as a function of maximum residual excitation, where it can be seen that the ratio of cross sections and the relative phase space for pn and d emission have roughly similar slopes. This comparison suggests that the increase in pn relative to d emission observed with increasing bombarding energy in all the investigated reactions may be understood in terms of a parallel relative increase in the number of states available to the pn successive evaporation.

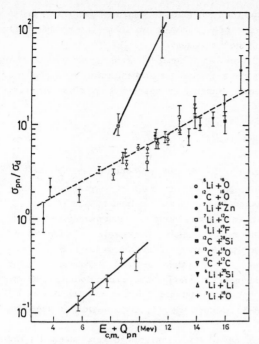

Figure 1. Experimental σ_{pn}/σ_d ratio values for various nuclear reactions as a function of maximum residual excitation

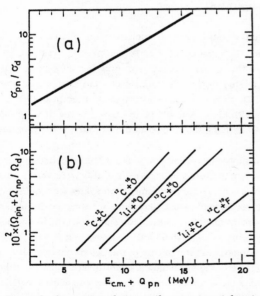

Figure 2. Comparison between the experimental systematics (a) shown in Figure 1 and calculated ratios of the number of levels available to pn + np emission over those available to d emission (b)

Apparently, the phase space available to multinucleon emission increases with energy faster than that available to the cluster emission simply because more nuclei, intermediate and residual, are involved in the former than in the latter decay.

Extensive Hauser-Feshbach calculations reproduced [6] remarkably well all the experimental data indicating that in all cases the cluster and multinucleon emissions proceed via evaporation from a compound nucleus, as opposed to a direct reaction mechanism.

3.2 Competition between p2n, dn and t evaporation

Measurements of the competition between p2n, dn and t evaporation have become recently available [2], for 6 heavy-ion-induced reactions, each at several bombarding energies, for target-projectile combinations resulting in compound nuclei 29<A<37.

The properties of the above triple competition were found to be commensurate to those of the pn versus d evaporation. In particular, an association of the relative yields for p2n, dn and t evaporation with the maximum excitation energy available to the A-3 ressidual nucleus, shown in Figure 3, demonstrates again a systematic trend, irrespectively of the interacting system. According to these systematics triton cluster emission dominates at low excitation while the multinucleon p2n evaporation dominates at higher energies. At about 7 MeV of maximum residual excitation the A-3 residual nucleus is predominantly produced via p2n evaporation.

Again standard Hauser-Feschback calculations reproduced very nicely [2] the experimental competition between p2n, dn and t evaporation.

3.3 Consideration of the evaporation competition

The competition between cluster and constituent-multinucleon evaporation does not seem to depend on the interacting system. Thus, systematic trends, shown in Figures 1 and 3, were recognized, according to which the relative cross sections for cluster versus multinucleon evaporations are simply related to the maximum excitation energy available to the commensurate residual nucleus. These regularities encourage an attempt to identify common properties of different clusters, although, as it will be discussed below, it is not as yet clear whether the same factors are underlying the competition associated with emission of more massive clusters.

Firstly, with respect to the nuclear physics implications, it should be emphasized that, although the experimental systematics demonstrated in Figures 1 and 3 were very nicely reproduced by the relevant calculations, the factors causing that systematic behavior unfortunately are not as yet explicity recognized. Most likely, an unexpected interplay between density of levels and angular momentum values, which can cancel out the individual characteristics of different interacting systems, is suggested by these systematics.

It is of course premature to try to understand the nature of the two deviations, shown in Figure 1, before understanding the nature of the systematics. Nevertheless, the facts that in the $^{12}C + ^{28}Si$ reaction the compound nucleus involved is a magic nucleus while the $^{6}Li + ^{6}Li$ is the lightest possible heavy-ion reaction may not be irrelevant.

Some comments with respect to common properties of the clusters is in order. Figures 1 and 3 clearly demonstrate that in both the evaporation of composite clusters is favored by

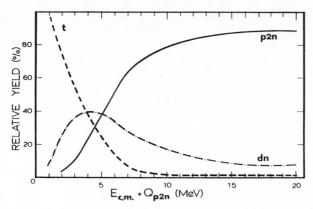

Figure 3. Systematics of the competition between p2n, dn and t evaporation

low energy while with increasing energy the evaporation of constituent nucleons becomes dominant. A more carefull comparison of Figures 1 and 3 suggests, however, that the dissociation of d to its constituents takes place faster than the dissociation of t. This most probably results from the larger binding energy of the latter cluster, the overcome of which at relatively low energies supresses more drastically the phase space available to the multinucleon p2n than in the case of pn evaporation. At higher energies binding energy differences clearly become irrelevant. It seems that about 2 MeV of energy available per emitted nucleon is required before the multinucleon evaporation becomes dominant.

4. Isoproduct competition in direct heavy-ion reactions

In several instances, such as in ^{6}Li- and ^{7}Li-induced reactions as well as on quasimolecular resonances, the competition between cluster and constituent multiparticle emission has been found to deviate significantly from that expected from pure evaporation. The significance of these deviations as well as the information they may convey either about the mechanism of the reactions or about the structural peculiarities of the involved nuclei are not yet conclusively evaluated. Nevertheless, here we shall discuss some interesting, although preliminary, conclusions concerning p2n, dn and t emission associated with direct alpha-transfer reactions induced by ^{7}Li beams.

The triton cluster versus constituent-multinucleon competition has been studied in 5 ^{7}Li-induced reactions on ^{12}C, ^{16}O, ^{19}F, ^{30}Si and ^{51}V. In all these reactions the competition between p2n, dn and t emission was found to deviate from that statistically expected.

The experimental behavior of the reaction ^{30}Si(^{7}Li, p2n/dn/t), which is demonstrated in Table 1 in juxtaposition with the statistical expectations, will help to discuss its essential features as influenced by the direct alpha-transfer mechanism expected to be present. The experimental data demonstrate that the ^{34}S residual nucleus is mainly produced via t emission, although it should be noted that the competing p2n and dn contributions are not at all negligible. The comparison of the experimental relative cross sections with those expected from pure evaporation clearly suggests that the experimental competition cannot be accounted for by pure evaporation. The experimental triton

Table 1. Relative cross sections for the production by p2n, dn and t emission of the ^{34}S residual nucleus in the $^7Li + {}^{30}Si$ reaction at the specified bombarding energies

E_{CM} (MeV)	Relative cross sections (%)					
	Experimental			Theoretical		
	σ_{p2n}	σ_{dn}	σ_t	σ_{p2n}	σ_{dn}	σ_t
13.0	10 ± 3	13 ± 3	77 ± 9	85.5	12.4	2.1
14.6	26 ± 4	20 ± 4	54 ± 5	86.6	12.0	1.4

contribution in particular exceeds by more than an order of magnitude that expected from evaporation, clearly indicating dominant presence of direct components in the $^{30}Si(^7Li,t)$ reaction at both bombarding energies.

The presence of multiparticle p2n and dn exit channels in the experimental measurements suggests on first inspection coexistence of direct and evaporation mechanisms in the $^{30}Si(^7Li, p2n/dn/t)$ reaction. If, however, the p2n and dn emissions in this reaction resulted from pure evaporation, their relative cross sections should have been correctly predicted by the relevant statistical calculations. The experimental multiparticle contributions, however, are in striking disagreement with the statistically expected. The σ_{p2n}/σ_{dn} ratio for instance assumes experimental values 0.8 ± 0.3 and 1.3 ± 0.3 at 13.0 and 14.6 MeV, respectively, compared to 6.9 and 7.2 expected from pure evaporation. Clearly, therefore, in this reaction not only the cluster but also the multiparticle emissions proceed predominantly via a non-statistical mechanism.

Apparently the nature as well as the energy dependence of the physical mechanisms coexisting in ^7Li-induced reactions are not at all clear, suggesting that these issues remain outstanding in spite of the opposite views more often than not expressed or implied in the relevant literature.

With respect to the energy dependence, it should be noted that the relative probabilities for multiparticle emission are significantly larger at 14.6 than at 13.0 MeV (Table 1). That information by itself, however, does not necessarily ensure an increasing contribution from non-statistical multiparticle emission with increasing bombarding energy since it can be equally well undersstood as due to an enchancement of the evaporation component against the contribution from the coexisting ^4He direct transfer mechanism. In fact such an anomalous energy dependence of the competition between direct and compound-nucleus mechanisms has been previously suggested for the reaction $^{12}C(^7Li,t)$ where a larger compound-nucleus component seems to exist at 38 than at 25 MeV bombarding energy [8].

In order to resolve the above question detailed excitation functions of the competition between p2n, dn and t emission were obtained. Such data associated with the $^{19}F(^9Li, p2n/dn/t)$ reaction are demonstrated in Figure 4 together with the theoretical contributions expected from pure evaporation. Rather unexpectedly the deviations of the experimental data from the statistically expected are the largest at the lower bombarding

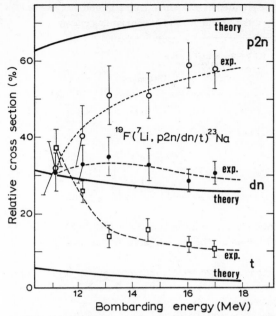

Figure 4. Experimental and statistically expected p2n, dn and t contributions

energies. As the bombarding energy increases experimental and theoretical contributions tend to be aligned.

If one as usually assumes that only compound-nucleus and direct ^{4}He- transfer mechanisms contribute, the conclusion suggested by Figure 4 is inevitable. The increasing bombarding energy favors the compound-nucleus component against the direct-reaction contribution. The rest of the investigated ^{7}Li-induced reactions demonstrate a similar behavior with that shown in Figure 4. Thus, either in ^{7}Li-induced reactions the competition between compound-nucleus and direct reaction mechanism has an anomalous energy dependence or an additional mechanism is present.

Preliminary measurements of absolute cross sections for the production of ^{23}Na in the above reaction suggest that a third mechanism must be present. This in all likelihood is associated with breakup of the projectile followed by ^{4}He transfer to the target nucleus.

Such breakup modes are not, at least conventionally, anticipated at the relatively low energies employed, although it should be noted that the experimental method used is especially suited to single out reaction events even of very low probability as long as these lead to the production of the A-3 residual nucleus.

5. Isoproduct competition associated with ^{4}He emission

Very little if anything is known about the next in complexity competition between 2p2n, dpn, dd, tp and ^{4}He emission leading to the same A-4 residual nucleus.

547

Table 2. Relative cross sections for the production by 2p2n, dd, dpn, tp and ^{4}He emission of the ^{24}Mg residual nucleus in the ^{12}C + ^{16}O reaction at 75 MeV bombarding energy

| Exit channel | Relative cross sections (%) | |
	Experimental	Theoretical
^{4}He	55 ± 10	6.6
tp	14 ± 3	37.0
dd+dpn	25 ± 5	44.5
2p2n	6 ± 2	11.4

Recently, however, that competition has been measured in the ^{12}C + ^{16}O reaction at bombarding energies well above the threshold for the disintegration of the alpha particle to its contituent nucleons.

Relevant experimental results obtained at 75 MeV bombarding energy are shown in Table 2 together with those expected from pure evaporation. Obviously theory and experiment are in a rather violent disagreement.

The above measurement corresponds to 10.6 MeV of energy available for excitation to the commensurate ^{24}Mg residual nucleus after 2p2n emission. According to the discussion presented in section 3.3 one qualitatively expects that at that excitation the composite ^{4}He evaporation would have been almost completely substituted by evaporation of the constituent multiparticles. While, however, that qualitative prediction is in rather good agreement with the results of the relevant statistical calculations (Table 2), this is not the case with the experimental data since the latter demosntrate an unexpectedly large probability for composite ^{4}He emission. Therefore, either the ^{12}C(^{16}O, α)^{24}Mg reaction does not proceed via evaporation from a compound nucleus, or otherwise the factors underlying the isoproduct competition associated with ^{4}He evaporation are not any more the same with those identified to influence the isoproduct competition in the case of d and t evaporation.

If the disagreement between theory and experiment observed in Table 2 were due to the direct character of the reaction, one would normally expect that disagreement to be caused primarily by the composite ^{4}He emission. This seems to be presently the case. Specifically, in Table 3 which demonstrates a comparison between theory and experiment of the same reaction, where however the composite ^{4}He contribution has been ignored in both theoretical and experimental data, a rather reasonable agreement between theory and experiment is observed, suggesting that the disaray previously observed in Table 2 is in fact almost exclusively caused by the unexpectedly strong composite ^{4}He component seen in the experiment.

Almost unanimous is the consensus in the literature that at the bombarding energies employed here the ^{12}C(^{16}O, α)^{24}Mg reaction proceeds via evaporation from a compound nucleus, with the singular exception of one investigation [9] which has suggested otherwise. The present preliminary results lend support to the latter view. Certain reservations stem from the opinion that the standard statistical calculations, which were proven extremely successful in reproducing the isoproduct competition associated with d ant t evaporation,

Table 3. Relative cross sections for the production by 2p2n, dd, dpn and tp emission of the ^{24}Mg residual nucleus in the $^{12}C + ^{16}O$ reaction aat 75 MeV bombarding energy

Exit channel	Relative cross sections (%)	
	Experimental	Theoretical
tp	32 ± 6	39.6
dd+dpn	56 ± 10	47.6
2p2n	12 ± 4	12.8

are not any more sufficient to describe the competition associated with ^4He evaporation. We are thus attempting to test the effect of various parameters which are not normally available in standard statistical calculations, such as deformation or even superdeformation which has been suggested [10] to characterize certain neighboring nuclei in this mass region.

Presently, nevertheless, the unexpectedly large relative cross section for composite ^4He emission observed in the reaction $^{12}C + ^{16}O$ remains outstanding.

6. Conclusions

Isoproduct competition, i.e. competition between cluster and constituent -multinucleon emission leading to the same residual nucleus, emerges as a rather *general characteristic* certainly of compound-nucleus and very probably of direct heavy-ion reactions.

Furthermore, it seems to be a rather *interesting characteristic* demonstrating unsuspected, thought-provoking, properties.

Finally, it constitutes a rather *useful characteristic* which can be implemented in the study of additional nuclear properties, such as the reaction mechanism, projectile breakup, utilization of Doppler-shift measurements as well as conditions for a meaningful comparison between theoretical and experimental cross-section values and the optimum experimental conditions for the discovery of virtual clusters.

Acknowledgement

I am indebted to my colleagues over the years E. Adamides, D. Bucurescu, E.N. Gazis, P. Kakanis, R.L. Kozub, C.J. Lister, J.W. Olness, A.D. Panagiotou, C.T. Papadopoulos and R. Vlastou for a rewarding collaboration and to my most recent students A.E. Aravantinos and G.P. Eleftheriades for their results, many of them unpublished, that I have included here. Finally, many discussions with S. Kossionides concerning material in this paper are gratefully acknowledged.

References

1   Xenoulis A.C., Aravantinos A.E., Lister C.J., Olness J.W., Kozub R.L.,
     Phys. Lett. **B106,** 461 (1981)
2   Xenoulis A.C., Aravantinos A.E., Eleftheriades G.P., Papadopoulos C.T., Gazis E.N.,
     Vlastou R., Nucl. Phys. **A516,** 108 (1990)
3   Moreto L.G., Wozniak G.J. , Nucl. Phys. **A488,** 337c (1988)
4   Xenoulis A.C., Gazis E.N., Kakanis P., Bucurescu D., Panagiotou A.D.,
     Phys. Lett. **B90,** 224(1980)
5   Gazis E.N., Papadopoulos C.T., Vlastou R., Xenoulis A.C., Phys. Rev. **C34,** 872 (1986)
6   Aravantinos A.E., Xenoulis A.C., Phys. Rev. **C35,** 1746 (1987)
7   Newton A.C., Can. J. Phys. **34,** 804 (1956)
8   Dennis L.C., Roy A., Frawley A.D., Kemper K.W., Nucl. Phys. **A359,** 455 (1981)
9   Bonetti R., Fioretto E., De Rosa A., Inglina G., Sanduli M., Phys. Rev. **34,** 1366 (1986)
10   Kolota J.J., Kryger R.A., DeYoung P.A., Prosser F.W., Phys. Rev. Lett. **61,** 1178 (1988)

Heavy-Cluster Transfer Reactions

A.T. Rudchik

Institute for Nuclear Research, 252028 Kiev, Ukraine

Abstract: The problems of heavy cluster transfer reactions and cluster spectroscopy for light nuclei are discussed.

Heavy-ion cluster transfer reactions have been sparsely studied for several reasons. and there are some problems concerning the mechanisms of many-nucleon transfer reactions. Such transfers can namely take place in many ways. Of these the statistical or direct one-step processe contributes fairly little. The direct heavy cluster transfer processes occupies as a rule only a small high-energy region. Only in special cases they give appreciable contributions to the experimental data.

For example, in the Li-spectra of the $^{12}C(^{14}N, Li)$ reactions at $E(^{14}N) = 130 MeV$ [1,2] the direct processes contributes at $E(Li) \sim 48 \pm 10 MeV$, the one-step Li evaporations at located at a narrow energy, whereas the multi-step statistical decay of excited ^{14}N in which Li is formed as residual nuclei occupies a wide energy region, cf. figure 1. In this reaction the formation of Li as residual nuclei from the decay of compound nuclei is possible only at very low-energy Li.

From this example one can conclude that for light nuclei the reaction mechanisms are very mixed. For a succesful identification of the reaction mechanism it is necessary to perform complex studies of spectra together with other data, such as angular distribution measurements, etc. In this way the study of heavy cluster problems may be possible as well.

As a second example, the analysis of $^6Li-$spectra from the $^{12}C(d,^6 Li)$ reaction at $E_d = 50 MeV$ [3], see figure 2, showed that the multi-step statistical processes give some contributions to the ^6Li cross section only for the 4^+ states of ^8Be. Indeed, the angular distributions were satisfactorily described by an $\alpha-$transfer process for the 0^+ and 2^+ states of ^8Be, as is shown in figure 3. For the 4^+ states of ^8Be, on the other hand, the calculated cross sections are partly lower than the experimental data [3].

Another problem is connected with the choice of the optical potentials in DWBA calculations. In figure 3 it was showed that absolute values of calculated cross sections strongly depend on the parametrization of the optical potential (full and dashed curves). This is typical for all heavy cluster transfer reactions [13]. It is a big obstacle for the development of heavy-

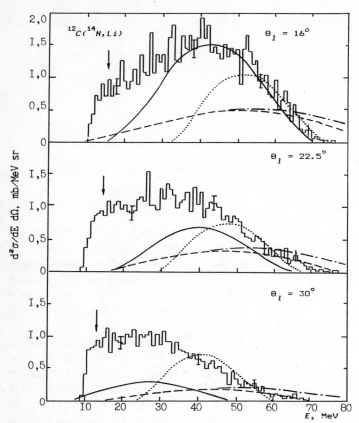

Fig. 1. Lithium spectra from the $^{12}C(^{14}N, Li)$ reactions at $E(N) = 130MeV$ [1,2] recorded in three angles. The dashed-dotted and dashed curves are the direct cluster transfer and two-cluster projectile break-up cross sections calculated with the Serber model [4,5] using the code SERBER [6]. The dotted curve is due to Hauser-Feshbach model calculations with the code LYBID [6] and the full curve is the cascade Monte Carlo calculations of light-particle evaporations from excited ^{14}N ($E^* \geq 20MeV$) with the code EVRESD [6]. The arrows show the region of Li residues from the cascade decays of the compound nucleus.

cluster spectroscopy. The situation is made easier when one of the reaction channels contains a light particle, because for such channels the optical potentials have been investigated over a wide energy range in more detail. At high energies the nuclear rainbow-scattering investigations can be used to select the appropriate potential.

The situation is especially difficult for cluster transfer reactions with radioactive products. In this case it is necessary to make big efforts to find a reasonable optical potential. One can hope that experiments with radioactive beams will help to solve this problem.

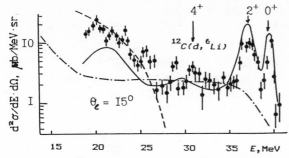

Fig. 2. The ^6Li spectrum from the $^{12}C(d,^6Li)$ reaction at $E = 50MeV$ [3]. The full curve represents FR DWBA calculations using doorway states [7,8], the dashed curve is the Hauser-Feshbach model calculation with the code LYBID [6], and the dashed-dotted curve is the Monte Carlo calculations of cascade light particle evaporations with the code EVRESD [6].

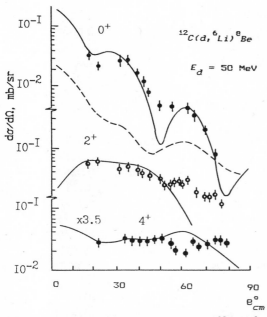

Fig. 3. Angular distributions of the $^{12}C(d,^6Li)$ cross section at $E = 50MeV$ [3]. The curves are FR DWBA calculations with the code LOLA [9] using the potentials A and α (full curve), and B and β (dashed curve). The optical potentials are given in Table 1.

Table 1. Optical potentials used in the present study.

Channel Name	V	W_s	W_D	r_v	a_v	r_w	a_w	r_c	Ref.
	MeV	MeV	MeV	fm	fm	fm	fm	fm	
$^{12}C + d$ A	83.1	0	15.1	1.15	0.80	1.22	0.75	1.20	[10]
B	73.2	0	11.0	1.25	0.70	1.26	0.65	1.30	[11]
$^8Be + {}^6Li$ α	148.0	0	7.2	1.47	0.65	1.47	0.65	1.47	[12]
β	190.0	7.0	7.0	1.28	0.70	2.50	0.90	1.25	[10]

Table 2. Spectroscopic data of the $^{12}C(d,^6Li)^8Be$ and $^{12}C(d,^8Be)^6Li$ reactions.

Meachanism	Cluster	$n_1l_1j_1 - n_2l_2j_2l$		A_s
Pick-up	α	$3S_0 - 2S_0$	0	0.8715
Heavy knock-on	^8Be	$3S_0 - 2S_0$	0	0.5032
Heavy pick-up	^6Li	$3S_1 - 2S_1$	0	0.3398
	$^6Li^\star_{2.185}$	$2D_1 - 1D_1$	0	-0.0593
			1	0.0890
			2	-0.0785
	$^6Li^\star_{4.31}$	$2D_1 - 1D_1$	0	-0.0593
			1	0.0636
			2	-0.0561
	$^6Li^\star_{5.7}$	$2D_1 - 1D_1$	0	-0.0254
			1	0.0382
			2	-0.0336
Knock-on	^6Li	$3S_1 - 2S_1$	0	0.5886
	$^6Li^\star_{2.185}$	$2D_1 - 1D_1$	0	-0.1028
			1	0.1542
			2	-0.1359
	$^6Li^\star_{4.31}$	$2D_1 - 1D_1$	0	-0.0734
			1	0.1101
			2	-0.0971
	$^6Li^\star_{5.7}$	$2D_1 - 1D_1$	0	-0.0441
			1	0.0661
			2	-0.0583

The third problem is connected with the calculations of heavy cluster spectroscopic factors. At present there are methods to carry out such calculations only for $1p-$shell nuclei [14]. These methods are realized in the code DESNA [15]. Thus there is now the full possibility to study heavy cluster transfer reactions with the FR DWBA model for $1p-$nuclei, but

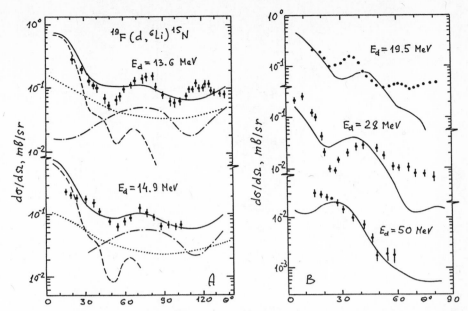

Fig. 4. Cross sections of the $^{19}F(d,^6Li)^{15}N$ reaction at $E = 13.6, 14.9, 19.5, 28$ and $50 MeV$. From refs. [19-23], respectively. Left panel: The dashed curves are FR DWBA cross sections for α transfer, dashed-dotted for ^6Li knock-on, the dotted curves are Hauser-Feshbach cross sections, and the full curves represent incoherent sums. Right panel: Full curves are FR DWBA cross sections for α transfer.

for $2s1d$−nuclei there is only partial progress in the calculations of heavy cluster amplitudes [13].

To understand the calculation problems in detail, let us look at the FR DWBA cross section for a transfer of a cluster x from the system $a = b + x$ to the system $B = A + x$ [13,16,17]. The differential cross section is given by the expression

$$\frac{d\sigma}{d\Omega} = \frac{1}{E_a E_b}\frac{k_b}{k_a}C_J \sum_{j_1 j_2 lm} |\sum_{l_1 l_2 s_x E_x} A_s(E_x) \sum_{L_b} \sqrt{\frac{2}{2L_b+1}} B_{lmL_b}^{j_1 j_2 S_x}(-1)^m \bar{P}_{L_b}^m|^2,$$
(1)

where

$$A_s(E_x) = S_{A+x}^{l_1 j_1 s_x}(E_x) \cdot S_{b+x}^{l_2 j_2 s_x}(E_x)\sqrt{2l+1}W(l_1 j_1 l_2 j_2; s_x l).$$
(2)

Here $S_{A+x}^{l_1 j_1 s_x}(E_x)$ and $S_{b+x}^{l_2 j_2 s_x}(E_x)$ are the spectroscopic factors of an x cluster in E_x excited states, and $B_{lmL_b}^{j_1 j_2 s_x}$ is the reaction amplitude which includes the amplitudes of the *direct* and the *exchange* processes.

From (1) and (2) one can conclude that heavy cluster spectroscopy is quantitatively very difficult, because the spectroscopic factorsget mixed in coherent sum, and consequently it is only possible to compare calculated

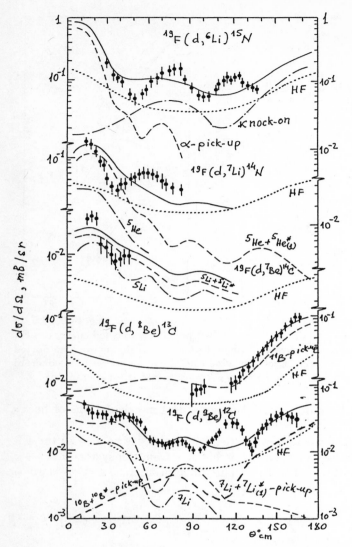

Fig. 5. Cross sections of $^{19}F(d, HI)$ reactions at $E_d = 13.6\,MeV$. The dashed-dotted curves are FR DWBA cross sections for transfer of clusters in their ground states, dashed curves for clusters in ground and excited states (coherent sum), dashed-crossed curves for ^6Li knock-on, the dotted curves are Hauser-Feshbach model calculations, and the full curves represent coherent sums.

and experimental cross sections, and to get some qualitative conclusions of cluster amplitudes. This is unfortunately the situation for many cases. However, A_s values can restrict the number of possibilities, and thus only a few cluster patterns can be realized. Also the $B-$amplitudes carry some restrictions. As a result, in many cases the heavy-cluster spectroscopy has resemblance to that of light clusters.

For example, the cluster spectroscopic data of the $^{12}C(d,^6 Li)^8 Be$ and $^{12}C(d,^8 Be)^6 Li$ reactions are presented in Table 2. One can see that the transfers of $\alpha, ^6 Li$ and $^8 Be$ clusters in ground states are preferred (A_s selection). But the $\alpha-$transfer gives its basic contribution to the calculated cross sections at $E_d = 13.6 MeV$ (B selection), see [13,18].

This example shows that in some reactions the studies of transferred heavy clusters is simlified due to the spectroscopic (A_s) and wave function integral (B) restrictions, but this is rather an exception than the rule. For instance, in the $^{19}F(d,^6 Li)$ reaction at low energies the knock-on of $^6 Li$ gives a significant contribution to the calculated cross section for large angles, cf. figure 4.

Next problem is related to the fact that the heavy clusters ($A > 4$) can be transferred in excited as well as in ground states. Figure 5 demonstrates this for the $^{19}F, (d, HI)$ reactions, where HI denotes the heavy ions $^7 Li, ^{7,8,9} Be$ and $^{10,11}B$. We can see that the transfer of $^5 He$ and $^{5,7} Li$ in the first excited states is comparable to the transfer of these clusters in the ground state. It is, however, possible to overcome these difficulties of the spectroscopic amplitudes are known.

From a careful analysis of many-nucleon transfer reactions one can conclude that many of them can be effectively used for the heavy-cluster studies.

References

1. O.A. Ponkratenko, O.F. Nemets, A.T. Rudchik: in Nuclear Spectroscopy and Nuclear Structure, Report of 38th Conference, Baku, 1988, p.364
2. O.A. Ponkratenko: Doctoral thesis, Institute for Nuclear Research, Kiev, 1991
3. O.Yu. Goryunov et al.: Ukr. Fiz. J. **34** 1144 (1989)
4. R. Serber: Phys. Rev. **72** 1008 (1947)
5. M.T. Magda, A. Pop, A. Sandulescu: JINR Preprint E7-85-64, Joint Institute of Nuclear Research, Dubna, 1985
6. O.A. Ponkratenko, A.T. Rudchik: INR Preprint KIYAI-88-44, Institute of Nuclear Research, Kiev, 1988
7. V.G. Kadmensky et al.: Ukr. Fiz. J. **33** 1309 (1988)
8. O.Yu. Goryunov et al.: Yad. Fiz. **49** 421 (1989)
9. R.M. deVries: Phys. Rev. C **8** 951 (1973)
10. K. Umeda, T. Yamaya, T. Suehiro et al.: Nucl. Phys. **A429** 88 (1984)
11. F. Hintenberger, G. Mairle, U. Schmidt-Rohr et al.: Nucl. Phys. **A111** 265 (1968)

12. J.E. Poling, E. Norbek, R.R. Carlson: Phys. Rev. C **13** 648 (1976)
13. O.F. Nemets, V.G. Neudachin, A.T. Rudchik, Yu.F. Smirnov, Yu.M. Tchuvil'sky: Nucleon associations in atomic nuclei and many nucleon transfer reactions, Naukova Dumka, Kiev, 1988
14. Yu.F. Smirnov, Yu.M. Tchuvil'sky: Phys. Rev. C **15** 84 (1977)
15. A.T. Rudchik, Yu.M. Tchuvil'sky: INR Preprint KIYAI-82-12, Institute of Nuclear Research, Kiev, 1982
16. N. Aystern, R.M. Drisko, E.C. Halbert, G.R. Satchler: Phys. Rev. B **133** 3 (1964)
17. N.S. Zelenskaya, I.B. Teplov: Bull. Akad. Sci. USSR, Phys. Ser. **41** 1709 (1977)
18. A.T. Rudchik, in Nuclear Reactions, Proceeding of the First Kiev International School on Nuclear Physics, 1990, p. 31-44
19. O.Yu. Goryunov et al.: Yad. Fiz. **22** 31 (1975)
20. L.G. Denes et al.:Phys. Rev. A **148** 1097 (1966)
21. H.H. Gotbrod et al.: Nucl. Phys. **A165** 240 (1971)
22. M. Bedjidian et al.: Nucl. Phys. **A189** 403 (1972)
23. o.Yu. Goryunov et al.: Yad. Fiz. **51** 929 (1990)
24. A.T. Rudchik, Doctoral Thesis, Institute of Nuclear Research, Kiev, 1983

Collisions Between Atomic Clusters

R. Schmidt and H.O. Lutz

Fakultät für Physik, Universität Bielefeld,
W-4800 Bielefeld, Fed. Rep. of Germany

Collisions between atomic clusters at incident energies from zero up
to several eV are theoretically investigated. The cross section of fusion
and other reaction channels are estimated by applying a reaction model
of nuclear physics in an appropriate manner.

It is found that the behaviour of the fusion cross section as function of the incident energy depends qualitatively on the charge of the
colliding clusters. The maximum fusion cross section may considerably
exceed the geometrical one provided the clusters are uncharged or unequal charged. The Coulomb barrier (or fusion barrier) of equal charged
clusters lies in the eV range but depends sensitively on the mass number of the clusters. Fusion represents the dominating reaction channel
as long as the incident energy is smaller than a critical value. At incident energies above this threshold binary, deep-inelastic-type collisions
are the dominating reaction channel in analogy to collisions between
atomic nuclei (heavy-ion collisions at low energies).

The model is applied also to collisions between macroscopic liquid
droplets. The calculated fusion cross section is compared to experimental data on droplet collisions, measured recently /1/, and found to be
in good agreement.

/1/ A. Brenn and A. Frohn, Exp. in Fluids 7 (1989) 441

Cluster Transfer at Very Low Energies

E. W. Schmid

Institut für Theoretische Physik, Universität Tübingen,
Auf der Morgenstelle 14, W-7400 Tübingen, Fed. Rep. of Germany

A three-cluster model for calculating nuclear transfer reactions at very low energies is discussed. Pauli effects enter by a special choice of effective two-cluster potentials. For solving the three-cluster equation a coupled channels method with open and (several) closed two-body channels is employed. Orthogonalization of channel spaces allows to study convergence with respect to the number of closed channels and it also allows to get physical insight into the details of the reaction process. The numerical treatment is discussed.

1. Introduction

Let me present a brief report on what our small group at Tübingen University (K. Bräuer, M. Walz, Z. Papp and myself) has been doing and what is planned for the near future. In the past our main concern was a systematic study of the effective interactions of fermion clusters. Orthogonalization of channel spaces allowed us to extend Feshbach's definition of effective two-body interactions to systems in which both partners are composite [1]. The desire to combine Pauli correctness of the resonating group model with the fitting power of the optical model led to the fish bone optical model [2]. It was seen that three-cluster forces arising from three-cluster Pauli exchange are considerably reduced by a renormalization of the wave function [3]. Cluster distortion (or Q-space, in Feshbach's language) leads to energy-dependence of the two-body effective interactions and, interrelated with this, to effective multibody forces. It became clear that energy-dependent two-body interactions cannot be embedded into multibody systems, and that this is not a question of mathematics but a result of missing Q-space information [4]. During our study of effective interactions we also got involved in the discussion on a short distance node in the nucleon-nucleon wave functionn. We found that such a node is either produced by nonorthogonality of multichannel test function spaces or by projecting out the interior region from a single channel space (and calling it Q-space), or by unitary off-shell transformation. Introducing a third (charged) body as a spectator and using the Coulomb potential as a yardstick for calibrating effective potentials led to the conclusion that the constituent quark model does not predict a short distance node [5].

Springer Series in Nuclear and Particle Physics **Clustering Phenomena in Atoms and Nuclei**
Editors: M. Brenner · T. Lönnroth · F.B. Malik   © Springer-Verlag Berlin, Heidelberg 1992

Finally we got tired of studying general properties of effective interactions and wanted to enter into practical applications. Looking around for good examples we got interested in nuclear reactions at very low energies. In the meantime we finished writing computer codes but we do not yet have results for realistic potentials. Nevertheless I'll give you a status report on our work.

Let's discuss things in terms of an example. Take the reaction $^7Li(p, \alpha)\alpha$ at several keV incident proton energy. What is so challenging about it? Well, at a few keV the cross section cannot be measured directly. There is no final parameter, like an overall potential strength, which can be adjusted. What is needed is a good treatment of the dynamics with an input which is insensitive to small changes of energy. There are tightly bound clusters and all open channels can be included explicitly in a calculation. At only a few keV the clusters meet and react in the nuclear surface where they are quasifree. But that's not all. Since the reaction has a high Q-value a deep inelastic process must also be present. This is seen immediately by looking at the inverse reaction: Two α-particles are coming in at an energy of $\sim$17 MeV. When they hit each other, this energy is converted into internal excitation energy of only one of the two α-clusters. The excited cluster carries a proton far out into the surface of the interaction region, from where the proton penetrates the barrier. Intuitively one should think that at least one closed channel with an excited α-cluster is needed to describe this process. And here comes another challenge: DWBA calculations [6] successfully describe the process as a one-step direct reaction. How can this be understood? Does nonorthogonality and overcompleteness of channel spaces make it unnecessary to introduce an intermediate channel?

In the following I'll define the three-cluster model which we employ for studying nuclear reactions at very low energies. I'll then discuss the method we use for solving the basic three-body equation.

2. Theory

The method we want to employ for calculating nuclear reaction cross sections at very low energies has the following ingredients: Definition of a three-cluster model, calibration of two-cluster effective interactions, construction of a test function space, setting up a two-body coupled channels equation, orthogonalization of channel spaces, numerical solution of the orthogonalized coupled channels equation and estimation of the achieved convergence.

2.1 Definition of the cluster model

The most severe restrictive assumption we make lies in the definition of the cluster model. We introduce three clusters and treat them as inert bodies. The clusters essentially interact via effective two-body potentials; a small three-body potential will be allowed for minor adjustments. All knowledge on the internal structure of the clusters enters by the effective potentials. An extension of the theory to more than three clusters is possible, but not considered at the moment. In the $^7Li(p,\alpha)\alpha$ example, the ^7Li ground state is approximated by an α-triton bound state, and one of the two α-particles in the final channel is a bound state of triton and proton.

2.2 Two-cluster effective interactions

As two-cluster effective potentials we are using potentials of the fish bone optical model [2]. With this choice we want to combine Pauli-correctness of the resonating group model with fitting power of the optical model. The potentials read

$$V_{eff} = V_{opt} - \sum_{\mu,\nu} |u_\mu\rangle\langle u_\mu|(T + V_{opt} - \epsilon)|u_\nu\rangle \overline{M}_{\mu\nu}\langle u_\nu| \tag{1}$$

with

$$\overline{M}_{\mu\nu} = \begin{cases} 1 & \text{if } \eta_\mu = 1 \quad \text{or } \eta_\nu = 1 \ , \\ \text{else:} \\ 1 - \sqrt{(1-\eta_\mu)/(1-\eta_\nu)} & \text{for } \mu \leq \nu \\ 1 - \sqrt{(1-\eta_\nu)/(1-\eta_\mu)} & \text{for } \mu \geq \nu \ . \end{cases} \tag{2}$$

The u_μ are eigenstates of the RGM norm kernel for eigenvalues η_μ, ϵ is a chosen negative or positive energy with a value outside of the range of physical interest. V_{opt} is a (local) optical potential with fitting parameters. The double sum rigorously represents the kinetic energy exchange part and the norm kernel part of the renormalized RGM interaction. It rigorously represents the potential energy exchange part when operating on Pauli forbidden states and approximately represents it when operating on partly Pauli forbidden states. When V_{eff} is used in a two-body Schrödinger equation the Pauli forbidden two-cluster states appear as redundant solutions at $E = \epsilon$.

For generating the RGM norm kernel eigenstates we are using the equal frequency harmonic oscillator limit of RGM, i.e. the u_μ are harmonic oscillator states. The fitting parameters of V_{opt} are fitted to the experimental two-cluster phase shifts with special emphasis on the energy region of interest.

With potentials given by (1,2) our basic three-cluster Schrödinger equation reads

$$H_{eff}\psi \equiv \Big[T + V_{eff}(1,2) + V_{eff}(2,3) + V_{eff}(3,1) + V_{opt}(1,2,3)\Big]\psi = E\psi \ . \tag{3}$$

We allow for a simple one-parameter three-body potential $V_{opt}(1,2,3)$ to adjust the reaction cross section to scattering experiments at energies E in the range 50 keV $< E <$ 200 keV. Our model is trustworthy only when $V_{opt}(1, 2, 3)$ turns out to be very small!

Eq. (3) has been the basis of our earlier studies when we used a Faddeev method to solve it [7]. We gave up this line of research because Faddeev methods become very tedious when two of the three bodies carry an electric charge. When all three bodies are charged, like in most cases of astrophysical interest, a Faddeev method does not even exist.

2.3 The test function space

We are using two-body function spaces for the open channels as well as for closed channels. The space of an open channel is spanned by a calculated bound state ϕ_i of a two cluster subsystem times a free relative motion function χ_i for the third cluster. The closed channels are spanned by chosen square integrable states ϕ_i of two cluster subsystems times free relative motion functions χ_i for the third cluster. Thus we have

$$\psi = \sum_{i=1}^{n} X(\phi_i \chi_i), \quad \delta_\psi = \sum_{i=1}^{n} X(\phi_i \delta \chi_i) \,. \tag{4}$$

The two-cluster bound states should not differ too much from the true eigenstates of the respective subsystem Hamiltonian. Otherwise convergence trouble will arise. The projection operator X projects out all Pauli forbidden subsystem states; it is a product of all operators of the form $(\mathbb{1} - |u\rangle\langle u|)$ with u being a Pauli forbidden two-cluster state.

2.4 The coupled channels equation

We project (3) onto the test function space (4) by

$$\langle \delta\psi|(H - E)|\psi\rangle = 0 \tag{5}$$

and get the coupled channels equation

$$\sum_{j=1}^{n} \langle \phi_i|X(H - E)X|\phi_j\rangle|\chi_j\rangle \equiv \sum_{j=1}^{n} (H_{ij} - E\,N_{ij})|\chi_j\rangle = 0 \,, \; i = 1, \ldots, n \,. \tag{6}$$

In (6) there is no antisymmetrizer because (different) clusters are distinguishable objects. Pauli effects coming from nucleon exchange between two clusters enter into (3) via the effective interaction potential (1). The calculation of H_{ij} and N_{ij} by (6) requires a folding procedure. This means that special care is needed for treating Pauli forbidden states. Folding mixes Pauli forbidden states with Pauli allowed states, and the ϵ-technique of (1) has to be modified. We are calculating H_{ij} and N_{ij} by the Talmi–Moshinsky–Tobocman method [8] and use ϵ only to mark the Pauli forbidden subsystem states appearing in the projection operator X.

Pauli effects coming from nucleon exchange between all three clusters are not taken into account explicitly because there is no *derived* three-body force; they justify, however, the inclusion of a small three-body potential $V_{opt}(1,2,3)$. For the Coulomb interaction we are using the double folding approximation without Pauli effect. The Talmi–Moshinsky–Tobocman method yields nonlocal folding potentials. We either extract a local part, or use the damping factor given by Tobocman [8].

2.5 Orthogonalization of channel spaces

As is well known the channel spaces (4) are nonorthogonal. If one is only interested in asymptotic quantities, like reaction cross sections, this feature is of little importance. Eq. (6) will give correct asymptotic results without orthogonalization of test function spaces. This is probably the reason why so few people are interested in orthogonalization.

Nonorthogonality of function spaces, however, has some unpleasant consequences. Intuition tells us that only the microscopic interaction can couple channels. But in the coupled channels equation (6) the coupling of channels comes from all sources: H_{ij} with $i \neq j$ has a contribution coming from the microscopic kinetic energy operator, from the microscopic potential, and even $E \cdot N_{ij}$ couples channels. Of course, there is a lot of cancellation, but our insight into what is really going on is blurred.

Intuition also tells us that equation (6) will accept only those elements of the test function space which are needed to describe the physical process. Nonorthogonality destroys our physical insight also here. As long as we do not orthogonalize the test function space, the solution of the coupled channels equation cannot tell us whether a certain closed channel is important or not. It cannot even tell us whether a reaction is a surface reaction or a reaction coming from the region of strong interpenetration. Whenever we want to understand what is going on in the interaction region, i.e. whenever we are asking for off-shell information, we have to orthogonalize function spaces.

An orthogonalization procedure has been presented earlier [1] and has already been successfully applied in practice [9]. The concept is simple because the procedure is an implementation of the Gram–Schmidt method. Just allow all states of the first channel, project out of the second channel all states which are already present in the first channel, project of the third channel whatever is already present in the first two channels, and so on. Also the numerical implementation is not difficult. It's a recursive sequence of matrix operations involving (N_{ij}) and (H_{ij}) of (6). However, it is a little lengthy to write things down on paper. Nevertheless I'll do it.

Orthogonalization leads from (6) to the coupled channels equation

$$\sum_{j=1}^{n}(H_{ij}^{(n)} - E\mathbb{1})|\chi_j^{(n)}\rangle = 0 , \quad (i = 1,\ldots,n) .\tag{7}$$

in which the overlap operator matrix (N_{ij}) does no longer appear; the upper index (n) indicates that all n channels have been orthogonalized.

We add an upper index (0) and a tilde to the operator matrices (H_{ij}) and (N_{ij}) appearing in (6) and proceed as follows [4]. We apply a recursion which runs from $k = 1$ to $k = n$. For every k we have the following steps:

$$\tilde{N}_{kk}^{(k-1)}|u_{k,\nu}\rangle = \eta_\nu|u_{k,\nu}\rangle ,\tag{8}$$

$$H_{kk}^{(k-1)} = \tilde{H}_{kk}^{(k-1)} + \sum_{\nu}|u_{k,\nu}\rangle\varepsilon\langle u_{k,\nu}| ,\tag{9a}$$

$$H_{ij}^{(k-1)} = \tilde{H}_{ij}^{(k-1)} \quad \text{for } i,j \neq k ,\tag{9b}$$

$$N_{kk}^{(k-1)} = \tilde{N}_{kk}^{(k-1)} + \sum_{\nu}|u_{k,\nu}\rangle \langle u_{k,\nu}| ,\tag{9c}$$

$$N_{ij}^{(k-1)} = \tilde{N}_{ij}^{(k-1)} \quad \text{for } i,j \neq k ,\tag{9d}$$

$$(\tilde{H}_{ij}^{(k)}) = (B^{(k-1)\dagger})(C^{(k-1)\dagger})(H_{ij}^{(k-1)})(C^{(k-1)})(B^{(k-1)}) ,\tag{10a}$$

$$(\tilde{N}_{ij}^{(k)}) = (B^{(k-1)\dagger})(C^{(k-1)\dagger})(N_{ij}^{(k-1)})(C^{(k-1)})(B^{(k-1)}) ,\tag{10b}$$

$$(B^{(k-1)}) =
\begin{pmatrix}
1 & & & & & & & 0 \\
& \ddots & & & & & & \\
& & 1 & & & & & \\
& & & (N_{kk}^{(k-1)})^{1/2} & -N_{kk+1}^{(k-1)} & \cdots & -N_{kn}^{(k-1)} & \\
& & & 1 & & & & \\
& & & & \ddots & & & \\
0 & & & & & & 1 &
\end{pmatrix} ,\tag{11a}$$

$$
(C^{(k-1)}) \; = \; \begin{pmatrix} 1 & & & & & & 0 \\ & \ddots & & & & & \\ & & 1 & & & & \\ & & & (N_{kk}^{(k-1)})^{-1} & & & \\ & & & & 1 & & \\ & & & & & \ddots & \\ 0 & & & & & & 1 \end{pmatrix} . \qquad (11b)
$$

One has to solve an eigenvalue equation (8) and detect zero eigenvalues. In (9a) and (9b) the sum runs over zero states, only. In the final equation (7) all zero states will appear as redundant solutions of the equation at $E = \epsilon$ with ϵ being a chosen unphysical energy. Eqs. (10a,b) are the actual recursion which leads from $(k-1)$ to k.

For our practical application we are representing all operators appearing in (8–11) in the (equal frequency) harmonic oscillator space which has already been introduced in (1) to describe Pauli effects. In each channel, only a finite number of states is affected by orthogonalization. Therefore the recursion (10a,b) is carried out with finite matrices. After completing orthogonalization one adds on the asymptotic part of the matrices and returns to the representation which one has had before, e.g. differential operators for kinetic energy, etc..

2.6 Numerical treatment

The computer software package needed for setting up eq. (6) and for carrying out orthogonalization has been presented a year ago at the Uzhgorod conference by Z. Papp and myself.

For solving the coupled channels equation (7) we are now having at our disposal two independent computer codes. The first one has been written by Z. Papp. It solves the coupled channels equation in form of a Lippmann–Schwinger equation in Sturmian function representation. The second code I have written myself with nostalgic pleasure (people who know that in 1960 I joined the group of Karl Wildermuth at Florida State University, knowing nothing about clusters but knowing how to program computers in machine language, will understand this remark). The code solves (7) in integro-differential form, with nonlocal potentials in harmonic oscillator representation. The Fox–Goodwin recursion is used to generate partial solutions from which the full solution is then constructed by superposition. This method is especially suited when a vector processor is used to generate the partial solutions simultaneously.

Since both codes became operative only about a month ago I can not yet present realistic results. K. Bräuer, while debugging my code, did a simplified $^7Li(p, \alpha)\alpha$ calculation. The code is running but it would be premature to quote results.

2.7 The question of convergence

Our starting equation (3) contains crucial model assumptions (three inert clusters interacting by energy-independent potentials). We are accepting this approximation and will test it at energies accessible to scattering experiments. In solving (3), however, we want to be as exact as possible.

Since we cannot solve (3) directly we had to go over from the three-body equation (3) to the two-body orthogonalized coupled channels equation (7). How much accuracy are we sacrificing by this second step? Here we are right at the center of few body physics: solving a three-body equation!

We are below the three-body breakup threshold and only very few two-body channels are open. Therefore the coupled channels ansatz (4) is adequate. Our codes can handle, in addition to open channels, several closed channels. Will we be able to achieve convergence?

Our way of orthogonalizing coupled channel spaces has one nice feature. With ψ from (4) and solutions $\chi_i^{(n)}$ from (7) we have

$$\langle \psi | \psi \rangle = \sum_{i=1}^{n} \langle \chi_i^{(n)} | \chi_i^{(n)} \rangle \quad \left(\neq \sum_{i=1}^{n} \langle \chi_i | \chi_i \rangle ! \right) . \tag{12}$$

Convergence of the sum over $\langle \chi_i^{(n)} | \chi_i^{(n)} \rangle$ has been studied in detail in case of one open channel and many closed ones [10]. The sum converges under certain mathematical conditions, which are always fulfilled in physics. I have no doubt that the proof can be extended to cases where more than one channel is open.

People have always been searching for variational principles which can be applied to scattering calculations. Eq. (12) looks rather tempting, in this respect. For a given normalization of the incident flux the norm sum of the closed channel wave functions will approach a certain value with $n \to \infty$. Will it approach this converged value from below? Unfortunately this is not necessarily true because

$$\langle \chi_i^{(n)} | \chi_i^{(n)} \rangle \neq \langle \chi_i^{(m)} | \chi_i^{(m)} \rangle , \quad i \leq n < m . \tag{13}$$

It is not meaningful, in the general case, to keep n fixed and to vary some non-linear parameter in (4) in order to reach a maximum of the norm sum of the closed channels. What is meaningful, however, is to increase the number of channels and to check whether the norm sum of the closed channels is already saturated and remains unchanged. In the latter case one has at least an indication that the achieved accuracy is good.

3. Summary

We have discussed a three-cluster model for calculating nuclear reaction processes at very low energies. Pauli effects enter into the three-cluster Schrödinger equation by a special choice of the effective interaction potentials. For solving the three-body equation a coupled channels method with open and (several) closed two-body channels is employed. Orthogonalization of channel spaces allows to study convergence with respect to the number of closed channels and it also allows to get physical insight into details of the reaction process. The numerical treatment uses the fish bone optical model for fitting two-cluster potentials, the Talmi–Moshinsky–Tobocman method with Pauli projection for setting up the coupled channels equation, a recursion of matrix operations for orthogonalization and two independent methods for solving the coupled channels equation.

References

[1] E.W. Schmid and G. Spitz, Z. Phys. **A321**, 581 (1985)
[2] E.W. Schmid, Z. Phys. **A297**, 105 (1980);
    E.W. Schmid, Z. Phys. **A302**, 311 (1981)
[3] E.W. Schmid, M. Orlowski, and Bao Cheng-guang, Z. Phys. **A308**, 237 (1982)
[4] S. Nakaichi-Maeda and E.W. Schmid, Phys. Rev. **C35**, 799 (1987)
[5] E.W. Schmid, A. Faessler, H. Ito, and G. Spitz, Few-Body Systems **5**, 45 (1988)
[6] G. Raimann, B. Bach, K. Grün, H. Herndl, H. Oberhummer, S. Engstler, C. Rolfs, H. Abele, R. Neu, and G. Staudt, Phys. Lett. **B249**, 191 (1990)
[7] K. Hahn, P. Doleschall, and E.W. Schmid, Phys. Rev. **C31**, 325 (1985)
[8] W. Tobocman, Nucl. Phys. **A357**, 293 (1981)
[9] G. Spitz and E.W. Schmid, Few-Body Systems **1**, 37 (1986)
[10] E.W. Schmid and H. Fiedeldey, Phys. Rev. **C39**, 2170 (1989)

Part VII

**Post-Conference
Thoughts**

Nuclear Versus Atomic Clusters[†]

B.R. Mottelson

Nordita, Blegdamsvej 17, DK-2100 Copenhagen, Denmark

Abstract: The purpose of this lecture will not be an attempt to summarize the contributions of this conference. This will be practically impossible because of the diverse field of participants. However, I shall say a few words about my own favourites in the field of nuclear clusters. Secondly, I shall have a look at atomic clusters from the perspectives of a nuclear physicist in order to describe some ideas suggested from nuclear physics. In this context it will be necessary to translate the languages, since there seems to be confusion in the use of the word "cluster". When nuclear physicists talk about clusters they refer to *subunits* within the nucleus while the cluster community speak of *collections of atoms* which are single entities themselves.

1. Nuclear Clusters

1.1 Cluster radioactivity

My first subject is the one of cluster radioactivity which represents a new form of large amplitude collective motion in the nucleus. Cluster radioactivity was theoretically suggested by Greiner and Sandulescu and experimentally observed by Rose and Jones in the spontaneous decay of ^{223}Ra into ^{209}Pb and a ^{14}C fragment. The branching ratio for this decay to the α-decay is only $\sim 8 \cdot 10^{-10}$. Since then there has been a development of the experimental techniques. Extensive studies by Price and collaborators have today resulted in 17 known examples of decays where the fragments range from ^{14}C to ^{34}Si. In some cases the branching ratios are as small as $\sim 10^{-16}$ and the limits are still being extended. These remarkable findings lead to an era of spectroscopy since it is found that the decay predominantly goes to excited states of the daughter nucleus. As discussed in the talk to this conference by Price, the observations point to the possibility of an analysis of the cluster formation process in terms of the available Nilsson orbits and the configurations of the initial and the final states.

Cluster radioactivity raises some very interesting questions that seemingly have been quite lightly passed over at this conference. Generally the

---

[†] This text was prepared by P. Manngård from B.M.'s lecture notes and a videotape of the lecture.

Springer Series in Nuclear and Particle Physics **Clustering Phenomena in Atoms and Nuclei**
Editors: M. Brenner · T. Lönnroth · F.B. Malik   © Springer-Verlag Berlin, Heidelberg 1992

small branching ratios have been recognized to be due to a penetration factor through a large potential barrier resulting in a small probability for the cluster to escape. However, there are some "deeper thruths" hidden in this explanation. If we calculate the energy of the orbits involved in the formation as a function of some generalized coordinate we will find that there are a number of avoided crossings. We note that there is a velocity depending on the potential barrier and the reduced mass associated with the decay process. This velocity dependence causes transitions to excited states if the process is "fast". Also, in some cases, because of symmetry reasons the orbitals do actually cross and thereby giving an increasing probability of decay into excited states. Finally, we have the case of doubly occupied orbits crossing. The transition probability in such cases should be proportional to the pairing energy squared.

1.2 Neutron halo

The discovery and study of a neutron halo in light nuclei with large neutron excess is a most remarkable and exiting subject that deserves to be discussed at this conference too. When a high energy (about 1 GeV/A) heavy particle collides with a light target, a large number of light fragments moving with essentially the same velocity as the projectile are created. These fragments can then be passed through a mass-separator and subsequently used as a projectile. As an example let us consider the fragment ^{11}Li colliding with ^{12}C and thereby producing ^{9}Li and two neutrons. When the mass of ^{11}Li is obtained from β-decay one finds that the neutron separation energy is only ~ 200 keV. Thus ^{11}Li is very weekly bound, but ^{10}Li is not bound at all! The potential is not strong enough to bind ^{9}Li + n but a pair of neutrons is bound, reflecting the importance of the pairing energy in nuclear physics. The small separation energy implies that the neutrons extend far out from the nucleus. This large extension can be directly seen from cross-section measurements. If one plots the radii of light nuclei versus mass number one finds that most of them follow the familiar $A^{1/3}$ dependence but ^{11}Li and ^{14}Be have interaction radii that are about 30 % larger than the others (see Tanihata et al., ref. [1]). This means that we have a new class of nuclei with a charge in the center and the neutrons extend far out. Furthermore, if one measures the momentum of the ^{9}Li recoil one obtains the momentum distribution of the neutrons that are kicked off. It is found that there are two distinct distributions of transverse momentum for the neutrons. First of all there is a broad momentum distribution corresponding to neutrons with the normal amount of momentum. On top of this one sees a very narrow one (factor three narrower) reflecting the expected low momentum distribution of such an extended system, cf. Kobayashi et al. (ref. [2]).

This class of nuclei are expected to have some very interesting properties. The collective motion of the neutrons with respect to the protons may have much lower frequency components than the dipole frequencies in normal nuclei. In this system the normal dipole energy would be ~ 20 MeV but the large neutron halo can give rise also to lower frequency components. In fact, theoretical calculations predict a strenght at low frequencies that is totally unknown in heavier nuclei. This is very important for many processes, i.e. the r-process in neutron rich nuclei. Here the rate of the process by which elements are formed is sensitive to the dipole capture probability of the neutrons from the continuum.

Experiments in progress are planned to establish the electric dipole response of these new nuclei, to study the n-n correlations and to measure the quadrupole moment of ^{11}Li. Further, the general significance of the r-process will be explored. These studies have a bearing on the general many-body phenomena of the "Efetov states" as well as on the important nuclear physics and astrophysical issues involved in defining the neutron drip line.

2. Atomic Clusters

Atomic clusters are a new and fascinating class of many body problem and I shall now explain how I see the field of atomic clusters. This new field has received a lot of interest for a number of reasons. First of all it constitutes a link between molecules and bulk solids. There are also similarities between these systems and atoms constituting many body systems of electrons held together by and moving in a potential generated by the positive charge. Especially for sufficiently large clusters of metallic atoms there are similarities to atomic nuclei. Here we are dealing with a delocalized motion of electrons in an approximatively constant potential, i.e. an independent particle system. However, there is a very interesting difference. Nuclear forces are predominantly attractive giving us a special scheme for discussing correlations and many body effects in approximatively degenerate sets of orbits occuring in that potential. On the other hand, forces between electrons are repulsive. It is therefore a challenge to try to understand how this changes the nature of the correlations and the collective features of these systems. Finally, atomic clusters show certain structural dependencies on the number of atoms or electrons participating. These define a number of regimes making the systems fascinating. Thus, we have an enormously rich and varied field to study, but I will confine myself here to metallic atoms, since the connections to nuclear physics are perhaps most intimate in these cases.

2.1 Periodicity

When the abundancies of clusters produced in an experiment are plotted as a function of particle number (cluster size), one recognizes that there are a number of distinct periodicities observed in the stabilities of the clusters.

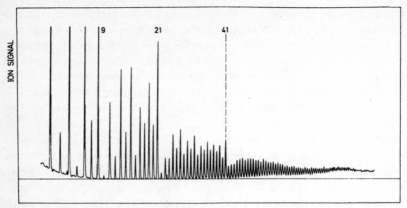

Fig. 1. Abundancies of clusters as a function of the number of atoms in the cluster. The system is Na_n^+ from Brechignac et al., J. Chem. Phys. **90** 1493 (1989)

Starting with small clusters one finds period two or odd-even, period four or quartet structures, shell structures (periods of 10's to 100's), supershells (periods of 1000's) and finally structures due to layer packing containing several tens of thousands of atoms. Figure 1 in P. Martins lecture to this conference shows a typical shell structure (8, 20, 40, 58) for neutral sodium atoms. One also sees the odd-even differences with the even species more abundant as well as the quartet structure (i.e. 26, 30, 34, 38). These periodicities and shells are seen for a number of species and are associated with the electrons and not with the atoms. That this is the case can be seen for Na_n^+ in Fig. 1., where the numbers of the shell structure are shifted by one unit (9, 21, 41) and the clusters containing an odd number of atoms are more abundant. It is worth noticing that the odd-even difference rapidly gets smaller for clusters containing more than about 40 atoms. This has also been observed for silver and other systems.

In sodium the shell structure associated with quantal shells extend up to about 1400 atoms, but between 1400 and 1900 the periodicity changes strikingly (see the lecture to this conference by P. Martin). There is still periodicity, but the period is about 2.5 times larger than for the quantal shells. These oscillations continue for systems involving up to more than 40000 atoms and correspond to packing of successive layers on an icosahedron. Martin has thereby demonstrated that, at sufficiently large numbers of atoms, these systems go over to a periodicity associated with the facetting of the system. However, if these systems are heated up (temperature increases from one hundred to several hundred Kelvin) by photons it is observed that the structures of layer packing disappear and are replaced by shell structures which are to be associated with quantal shells.

Thus, we have a number of remarkable periodocities associated with different kinds of structures. Now I will try to say how far we are from

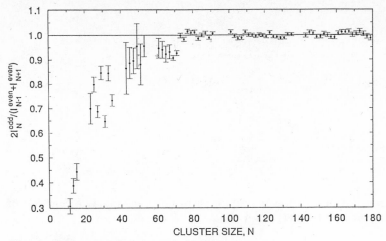

Fig. 2. Odd-even effect in atomic clusters.

understanding them and how that relates to the structural features of the systems. Starting with the simplest, i.e. odd-even differences, I remind that in nuclear physics the extra stability of the even systems is a well-known theme with far-reaching consequences for the nuclear structure. The extra binding of even systems as compared to the odd ones adds a term to the systematics of the binding energy, that is called the pairing energy. This effect can be parametrized by an average binding energy difference $\Delta(A)$ between the ground states of systems with even numbers of particles as compared with those with an odd number. In a limited region of the periodic table known binding energies are fitted and the pairing parameter (Δ) can be measured from the systematic differences between even and odd nuclei in that mass region. It is found to vary approximately as $A^{-1/2}$, where A is the number of particles. For atomic clusters the evidence of pairing is obtained from the measured abundancies. In order to see the systematics the ratio of the odd species to the average of the two neighbouring even ones is plotted as a function of cluster size in Fig. 2.

There is a major effect in light systems which tends to disappear already for clusters containing 40 - 50 atoms. We shall attempt to interpret the odd/even abundance effects in clusters in terms of an average pairing parameter Δ, and to interpret its variation with the size of the cluster. We are, however faced with a problem when trying to extract the parameter from data because these systems are at non-zero temperature. We have to consider the termodynamics of how the single particle levels are occupied as well as energy spacings. Let us try the simplest possible model with uniformly spaced one-particle levels from a potential model with a twofold degeneracy, i.e. spin up-spin down. Filling up the levels in this model we find that even systems in their ground states are systematically lower in en-

575

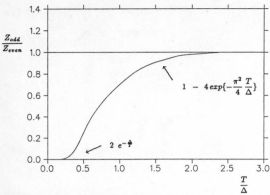

Fig. 3. The ratio of odd and even cluster partition functions as a function of temperature divided by the parameter Δ.

ergy by a term which has the structure of a pairing term and the difference is just one quarter of the distance between the levels. By measuring the favouring of odd or even we can estimate the magnitude of the parameter. In the Fermi gas picture the spacing of levels is $\frac{4}{3}$ times the Fermi- energy divided by the number of particles, and therefore the parameter is $\frac{1}{3}$ times the Fermi-energy divided by the number of particles. For sodium this gives approximately the parameter 1 eV divided by the number of particles. This picture is a simplified version of the pioneering work of Kubo (ref. 4) that he introduced several years ago when discussing the effects of finite size on the electronic properties of small metallic particles.

We should now consider the effects of non-zero temperature. We must then calculate the partition functions for odd and even systems and compare them. In Fig. 3 the odd partition function divided by the even is plotted as a function of temperature divided by the parameter Δ. We see that, if the temperature is very low, compared to this parameter, there are almost no odd systems produced and the evens are favoured by an exponential factor involving the ratio of the gap to the temperature. At higher temperatures the odd-even difference decays exponentially as the inverse ratio of temperature over gap and as temperature gets larger we are averaging over many levels and get equal abundancies of odd and even systems. When the temperature is comparable to the level spacing the odd and even abundancies are equal except for terms that go to zero exponentially in the ratio T/Δ, and since Δ is proportional to N^{-1} we can see why the odd-even difference disappears quite suddenly at a characteristic value of N.

However, this simple picture is somewhat spoiled by the fact that levels are not equidistant. If the statistics of level spacing is included it is possible to fit the experimental curve with a curve which is determined by the ratio of Δ/T. If the temperature is known we then get the gap approximately

1.4 eV divided by the number of particles. This seems good considering the simple estimate, but, a very important fact has been neglected! These electrons that go into twofold occupation of the levels close to the Fermi surface have an especially strong Coulomb interaction with each other, because of the exchange interaction. We thus have a repulsive effect of the Coulomb exchange term that tends to reduce the magnitude of the odd-even differences. This reduction, as worked out by Snider and Sorbello (ref. [5]), depends on the ratio of the Wigner-Seitz radius to the atomic unit of length (Bohr radius of hydrogen), and in the case of sodium this effect cancels about $\frac{2}{3}$ of the odd/even energy differences in the ground state. We thus get an odd-even difference of about $\frac{1}{3}$ eV per number of particles which is more than a factor three smaller than observed. I would like to present this as the first evidence that somewhere in our picture there is an effective attraction between the electrons that counteracts this exchange term and thereby reducing this repulsive interaction by a major amount.

About the quartets I will only say, since they are sometimes observed and sometimes not, that our present understanding is that they are associated with the existence of residual axial symmetry. If the electron orbit has a finite orbital angular momentum about a symmetry axis then time reversal gives a fourfold degeneracy of that orbit and thereby produces quartets. This is an interesting topic containing information for example on differences between singlet and triplet states.

2.2 Shell Structure

We have seen here pictures showing these marvellous periods and the fact that these shells extend to very large quantum numbers in clusters, see P. Martin and S. Bjørnholm at this conference. This gives the nuclear physicist an opportunity to test ideas discussed in nuclear physics for more than ten years, ideas that have not been possible to test simply because nuclear systems are not big enough. What do we know about shell structure? It is a kind of bunchiness in the spectrum of single particle levels. When we fill all the levels in a bunch we get a very stable system whereas the next particle will go into a level in the next bunch at considerably higher energy and therefore produce much less stable systems. This is the picture of most textbooks, where the Schrödinger equation is solved in some potential well with the correct size to obtain the observed levels. However, this is no explanation, but merely tells us that we know how to solve the Schrödinger equation. Instead we should ask these questions. How much bunching is there? What is the periodicity of the bunching? If we go to the limit of very large systems, how much bunching would be left? What are the orbits that appear in a particular bunch and what other consequencies does this have for the structure of these systems? The bunchiness needs an explanation that is very seldom considered. The answer to these questions was provided more than ten years ago by Balian and Bloch [3]. I will try to explain it in

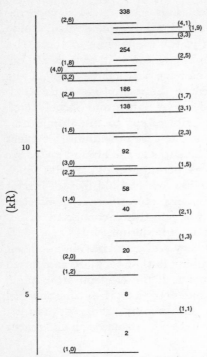

Fig. 4. Level bunching in a spherical square well. The numbers correspond to radial and angular quantum numbers, (n, l).

a simplified way that I like. Figure 4 shows the levels in a spherical square well potential. The levels come in bunches and we see the familiar numbers separating them. We see that the origin of the bunching is associated with the possibility of a compensation between the energy associated with changing the radial quantum number, n, by some integer number and changing the angular quantum number, l, by some corresponding integer number.

As an example the bunch between 8 and 20 involves the first orbit of angular momentum two, the 1d in nuclear physics notation, and the second orbit of angular momentum zero, the 2s. The approximative degeneracy of these two orbits is then the origin of the bunch that defines this closed shell. That is, increasing l by two units and decreasing n by one unit, we get an approximately degenerate pair of orbits. Similarly, when we come to larger shells, for example between 138 and 186, we have the first orbit of angular momentum seven and the second orbit of angular momentum four or we increase n by one unit and decrease l by three units. This kind of bunchiness from the compensation can be interpreted by recognizing that the derivative of the hamiltonian with respect to the action is classically the period of the motion in the corresponding direction. The ratio of the derivative of the energy eigenvalues with respect to radial and angular quantum numbers,

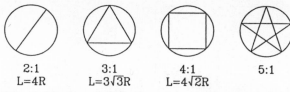

2:1	3:1	4:1	5:1
L=4R	L=3√3R	L=4√2R	

Fig. 5. Motion pattern given by combinations of the radial and angular quantum numbers in the relation n:l indicated.

$(\partial\epsilon/\partial n : \partial\epsilon/\partial l)$ will be a:b with a and b integer numbers. This means that the ratio of the radial and angular periods is in the ratio of integers and this then means that the classical motion forms a closed periodic orbit after a radial and b angular oscillations. The simplest such motions are shown in Fig. 5. We see (far left) the simplest motion with n:l = 2:1, called the pendulating motion, i.e. two radial oscillations during one angular oscillation, and three other more complicated motions with n:l indicated in the figure.

Associated with each of these closed periodic orbits is a bunchiness in the single particle spectrum that will be recognized as a shell. The occurence of approximative degeneracy in the spectrum leading to an increase in level density will have an important effect on the stability of the corresponding quantal system. The simplest motion (2:1) is very important for nuclear physics where quantum numbers are fairly small, but for very large systems it is relatively unimportant. This is because the degeneracy of the pendulating motion only involves two degrees of freedom (the number of coordinates needed to specify a direction in space), whereas all the other orbits are planar and their degeneracy involve three degrees of freedom. Therefore these planar motions are fundamentally much more important than the linear motion that involves the lowest frequency. The remarkable thing about this whole scheme is that for systems up to several thousand constituents the square well potential is dominated by the triangel and the square motions. This is shown beautifully in Fig. 6 taken from Balian and Bloch's paper [3] in which they plot the level density smeared out over an interval small compared to the separation of shells but large enough to overlap the different levels within the bunch so we don't get confused with the tremendous complexity of the total spectrum.

We see that indeed this oscillation continues up to systems containing several thousand, i.e. to large quantum numbers. The amplitude of the oscillation can be very easily estimated from the simple semiclassical geometry sketched above. We find that the amplitude of these oscillations is increasing as the one third power of the number of particles, while the total level density is increasing as the one half power. The ratio of oscillation to background is decreasing but only at the very small rate of the one sixth power of the number of particles. Figure 6 also shows that there is beating in the spectrum. This is associated with the fact that the periods of the square

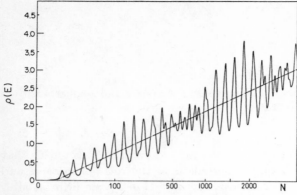

Fig. 6. Smoothed level density as a function of the number of constituents [3].

and the triangle differ by about 10 % and so we see about ten oscillations for each one of these beats. When they get out of phase with respect to each other we have a small shell structure effect and correspondingly a large effect when they are in phase.

This shell structure has been identified by the experimental techniques of both Martin and Bjørnholm. A theoretical calculation of the effect of single particle orbits in a Wood-Saxon potential shows a very suggestive correspondence with the experimental abundancies of Bjørnholm. (See Fig. 5. in Bjørnholm's lecture). Although the details of the potential are not right, further development of both theoretical and experimental studies of these systems will shed light on the structure of the one particle potential in these systems.

Now there is a lot of questions that are opened. Very interesting will be the questions
— what happens in the middle of the shell,
— what is the effect of these numbers on the fact that these systems are after all made of ions and therefore they are not completely liquid,
— what about the irregularities in the surface,
— what about at low temperature where there are crystal structures,
and so on.
These are question that should be answered in the future.

References

1. Tanihata, I. et al.: Phys. Lett. **160B** 380 (1985); Phys. Rev. Lett. **55** 2676 (1985)
2. Kobayashi, T., Yamakawa, O., Omata, K., Sugimoto, K., Shimoda, T., Tanihata, I.: Phys. Rev. Lett. **60** 2599 (1988)
3. Balian, R., Bloch, C.: Ann. Phys. **69** 76 (1972)
4. Kubo, R.: J. Phys. Soc. Jap. **17** 975 (1962)
5. Snider, D.R., Sorbello, R.S.: Surface Sci. **143** 204 (1984)

Index of Contributors

Printing: Mercedesdruck, Berlin
Binding: Buchbinderei Lüderitz & Bauer, Berlin